Shop Manual for

Automotive Suspension
and Steering Systems

Second Edition

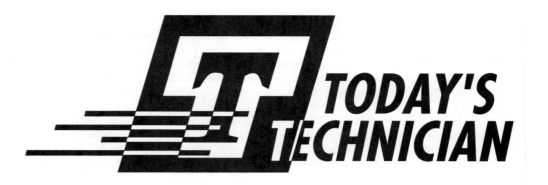

TODAY'S TECHNICIAN

Shop Manual for
Automotive Suspension and Steering Systems

Second Edition

Don Knowles

Knowles Automotive Training
Moose Jaw, Saskatchewan
CANADA

Jack Erjavec

Series Advisor
Columbus State Community College
Columbus, Ohio

DELMAR

™
THOMSON LEARNING

Africa • Australia • Canada • Denmark • Japan • Mexico • New Zealand • Philippines
Puerto Rico • Singapore • Spain • United Kingdom • United States

NOTICE TO THE READER

Cover Design: Cheri Plasse

Delmar Staff

Publisher: Alar Elken
Acquisitions Editor: Vernon R. Anthony
Developmental Editor: Catherine A. Wein
Project Editor: Megeen Mulholland

Production Manager: Mary Ellen Black
Production Coordinator: Karen Smith
Art and Design Coordinator: Cheri Plasse
Editorial Assistant: Betsy Hough

Printed in the United States of America
4 5 6 7 8 9 10 QPD 03 02 01

For more information, contact Delmar, 3 Columbia Circle, PO Box 15015, Albany, NY 12212-0515; or find us on the World Wide Web at http://www.delmar.com

International Division List

Asia
Thomson Learning
60 Albert Street, #15-01
Albert Complex
Singapore 189969
Tel: 65 336 6411
Fax: 65 336 7411

Australia/New Zealand:
Nelson/Thomson Learning
102 Dodds Street
South Melbourne, Victoria 3205
Australia
Tel: 61 39 685 4111
Fax: 61 39 685 4199

Latin America:
Thomson Learning
Seneca, 53
Colonia Polanco
11560 Mexico D.F. Mexico
Tel: 525-281-2906
Fax: 525-281-2656

Spain:
Thomson Learning
Calle Magallanes, 25
28015-MADRID
ESPANA
Tel: 34 91 446 33 50
Fax: 34 91 445 62 18

Japan:
Thomson Learning
Palaceside Building 5F
1-1-1 Hitotsubashi, Chiyoda-ku
Tokyo 100 0003 Japan
Tel: 813 5218 6544
Fax: 813 5218 6551

UK/Europe/Middle East
Thomson Learning
Berkshire House
168-173 High Holborn
London
WC1V 7AA United Kingdom
Tel: 44 171 497 1422
Fax: 44 171 497 1426

Canada:
Nelson/Thomson Learning
1120 Birchmount Road
Scarborough, Ontario
Canada M1K 5G4
Tel: 416-752-9100
Fax: 416-752-8102

Library of Congress Cataloging-in-Publication Data
Knowles, Don.
    Classroom Manual and shop manual for automotive suspension and steering systems / Don
Knowles. — 2nd ed.
        p. cm. — (Today's technician)
    Includes bibliographical references and index.
    ISBN 0-8273-8649-4
        1. Automobiles — Springs and suspension — Maintenance and repair.
    2. Automobiles — Steering-gear — Maintenance and repair. I. Title.
    II. Series.
    TL257.K594 1998   98-34219
    629.2'43—dc21   CIP

CONTENTS

Photo Sequences

Job Sheets

1. Demonstrate Proper Lifting Procedures

2. Locate and Inspect Shop Safety Equipment

3. Shop Housekeeping Inspection

4. Raise a Car with a Floor Jack and Support It on Safety Stands

5. Follow the Proper Procedures to Hoist a Car

6. Determine the Availability and Purpose of Suspension and Steering Tools

7. Service Integral Wheel Bearing Hubs

8. Diagnose Wheel Bearings

9. Clean, Lubricate, Install, and Adjust Nonsealed Wheel Bearings

10. Tire Demounting and Mounting

11. Tire and Wheel Runout Measurement

12. Off-Car Wheel Balancing

13. On-Car Wheel Balancing

14. Remove Strut and Spring Assembly and Disassemble Spring and Strut

15. Assemble Strut and Spring Assembly and Install Strut and Spring Assembly in the Car

16. Install Strut Cartridge Off-Car

17. Measuring Lower Ball Joint Vertical and Radial Movement, Short-and-Long Arm Suspension Systems

18. Ball Joint Replacement

19. Steering Knuckle Removal, MacPherson Strut Front Suspension

20. Remove and Service Rear Suspension Strut and Coil Spring

21. Remove Rear Suspension Lower Control Arm and Ball Joint

22. Install Rear Suspension Lower Control Arm and Ball Joint Assembly

23. Electronic Air Suspension System Diagnosis (Lincoln Town Car)

24. Adjust Trim Height, Rear Load-Levelling Air Suspension System

25. Diagnose Automatic Ride Control (ARC) System (Ford Explorer or Mountaineer)

26. Remove and Replace Air Bag Inflator Module and Steering Wheel

27. Remove and Replace Steering Column

28. Diagnose, Remove, and Replace Idler Arm

29. Remove and Replace Outer Tie-Rod End, Parallelogram Steering Linkage

30. Draining and Flushing Power Steering Systems

31. Testing Power Steering Pump Pressure

PREFACE

Thanks to the support the *Today's Technician* series has received from those who teach automotive technology, Delmar Publishers is able to live up to its promise to provide new editions every three years. We have listened to our critics and our fans and present this new revised edition. By revising our series every three years, we can and will respond to changes in the industry, changes in the certification process, and the ever-changing needs of those who teach automotive technology.

The *Today's Technician* series by Delmar Publishers features textbooks that cover all mechanical and electrical systems of automobiles and light trucks. Principal titles correspond to the eight major areas of ASE (National Institute for Automotive Service Excellence) certification. Additional titles include remedial skills and theories common to all of the certification areas and advanced or specialized subject areas that reflect the latest technological trends.

Each title is divided into two manuals: a Classroom Manual and a Shop Manual. Dividing the material into two manuals provides the reader with the information needed to begin a successful career as an automotive technician without interrupting the learning process by mixing cognitive and performance-based learning objectives.

Each Classroom Manual contains the principles of operation for each system and subsystem. It also discusses the design variations used by different manufacturers. The Classroom Manual is organized to build upon basic facts and theories. The primary objective of this manual is to allow the reader to gain an understanding of how each system and subsystem operates. This understanding is necessary to diagnose the complex automobile systems.

The understanding acquired by using the Classroom Manual is required for competence in the skill areas covered in the Shop Manual. All the high priority skills, as identified by ASE, are explained in the Shop Manual. The Shop Manual also includes step-by-step instructions for diagnostic and repair procedures. Photo Sequences are used to illustrate many of the common service procedures. Other common procedures are listed and are accompanied with line drawings and photographs that allow the reader to visualize and conceptualize the finest details of the procedure. The Shop Manual also contains the reasons for performing the procedures, as well as when that particular service is appropriate.

The two manuals are designed to be used together and are arranged in corresponding chapters. Not only are the chapters in the manuals linked together, the contents of the chapters are also linked. Both manuals contain clear and thoughtfully selected illustrations. Many of the illustrations are original drawings or photos prepared for inclusion in this series. This means that the art is a vital part of each manual.

The page layout remains the same. The main body of the text includes all of the "need-to-know" information and illustrations. In the side margins are many of the special features of the series. These are provided to the side of the text so they do not break the flow of the text, making it easier to read and follow.

Highlights of This Edition—Classroom Manual

The complete text was updated to include the latest technology. Chapter 8 on computer-controlled suspension systems was rewritten to include information on the latest integrated antilock brake systems (ABS), acceleration slip regulation (ASR), and vehicle dynamic control (VDC) systems. This chapter also includes the latest computer-controlled suspension systems on sport utility vehicles (SUVs). The latest electronic power steering systems are included in Chapter 12 on rack and pinion steering gears. Chapters 15 and 16 on wheel alignment were rewritten to emphasize four wheel alignment theory and the latest electronic wheel alignment equipment.

Highlights of This Edition—Shop Manual

The complete text was updated to include the latest diagnostic and service procedures. Chapter 8 on computer-controlled suspension systems was rewritten to include diagnostic and service procedures on the latest suspension systems such as the commputer-controlled systems on sport utility vehicles (SUVs). The latest diagnostic and service procedures on electronic power steering systems are included in Chapter 12 on rack and pinion steering gear diagnosis and service. Chapters 15 and 16 on wheel alignment were rewritten to emphasize four wheel alignment procedures with the latest electronic wheel alignment equipment.

Classroom Manual

To stress the importance of safe work habits, the Classroom Manual dedicates one full chapter to safety. Included in this chapter are common safety practices, safety equipment, and safe handling of hazardous materials and wastes. This includes information on MSDS sheets and OSHA regulations. Other features of this manual include:

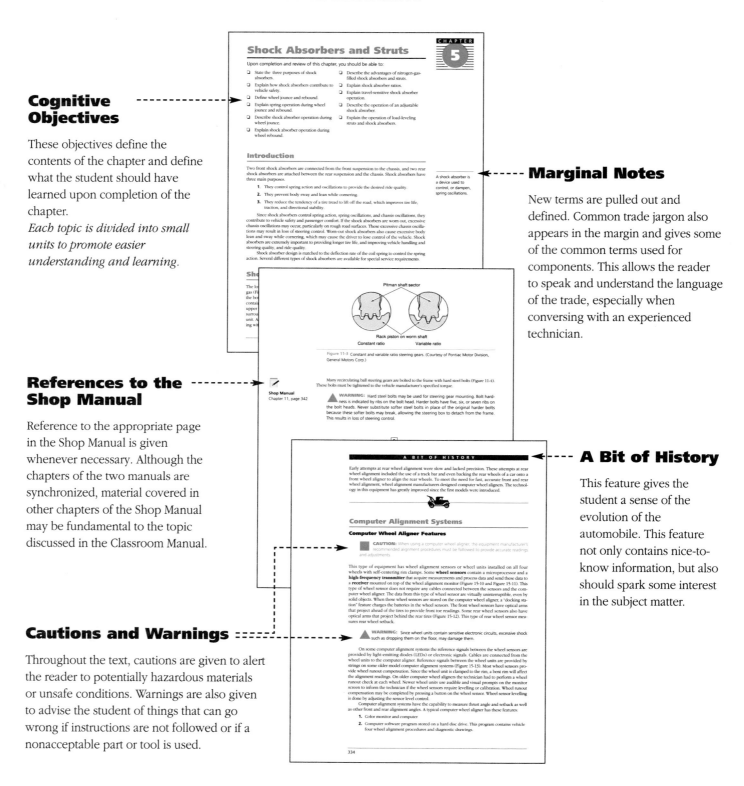

Cognitive Objectives

These objectives define the contents of the chapter and define what the student should have learned upon completion of the chapter.

Each topic is divided into small units to promote easier understanding and learning.

References to the Shop Manual

Reference to the appropriate page in the Shop Manual is given whenever necessary. Although the chapters of the two manuals are synchronized, material covered in other chapters of the Shop Manual may be fundamental to the topic discussed in the Classroom Manual.

Cautions and Warnings

Throughout the text, cautions are given to alert the reader to potentially hazardous materials or unsafe conditions. Warnings are also given to advise the student of things that can go wrong if instructions are not followed or if a nonacceptable part or tool is used.

Marginal Notes

New terms are pulled out and defined. Common trade jargon also appears in the margin and gives some of the common terms used for components. This allows the reader to speak and understand the language of the trade, especially when conversing with an experienced technician.

A Bit of History

This feature gives the student a sense of the evolution of the automobile. This feature not only contains nice-to-know information, but also should spark some interest in the subject matter.

Summaries

Each chapter concludes with summary statements that contain the important topics of the chapter. These are designed to help the reader review the contents.

Review Questions

Short answer essay, fill-in-the-blank, and multiple-choice type questions follow each chapter. These questions are designed to accurately assess the student's competence in the stated objectives at the beginning of the chapter.

The J1930 List of Terminology

Located in the appendix, this list serves as a reference to the acceptable industry terms as defined by SAE.

Terms to Know

A list of new terms appears next to the Summary. Definitions for these terms can be found in the Glossary at the end of the manual.

Third, employers are responsible for maintaining permanent files regarding hazardous materials. These files must include information on hazardous materials in the shop, proof of employee training programs, and information about accidents such as spills or leaks of hazardous materials. The employer's files must also include proof that employees' requests for hazardous material information such as MSDS have been met. A general right-to-know compliance procedure manual must be maintained by the employer.

Shop Manual
Chapter 1, page 6

Summary

- The United States Occupational Safety and Health Act of 1970 assures safe and healthful working conditions, and authorizes enforcement of safety standards.
- Many hazardous materials and conditions can exist in an automotive shop, including flammable liquids and materials, corrosive acid solutions, loose sewer covers, caustic liquids, high-pressure air, frayed electrical cords, hazardous waste materials, carbon monoxide, improper clothing, harmful vapors, high noise levels, and spills on shop floors.
- Material safety data sheets (MSDS) provide information regarding hazardous materials, labeling, and handling.
- The danger regarding hazardous conditions and materials may be avoided by applying the necessary safety precautions. These precautions include all areas of safety such as personal safety, gasoline safety, housekeeping safety, general shop safety, fire safety, and hazardous waste handling safety.
- The automotive shop must supply the necessary shop safety equipment, and all shop personnel must be familiar with the location and operation of this equipment. Shop safety equipment includes gasoline safety cans, steel storage cabinets, combustible material containers, fire extinguishers, eyewash fountains, safety glasses and face shields, first-aid kits, and hazardous waste disposal containers.

Review Questions

Short Answer Essays

1. Explain the purposes of the Occupational Safety and Health Act.
2. Define 12 shop hazards, and explain why each hazard is dangerous.
3. Describe five steps that are necessary for personal protection in the automotive shop.
4. Explain why smoking is dangerous in the shop.
5. Describe the danger in drug or alcohol use in the shop.
6. Explain three safety precautions related to electrical safety in the shop.
7. Define six essential safety precautions regarding gasoline handling.
8. Describe five steps required to provide housekeeping safety in the shop.

Terms to Know

Corrosive
Environmental Protection Agency (EPA)
Hazard Communication Standard
Ignitible
Material safety data sheets (MSDS)
Multipurpose dry chemical fire extinguisher
Occupational Safety and Health Act (OSHA)
Reactive
Resource Conservation and Recovery Act (RCRA)
Right-to-know laws
Toxic
Workplace hazardous materials information systems (WHMIS)

- Flutes on seal lips provide a pumping action to direct oil back into a housing.
- Bearing hub units are compact compared to bearings that are mounted in the wheel hub. This compactness makes hub bearing units suitable for FWD cars.
- Some bearing hub units are bolted to the steering knuckle; other bearing hub units are pressed into the steering knuckle.
- Some steering knuckles contain two separate tapered roller bearings.
- Rear axle bearings are mounted between the drive axles and the housing on rear wheel drive cars.

Review Questions

Short Answer Essays

1. Define a radial bearing load.
2. Define a thrust bearing load and give another term for this type of load.
3. Explain an angular bearing load.
4. Describe the main parts of a bearing, including the location and purpose of each part.
5. Describe the difference between a maximum-capacity cylindrical ball bearing and an ordinary cylindrical ball bearing.
6. Explain the design and purpose of bearing seals.
7. Explain the purpose of the sealer on the outside surface of a seal housing.
8. Describe the advantage of a bearing hub unit compared to bearings that were mounted in the wheel hub.
9. Explain how the proper bearing endplay adjustment is obtained when two tapered roller bearings are mounted in the steering knuckle.
10. Describe two different types of rear axle bearings on rear-wheel-drive cars.

Fill-in-the-Blanks

1. A bearing may be described as a component that supports a _____, _____, or _____.
2. A bearing is designed to support a load and _____ _____.
3. An angular bearing load is applied at an angle between the _____ and the _____.
4. A cylindrical ball bearing is designed primarily to withstand _____ loads.
5. A bearing shield is attached to the _____ bearing race.
6. Lubrication and proper _____ adjustment are important on tapered roller bearings.
7. A needle roller bearing is not designed to carry _____ loads.
8. A springless seal may be used to seal a _____ lubricant into a hub.

59

SAE J1930 Revised SEP95

TABLE 1—CROSS REFERENCE AND LOOK UP (CONTINUED)

Existing Usage	Acceptable Usage	Acceptable Acronized Usage
ALT (Alternator)	Generator	GEN
Alternator	Generator	GEN
Ambient Air Temperature	Ambient Air Temperature	AAT
AM1 (Air Management 1)	Secondary Air Injection Bypass [1]	AIR Bypass [1]
AM2 (Air Management 2)	Secondary Air Injection Diverter [1]	AIR Diverter [1]
APP (Accelerator Pedal Position)	Accelerator Pedal Position [1]	APP [1]
APS (Absolute Pressure Sensor)	Barometric Pressure Sensor [1]	BARO Sensor [1]
ATS (Air Temperature Sensor)	Intake Air Temperature Sensor [1]	IAT Sensor [1]
Automatic Transaxle	Automatic Transaxle [1]	A/T [1]
Automatic Transmission	Automatic Transmission [1]	A/T [1]
B+ (Battery Positive Voltage)	Battery Positive Voltage	B+
Backpressure Transducer	Exhaust Gas Recirculation Backpressure Transducer [1]	EGR Backpressure Transducer [1]
BARO (Barometric Pressure)	Barometric Pressure [1]	BARO [1]
Barometric Pressure Sensor	Barometric Pressure Sensor [1]	BARO Sensor [1]
Battery Positive Voltage	Battery Positive Voltage	B+
BLM (Block Learn Memory)	Long Term Fuel Trim [1]	Long Term FT [1]
BLM (Block Learn Multiplier)	Long Term Fuel Trim [1]	Long Term FT [1]
BLM (Block Learn Matrix)	Long Term Fuel Trim [1]	Long Term FT [1]
Block Learn Integrator	Long Term Fuel Trim [1]	Long Term FT [1]
Block Learn Matrix	Long Term Fuel Trim [1]	Long Term FT [1]
Block Learn Memory	Long Term Fuel Trim [1]	Long Term FT [1]
Block Learn Multiplier	Long Term Fuel Trim [1]	Long Term FT
BP (Barometric Pressure) Sensor	Barometric Pressure Sensor [1]	BARO Sensor [1]
BPP (Brake Pedal Position)	Brake Pedal Position [1]	BPP [1]
Brake Pressure	Brake Pressure	Brake Pressure
Brake Pedal Position	Brake Pedal Position [1]	BPP [1]
C3I (Computer Controlled Coil Ignition)	Electronic Ignition [1]	EI [1]
CAC (Charge Air Cooler)	Charge Air Cooler [1]	CAC [1]
Calculated Load Value	Calculated Load Value	LOAD
Camshaft Position	Camshaft Position [1]	CMP [1]
Camshaft Position Actuator	Camshaft Position Actuator [1]	CMP Actuator [1]
Camshaft Position Controller	Camshaft Position Actuator [1]	CMP Actuator [1]
Camshaft Position Sensor	Camshaft Position Sensor [1]	CMP Sensor [1]
Camshaft Sensor	Camshaft Position Sensor [1]	CMP Sensor [1]
Camshaft Timing Actuator	Camshaft Position Actuator [1]	CMP Actuator [1]
Canister	Canister [1]	Canister [1]
Canister	Evaporative Emission Canister	EVAP Canister [1]
Canister Purge	Evaporative Emission Canister Purge [1]	EVAP Canister Purge [1]
Canister Purge Vacuum Switching Valve	Evaporative Emission Canister Purge Valve [1]	EVAP Canister Purge Valve [1]
Canister Purge Valve	Evaporative Emission Canister Purge Valve [1]	EVAP Canister Purge Valve [1]
Canister Purge VSV (Vacuum Switching Valve)	Evaporative Emission Canister Purge [1]	EVAP Canister Purge [1]
CANP (Canister Purge)	Evaporative Emission Canister Purge [1]	EVAP Canister Purge [1]
CARB (Carburetor)	Carburetor [1]	CARB [1]
Carburetor	Carburetor [1]	CARB [1]
Catalytic Converter Heater	Catalytic Converter Heater	Catalytic Converter Heater
CCC (Converter Clutch Control)	Torque Converter Clutch [1]	TCC [1]
CCO (Converter Clutch Override)	Torque Converter Clutch [1]	TCC [1]
CCS (Coast Clutch Solenoid)	Coast Clutch Solenoid	CCS

373

Shop Manual

To stress the importance of safe work habits, the Shop Manual also dedicates one full chapter to safety. Other important features of this manual include:

Performance Objectives

These objectives define the contents of the chapter and define what the student should have learned upon completion of the chapter. These objectives also correspond with the list of required tasks for ASE certification. *Each ASE task is addressed.*

Although this textbook is not designed to simply prepare someone for the certification exams, it is organized around the ASE task list. These tasks are defined generically when the procedure is commonly followed and specifically when the procedure is unique for specific vehicle models. Imported and domestic model automobiles and light trucks are included in the procedures.

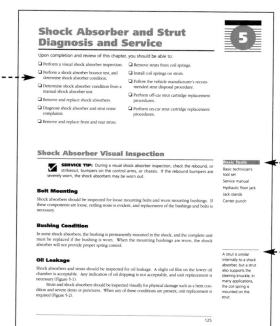

Tools Lists

Each chapter begins with a list of the Basic Tools needed to perform the tasks included in the chapter. Whenever a Special Tool is required to complete a task, it is listed in the margin next to the procedure.

Marginal Notes

New terms are pulled out and defined. Common trade jargon also appears in the margin and gives some of the common terms used for components. This allows the reader to speak and understand the language of the trade, especially when conversing with an experienced technician.

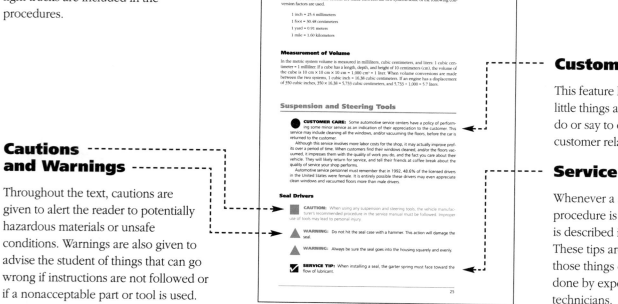

Customer Care

This feature highlights those little things a technician can do or say to enhance customer relations.

Service Tips

Whenever a special procedure is appropriate, it is described in the text. These tips are generally those things commonly done by experienced technicians.

Cautions and Warnings

Throughout the text, cautions are given to alert the reader to potentially hazardous materials or unsafe conditions. Warnings are also given to advise the student of things that can go wrong if instructions are not followed or if a nonacceptable part or tool is used.

References to the Classroom Manual

Reference to the appropriate page in the Classroom Manual is given whenever necessary. Although the chapters of the two manuals are synchronized, material covered in other chapters of the Classroom Manual may be fundamental to the topic discussed in the Shop Manual.

Photo Sequences

Many procedures are illustrated in detailed Photo Sequences. These detailed photographs show the students what to expect when they perform particular procedures. They also can provide a student a familiarity with a system or type of equipment, which the school may not have.

Case Studies

Case Studies concentrate on the ability to properly diagnose the systems. Each chapter ends with a case study in which a vehicle has a problem, and the logic used by a technician to solve the problem is explained.

ASE Style Review Questions

Each chapter contains ASE style review questions that reflect the performance objectives listed at the beginning of the chapter. These questions can be used to review the chapter as well as to prepare for the ASE certification exam.

Terms to Know

Terms in this list can be found in the Glossary at the end of the manual.

Rear Suspension Tie Rod Inspection and Replacement

Rear tie rods should be inspected for worn grommets, loose mountings, and bent conditions. Loose tie-rod bushings or a bent tie rod will change the rear wheel tracking and result in reduced directional stability. Worn tie-rod bushings also cause a rattling noise on road irregularities. When the rear tie rod is replaced, remove the front and rear rod mounting nuts. The lower control arm or rear axle may have to be pried rearward to remove the tie rod. Check the tie-rod grommets and mountings for wear, and replace parts as required. When the tie rod is reinstalled, tighten the mounting bolts to specifications, and check the rear wheel toe.

Classroom Manual
Chapter 7, page 156

Guidelines for Servicing Rear Suspension Systems

1. A squeaking noise may be caused by a suspension bushing, or a defective strut or shock absorber.
2. A rattling noise may be caused by worn suspension bushings, loose shock absorber or strut bushings or mounts, broken springs or spring insulators, or defective struts or shock absorbers.
3. Excessive body sway or roll, when one wheel strikes a road irregularity, may be caused by a defective stabilizer bar or bushings.
4. Excessive lateral movement of the rear chassis may be caused by a defective track bar or bushings.
5. Reduced curb riding height causes harsh riding.
6. Reduced curb riding height may be caused by sagged springs, worn control arm bushings, or bent control arms.
7. Sagged rear springs cause excessive steering effort and rapid steering wheel return after a turn.
8. When the coil spring is mounted on the rear strut, never loosen the upper strut nut until the spring is compressed with a spring compressing tool.
9. If a coil spring has a vinyl coating, the spring should be taped in the compressing tool contact areas to prevent chipping the coating.
10. On some independent rear suspension systems, a ball joint connects the lower control arm to the knuckle. Many of these ball joints have a conventional wear indicator and a press fit in the control arm.
11. When some lower control arms are installed, the control arm bolts should be tightened with the normal vehicle weight supp...
12. Suspension adjustment link studs sho...
13. A nut should never be loosened to in...
14. A broken leaf-spring center bolt may... improper tracking and reduced direct...

CASE STUDY

A customer complained about steering pull... also said the problem had just occurred in th...

208

Photo Sequence 5
Typical Procedure for Measuring Front and Rear Curb Riding Height

PS-1 Check the trunk for extra weight. PS-2 Check the tires for normal inflation pressure. PS-3 Park the car on a level shop floor or alignment track.

PS-4 Find the vehicle manufacturer's specified curb riding height measurement locations in the service manual. PS-5 Measure and record the right front curb riding height. PS-6 Measure and record the left front curb riding height.

PS-7 Measure and record the right rear curb riding height. PS-8 Measure and record the left rear curb riding height. PS-9 Compare the measurement results to the specified curb riding height in the service manual.

CASE STUDY 2

A technician was installing a rear axle bearing and seal on a small rear-wheel-drive car. The technician raised the vehicle on a lift to a comfortable working height and proceeded to pull the rear axle with a slide hammer-type puller. This technician was very strong and weighed about 250 pounds. The rear axle was extremely tight, and the technician operated the slide hammer puller with all the strength he had. Each time he operated the slide hammer, the rear axle bounced sideways on the lift. Suddenly the car moved sideways so much that the car fell off the lift and over on its side. Fortunately, the technician was able to jump out of the way of the falling vehicle, but the vehicle body suffered considerable damage.

The technician learned to avoid service operations on a lift that may cause the car to move on the lift. However, if such service operations must be performed, the car should be chained to the lift.

Terms to Know

Asbestos parts washer	Particulate emissions	Sulfuric acid
Carbon monoxide	Shop layout	

ASE Style Review Questions

1. While discussing shop layout and safety:
   Technician A says the location of each piece of safety equipment should be clearly marked.
   Technician B says some shops have a special tool room where special tools are located.
   Who is correct?
   A. A only   C. Both A and B
   B. B only   D. Neither A nor B

2. Breathing carbon monoxide may cause:
   A. arthritis.   C. impaired vision.
   B. cancer.   D. headaches.

3. While discussing shop rules:
   Technician A says breathing asbestos dust may cause heart defects.
   Technician B says oily rags should be stored in uncovered trash containers.
   Who is correct?
   A. A only   C. Both A and B
   B. B only   D. Neither A nor B

4. All these shop rules are correct EXCEPT:
   A. USC tools may be substituted for metric tools.
   B. Foot injuries may be caused by loose sewer covers.
   C. Hands should be kept away from electric-drive cooling fans.
   D. Power tools should not be left running and unattended.

5. While discussing air quality:
   Technician A says a restricted chimney on a shop furnace may cause carbon monoxide gas in the shop.
   Technician B says monitors are available to measure the level of carbon monoxide in the shop air.
   Who is correct?
   A. A only   C. Both A and B
   B. B only   D. Neither A nor B

14

9. A car experiences excessive body sway when cornering, but there is no abnormal noise in the suspension. The most likely cause of this problem is:
A. stabilizer bar bushing is missing.
B. a weak stabilizer bar.
C. stabilizer bar grommets worn out.
D. a broken stabilizer bar.

10. The steering on a vehicle pulls to the left while driving straight ahead. The most likely cause of this problem is:
A. the left front strut rod is bent.
B. the left lower ball joint is worn.
C. worn front stabilizer bushings on the left side.
D. reduced curb riding height on both sides of the front suspension.

ASE Challenge Questions

At the end of each technical chapter of the Shop Manual are challenging ASE style questions. The questions are written with the same criteria and rigor as actual ASE certification questions. These questions are designed to make the reader think and use critical thinking skills, along with the knowledge gained by studying the chapter.

Job Sheets

Located at the end of each chapter, the Job Sheets provide a format for students to perform procedures covered in the chapter. A reference to the ASE Task addressed by the procedure is referenced on the Job Sheet.

ASE Challenge Questions

1. The customer says her 1987 Blazer pulls to the left. A cursory check of the vehicle shows that the right front tire height is too low. The next step to diagnose the condition would be:
A. check the ball joints.
B. check the torsion bar.
C. check strut rod bushings.
D. check stabilizer bar bushings.

2. A customer says she drives her front-wheel-drive car has a steering wander problem on irregular road surfaces. All of the following could cause this problem EXCEPT:
A. worn stabilizer bar bushings.
B. worn ball joints.
C. worn strut rod bushings.
D. worn tie-rod ends.

3. While diagnosing a customer's complaint of steering instability on a MacPherson strut front suspension with wear indicator type ball joints, an inspection of the ball joints shows the grease fittings to be solid. Technician A says no movement of the fitting means the ball joint is good. Technician B says ball joint wear may not be apparent with the weight on the joint.
Who is correct?
A. A only   C. Both A nad B
B. B only   D. Neither A nor B

4. A customer says her rear-wheel-drive car has excessive body sway while cornering. In diagnosing this problem, which of the following components should you check first?
A. Control arm bushings
B. Strut rod bushings
C. Stabilizer bar bushings
D. Shock absorber bushings

5. Technician A says a bent strut rod can cause a car to pull to one side. Technician B says a deteriorated strut rod bushing can cause steering and braking problems.
Who is correct?
A. A only   C. Both A and B
B. B only   D. Neither A nor B

181

Job Sheet 14
(14)

Name _____ Date _____

Remove Strut and Spring Assembly and Disassemble Spring and Strut

Upon completion of this job sheet, you should be able to remove a strut and spring assembly from a car and disassemble the spring and strut.

ASE Correlation

This job sheet is related to the ASE Suspension and Steering Task List content area: B. Suspension Systems Diagnosis and Repair, 1. Front Suspensions, Task: 6. Inspect and replace front suspension system coil springs and spring insulators (silencers).

Tools and Materials

Front-wheel-drive car
Floor jack
Safety stands
Coil spring compressor

Vehicle make and model year _____
Vehicle VIN _____

Procedure

1. With the vehicle parked on the shop floor, perform a strut bounce test.
   Based on the bounce test results, state the strut condition and give the reason for your

jack. If a floor jack is used to raise the vehicle,
under the chassis so the lower control arms and
floor jack from under the vehicle.
stands? yes _____ no _____

system (ABS) wheel-speed sensor wire from
amps may have to be removed from the strut.
or wire disconnected? yes _____ no _____

145

Diagnostic Chart

Chapters include detailed diagnostic charts linked with the appropriate ASE task. These charts list common problems and most probable causes. They also list a page reference in the Classroom Manual for better understanding of the system's operation and a page reference in the Shop Manual for details on the procedure necessary for correcting the problem.

Table 9-1 NATEF and ASE TASK

Diagnose steering column noises, looseness, and binding problems (including tilt mechanisms); determine needed repairs.

Problem Area	Symptoms	Possible Causes	Classroom Manual	Shop Manual
NOISES	Rattling on road irregularities	Loose, worn coupling or universal joints	229	280
	Chunk while tilting	1. Worn universal joint or spherical bearing	229	280
		2. Worn tilt bumper	232	273
BINDING	Excessive steering effort	1. Column misaligned	233	283
		2. Improperly installed or defective dust shield	238	283
		3. Worn or defective column bearings	233	284
LOOSENESS	Excessive steering wheel freeplay	1. Worn universal joints or flexible coupling	229	280
		2. Worn tilt spherical bearing or universal joint	233	284
		3. Defective column bearings	233	284

Table 9-2 NATEF and ASE TASK

Inspect and replace steering shaft U-joint(s), flexible coupling(s), collap... steering wheels (includes steering wheels with air bags and/or other st... controls and components).

Problem Area	Symptoms	Possible Causes
NOISE	Rattling on road irregularities	Worn or loose flexible coupling or universal joint
STEERING CONTROL	Excessive steering wheel freeplay	Worn or loose flexible coupling or universal joint
	Improper steering wheel position	Collapsed column
	Damaged steering wheel	Frontal collision
AIR BAG DEPLOYMENT	Failure to deploy	Defective clock spring

298

ASE Practice Examination

A 50 question ASE practice exam, located in the appendix, is included to test students on the content of the complete Shop Manual.

APPENDIX A

ASE PRACTICE EXAMINATION

1. After new tires and new alloy rims are installed on a sports car, the owner complains about steering wander and steering pull in either direction while braking.
Technician A says there may be brake fluid on the front brake linings.
Technician B says the replacement rims may have a different offset than the original rims.
Who is correct?
A. Technician A   C. Both A and B
B. Technician B   D. Neither A nor B

2. Technician A says when a vehicle pulls to one side, the problem will not be caused by the manual steering gear.
Technician B says when an unbalanced power steering gear valve causes a vehicle to pull to one side the steering effort will be very light in the direction of the pull and normal or heavier in the opposite direction.
Who is correct?
A. Technician A   C. Both A and B
B. Technician B   D. Neither A nor B

3. The outside edge of the left front tire on a rear wheel drive car is badly scalloped.
Technician A says the cause could be worn balls joints.
Technician B says the cause could be incorrect tire pressure.
Who is correct?
A. Technician A   C. Both A and B
B. Technician B   D. Neither A nor B

4. The owner of a large rear wheel drive sedan says the front tires squeal loudly during low-speed turns. The most probable cause of this condition is:
A. excessive positive camber.
B. negative caster adjustment.
C. improper steering axis inclination (S.A.I.).
D. improper turning angle.

5. A mini pickup has a severe shudder when the vehicle is started from a stop with a load in the bed.
Technician A says the problem may be worn spring eyes.
Technician B says the problem may be axle torque wrap-up.
Who is correct?
A. Technician A   C. Both A and B
B. Technician B   D. Neither A nor B

6. A cyclic noise ("moaning," "whining," or "howling") that changes pitch with road speed and is present whenever the vehicle is in motion may be caused by any of the following EXCEPT:
A. worn differential gears.
B. rear axle bearings.
C. incorrect driveshaft runout.
D. off-road tire tread pattern.

7. Technician A says hard steering may be caused by low hydraulic pressure due to a stuck flow control valve in the pump.
Technician B says hard steering may be caused by low hydraulic pressure due to a worn steering gear piston ring or housing bore.
Who is correct?
A. Technician A   C. Both A and B
B. Technician B   D. Neither A nor B

8. Wheels and tires on a pickup truck were changed from a standard 14-inch to standard 15-inch light truck rims. The first time the brakes were applied, the truck shook and shuddered. When the 15-inch wheels were replaced by the 14-inch wheels, braking was uneventful.
Technician A says the 15-inch rim is one inch wider which causes the brakes to grab.
Technician B says the additional inch diameter increases braking leverage overloading worn suspension bushings.
Who is correct?
A. Technician A   C. Both A and B
B. Technician B   D. Neither A nor B

585

Instructor's Guide

The Instructor's Guide is provided free of charge as part of the *Today's Technician* series of automotive technology textbooks. It contains Lecture Outlines, Answers to Review Questions, a Pretest and a Test Bank including ASE style questions.

Classroom Manager

The complete ancillary package is designed to aid the instructor with classroom preparation and provide tools to measure student performance. For an affordable price, this comprehensive package contains:

Instructor's Guide
200 Transparency Masters
Answers to Review Questions

Lecture Outlines and Lecture Notes
Printed and Computerized Test Bank
Laboratory Worksheets and Practicals

Reviewers

I would like to extend a special thanks to those who saw things I overlooked and for their contributions:

Randall K. Bennett
Stark Technical College
North Canton, OH

Dana K. Brewer
Wyoming Technical Institute
Laramie, WY

Ricardo Casas
Western Technical Institute
El Paso, TX

Richard Diklich
Longview Community College
Lee's Summit, MO

Dr. Roger Donovan
Illinois Central College
East Peoria, IL

Thomas J. Fitch
Monroe Community College
Rochester, NY

Laurence S. Gaff
TAD Technical Institute
Chelsea, MA

Anthony Hoffman
Arizona Western College
Yuma, AZ

Ernesto Leyva
Western Technical Institute
El Paso, TX

Norris Martin
Texas State Technical College—Waco
Waco, TX

John E. Wood
Ranken Technical College
St. Louis, MO

Contributing Companies

I would also like to thank these companies who provided technical information and art for this edition:

AMMCO Tools
American Honda Motor Company, Inc.
Automotive Diagnostics, A Division of SPX Corporation
Bada Company
Chrysler Corporation
Cooper Industries, Inc.

Contributing Companies (continued)

CRC Industries, Inc.
Deere and Company
DuPont Automotive Products
Federal-Mogul Corporation
Ford Motor Company
General Fire Extinguisher Corporation
Hunter Engineering Company
Kleer-Flo Company
Lincoln, St. Louis
L.S. Starrett Company
Mac Tools, Inc.
Moog Automotive, Inc.
National Institute of Automotive Service Excellence (ASE)
Nissan Motors
OTC Division, SPX Corporation
Perfect Circle Dana
Sears Industrial Tools
The Sherwin Williams Company
Sealed Power Corporation
SKF Automotive Products
Snap-on Tools Corporation
Society of Automotive Engineers, Inc. (SAE)
Specialty Products Company
Timken Company
Western Emergency Equipment

Portions of materials contained herein have been reprinted with permission of General Motors Corporation, Service Technology Group.

Safety Practices

Upon completion and review of this chapter, you should be able to:

❏ List three requirements for shop layout and explain why these requirements are important.

❏ Observe all shop rules when working in the shop.

❏ Operate vehicles in the shop according to shop driving rules.

❏ Observe all shop housekeeping rules.

❏ Follow the necessary procedures to maintain satisfactory shop air quality.

❏ Observe all personal safety precautions while working in the automotive shop.

❏ Demonstrate proper lifting procedures and precautions.

❏ Demonstrate proper vehicle lift operating and safety procedures.

❏ Observe all safety precautions when hydraulic tools are used in the automotive shop.

❏ Follow safety precautions regarding the use of power tools.

❏ Demonstrate proper safety precautions while using compressed-air equipment.

❏ Follow safety precautions for using cleaning equipment in the automotive shop.

Shop Layout

There are many different types of shops in the automotive service industry, including:

New car dealer

Independent repair shop

Specialty shop

Service station

Fleet shop

 CAUTION: Always know the location of all safety equipment in the shop, and be familiar with the operation of this equipment.

The **shop layout** in any shop is important to maintain shop efficiency and contribute to safety. Shop layout includes bays for various types of repairs, space for equipment storage, and office locations. Most shops have specific bays for certain types of work, such as electrical repair, wheel alignment and tires, and machining (Figure 1-1). Safety equipment such as fire extinguishers, first-aid kits, and eyewash fountains must be in easily accessible locations, and the location of each piece of safety equipment must be clearly marked. Areas such as the parts department and the parts cleaning area must be located so they are easily accessible from all areas of the shop. The service manager's office should also be centrally located. All shop personnel should familiarize themselves with the shop layout, especially the location of safety equipment. If you know the exact fire extinguisher locations, you may get an extinguisher into operation a few seconds faster. Those few seconds could make the difference between a fire that is quickly extinguished and one that gets out of control, causing extensive damage and personal injury!

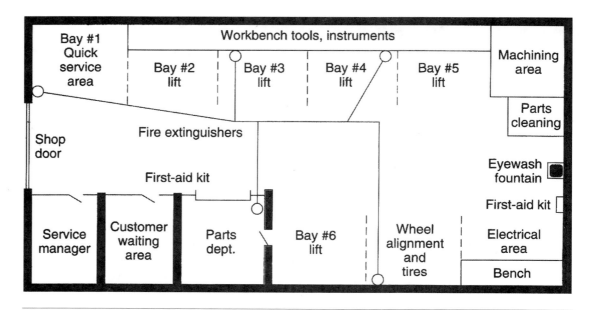

Figure 1-1 Typical shop layout.

The tools and equipment required for a certain type of work are stored in that specific bay. For example, the equipment for electrical and electronic service work is stored in the bay allotted to that type of repair. When certain bays are allotted to specific types of repair work, unnecessary equipment movement is eliminated. Each technician has his or her own tools on a portable roll cabinet that is moved to the vehicle being repaired. Special tools are provided by the shop, and these tools may be located on tool boards attached to the wall. Other shops may have a tool room where special tools are located. Adequate workbench space must be provided in those bays where bench work is required.

Shop Rules

> **WARNING:** Shop rules, vehicle operation in the shop, and shop housekeeping are serious business. Each year a significant number of technicians are injured and vehicles are damaged as a result of disregarding shop rules, careless vehicle operation, and sloppy housekeeping.

The application of some basic shop rules helps prevent serious, expensive accidents. Failure to comply with shop rules may cause personal injury or expensive damage to vehicles and shop facilities. It is the responsibility of the employer and all shop employees to make sure that shop rules are understood and followed until these rules become automatic habits. The following basic shop rules should be followed.

1. Always wear safety glasses and other protective equipment that is required by a service procedure. For example, an **asbestos parts washer** must be used to avoid breathing asbestos dust into the lungs (Figure 1-2). *Asbestos dust is a known cause of lung cancer.* This dust is encountered in manual transmission clutch facings and brake linings.

2. Tie long hair securely behind your head, and do not wear loose or torn clothing.

Figure 1-2 Brake assembly washer prevents asbestos dust. (Courtesy of Kleer-Flo Company)

3. Do not wear rings, watches, or loose hanging jewelry. If jewelry such as a ring, metal watch band, or chain makes contact between an electrical terminal and ground, the jewelry becomes extremely hot, resulting in severe burns.

4. Set the parking brake when working on a vehicle. If the vehicle has an automatic transmission, place the gear selector in park unless a service procedure requires another selector position. When the vehicle is equipped with a manual transmission, position the gear selector in neutral with the engine running, or in reverse with the engine stopped.

5. Always connect a shop exhaust hose to the vehicle tailpipe and be sure the shop exhaust fan is running. If it is absolutely necessary to operate a vehicle without a shop exhaust pipe connected to the tailpipe, open the large shop door to provide adequate ventilation. **Carbon monoxide** in the vehicle exhaust may cause severe headaches and other medical problems. High concentrations of carbon monoxide may result in death!

6. Keep hands, clothing, and wrenches away from rotating parts such as cooling fans. Remember that electric-drive fans may start turning at any time, even with the ignition off.

7. Always leave the ignition switch off unless a service procedure requires another switch position.

8. Do not smoke in the shop. If the shop has designated smoking areas, smoke only in these areas.

9. Store oily rags and other discarded combustibles in covered metal containers designed for this purpose.

10. Always use the wrench or socket that fits properly on the bolt. Do not substitute metric for English wrenches, or vice versa.

Carbon monoxide is a poisonous gas, and when inhaled it may cause headaches, nausea, ringing in the ears, fatigue, and heart flutter. In strong concentrations, it causes death.

11. Keep tools in good condition. For example, do not use a punch or chisel with a mushroomed end. When struck with a hammer, a piece of the mushroomed metal could break off, resulting in severe eye or other injury.

12. Do not leave power tools running and unattended.

13. Serious burns may be prevented by avoiding contact with hot metal components, such as exhaust manifolds, other exhaust system components, radiators, and some air conditioning hoses.

14. When a lubricant such as engine oil is drained, always use caution because the oil could be hot enough to cause burns.

15. Prior to getting under a vehicle, be sure the vehicle is placed securely on safety stands.

16. Operate all shop equipment, including lifts, according to the equipment manufacturer's recommended procedure. Do not operate equipment unless you are familiar with the correct operating procedure.

17. Do not run or engage in horseplay in the shop.

18. Obey all state and federal fire, safety, and environmental regulations.

19. Do not stand in front of or behind vehicles.

20. Always place fender covers and a seat cover on a customer's vehicle before working on the car.

21. Inform the shop foreman of any safety dangers and suggestions for safety improvement.

Vehicle Operation

When driving a customer's vehicle, certain precautions must be observed to prevent accidents and maintain good customer relations.

1. Prior to driving a vehicle, make sure the brakes are operating and fasten the safety belt.

2. Check to be sure there is no person or object under the car before you start the engine.

3. If the vehicle is parked on a lift, be sure the lift is fully down and the lift arms, or components, are not in contact with the vehicle chassis.

4. Check to see if there are any objects directly in front of or behind the vehicle before driving away.

5. Always drive slowly in the shop, and watch carefully for personnel and other moving vehicles.

6. Make sure the shop door is up high enough so there is plenty of clearance between the top of the vehicle and the door.

7. Watch the shop door to be certain that it is not coming down as you attempt to drive under the door.

8. If a road test is necessary, obey all traffic laws, and never drive in a reckless manner.

9. Do not squeal tires when accelerating or turning corners.

If the customer observes that service personnel take good care of his or her car by driving carefully and installing fender and seat covers, the service department image is greatly enhanced in the customer's eyes. These procedures impress upon the customer that shop personnel respect the car. Conversely, if grease spots are found on upholstery or fenders after service work is completed, the customer will probably think the shop is careless, not only in car care, but also in service work quality.

Housekeeping

CUSTOMER CARE: When a customer sees that you are concerned about his or her vehicle, and that you operate a shop with excellent housekeeping habits, the customer will be impressed and will likely keep returning for service.

Careful housekeeping habits prevent accidents and increase worker efficiency. Good housekeeping also helps impress upon the customer that quality work is a priority in this shop. Follow these housekeeping rules:

1. Keep aisles and walkways clear of tools, equipment, and other items.
2. Be sure all sewer covers are securely in place.
3. Keep floor surfaces free of oil, grease, water, and loose material.
4. Proper trash containers must be conveniently located, and these containers should be emptied regularly.
5. Access to fire extinguishers must be unobstructed at all times, and fire extinguishers should be checked for proper charge at regular intervals.

SERVICE TIP: When you are finished with a tool, never set it on the customer's car. After using a tool, the best place for it is in your tool box or on the workbench. Many tools have been lost by leaving them on customer's vehicles.

6. Tools must be kept clean and in good condition.
7. When not in use, tools must be stored in their proper location.
8. Oily rags and other combustibles must be placed in proper covered containers.
9. Rotating components on equipment and machinery must have guards, and all shop equipment should have regular service and adjustment schedules.
10. Benches and seats must be clean.
11. Keep parts and materials in their proper location.
12. When not in use, creepers must not be left on the shop floor. Creepers should be stored in a specific location.
13. The shop should be well lighted, and all lights should be in working order.
14. Frayed electrical cords on lights or equipment must be replaced.
15. Walls and windows should be cleaned regularly.
16. Stairs must be clean, well lighted, and free of loose material.

If these housekeeping rules are followed, the shop will be a safer place to work, and customers will be impressed with the appearance of the premises.

> Excellent housekeeping involves general shop cleanliness, proper shop safety equipment in good working condition, and the proper maintenance of all shop equipment and tools.

Air Quality

 CAUTION: Never run a vehicle's engine inside the shop without an exhaust hose connected to the tailpipe.

Vehicle exhaust contains small amounts of carbon monoxide, which is a poisonous gas. Strong concentrations of carbon monoxide may be fatal. All shop personnel are responsible for air quality in the shop. Shop management is responsible for an adequate exhaust system to remove exhaust fumes from the maximum number of vehicles that may be running in the shop at the same time.

Technicians should never run a vehicle in the shop unless a shop exhaust hose is installed on the tailpipe of the vehicle. The exhaust fan must be switched on to remove exhaust fumes.

If shop heaters or furnaces have restricted chimneys, they release carbon monoxide emissions into the shop air. Therefore, chimneys should be checked periodically for restriction and proper ventilation.

Monitors are available to measure the level of carbon monoxide in the shop. Some of these monitors read the amount of carbon monoxide present in the shop air; others provide an audible alarm if the concentration of carbon monoxide exceeds the danger level.

Diesel exhaust contains some carbon monoxide, but **particulate emissions** are also present in the exhaust from these engines. Particulates are basically small carbon particles that can be harmful to the lungs.

The **sulfuric acid** solution in car batteries is a corrosive, poisonous liquid. If a battery is charged with a fast charger at a high rate for a period of time, the battery becomes hot, and the sulfuric acid solution begins to boil. Under this condition, the battery may emit a strong sulfuric acid smell and these fumes may be harmful to the lungs. If this happens, the battery charger should be turned off or the charging rate should be reduced considerably.

 WARNING: When an automotive battery is charged, hydrogen gas and oxygen gas escape from the battery. If these gases combine they form water, but hydrogen gas by itself is explosive. While a battery is charging, sparks, flames, and other sources of ignition must not be allowed near the battery.

WARNING: Breathing asbestos dust must be avoided because this dust is a known contributor to lung cancer.

Some automotive clutch facings and brake linings contain asbestos. Never use an air hose to blow dirt from these components, because this action disperses asbestos dust into the shop where it may be inhaled by technicians and other people in the shop. An asbestos parts washer or a vacuum cleaner with special attachments must be used to clean the dust from these components.

Even though technicians take every precaution to maintain air quality in the shop, some undesirable gases may still get in the air. For example, exhaust manifolds may get oil on them during an engine overhaul. When the engine is started and these manifolds get hot, the oil burns off the manifolds and pollutes the shop air with oil smoke. Adequate shop ventilation must be provided to take care of this type of air contamination.

Personal Safety

Personal safety is the responsibility of each technician in the shop. Always follow these safety practices:

1. Always use the correct tool for the job. If the wrong tool is used, it may slip and cause injury.

2. Follow the vehicle manufacturer's recommended service procedures.

3. Always wear eye protection, such as safety glasses or a face shield (Figure 1-3).

4. Wear protective gloves when cleaning parts in hot or cold tanks, and when handling hot parts such as exhaust manifolds.

Figure 1-3 Proper eye protection must be worn in the shop. (Courtesy of Mac Tools, Inc.)

5. Do not smoke when working on a vehicle. A spark from a cigarette or lighter may ignite flammable materials in the work area.

6. When working on a running engine, keep hands and tools away from rotating parts. Remember that electric-drive fans may start turning at any time.

7. Do not wear loose clothing, and keep long hair tied behind your head. Loose clothing or long hair is easily entangled in rotating parts.

8. Wear safety shoes or boots.

9. Do not wear watches, jewelry, or rings when working on a vehicle. Severe burns occur when jewelry makes contact between an electric terminal and ground. Rings may catch on an object resulting in painful injury.

10. Always place a shop exhaust hose on the vehicle tailpipe if the engine is running and be sure the exhaust fan is running. Carbon monoxide in the vehicle exhaust can be harmful or fatal.

11. Be sure that the shop has adequate ventilation. Carbon monoxide is odorless; do not expect to be able to smell it.

12. Make sure the work area has adequate lighting.

13. Use trouble lights with steel or plastic cages around the bulb. If an unprotected bulb breaks, it may ignite flammable materials in the area.

14. When servicing a vehicle always apply the parking brake. Place the transmission in park with an automatic transmission, or neutral with a manual transmission.

15. Avoid working on a vehicle parked on an incline.

16. Never work under a vehicle unless the vehicle chassis is supported securely on safety stands.

17. When one end of a vehicle is raised, place wheel chocks on both sides of the wheels remaining on the floor.

18. Be sure that you know the location of shop first-aid kits and eyewash fountains.

19. Familiarize yourself with the location of all shop fire extinguishers.

20. Do not use any type of open flame heater to heat the work area.

21. Collect oil, fuel, brake fluid, and other liquids in the proper safety containers.

22. Use only approved cleaning fluids and equipment. *Do not use gasoline to clean parts.*

23. Obey all state and federal safety, fire, and hazardous material regulations.

24. Always operate equipment according to the equipment manufacturer's recommended procedure.

25. Do not operate equipment unless you are familiar with the correct operating procedure.

26. Do not leave running equipment unattended.

27. Do not use electrical equipment, including trouble lights, with frayed cords.

28. Be sure the safety shields are in place on rotating equipment.

29. Before operating electric equipment, be sure the power cord has a ground connection.

30. When working in an area where noise levels are extreme, wear ear plugs or covers.

31. All shop equipment should have regular maintenance.

Lifting and Carrying

Many automotive service jobs require heavy lifting. Know your maximum weight lifting ability and do not attempt to lift more than this weight. If a heavy part exceeds your weight lifting ability, have a coworker help with the lifting job. Follow these steps when lifting or carrying an object:

1. If the object is going to be carried, be sure your path is free from loose parts or tools.

2. Position your feet close to the object; position your back reasonably straight for proper balance.

3. Your back and elbows should be kept as straight as possible. Continue to bend your knees until your hands reach the best lifting location on the object.

4. Be certain the container is in good condition. If a container falls apart during the lifting operation, parts may drop out of the container and result in foot injury or part damage.

5. Maintain a firm grip on the object; do not attempt to change your grip while lifting is in progress.

6. Straighten your legs to lift the object and keep the object close to your body. Use leg muscles rather than back muscles (Figure 1-4).

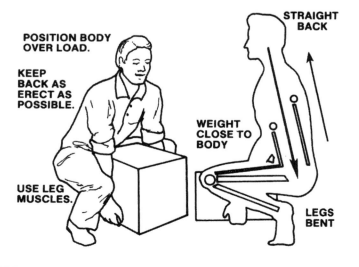

Figure 1-4 Use your leg muscles never your back to lift heavy objects.

7. If you have to change direction of travel, turn your whole body. Do not twist.

8. Do not bend forward to place an object on a workbench or table. Position the object on the front surface of the workbench and slide it back. Do not pinch your fingers under the object while setting it on the front of the bench.

9. If the object must be placed on the floor or a low surface, bend your legs to lower the object. Do not bend your back forward, because this movement strains back muscles.

10. When a heavy object must be placed on the floor, place suitable blocks under the object to prevent jamming your fingers.

Shop Manual
Chapter 1, page 2

Hand Tool Safety

Many shop accidents are caused by improper use and care of hand tools. Follow these safety steps when working with hand tools:

1. Maintain tools in good condition and keep them clean. Worn tools may slip and result in hand injury. If a hammer with a loose head is used, the head may fly off and cause personal injury or vehicle damage. If your hand slips off a greasy tool, it may cause some part of your body to hit the vehicle, causing injury.

2. Using the wrong tool for the job may damage the tool, fastener, or your hand, if the tool slips. If you use a screwdriver as a chisel or pry bar, the blade may shatter causing serious personal injury.

3. Use sharp pointed tools with caution. Always check your pockets before sitting on the vehicle seat. A screwdriver, punch, or chisel in the back pocket may put an expensive tear in the upholstery. Do not lean over fenders with sharp tools in your pockets.

4. Tools that are intended to be sharp should be kept sharp. A sharp chisel, for example, will do the job faster with less effort.

Lift Safety

 WARNING: Do not raise a vehicle on a lift if the vehicle weight exceeds the maximum capacity of the lift.

 WARNING: When a vehicle is raised on a lift, the vehicle must be raised high enough to allow engagement of the lift locking mechanism.

Special precautions and procedures must be followed when a vehicle is raised on a lift. Follow these steps for a hoist lift:

1. Always be sure the lift is completely lowered before driving a vehicle on or off the lift.

2. Do not hit or run over lift arms and adaptors when driving a vehicle on or off the lift. Have a coworker guide you when driving a vehicle onto the lift. Do not stand in front of a lift with the car coming toward you.

3. Be sure the lift pads contact the car manufacturer's recommended lifting points shown in the service manual. If the proper lifting points are not used, components under the vehicle such as brake lines or body parts, may be damaged. Failure to use the recommended lifting points may cause the vehicle to slip off the lift, resulting in severe vehicle damage and personal injury.

A lift may be referred to as a hoist.

4. Before a vehicle is raised or lowered, close the doors, hood, and trunk lid.

5. When a vehicle has been lifted a short distance off the floor, stop the lift and check the contact between the hoist lift pads and the vehicle to be sure the lift pads are still on the recommended lifting points.

6. When a vehicle has been raised, be sure the safety mechanism is in place to prevent the lift from dropping accidentally.

7. Prior to lowering a vehicle, always make sure there are no objects, tools, or people under the vehicle.

8. Do not rock a vehicle on a lift during a service job.

9. When a vehicle is raised, removal of some heavy components may cause vehicle imbalance. For example, since front-wheel-drive cars have the engine and transaxle at the front of the vehicle, these cars have most of their weight on the front end. Removing a heavy rear-end component on these cars may cause the back end of the car to rise off the lift. If this happens the vehicle could fall off the lift!

10. Do not raise a vehicle on a lift with people in the vehicle.

11. When raising pickup trucks and vans on a lift, remember these vehicles are higher than a passenger car. Be sure there is adequate clearance between the top of the vehicle and the shop ceiling, or components under the ceiling.

12. *Do not raise a four-wheel-drive vehicle with a frame contact lift because this may damage axle joints.*

13. Do not operate a front-wheel-drive vehicle that is raised on a frame contact lift. This may damage the front drive axles.

Hydraulic Jack and Safety Stand Safety

 WARNING: Always make sure the safety stand weight capacity rating exceeds the vehicle weight that you are planning to raise.

 WARNING: Never lift a vehicle with a floor jack if the weight of the vehicle exceeds the rated capacity of the jack.

Accidents involving the use of floor jacks and safety stands may be avoided if these safety precautions are followed:

1. Never work under a vehicle unless safety stands are placed securely under the vehicle chassis and the vehicle is resting on these stands (Figure 1-5).

2. Prior to lifting a vehicle with a floor jack, be sure that the jack lift pad is positioned securely under a recommended lifting point on the vehicle. Lifting the front end of a vehicle with the jack placed under a radiator support may cause severe damage to the radiator and support.

3. Position the safety stands under a strong chassis member such as the frame or axle housing. The safety stands must contact the vehicle manufacturer's recommended lifting points.

4. Since the floor jack is on wheels, the vehicle and safety stands tend to move as the vehicle is lowered from a floor jack onto safety stands. Always be sure the safety stands

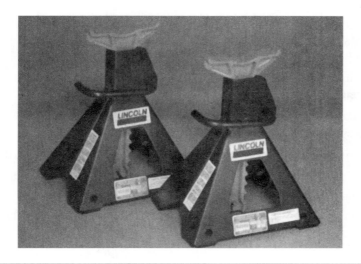

Figure 1-5 Safety stands. (Courtesy of Lincoln, St. Louis)

remain under the chassis member during this operation, and be sure the safety stands do not tip. All the safety stand legs must remain in contact with the shop floor.

5. When the vehicle is lowered from the floor jack onto safety stands, remove the floor jack from under the vehicle. Never leave a jack handle sticking out from under a vehicle. Someone may trip over the handle and injure himself or herself.

Power Tool Safety

Power tools use electricity, shop air, or hydraulic pressure as a power source. Careless operation of power tools may cause personal injury or vehicle damage. Follow these steps for safe power tool operation:

1. Do not operate power tools with frayed electrical cords.
2. Be sure the power tool cord has a proper ground connection.
3. Do not stand on a wet floor while operating an electric power tool.
4. Always unplug an electric power tool before servicing the tool.
5. Do not leave a power tool running and unattended.
6. When using a power tool on small parts, do not hold the part in your hand. The part must be secured in a bench vise or with locking pliers.
7. Do not use a power tool on a job where the maximum capacity of the tool is exceeded.
8. Be sure that all power tools are in good condition; always operate these tools according to the tool manufacturer's recommended procedure.
9. Make sure all protective shields and guards are in position.
10 Maintain proper body balance while using a power tool.
11. Always wear safety glasses or a face shield.
12. Wear ear protection.
13. Follow the equipment manufacturer's recommended maintenance schedule for all shop equipment.

14. Never operate a power tool unless you are familiar with the tool manufacturer's recommended operating procedure. Serious accidents occur from improper operating procedures.

15. Always make sure that the wheels are securely attached and in good condition on the electric grinder.

16. Keep fingers and clothing away from grinding and buffing wheels. When grinding or buffing a small part, hold the part with a pair of locking pliers.

17. Always make sure the sanding or buffing disc is securely attached to the sander pad.

18. Special heavy-duty sockets must be used on impact wrenches. If ordinary sockets are used on an impact wrench, they may break and cause serious personal injury.

19. Never operate an air chisel unless the tool is securely connected to the chisel with the proper retaining device.

20. Never direct a blast of air from an air gun against any part of your body. If air penetrates the skin and enters the bloodstream, it may cause very serious health problems and even death.

Compressed-Air Equipment Safety

The shop air supply contains high pressure air in the shop compressor and air lines. Serious injury or property damage may result from careless operation of compressed-air equipment. Follow these steps to improve safety.

1. Safety glasses or a face shield should be worn for all shop tasks, including those tasks involving the use of compressed-air equipment.

2. Wear ear protection when using compressed-air equipment.

3. Always maintain air hoses and fittings in good condition. If an end suddenly blows off an air hose, the hose will whip around, possibly causing personal injury.

4. Do not direct compressed air against the skin. This air may penetrate the skin, especially through small cuts or scratches. If air penetrates the skin and enters the bloodstream, it can be fatal or cause serious health complications. Use only air gun nozzles approved by Occupational Safety and Health Act (OSHA).

5. Do not use an air gun to blow debris off clothing or hair.

6. Do not clean the workbench or floor with compressed air. This action may blow very small parts against your skin or into your eye. Small parts blown by compressed air may also cause vehicle damage. For example, if the car in the next stall has the air cleaner removed, a small part may find its way into the carburetor or throttle body. When the engine is started, this part will likely be pulled into the cylinder by engine vacuum, and the part will penetrate through the top of a piston.

7. Never spin bearings with compressed air because the bearing will rotate at extremely high speed. This may damage the bearing or cause it to disintegrate, causing personal injury.

8. All pneumatic tools must be operated according to the tool manufacturer's recommended operating procedure.

9. Follow the equipment manufacturer's recommended maintenance schedule for all compressed-air equipment.

Tools operated by air pressure may be referred to as pneumatic tools.

Shop Manual
Chapter 1, page 13

Guidelines for Safety Practices

1. Technicians must be familiar with shop layout, especially the location of safety equipment. This knowledge provides a safer, more efficient shop.

2. Shop rules must be observed by everyone in the shop to provide adequate shop safety, personal health protection, and vehicle protection.

3. The application of driving rules in the shop increases safety, protects customer vehicles and shop property, and improves the shop image in the eyes of the customer.

4. When good housekeeping habits are developed, shop safety is improved, worker efficiency is increased, and customers are impressed.

5. If some basic rules are followed to maintain shop air quality, the personal health of shop employees is improved.

6. Personal safety is the responsibility of everyone in the shop.

7. When lifting heavy objects, always bend your knees rather than your back.

8. Many shop accidents are caused by the improper use of hand tools.

9. Never exceed the rated capacity of a hydraulic press, hydraulic jack, vehicle lift, or safety stands.

10. When raising a vehicle with a lift or floor jack, always be sure the lifting equipment is contacting the vehicle on the manufacturer's recommended lifting points.

11. After a vehicle is raised on a lift, be sure the lift locking mechanism is in place.

12. Never operate electrical equipment with a frayed cord or without a ground wire.

13. Never direct a blast of compressed air against human flesh.

CASE STUDY 1

A technician raised a Grand Marquis on a lift to perform an oil and filter change and a chassis lubrication. This lift was a twin post-type with separate front and rear lift posts. On this type of lift, the rear wheels must be positioned in depressions in the floor to position the rear axle above the rear lift arm. Then the front lift post and arms must be moved forward or rearward to position the front lift arms under the front suspension. The front lift arms must also be moved inward or outward so they are lifting on the vehicle manufacturer's specified lifting points.

The technician carefully positioned the front lift post and arms properly, but he forgot to check the position of the rear tires in the floor depressions. The car was raised on the lift, and the technician proceeded with the service work. Suddenly there was a loud thump, and the rear of the car bounced up and down! The rear lift arms were positioned against the floor of the trunk rather than on the rear axle, and the lift arms punched through the floor of the trunk, narrowly missing the fuel tank. The technician was extremely fortunate the car did not fall off the lift resulting in more severe damage. If the rear lift arms had punctured the fuel tank, a disastrous fire could have occurred! Luckily, these things did not happen.

The technician learned a very important lesson about lift operation. Always follow all the recommended procedure in the lift operator's manual! The trunk floor was repaired at no cost to the customer, and fortunately, the shop and the vehicle escaped without major damage.

A technician was installing a rear axle bearing and seal on a small rear-wheel-drive car. The technician raised the vehicle on a lift to a comfortable working height and proceeded to pull the rear axle with a slide hammer-type puller. This technician was very strong and weighed about 250 pounds. The rear axle was extremely tight, and the technician operated the slide hammer puller with all the strength he had. Each time he operated the slide hammer, the rear axle bounced sideways on the lift. Suddenly the car moved sideways so much that the car fell off the lift and over on its side. Fortunately, the technician was able to jump out of the way of the falling vehicle, but the vehicle body suffered considerable damage.

The technician learned to avoid service operations on a lift that may cause the car to move on the lift. However, if such service operations must be performed, the car should be chained to the lift.

Terms to Know

Asbestos parts washer

Carbon monoxide

Particulate emissions

Shop layout

Sulfuric acid

ASE Style Review Questions

1. While discussing shop layout and safety:
   Technician A says the location of each piece of safety equipment should be clearly marked.
   Technician B says some shops have a special tool room where special tools are located.
   Who is correct?
   A. A only
   B. B only
   C. Both A and B
   D. Neither A nor B

2. Breathing carbon monoxide may cause:
   A. arthritis.
   B. cancer.
   C. impaired vision.
   D. headaches.

3. While discussing shop rules:
   Technician A says breathing asbestos dust may cause heart defects.
   Technician B says oily rags should be stored in uncovered trash containers.
   Who is correct?
   A. A only
   B. B only
   C. Both A and B
   D. Neither A nor B

4. All these shop rules are correct EXCEPT:
   A. USC tools may be substituted for metric tools.
   B. Foot injuries may be caused by loose sewer covers.
   C. Hands should be kept away from electric-drive cooling fans.
   D. Power tools should not be left running and unattended.

5. While discussing air quality:
   Technician A says a restricted chimney on a shop furnace may cause carbon monoxide gas in the shop.
   Technician B says monitors are available to measure the level of carbon monoxide in the shop air.
   Who is correct?
   A. A only
   B. B only
   C. Both A and B
   D. Neither A nor B

6. Some of the concerns about air quality safety in an automotive shop are that:
   A. automotive batteries contain hydrochloric acid.
   B. a battery gives off hydrocarbon gas during the charging process.
   C. shop furnaces must be properly vented to reduce carbon monoxide emissions.
   D. diesel exhaust contains sulfur dioxide.

7. While discussing personal safety:
   Technician A says rings and jewelry may be worn in the automotive shop.
   Technician B says some electric-drive cooling fans may start turning at any time.
   Who is correct?
   A. A only
   B. B only
   C. Both A and B
   D. Neither A nor B

8. While lifting heavy objects in the automotive shop:
   A. bend your back to pick up the heavy object.
   B. place your feet as far as possible from the object.
   C. bend forward to place the object on the workbench.
   D. straighten your legs to lift an object off the floor.

9. While discussing power tool safety:
   Technician A says an electric power tool cord does not require a ground.
   Technician B says frayed electric cords should be replaced.
   Who is correct?
   A. A only
   B. B only
   C. Both A and B
   D. Neither A nor B

10. While operating hydraulic equipment safely in the automotive shop remember that:
    A. safety stands have a maximum weight capacity.
    B. the driver's door should be open when raising a vehicle on a lift.
    C. a lift does not require a safety mechanism to prevent lift failure.
    D. four-wheel-drive vehicles should be lifted on a frame contact lift.

Job Sheet 1

Name _____ Date _____

Demonstrate Proper Lifting Procedures

Upon completion of this job sheet, you should be able to follow the proper procedure when lifting heavy objects.

Tools and Materials

A heavy object with a weight that is within the weight lifting capability of the technician.

Procedure

Have various members of the class demonstrate proper weight lifting procedures by lifting an object off the shop floor and placing it on the workbench. Other class members are to observe and record any improper weight lifting procedures.

1. If the object is to be carried, be sure your path is free from loose parts or tools.

2. Position your feet close to the object and position your back reasonably straight for proper balance.

3. Your back and elbows should be kept as straight as possible. Continue to bend your knees until your hands reach the best lifting location on the object to be lifted.

4. Be certain the container is in good condition. If a container falls apart during the lifting operation, parts may drop out of the container and result in foot injury or part damage.

5. Maintain a firm grip on the object; do not attempt to change your grip while lifting is in progress.

6. Straighten your legs to lift the object, keeping the object close to your body. Use leg muscles rather than back muscles.

7. If you have to change direction of travel, turn your whole body. Do not twist.

8. Do not bend forward to place an object on a workbench or table. Position the object on the front surface of the workbench and slide it back. Do not pinch your fingers under the object while setting it on the front of the bench.

9. If the object must be placed on the floor or a low surface, bend your legs to lower the object. Do not bend your back forward, because this movement strains back muscles.

10. When a heavy object must be placed on the floor, put suitable blocks under the object to prevent jamming your fingers under the object.

11. List any improper weight lifting procedures observed during the weight lifting demonstrations.

    1. _____
    2. _____
    3. _____
    4. _____
    5. _____
    6. _____

☑ **Instructor Check** _____

Job Sheet 2

Name _____ Date _____

Locate and Inspect Shop Safety Equipment

Upon completion of this job sheet, you should be familiar with the location of shop safety equipment and know if this equipment is serviced properly.

Tools and Materials

None

Procedure

1. *Fire extinguishers:* Are the fire extinguishers tagged to indicate they have been checked or serviced recently? yes _____ no _____

   Draw a basic diagram of the shop layout in the space below and mark the locations of the fire extinguishers and fire exits.

2. *Eyewash fountain or shower:* Is the eyewash fountain or shower operating properly?
   yes _____ no _____
   Mark the location of the eyewash fountain or shower on the shop layout diagram in step 1.

3. *First-aid kits:* Are the first-aid kits properly stocked with supplies?

   yes _____ no _____

   Mark the location of the first-aid kits on the shop layout diagram in step 1

4. *Electrical shut-off box:* Mark the location of the shop electrical shut-off box on the shop layout diagram in step 1.

5. Are the trash containers equipped with proper covers?

   yes _____ no _____

   Mark the location of the trash containers on the shop layout diagram in step 1.

6. *Metal storage cabinet:* Mark the location of metal storage cabinet(s) for combustible materials on the shop layout diagram in step 1.

☑ **Instructor Check** _____

Job Sheet 3

Name _____ Date _____

Shop Housekeeping Inspection

Upon completion of this job sheet, you should be able to apply shop housekeeping rules in your shop.

Tools and Materials

None

Procedure

When another class of Automotive Technology students is working in the shop, evaluate their shop housekeeping procedures using the sixteen shop housekeeping procedures. List all the improper shop housekeeping procedures that you observed in the space provided at the end of the job sheet.

1. Keep aisles and walkways clear of tools, equipment, and other items.
2. Be sure all sewer covers are securely in place.
3. Keep floor surfaces free of oil, grease, water, and loose material.
4. Proper trash containers must be conveniently located, and these containers should be emptied regularly.
5. Access to fire extinguishers must be unobstructed at all times; fire extinguishers should be checked for proper charge at regular intervals.

> **WARNING:** When you are finished with a tool never set it on the customer's car. After using a tool, the best place for it is in your tool box or on the workbench. Many tools have been lost by leaving them on customer's cars.

6. Tools must be kept clean and in good condition.
7. When not in use tools must be stored in their proper location.
8. Oily rags and other combustibles must be placed in proper, covered metal containers.
9. Rotating components on equipment and machinery must have guards. All shop equipment should have regular service and adjustment schedules.
10. Benches and seats must be kept clean.
11. Keep parts and materials in their proper location.
12. When not in use creepers, must not be left on the shop floor. Creepers should be stored in a specific location.
13. The shop should be well lighted and all lights should be in working order.
14. Frayed electrical cords on lights or equipment must be replaced.
15. Walls and windows should be cleaned regularly.
16. Stairs must be clean, well lighted, and free of loose material.

Observed improper shop housekeeping procedures:

1. _____
2. _____
3. _____
4. _____
5. _____
6. _____

✔ **Instructor Check** _____

Tools and Shop Practices

Upon completion and review of this chapter, you should be able to:

❏ Perform automotive measurements using the United States Customary (USC) system and the international system (SI) of weights and measures.

❏ Explain two measurements that may be performed with a dial indicator during suspension and steering system service.

❏ Describe the basic purpose of a coil spring compressing tool.

❏ Describe how a plumb bob may be used when performing suspension and steering service.

❏ Follow the recommended procedure while operating hydraulic tools, such as presses, floor jacks, and vehicle lifts to perform automotive service tasks.

❏ Fulfill employee obligations when working in the shop.

❏ Accept job responsibility for each job completed in the shop.

❏ Describe the ASE technician testing and certification process including the eight areas of certification.

Measuring Systems

Two systems of weight and measures are commonly used in the United States. One system is the **United States customary (USC) system,** which is commonly referred to as the English system. Well-known measurements for length in the USC system are the inch, foot, yard, and mile. In this system, the quart and gallon are common measurements for volume, and ounce, pound, and ton are measurements for weight. A second system of weights and measures is called the **international system (SI),** which is often referred to as the metric system.

In the USC system the basic linear measurement is the yard, whereas the corresponding linear measurement in the metric system is the meter. Each unit of measurement in the metric system is related to the other metric units by a factor of 10. Thus every metric unit can be multiplied or divided by 10 (or 100 or 1,000) to obtain larger units (multiples), or smaller units (submultiples). For example, the meter may be divided by 100 to obtain centimeters (1/100 meter), or millimeters (1/1,000 meter).

The U.S. government passed the Metric Conversion Act in 1975 in an attempt to move American industry and the general public to accept and adopt the metric system. The automotive industry has adopted the metric system, and in recent years most bolts, nuts, and fittings on vehicles have been changed to metric. During the early 1980s some vehicles had a mix of English and metric bolts. Imported vehicles have used the metric system for many years. Although the automotive industry has changed to the metric system, the general public in the United States has been slow to convert from the USC system to the metric system. One of the factors involved in this change is cost. What would it cost to change every highway distance and speed sign in the United States to read kilometers? Probably hundreds of millions, or even billions, of dollars.

The United States Customary (USC) system of weights and measures is sometimes referred to as the English system.

The international system (SI) of weights and measures is called the metric system.

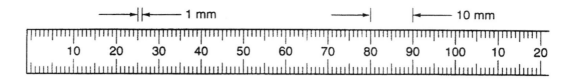

Figure 2-1 Metric tape graduated in millimeters.

Service technicians must be able to work with both the USC and the metric system. One meter (m) in the metric system is equal to 39.37 inches (in) in the USC system. A metric tape measure may be graduated in millimeters, and 10 millimeters = 1 centimeter (Figure 2-1). Some common equivalents between the metric and USC systems are these:

1 meter (m) = 39.378 inches (in)

1 centimeter (cm) = 0.3937 inch

1 millimeter (mm) = 0.03937 inch

1 inch = 2.54 cm

1 inch = 25.4 mm

In the USC system, phrases such as 1/8 of an inch are used for measurements. The metric system uses a set of prefixes. For example, in the word kilometer the prefix *kilo* means 1,000, indicating that there are 1,000 meters in a kilometer. Common prefixes in the metric system are:

NAME	SYMBOL	MEANING
mega	M	one million
kilo	k	one thousand
hecto	h	one hundred
deca	da	ten
deci	d	one tenth of
centi	c	one hundredth of
milli	m	one thousandth of
micro	µ	one millionth of

Measurement of Mass

In the metric system mass is measured in grams, kilograms, or tonnes: 1,000 grams (g) = 1 kilogram (kg). In the USC system mass is measured in ounces, pounds, or tons. When converting pounds to kilograms, 1 pound = .453 kilograms.

Measurement of Length

In the metric system length is measured in millimeters, centimeters, meters, or kilometers: 10 millimeters (mm) = 1 centimeter (cm). In the USC system length is measured in inches, feet, yards, or miles. When distance conversions are made between the two systems some of the following conversion factors are used.

> 1 inch = 25.4 millimeters
>
> 1 foot = 30.48 centimeters
>
> 1 yard = 0.91 meters
>
> 1 mile = 1.60 kilometers

Measurement of Volume

In the metric system volume is measured in milliliters, cubic centimeters, and liters: 1 cubic centimeter = 1 milliliter. If a cube has a length, depth, and height of 10 centimeters (cm), the volume of the cube is 10 cm × 10 cm × 10 cm = 1,000 cm³ = 1 liter. When volume conversions are made between the two systems, 1 cubic inch = 16.38 cubic centimeters. If an engine has a displacement of 350 cubic inches, 350 × 16.38 = 5,733 cubic centimeters, and 5,733 ÷ 1,000 = 5.7 liters.

Suspension and Steering Tools

CUSTOMER CARE: Some automotive service centers have a policy of performing some minor service as an indication of their appreciation to the customer. This service may include cleaning all the windows, and/or vacuuming the floors, before the car is returned to the customer.

Although this service involves more labor costs for the shop, it may actually improve profits over a period of time. When customers find their windows cleaned, and/or the floors vacuumed, it impresses them with the quality of work you do, and the fact you care about their vehicle. They will likely return for service, and tell their friends at coffee break about the quality of service your shop performs.

Automotive service personnel must remember that in 1992, 48.6% of the licensed drivers in the United States were female. It is entirely possible these drivers may even appreciate clean windows and vacuumed floors more than male drivers.

Seal Drivers

 CAUTION: When using any suspension and steering tools, the vehicle manufacturer's recommended procedure in the service manual must be followed. Improper use of tools may lead to personal injury.

 WARNING: Do not hit the seal case with a hammer. This action will damage the seal.

 WARNING: Always be sure the seal goes into the housing squarely and evenly.

SERVICE TIP: When installing a seal, the garter spring must face toward the flow of lubricant.

Figure 2-2 Seal drivers. (Courtesy of Snap-on Tools Corp.)

Seal drivers are designed to fit squarely against the seal case and inside the seal lip. A soft hammer is used to tap the seal driver and drive the seal straight into the housing. Some tool manufacturers market a seal driver kit with drivers to fit many common seals (Figure 2-2).

Bearing Pullers

A variety of **bearing pullers** are available to pull different sizes of bearings in various locations (Figure 2-3). Some bearing pullers are slide hammer-type, whereas others are screw-type.

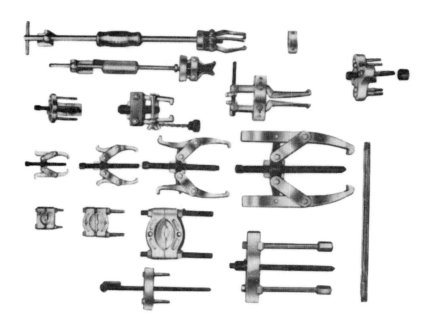

Figure 2-3. Bearing pullers (Courtesy of Snap-on Tools Corp.)

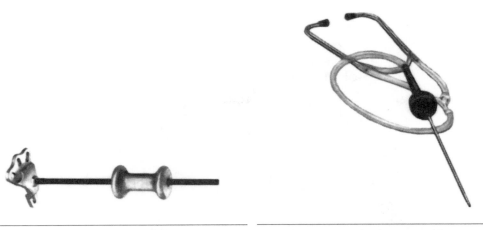

Figure 2-4. Rear axle puller. (Courtesy of Mac Tools, Inc.)

Figure 2-5. Stethoscope. (Courtesy of Snap-on Tools Corp.)

Axle Pullers

Axle pullers are used to pull rear axles in rear-wheel-drive vehicles. Most rear axle pullers are slide hammer-type (Figure 2-4).

Stethoscope

A **stethoscope** is used to amplify sound and diagnose the source of noises such as bearing noise. The stethoscope pickup is placed on the suspected noise source, and the ends of the two arms are placed in the technician's ears (Figure 2-5).

Front Bearing Hub Tool

Front bearing hub tools are designed to remove and install front wheel bearings on front-wheel-drive cars (Figure 2-6). These bearing hub tools are usually designed for a specific make of vehicle.

Dial Indicator

 SERVICE TIP: Dial indicators must be kept clean and dry for accuracy and long life.

Dial indicators are used for measuring in many different locations. In the suspension and steering area, dial indicators are used for measurements such as tire runout and ball joint movement (Figure 2-7). Dial indicators have many different attaching devices to connect the indicator to the component to be measured.

Tire Tread Depth Gauge

A **tire tread depth gauge** measures tire tread depth. This measurement should be taken at three or four locations around the tire's circumference to obtain an average tread depth (Figure 2-8). This gauge is essential when making tire warranty adjustments.

Tire Changer

 CAUTION: Never operate a tire changer until you are familiar with its operation. This action may lead to serious personal injury. Your instructor will explain and

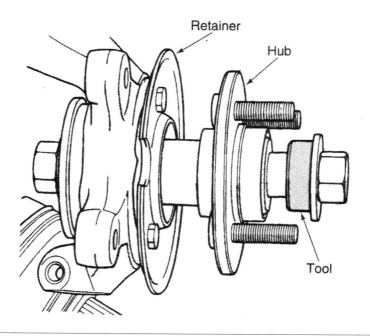

Retainer

Hub

Tool

Figure 2-6 Front bearing hub tool. (Courtesy of Chrysler Corp.)

demonstrate the operation of this equipment. You should read the equipment operator's manual and use this equipment under the instructor's supervision until you are familiar with it.

Tire changers are used to demount and mount tires (Figure 2-9). These changers may be used on most common tire sizes. There are a wide variety of tire changers available, and each one has somewhat different operating procedures. Always follow the procedure in the equipment operator's manual, and the directions provided by your instructor.

Figure 2-7 Dial indicator for measuring ball joint wear and wheel runout. (Courtesy of Mac Tools, Inc.)

Figure 2-8 Tire tread depth gauge. (Courtesy of Mac Tools, Inc.)

Wheel Balancer

CAUTION: Using a wheel balancer before you are familiar with its operation may result in serious personal injury and property damage.

Electronic wheel balancers are used in most automotive repair shops (Figure 2-10). Do not attempt to use this equipment until you have studied wheel balance theory. Your instructor will explain and demonstrate the use of this equipment before you attempt to use it. This equipment should be used under the supervision of your instructor until you are familiar with it.

Machinist's Rule

A **machinist's rule** is used to perform many accurate measurements in the shop. Most machinists rules are graduated in inches and millimeters (Figure 2-11).

Figure 2-9 Typical tire changer.

Figure 2-10 Typical electronic wheel balancer. (Courtesy of Hunter Engineering Co.)

Cat #	Length	Width	Graduations Inches Front	Back	Type	Wt. (lbs)
9-GHCF645	6"	1/2"	10, 100	32, 64	FLEX	0.2
9-GHCF1265	12"	1"	10, 100	32, 64	RIGID	0.4
9-GHCF1865	18"	1 1/8"	10, 100	32, 64	RIGID	0.6
9-GHCF2465	21"	1 1/8"	10, 100	32, 64	RIGID	0.8
9-GHCF667ME	6"	3/4"	32, 64	mm, .5mm	RIGID	0.2
9-GHCF1247M	12"	1/2"	32, 64	mm, .5mm	FLEX	0.4

Cat #	Length	Width	Graduations Inches Front	Back	Type	Wt. (lbs)
9-GH3001	6"	15/32"	32, 64	Dec. Eq.	FLEX	0.2
9-GH300ME	6"	15/32"	mm, 64	Metric Eq.	FLEX	0.2
9-GH311ME	6"	3/4"	16, 32	–	FLEX	0.2
9-GHCF616	6"	15/32"	10, 100	32, 64	FLEX	0.2
9-GH616	6"	15/32"	10, 100	32, 64	FLEX	0.2
9-GHCF676	6"	3/4"	8, 16	32, 64	RIGID	0.2
9-GHCF1227ME	12"	1"	32, 64	mm, .5mm	RIGID	0.4
9-GH1210	12"	1"	16, 32	32, 64	CENTER FINDING RIGID	0.2

Figure 2-11 Machinist's rule. (Courtesy of Sears Industrial Tools)

Ball Joint Removal and Pressing Tools

Ball joint removal and pressing tools are designed to remove and replace pressed-in ball joints on front suspension systems (Figure 2-12). The size of the removal and pressing tool must match the size of the ball joint.

Tie-Rod End and Ball Joint Puller

Some car manufacturer's recommend a **tie-rod end and ball joint puller** to remove tie-rod ends and pull ball joint studs from the steering knuckle (Figure 2-13).

Control Arm Bushing Tools

A variety of **control arm bushing tools** are available to remove and replace control arm bushings. The bushing removal and installation tool must match the size of the control arm bushing (Figure 2-14).

Coil Spring Compressor Tool

CAUTION: There is a tremendous amount of energy in a compressed coil spring. Never disconnect any suspension component that will suddenly release this tension. This action may result in serious personal injury and vehicle or property damage.

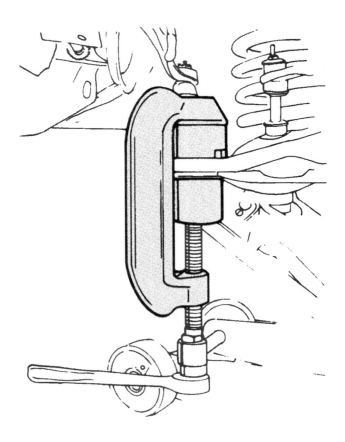

Figure 2-12 Ball joint removal and pressing tools. (Courtesy of Chevrolet Motor Division, General Motors Corp.)

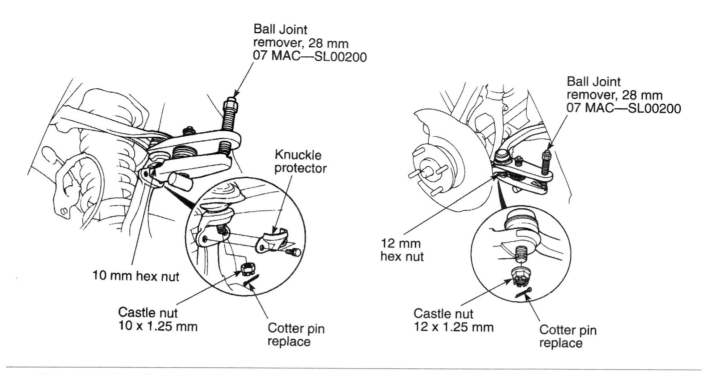

Figure 2-13 Tie-rod end and ball joint puller. (Courtesy of American Honda Motor Co., Inc.)

⚠ **WARNING:** The vehicle and equipment manufacturers' recommended procedures must be followed for each type of spring compressor tool.

There are many types of **coil spring compressor tools** available to the automotive service industry (Figure 2-15). These tools are designed to compress the coil spring and hold it in the compressed position while removing the strut from the coil spring, or performing other suspension work. Various types of spring compressor tools are required on different types of front suspension systems. While using spring compressor tools, the vehicle manufacturer's and equipment manufacturer's recommended procedure must be followed.

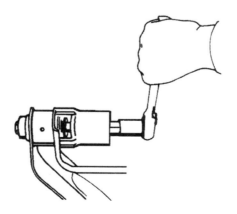

Figure 2-14 Control arm bushing removal and replacement tools. (Courtesy of Chevrolet Motor Division, General Motors Corp.)

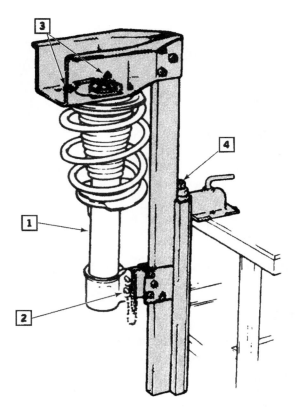

1	Strut assembly
2	Install locking pins through strut assembly
3	Tighten nuts till flush with strut compressor
4	Compressor forcing screw

Figure 2-15. Coil spring compressor tool. (Courtesy of Oldsmobile Motor Division, General Motors Corp.)

Power Steering Pressure Gauge

CAUTION: The power steering pump delivers extremely high pressure during the pump pressure test. Always follow the recommended test procedure in the vehicle manufacturer's service manual to avoid personal injury during this test.

A **power steering pressure gauge** is used to test the power steering pump pressure (Figure 2-16). Since the power steering pump delivers extremely high pressure during this test, the recommended procedure in the vehicle manufacturer's service manual must be followed.

Pitman Arm Puller

CAUTION: Never strike a puller with a hammer when it is installed and tightened. This action may result in personal injury.

A **pitman arm puller** is a heavy-duty puller designed to remove the pitman arm from the pitman shaft (Figure 2-17).

Figure 2-16 Power steering pressure gauge. (Courtesy of Mac Tools, Inc.)

Figure 2-17 Pitman arm puller. (Courtesy of Mac Tools, Inc.)

Vacuum Hand Pump

A **vacuum hand pump** creates a vacuum to test vacuum operated components and hoses (Figure 2-18).

Turning Radius Gauge

Turning radius gauge turn tables are placed under the front wheels during a wheel alignment. The top plate in the turning radius gauge rotates on the bottom plate to allow the front wheels to be turned during a wheel alignment. A degree scale and a pointer on the gauge indicate the number of degrees the front wheels are turned (Figure 2-19). If the car has four wheel steering (4WS), the turning radius gauges are also placed under the rear wheels during a wheel alignment.

Plumb Bob

A **plumb bob** is a metal weight with a tapered end (Figure 2-20). This weight is suspended on a string. Plumbers use a plumb bob to locate pipe openings directly below each other at the top and bottom of partitions. Some vehicle manufacturers recommend checking vehicle frame measurements with a plumb bob.

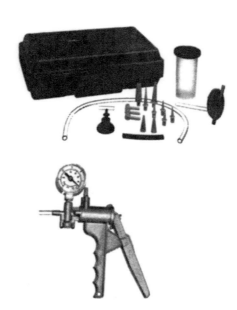

Figure 2-18 Vacuum hand pump. (Courtesy of Mac Tools, Inc.)

Figure 2-19 Turning radius gauge. (Courtesy of American Honda Motor Co., Inc.)

Tram Gauge

A **tram gauge** is a long, straight graduated bar with an adjustable pointer at each end (Figure 2-21). The tram gauge is used for performing frame and body measurements.

Magnetic Wheel Alignment Gauge

 SERVICE TIP: Magnetic wheel alignment gauge mounting surfaces must be clean with no metal burrs.

Each **magnetic wheel alignment gauge** contains a strong magnet that holds the gauge securely on the front wheel hubs. The magnetic wheel alignment gauge is capable of measuring some of the front suspension alignment angles (Figure 2-22).

Rim Clamps

When the wheel hub is inaccessible to the magnetic alignment gauge, an adjustable **rim clamp** may be attached to each front wheel. The magnetic gauges may be attached to the rim clamp (Figure 2-23). Rim clamps are also used on computer wheel aligners.

Figure 2-20 Plumb bob. (Courtesy of Sears Industrial Tools)

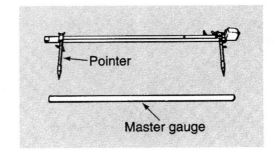

Pointer

Master gauge

Figure 2-21 Tram gauge. (Reprinted with permission)

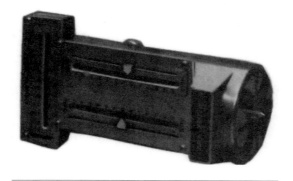

Figure 2-22 Magnetic wheel alignment gauge. (Courtesy of Snap-on Tools Corp.)

Figure 2-23 Rim clamps. (Courtesy of Mac Tools, Inc.)

Brake Pedal Jack

A **brake pedal jack** must be installed between the front seat and the brake pedal to apply the brakes while checking some front wheel alignment angles (Figure 2-24).

Tie-Rod Sleeve Adjusting Tool

 WARNING: Do not use anything except a tie rod adjusting tool to adjust the tie-rod sleeves. Tools such as a pipe wrench will damage the sleeves.

A **tie-rod sleeve adjusting tool** is required to rotate the tie-rod sleeves and perform some front wheel adjustments (Figure 2-25).

Steering Wheel Locking Tool

A **steering wheel locking tool** is required to lock the steering wheel while performing some front suspension service (Figure 2-26).

Toe Gauge

A **toe gauge** is a long, straight, graduated bar that may be used to measure front wheel toe (Figure 2-27).

Track Gauge

Some **track gauges** use a fiber-optic alignment system to measure front wheel toe and to determine if the rear wheels are tracking directly behind the front wheels. The front and rear fiber-optic gauges may be connected to the wheel hubs or to rim clamps attached to the wheel rims (Figure 2-28). A remote light source in the main control is sent through fiber-optic cables to the wheel gauges. A strong light beam between the front and rear wheel units informs the technician if the rear wheel tracking is correct.

Figure 2-24 Brake pedal jack. (Courtesy of Mac Tools, Inc.)

Figure 2-25 Tie-rod sleeve adjusting tool. (Courtesy of Mac Tools, Inc.)

Figure 2-26 Steering wheel locking tool. (Courtesy of Mac Tools, Inc.)

Figure 2-27 Toe gauge. (Courtesy of Mac Tools, Inc.)

Computer Wheel Aligner

Many automotive shops are equipped with a **computer wheel aligner** (Figure 2-29). These wheel aligners perform all front and rear wheel alignment angles quickly and accurately.

Scan Tool

There are a variety of **scan tools** available for diagnosing automotive computer systems (Figure 2-30). These testers will obtain fault codes and perform other diagnostic functions on computer-controlled suspension systems.

Bench Grinder

CAUTION: Always wear a face shield when using a bench grinder. Failure to observe this precaution may cause personal injury.

CAUTION: When grinding small components on a grinding wheel, wire brush wheel, or buffing wheel, always hold these components with a pair of vise grips to avoid injury to fingers and hands.

CAUTION: Grinding and buffing wheels on bench grinders must be mounted on the grinder according to the instructions provided by the manufacturer of the bench grinder. Grinding and buffing wheels must be retained with the manufacturer's specified washers and nuts, and the retaining nuts must be tightened to the specified torque. Personal injury may occur if grinding and buffing wheels are not properly attached to the bench grinder.

Figure 2-28 Fiber-optic track and toe gauge. (Courtesy of Snap-on Tools Corp.)

Figure 2-29 Computer wheel aligner. (Courtesy of Hunter Engineering Company)

Bench grinders usually have a grinding wheel and a wire wheel brush driven by an electric motor (Figure 2-31). The grinding wheel (item G, Figure 2-32) may be replaced with a grinding disc containing several layers of synthetic material (item K, Figure 2-32). A buffing wheel may be used in place of the wire wheel brush. The grinding wheel may be used for various grinding jobs and deburring. A buffing wheel is most commonly used for polishing.

Most bench grinders have grinding wheels and wire brush wheels that are 6 to 10 in (15.24 to 25.4 cm) in diameter. Bench grinders must be securely bolted to the workbench.

Figure 2-30 Scan tool for diagnosing computer-controlled suspension systems. (Courtesy of OTC Division, SPX Corp.)

Figure 2-31 Bench grinders. (Courtesy of Snap-on Tools Corp.)

Figure 2-32 Bench grinder attachments. (Courtesy of Snap-on Tools Corp.)

Hydraulic Pressing and Lifting Equipment

Hydraulic Press

 WARNING: When operating a hydraulic press always be sure that the components being pressed are supported properly on the press bed with steel supports.

CAUTION: When using a hydraulic press, never operate the pump handle if the pressure gauge exceeds the maximum pressure rating of the press. If this pressure maximum is exceeded, some part of the press may suddenly break and cause severe personal injury.

When two components have a tight precision fit between them, a **hydraulic press** is used to either separate these components or press them together. The hydraulic press rests on the shop floor, and an adjustable steel beam bed is retained to the lower press frame with heavy steel pins. A hydraulic cylinder and ram is mounted on the top part of the press with the ram facing downward toward the press bed (Figure 2-33). The component being pressed are placed on the press bed with appropriate steel supports. A hand-operated hydraulic pump is mounted on the side of the press. When the handle is pumped, hydraulic fluid is forced into the cylinder, and the ram is extended against the component on the press bed to complete the pressing operation. A pressure

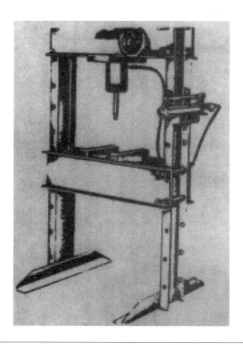

Figure 2-33 Hydraulic press. (Courtesy of Mac Tools, Inc.)

Figure 2-34 Hydraulic floor jack. (Courtesy of Mac Tools, Inc.)

gauge on the press indicates the pressure applied from the hand pump to the cylinder. The press frame is designed for a certain maximum pressure, and this pressure must not be exceeded during hand pump operation.

Floor Jack

WARNING: The maximum lifting capacity of the floor jack is usually written on the jack decal. Never lift a vehicle that exceeds the jack lifting capacity. This action may cause the jack to break or collapse, resulting in vehicle damage or personal injury.

A **floor jack** is a portable unit mounted on wheels. The lifting pad on the jack is placed under the chassis of the vehicle, and the jack handle is operated with a pumping action (Figure 2-34). This jack handle operation forces fluid into a hydraulic cylinder in the jack, and this cylinder extends to force the jack lift pad upward and lift the vehicle. Always be sure that the lift pad is positioned securely under one of the car manufacturer's recommended lifting points. To release the hydraulic pressure and lower the vehicle, the handle or release lever must be turned slowly. Do not leave the jack handle where someone can trip over it.

Lift

WARNING: Always be sure that the lift arms are securely positioned under the car manufacturer's recommended lifting points before raising the vehicle. These lifting points are shown in the service manual.

WARNING: The maximum capacity of the vehicle lift is placed on an identification plate. Never lift a vehicle that is heavier than the maximum capacity of the lift.

Figure 2-35. Lifts are used to raise a vehicle. (Courtesy of Snap-on Tools Corp.)

A **vehicle lift** is used to raise a vehicle so the technician can work under the vehicle. The lift arms must be placed under the car manufacturer's recommended lifting points prior to raising a vehicle. Twin posts are used on some lifts, whereas other lifts have a single post (Figure 2-35). Some lifts have an electric motor, which drives a hydraulic pump to create fluid pressure and force the lift upward. Other lifts use air pressure from the shop air supply to force the lift upward. If shop air pressure is used for this purpose, the air pressure is applied to fluid in the lift cylinder. A control lever or switch is placed near the lift. The control lever supplies shop air pressure to the lift cylinder, and the switch turns on the lift pump motor. Always be sure that the safety lock is engaged after the lift is raised. When the safety lock is released, a release lever is operated slowly to lower the vehicle.

Cleaning Equipment Safety and Environmental Considerations

Cleaning Equipment Safety

All technicians are required to clean parts during their normal work routines. Face shields and protective gloves must be worn while operating cleaning equipment. In most states, environmental regulations require that the runoff from steam cleaning must be contained in the steam cleaning system. This runoff cannot be dumped into the sewer system. Since it is expensive to contain this runoff in the steam cleaning system, the popularity of steam cleaning has decreased. The solution in hot and cold cleaning tanks may be caustic, and contact between this solution and skin or eyes must be avoided. Parts cleaning often creates a slippery floor, and care must be taken when walking in the parts cleaning area. The floor in this area should be cleaned frequently. When the cleaning solution in hot or cold cleaning tanks is replaced, environmental regulations require that the old solution be handled as hazardous waste. Use caution when placing aluminum or aluminum alloy parts in a cleaning solution. Some cleaning solutions will damage these components. Always follow the cleaning equipment manufacturer's recommendations.

Parts Washers with Electromechanical Agitation

Some parts washers provide electromechanical agitation of the parts to provide improved cleaning action (Figure 2-36). These parts washers may be heated with gas or electricity. Various water-based hot tank cleaning solutions are available depending on the type of metals being cleaned. For example, Kleer-Flo Greasoff® number 1 powdered detergent is available for cleaning iron and steel. Non-heated electromechanical parts washers are also available, and these washers use cold cleaning solutions such as Kleer-Flo Degreasol® formulas.

Many cleaning solutions, such as Kleer-Flo Degreasol® 99R contain no ingredients listed as hazardous by the Environmental Protection Agency's RCRA Act. This cleaning solution is a blend of sulfur-free hydrocarbons, wetting agents, and detergents. Degreasol® 99R does not contain aromatic or chlorinated solvents, and it conforms to California's Rule 66 for clean air. Always use the cleaning solution recommended by the equipment manufacturer.

Cold Parts Washer with Agitaton Immersion Tank

Some parts washers have an agitator immersion chamber under the shelves that provides thorough parts cleaning. Folding work shelves provide a large upper cleaning area with a constant flow of solution from the dispensing hose (Figure 2-37). This cold parts washer operates on Degreasol® 99R cleaning solution.

Aqueous Parts Cleaning Tank

The aqueous parts cleaning tank uses a water-based environmentally friendly cleaning solution, such as Greasoff® 2, rather than traditional solvents. The immersion tank is heated and agitated for effective parts cleaning (Figure 2-38). A sparger bar pumps a constant flow of cleaning solution

Figure 2-36 Parts washer with electromechanical agitation. (Courtesy of Kleer-Flo Company)

Figure 2-37 Cold parts washer with agitated immersion tank. (Courtesy of Kleer-Flo Company)

Figure 2-38 Aqueous parts cleaning tank. (Courtesy of Kleer-Flo Company)

across the surface to push floating oils away, and an integral skimmer removes these oils. This action prevents floating surface oils from redepositing on cleaned parts.

Employer and Employee Obligations

When you begin employment, you enter into a business agreement with your employer. A business agreement involves an exchange of goods or services that have value. Although the automotive technician may not have a written agreement with his or her employer, the technician exchanges time, skills, and effort for money paid by the employer. Both the employee and the employer have obligations. The automotive technician's obligations include the following:

1. *Productivity*: As an automotive technician you have a responsibility to your employer to make the best possible use of time on the job. Each job should be done in a reasonable length of time. Employees are paid for their skills, effort, and time.

2. *Quality*: Each repair job should be a quality job! Work should never be done in a careless manner. Nothing improves customer relations like quality workmanship.

3. *Teamwork*: The shop staff are a team, and everyone including technicians and management personnel are team members. You should cooperate with and care about other team members. Each member of the team should strive for harmonious relations with fellow workers. Cooperative teamwork helps improve shop efficiency, productivity, and customer relations. Customers may be "turned off" by bickering between shop personnel.

4. *Honesty*: Employers and customers expect and deserve honesty from automotive technicians. Honesty creates a feeling of trust among technicians, employers, and customers.

5. *Loyalty*: As an employee, you are obliged to act in the best interests of your employer, both on and off the job.

6. *Attitude*: Employees should maintain a positive attitude at all times. As in other professions, automotive technicians have days when it may be difficult to maintain a positive attitude. For example, there will be days when the technical problems on a certain vehicle are difficult to solve. However, a negative attitude certainly will not help the situation! A positive attitude has a positive effect on the job situation as well as on the customer and employer.

7. *Responsibility*: You are responsible for your conduct on the job and your work-related obligations. These obligations include always maintaining good workmanship and customer relations. Attention to details such as always placing fender and seat covers on customer vehicles prior to driving or working on the vehicle greatly improve customer relations.

8. *Following directions*: All of us like to do things "our way." Such action, however, may not be in the best interests of the shop, and as an employee you have an obligation to follow the supervisor's directions.

9. *Punctuality and regular attendance*: Employees have an obligation to be on time for work, and to be regular in attendance on the job. It is very difficult for a business to operate successfully if it cannot count on its employees to be on the job at the appointed time.

10. *Regulations*: Automotive technicians should be familiar with all state and federal regulations pertaining to their job situation, such as the Occupational Safety and Health Act (OSHA) and hazardous waste disposal laws. In Canada, employees should be familiar with workplace hazardous materials information systems (WHMIS).

Employer to employee obligations include:

1. *Wages*: The employer has a responsibility to inform the employee regarding the exact amount of financial remuneration they will receive and when they will be paid.

2. *Fringe benefits*: A detailed description of all fringe benefits should be provided by the employer. These benefits may include holiday pay, sickness and accident insurance, and pension plans.

3. *Working conditions*: A clean, safe workplace must be provided by the employer. The shop must have adequate safety equipment and first-aid supplies. Employers must be

certain that all shop personnel maintain the shop area and equipment to provide adequate safety and a healthy workplace atmosphere.

4. *Employee instruction*: Employers must provide employees with clear job descriptions, and be sure that each worker is aware of his or her obligations.

5. *Employee supervision*: Employers should inform their workers regarding the responsibilities of their immediate supervisors and other management personnel.

6. *Employee training*: Employers must make sure that each employee is familiar with the safe operation of all the equipment that they are required to use in their job situation. Since automotive technology is changing rapidly, employers should provide regular update training for their technicians. Under the right-to-know laws, employers are required to inform all employees about hazardous materials in the shop. Employees should be familiar with material safety data sheets (MSDS), which detail the labeling and handling of hazardous waste, and the health problems if exposed to hazardous waste.

Job Responsibilities

An automotive technician has specific responsibilities regarding each job performed on a customer's vehicle. These job responsibilities include:

1. Do every job to the best of your ability. There is no place in the automotive service industry for careless workmanship! Automotive technicians and students must realize they have a very responsible job. During many repair jobs you, as a student or technician working on a customer's vehicle, actually have the customer's life and the safety of his or her vehicle in your hands. For example, if you are doing a brake job and leave the wheel nuts loose on one wheel, that wheel may fall off the vehicle at high speed. This could result in serious personal injury for the customer and others, plus extensive vehicle damage. If this type of disaster occurs, the individual who worked on the vehicle and the shop may be involved in a very expensive legal action. As a student or technician working on customer vehicles, you are responsible for the safety of every vehicle that you work on! Even when careless work does not create a safety hazard, it leads to dissatisfied customers who often take their business to another shop. Nobody benefits when that happens.

2. Treat customers fairly and honestly on every repair job. Do not install parts that are unnecessary to complete the repair job.

3. Use published specifications; do not guess at adjustments.

4. Follow the service procedures in the service manual provided by the vehicle manufacturer or an independent manual publisher.

5. When the repair job is completed, always be sure the customer's complaint has been corrected.

6. Do not be too concerned with work speed when you begin working as an automotive technician. Speed comes with experience.

National Institute for Automotive Service Excellence (ASE) Certification

The **National Institute for Automotive Service Excellence (ASE)** has provided voluntary testing and certification of automotive technicians on a national basis for many years. The image of the automotive service industry has been enhanced by the ASE certification program. More than

265,000 ASE certified automotive technicians now work in a wide variety of automotive service shops. ASE provides certification in eight areas of automotive repair:

1. Engine repair

2. Automatic transmissions/transaxles

3. Manual drive train and axles

4. Suspension and steering

5. Brakes

6. Electrical systems

7. Heating and air conditioning

8. Engine performance

A technician may take the ASE test and become certified in any or all of the eight areas. When a technician passes an ASE test in one of the eight areas, an Automotive Technician's shoulder patch is issued by ASE. If a technician passes all eight tests he or she receives a Master Technician's shoulder patch (Figure 2-39). Retesting at five-year intervals is required to remain certified.

The certification test in each of the eight areas contains 40 to 80 multiple-choice questions. The test questions are written by a panel of automotive service experts from various areas of automotive service, including automotive instructors, service managers, automotive manufacturers' representatives, test equipment representatives, and certified technicians. The test questions are pretested and checked for quality by a national sample of technicians. On an ASE certification test, approximately 45 percent to 50 percent of the questions are Technician A and Technician B format, and the multiple-choice format is used in 40 percent to 45 percent of the questions. Less than 10 percent of ASE certification questions are an EXCEPT format, where the technician selects one incorrect answer out of four possible answers. ASE regulations demand that each technician must have two years of working experience in the automotive service industry prior to taking a certification test or tests. However, relevant formal training may be substituted for one year of working experience. Contact ASE for details regarding this substitution. The contents of the Suspension and Steering test are listed in Table 2-1.

ASE also provides certification tests in automotive specialty areas such as Advanced Engine Performance Specialist; Alternate Fuels Light Vehicle Compressed Natural Gas; Parts Specialist; Machinist, Cylinder Head Specialist; Machinist, Cylinder Block Specialist; and Machinist, Assembly Specialist.

Shops that employ ASE certified technicians display an official **ASE blue seal of excellence.** This blue seal increases the customer's awareness of the shop's commitment to quality service, and the competency of certified technicians.

Shop Manual
Chapter 2, page 20

Figure 2-39. ASE certification shoulder patches worn by automotive technicians and master technicians. (Courtesy of National Institute of Automotive Service Excellence [ASE])

TABLE 2-1 SUSPENSION AND STEERING TEST SUMMARY

Content Area	Questions in Test		Percentage of Test
A. Steering Systems Diagnosis and Repair	9		22.5%
1. Steering Columns and Manual Steering Gears		(2)	
2. Power-Assisted Steering Units		(4)	
3. Steering Linkage		(3)	
B. Suspension Systems Diagnosis and Repair	13		32.5%
1. Front Suspensions		(6)	
2. Rear Suspensions		(5)	
3. Miscellaneous Service		(2)	
C. Wheel Alignment Diagnosis, Adjustment, and Repair	13		32.5%
D. Wheel and Tire Diagnosis and Repair	5		12.5%
Total	40		100.0%

(Courtesy of National Institute for Automotive Service Excellence [ASE])

Guidelines for Tools and Shop Practices

1. Two systems of measurement in common use are the United States customary (USC) and international system (SI).

2. The SI system may be called the metric system.

3. In the metric system the units can be divided or multiplied by 10.

4. A seal driver should be used to install a seal rather than striking the seal case with a hammer.

5. Never operate any type of equipment unless you are familiar with the proper operating procedure.

6. A tremendous amount of energy is stored in a compressed coil spring. Always follow the tool manufacturer's and vehicle manufacturer's recommended procedure when using a spring compressor tool. Never disconnect any suspension component that would suddenly release the spring tension.

7. A power steering pump delivers extremely high pressure under certain test conditions. Always follow the vehicle manufacturer's recommended procedure during the pump pressure test.

8. Always wear eye protection and protective gloves when cleaning parts in any type of cleaning solution.

9. When employers and employees accept and fulfill their obligations, personal relationships and general attitudes are greatly improved, shop productivity is increased, and customer relations are improved.

10. If a technician accepts certain job responsibilities, job quality improves, and customer satisfaction increases.

11. ASE technician certification improves the quality of automotive repair and improves the image of the profession.

CASE STUDY 1

A technician was removing and replacing the alternator of a General Motors car. After installing the replacement alternator and connecting the alternator battery wire, the technician proceeded to install the alternator belt. The rubber boot was still removed from the alternator battery terminal. While installing this belt the technician's wristwatch expansion bracelet made electrical contact from the alternator battery terminal to ground on the alternator housing. Even though the alternator battery wire is protected with a fuse link (which melted), a high current flowed through the wristwatch bracelet. This heated the bracelet to a very high temperature and severely burned the technician's arm.

The technician forgot two safety rules:

1. Never wear jewelry, such as watches and rings, while working in an automotive shop.

2. Before performing electrical work on a vehicle, disconnect the negative battery cable. If the vehicle is equipped with an air bag, wait the specified time after this cable is disconnected.

CASE STUDY 2

A technician was removing and replacing the starter motor with the solenoid mounted on top of the motor. The technician began removing the battery cable from the solenoid, and the ring on one of his fingers made contact between the end of the wrench and the engine block. The current flow through the ring was so high that the positive battery terminal melted out of the battery. The technician's finger was so badly burned that a surgeon had to cut the ring from his finger and repair the finger.

This technician forgot to disconnect the negative battery cable before working on the vehicle!

CASE STUDY 3

A technician had just replaced the engine in a Ford vehicle, and he was performing final adjustments such as timing and air-fuel mixture. In this shop, the cars were parked in the work bays at an angle on both sides of the shop. With the engine running at fast idle the automatic transmission suddenly slipped into reverse. The car went backward across the shop and collided with a car in one of the electrical repair bays. Both vehicles were damaged to a considerable extent. Fortunately, no personnel were injured.

This technician forgot to apply the parking brake while working on the vehicle!

Terms to Know

ASE blue seal of excellence	Bearing pullers	Computer wheel aligner
Axle pullers	Brake pedal jack	Control arm bushing tools
Ball joint removal and pressing tools	Coil spring compressor tools	Dial indicators

Electronic wheel balancers

Floor jack

Hydraulic press

International system (SI)

Machinist's rule

Magnetic wheel alignment gauge

National Institute for Automotive
    Service Excellence (ASE)

Pitman arm puller

Plumb bob

Power steering pressure gauge

Rim clamps

Scan tools

Seal drivers

Steering wheel locking tool

Stethoscope

Tie-rod end and ball joint puller

Tie-rod sleeve adjusting tool

Tire changer

Tire tread depth gauge

Toe gauge

Track gauges

Tram gauge

Turning radius gauge turn tables

United States Customary (USC) system

Vacuum hand pump

Vehicle lift

ASE Style Review Questions

1. While discussing systems of weights and measures:
   Technician A says the international system (SI) is called the metric system.
   Technician B says every unit in the metric system can be divided or multiplied by 10.
   Who is correct?
   A. A only
   B. B only
   C. Both A and B
   D. Neither A nor B

2. In the metric system:
   A. 1 liter is equal to 1,000 cm3.
   B. 1 mile is equal to 1.8 kilometers.
   C. 1 inch is equal to 2.54 millimeters.
   D. 1 meter is equal to 36.37 inches.

3. While discussing steering and suspension tools:
   Technician A says there is no energy in a compressed coil spring.
   Technician B says that a coil spring compressor tool must be used to compress the coil spring before removing any suspension component that would release the spring tension.
   Who is correct?
   A. A only
   B. B only
   C. Both A and B
   D. Neither A nor B

4. While discussing shop cleaning equipment safety:
   Technician A says hot tanks contain caustic solutions.
   Technician B says some metals such as aluminum dissolve in hot tanks.
   Who is correct?
   A. A only
   B. B only
   C. Both A and B
   D. Neither A nor B

5. While discussing employer and employee responsibilities:
   Technician A says employers are required to inform their employees about hazardous materials in the shop.
   Technician B says that employers have no obligation to inform their employees about the safe operation of shop equipment.
   Who is correct?
   A. A only
   B. B only
   C. Both A and B
   D. Neither A nor B

6. While discussing turning radius gauges:
   Technician A says these gauges are placed under the front wheels during a wheel alignment.
   Technician B says these gauges contain a degree scale that indicates how much the wheels are turned.
   Who is correct?
   A. A only
   B. B only
   C. Both A and B
   D. Neither A nor B

7. All these statements about suspension and steering tools are true EXCEPT:
   A. A plumb bob may be used for frame measurement.
   B. A tram gauge may be used to measure front wheel alignment.
   C. A scan tool may be used to diagnose electronically controlled suspension systems.
   D. A brake pedal jack may be used to apply the brakes during a wheel alignment.

8. When using a vehicle lifts:
   A. the lift arms may be positioned on any strong chassis member.
   B. the safety catch must be engaged after the vehicle is raised on a lift.
   C. the vehicle hood should be open when raising a vehicle on a lift.
   D. the vehicle weight may exceed the maximum capacity of the lift.

9. When taking ASE certification tests:
   A. a technician must pass 4 of the 8 ASE certification tests to receive a master technician's shoulder patch.
   B. each ASE certification test contains 35 to 45 questions.
   C. an ASE specialty test is available in Advanced Engine Performance Specialist.
   D. retesting at 3-year intervals is required to maintain ASE certification.

10. When taking ASE certification tests:
    A. a technician must have four years of automotive repair experience before taking an ASE certification test.
    B. ASE may allow relevant training to be substituted for one year of work experience.
    C. on any ASE certification test 90 percent of the questions are Technician A and Technician B type.
    D. employers who employ ASE certified technicians may display a green ASE seal of excellence.

Job Sheet 4

4

Name _____ Date _____

Raise a Car With a Floor Jack and Support It on Safety Stands

Upon completion of this job sheet, you should be able to raise the front and rear of a vehicle with a floor jack and support the vehicle on safety stands.

Tools and Materials

A car
Hydraulic floor jack
Four safety stands

Vehicle Make and Model Year _____

Vehicle VIN _____

This job sheet must be completed under the supervision of your instructor!

Procedure

1. Determine the proper vehicle floor jack lifting points in the car manufacturer's service manual.

    Proper front lifting point _____
    Proper rear lifting point _____

2. Be sure the car is parked on a level shop floor surface with the parking brake applied.

3. Place the floor jack under the proper lifting point on the front of the vehicle, and raise the floor jack pad until it contacts the lifting point.

4. Release the parking brake and be sure the transmission is in neutral. Is the parking brake released?

    Is the transmission in neutral?
    Instructor check _____

5. Operate the floor jack and raise the vehicle to the desired height. Place safety stands under the proper support points on the vehicle chassis or suspension.

6. Very slowly operate the release lever on the hydraulic floor jack to slowly lower the vehicle until the support points lightly contact the safety stands. Stop lowering the floor jack.

7. Be sure the safety stands contact the proper support points on the vehicle, and check to be sure all the safety stand legs contact the shop floor evenly.

    Are the safety stands contacting the proper support points on the vehicle? Are all safety stand legs contacting the floor evenly?
    Instructor check _____

8. If the answer to both questions in step 7 is yes, very slowly operate the release lever on the hydraulic floor jack to slowly lower the vehicle until the vehicle weight is completely supported on the safety stands. Be sure all the safety stand legs are contacting the shop floor evenly, and then remove the floor jack.

9. Place the floor jack under the proper lifting point on the rear of the vehicle, and raise the floor jack pad until it contacts the lifting point.

10. Operate the floor jack and raise the vehicle to the desired height. Place safety stands under the proper support points on the rear of the vehicle chassis or suspension.

    Are the safety stands properly positioned? yes _____ no _____
    Instructor check _____

11. Very slowly operate the release lever on the hydraulic floor jack to slowly lower the vehicle until the rear support points lightly contact the safety stands. Stop lowering the floor jack.

12. Be sure the safety stands contact the proper rear support points on the vehicle, and check to be sure all the safety stand legs contact the shop floor evenly.

    Are the safety stands contacting the proper support points on the vehicle?
    yes _____ no _____
    Are all the safety stand legs contacting the floor evenly? yes _____ no _____
    Instructor check _____

13. If the answer to both questions in step 12 is yes, very slowly operate the release lever on the hydraulic floor jack to slowly lower the rear of the vehicle until the vehicle weight is completely supported on the safety stands. Be sure all the safety stand legs are contacting the shop floor evenly, then remove the floor jack.

✓ Instructor Check _____

Job Sheet 5

Name _____ Date _____

Follow the Proper Procedure to Hoist a Car

Upon completion of this job sheet, you should be able to raise and lower a car on a hoist.

Tools and Materials

Car
Lift with enough capacity to hoist the car

Vehicle Make and Model Year _____
Vehicle VIN _____

Procedure

1. Always be sure the lift is completely lowered before driving the car on the lift.

   List the types of lift(s) in your shop, and provide the location of the lift controls

   _____
   _____
   _____
   _____

   ▲ **WARNING:** Do not raise a four-wheel-drive vehicle with a frame contact lift because this may damage axle joints.

2. Do not hit or run over lift arms and adaptors when driving a car on the lift.

3. Have a coworker guide you when driving a car onto the lift.

   ■ **CAUTION:** Do not stand in front of a lift with the car coming toward you. This action may result in personal injury.

4. Be sure the lift pads on the lift are contacting the car manufacturer's recommended lifting points shown in the service manual.

   Is the vehicle properly positioned on lift?

   Recommended front lifting points: right side _____

   left side _____

   Recommended rear lifting points: right side _____

   left side _____

   Are all four lift pads contacting the recommended lifting points?

   Instructor check _____

CAUTION: If the proper lifting points are not used, components under the vehicle such as brake lines or body parts may be damaged. Failure to use the recommended lifting points may cause the car to slip off the lift resulting in severe vehicle damage and personal injury.

5. Be sure the doors and hood are closed, and be sure there are no people in the car.

6. When a car is lifted a short distance off the floor, stop the lift and check the contact between the lift pads and the car chassis to be sure the lift pads are still on the recommended lifting points.

7. Be sure there is adequate clearance between the top of the vehicle and the shop ceiling, or components under the ceiling.

8. When a car is raised on a lift, be sure the safety mechanism is in place to prevent the lift from dropping accidentally.

   Is the lift safety mechanism in place?   yes _____ no _____

   Instructor check _____

   List one precaution that should be observed when a front-wheel-drive vehicle is raised on a lift, and explain the reason for this precaution.

   _____

   _____

   _____

9. Prior to lowering a car on a lift, always make sure there are no objects, tools, or people under the vehicle.

   CAUTION: Do not rock a car on a lift during a service job. This action may cause the car to fall off the lift resulting in personal injury and vehicle damage.

   CAUTION: When a car is raised on a lift, removal of some heavy components may cause car imbalance on the lift. This action may cause the car to fall off the lift resulting in personal injury and vehicle damage.

10. Be sure the lift is lowered completely and no lift components are contacting the vehicle before backing the vehicle off the lift.

✓ **Instructor Check** _____

Job Sheet 6

Name _____ Date _____

Determine the Availability and Purpose of Suspension and Steering Tools

Tools and Materials

None

Procedure

Locate the following tools in your shop or tool room, and explain the purpose of each tool.

1. Stethoscope: available? yes _____ no _____

   Location _____

   Purpose _____

   _____

2. Dial indicator: available? yes _____ no _____

   Location _____

   Purpose _____

   _____

3. Ball joint removal and pressing tools: available? yes _____ no _____

   Location _____

   Purpose _____

   _____

4. Coil spring compressing tool: available? yes _____ no _____

   Location _____

   Purpose _____

   _____

5. Power steering pressure gauge: available? yes _____ no _____

   Location _____

   Purpose _____

   _____

6. Vacuum hand pump: available? yes _____ no _____

   Location _____

   Purpose _____

   _____

7. Turning radius gauge: available? yes _____ no _____

Location _____

Purpose _____

_____

8. Plumb bob: available? yes _____ no _____

Location _____

Purpose _____

_____

9. Tram gauge: available? yes _____ no _____

Location _____

Purpose _____

_____

10. Brake pedal jack: available? yes _____ no _____

Location _____

Purpose _____

_____

11. Steering wheel locking tool: available? yes _____ no _____

Location _____

Purpose _____

_____

12. Scan tool: available? yes _____ no _____

Location _____

Purpose _____

_____

✓ **Instructor Check** _____

Wheel Bearing and Seal Service

Upon completion and review of this chapter, you should be able to:

❏ Diagnose bearing defects.
❏ Clean and repack wheel bearings.
❏ Reassemble and adjust wheel bearings
❏ Remove and replace wheel bearing seals.
❏ Diagnose wheel bearing problems on the vehicle.

❏ Diagnose problems in wheel bearing hub units.
❏ Remove and replace front drive axles.
❏ Remove and replace wheel bearing hub units.
❏ Remove and replace rear axle bearings on rear-wheel-drive cars.

Basic Tools

Basic technician's tool set

Service manual

Inch-pound torque wrench

Foot-pound torque wrench

Fine-toothed round and flat files

Wheel bearing grease

Differential lubricant

Bearing galling refers to metal smears on the ends of the rollers.

Bearing abrasive step wear is a fine circular wear pattern on the ends of the rollers.

Bearing etching appears as a loss of material on the bearing rollers and races. Bearing surfaces are gray or grayish black.

Bearing indentations are surface depressions on the rollers and races.

Classroom Manual
Chapter 3, page 43

Diagnosis of Bearing Defects

Bearings are designed to provide long life, but there are many causes of premature bearing failure. If a bearing fails, the technician must decide if the bearing failure was caused by normal wear, or if the bearing failed prematurely. For example, if a front wheel bearing fails on a car that is one year old with an original odometer reading of 15,000 miles (24,000 kilometers), experience tells us the bearing failure is premature because front wheel bearings normally last for a much longer mileage period. Always listen to the customer's complaints, and obtain as much information as possible from the customer. Ask the customer specific questions about abnormal or unusual vehicle noises and operation. If a bearing fails prematurely, there must be some cause for the failure. The causes of premature bearing failure are:

1. Lack of lubrication
2. Improper type of lubrication
3. Incorrect endplay adjustment (where applicable)
4. Misalignment of related components, such as shafts or housings
5. Excessive bearing load
6. Improper installation or service procedures
7. Excessive heat
8. Dirt or contamination

When a bearing fails prematurely, the technician must correct the cause of this failure to prevent the new bearing from failing. The types of bearing failures and the necessary corrective service procedures are provided in Figure 3-1 and Figure 3-2. **Bearing fatigue spalling** appears as flaking of surface metal on bearing rollers and races. **Bearing brinelling** shows up as straight-line indentations on the races and rollers. **Bearing smears** appear as metal loss in a circular, blotched pattern around the bearing races and rollers. **Bearing frettage** shows up as a fine, corrosive wear pattern around the bearing races and rollers. This wear pattern is circular on the races.

The first indication of bearing failure is usually a howling noise while the bearing is rotating. The howling noise will likely vary depending on the bearing load. A front wheel bearing usually provides a more noticeable howl when the vehicle is turning a corner, because this places additional thrust load on the bearing. A defective rear axle bearing usually provides a howling noise that is more noticeable at lower speeds. The howling noise is more noticeable when driving on a narrow street with buildings on each side, because the noise vibrates off the nearby buildings. A rear axle

TAPERED ROLLER BEARING DIAGNOSIS

Consider the following factors when diagnosing bearing condition:
1. General condition of all parts during disassembly and inspection.
2. Classify the failure with the aid of the illustrations.
3. Determine the cause.
4. Make all repairs following recommended procedures.

ABRASIVE ROLLER WEAR

Pattern on races and rollers caused by fine abrasives.
Clean all parts and housings, check seals and bearings and replace if leaking, rough or noisy.

GALLING

Metal smears on roller ends due to overheat, lubricant failure or overload. Replace bearing, check seals and check for proper lubrication.

BENT CAGE

Cage damaged due to improper handling or tool usage. Replace bearing.

ABRASIVE STEP WEAR

Pattern on roller ends caused by fine abrasives.
Clean all parts and housings, check seals and bearings and replace if leaking, rough or noisy.

ETCHING

Bearing surfaces appear gray or grayish black in color with related etching away of material usually at roller spacing.
Replace bearings, check seals, and check for proper lubrication.

BENT CAGE

Cage damaged due to improper handling or tool usage.
Replace bearing.

INDENTATIONS

Surface depressions on race and rollers caused by hard particles of foreign material.
Clean all parts and housings. Check seals and replace bearings if rough or noisy.

MISALIGNMENT

Outer race misalignment due to foreign object.
Clean related parts and replace bearing. Make sure races are properly sealed..

Figure 3-1 Bearing failures and corrective procedures. (Courtesy of Chevrolet Motor Division, General Motors Corp.)

bearing noise is present during acceleration and deceleration, because the vehicle weight places a load on the bearing regardless of the operating condition. The rear axle bearing noise may be somewhat more noticeable during deceleration because there is less engine noise at that time.

FATIGUE SPALLING

Flaking of surface metal resulting from fatigue.
Replace bearing, clean all related parts.

STAIN DISCOLORATION

Discoloration can range from light brown to black caused by incorrect lubricant or moisture.
Re-use bearings if stains can be removed by light polishing or if no evidence of overheating is observed.
Check seals and related parts for damage.

FRETTAGE

Corrosion set up by small relative movement of parts with no lubrication. Replace bearing. Clean related parts. Check seals and check for proper lubrication.

CAGE WEAR

Wear around outside diameter of cage and roller pockets caused by abrasive material and inefficient lubrication.
Clean related parts and housings.
Check seals and replace bearings

BRINELLING

Surface indentations in raceway caused by rollers either under impact loading or vibration while the bearing is not rotating.
Replace bvearing if rough or noisy

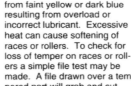

HEAT DISCOLORATION

Heat discoloration can range from faint yellow or dark blue resulting from overload or incorrect lubricant. Excessive heat can cause softening of races or rollers. To check for loss of temper on races or rollers a simple file test may be made. A file drawn over a tempered part will grab and cut metal, whereas, a file drawn over a hard part will glide readily with no metal cutting. Replace bearings if overheating damage is indicated.
Check seals and other parts.

SMEARS

Smearing of metal due to slippage. Slippage can be caused by poor fits, lubrication, overheating, overloads or handling damage.
Replace bearings, clean related parts and check for proper fit and lubrication.

CRACKED INNER RACE

Race cracked due to improper fit, cocking, or poor bearing seats.
Replace bearing and correct bearing seats.

Figure 3-2 Bearing failures and corrective procedures, continued. (Courtesy of Chevrolet Motor Division, General Motors Corp.)

Service and Adjustment of Tapered Roller Bearing-Type Wheel Bearings

Cleaning Bearings

 CAUTION: Always wear safety goggles when working in the shop.

CAUTION: Do not aim compressed air directly against any part of your body. Compressed air can penetrate human flesh and enter the blood stream with very serious consequences.

CAUTION: Do not spin the bearing at high speed with compressed air. Bearing damage or disintegration may result. Bearing disintegration may cause serious personal injury.

CAUTION: Never strike a bearing with a ball peen hammer. This action will damage the bearing and the bearing may shatter, causing severe personal injury.

WARNING: Never wash a sealed bearing or a bearing that is shielded on both sides because solvent may enter the bearing and destroy the lubricant in the bearing, resulting in very short bearing life.

Two separate tapered roller bearings are used in the front wheel hubs of many rear-wheel-drive cars. The rear wheel hubs on some front-wheel-drive cars have the same type of bearings. Similar service and adjustment procedures apply to these tapered roller bearings.

These bearings should be cleaned, inspected, and packed with wheel bearing grease at the vehicle manufacturer's recommended service intervals. Pry the grease seal out of the inner hub opening with a large flat-blade screwdriver, and discard the seal. This seal should always be replaced when the bearings are serviced. Do not attempt to wash sealed bearings or bearings that are shielded on both sides. If a bearing is sealed on one side, it may be washed in solvent and repacked with grease.

Bearings may be placed in a tray and lowered into a container of clean solvent. A brush may be used to remove old grease from the bearing (Figure 3-3). The bearings may be dried with compressed air after the cleaning operation. Be sure the shop air supply is free from moisture, which causes rust formation in the bearing. After all the old grease has been cleaned from the bearing, rinse the bearing in clean solvent and dry it thoroughly with compressed air.

When bearing cleaning is completed, bearings should be inspected for the defects illustrated in Figures 3-1 and 3-2. If any of these conditions are present on the bearing, replacement is necessary. Tapered roller bearings and their matching outer races must be replaced as a set. If the bearing installation is not done immediately, cover the bearings with a protective lubricant and wrap them in waterproof paper (Figure 3-4). Be sure to identify the bearings, or lay them in order, so you reinstall them in their original location. Do not clean bearings or races with paper towels. If you are

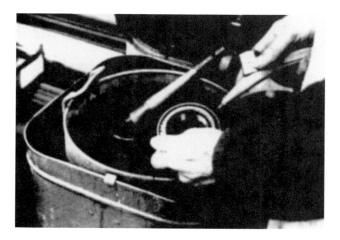

Figure 3-3 Cleaning a bearing with solvent. (Courtesy of the Timken Company)

Figure 3-4 Wrapping a bearing in waterproof paper. (Courtesy of the Timken Company)

using a shop towel for this purpose, be sure it is lint free. Lint from shop towels or paper towels may contaminate the bearing. Bearing races and the inner part of the wheel hub should be thoroughly cleaned with solvent and dried with compressed air. Inspect the seal mounting area in the hub for metal burrs. Remove any burrs with a fine round file.

Bearing races must be replaced if any of the defects described in Figures 3-1 and 3-2 are found. The proper bearing race driving tool must be used to remove the bearing races (Figure 3-5). If a driver is not available for the bearing races, a long brass punch and hammer may be used to drive the races from the hub. When a hammer and punch are used for this purpose, be careful not to damage the hub inner surface with the punch.

The new bearing races should be installed in the hub with the correct bearing race driving tool (Figure 3-6). When bearings and races are replaced, be sure they are the same as the original bearings. The part numbers should be the same on the old bearings and the replacement bearings.

Inspect the bearing and seal mounting surfaces on the spindle. Small metal burrs may be removed from the spindle with a fine-toothed file. If the spindle is severely scored in the bearing or seal mounting areas, spindle replacement is necessary.

Special Tools

Bearing driver

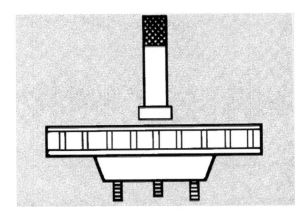

Figure 3-5 Bearing race removal. (Courtesy of Chevrolet Motor Division, General Motors Corp.)

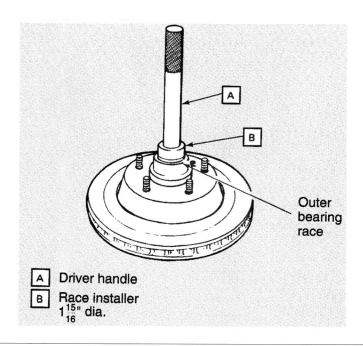

A | Driver handle
B | Race installer $1\frac{15}{16}''$ dia.

Figure 3-6 Bearing race installation. (Courtesy of Chevrolet Motor Division, General Motors Corp.)

Bearing Lubrication and Assembly

 WARNING: Cleanliness is very important during wheel bearing service. Always maintain cleanliness of hands, tools, work area, and all related bearing components. One small piece of dirt in a bearing will cause bearing failure.

 WARNING: Always keep grease containers covered when not in use. Uncovered grease containers are easily contaminated with dirt and moisture.

After the bearings and races have been cleaned and inspected, the bearings should be packed with grease. Always use the vehicle manufacturer's specified wheel bearing grease. Vehicle manufacturers usually recommend a lithium-based wheel bearing grease. Place a lump of grease in the palm of one hand and grasp the bearing in the other hand. Force the widest edge of the bearing into the lump of grease, and squeeze the grease into the bearing. Continue this process until grease is forced into the bearing around the entire bearing circumference. Place a coating of grease around the outside of the rollers, and apply a light coating of grease to the races. A bearing packing tool may be used to force grease into the bearings rather than using the hand method (Figure 3-7).

 SERVICE TIP: When a lip seal is installed, the garter spring should always face toward the flow of lubricant.

Special Tools

Seal driver

Place some grease in the wheel hub cavity and position the inner bearing in the hub (Figure 3-8). Check the fit of the new bearing seal on the spindle and in the hub. The seal lip must fit

Figure 3-7 Mechanical wheel bearing packer. (Courtesy of the Timken Company)

snugly on the spindle, and the seal case must fit properly in the hub opening. The part number on the old seal and the replacement seal should be the same. Be sure the seal is installed in the proper direction with the garter spring and higher part of the lip toward the lubricant in the hub. The new inner bearing seal must be installed in the hub with a suitable seal driver (Figure 3-9).

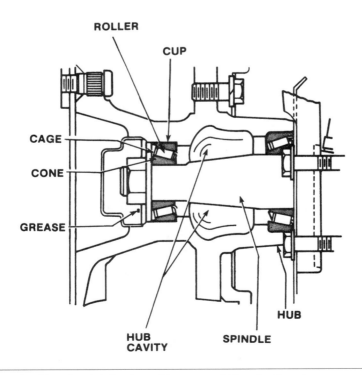

Figure 3-8 Wheel bearing lubrication. (Courtesy of Chrysler Corp.)

Figure 3-9 Seal installation. (Courtesy of Pontiac Motor Division, General Motors Corp.)

Place a light coating of wheel bearing grease on the spindle and slide the hub assembly onto the spindle. Install the outer wheel bearing and be sure there is adequate lubrication on the bearing and race. Be sure the washer and nut are clean and install these components on the spindle (Figure 3-10). Tighten the nut until it is finger tight.

Photo Sequence 1 shows a typical procedure for adjusting rear wheel bearings on a front-wheel-drive car.

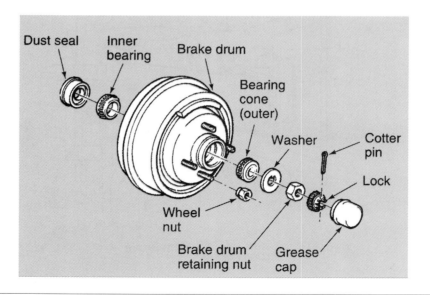

Figure 3-10 Installation of wheel bearings and related components. (Courtesy of Chrysler Corp.)

Photo Sequence 1
Typical Procedure for Adjusting Rear Wheel Bearings on a Front-Wheel-Drive Car

P1-1 Always make sure the car is positioned safely on a lift before working on the vehicle.

P1-2 Remove the dust cap from the wheel hub.

P1-3 Remove the cotter pin and nut retainer from the bearing adjusting nut.

P1-4 Tighten the bearing adjusting nut to 17 to 25 ft-lbs.

P1-5 Loosen the bearing adjusting nut one-half turn.

P1-6 Tighten the bearing adjusting nut to 10 to 15 in-libs.

P1-7 Position the adusting nut retainer over the adjusting nut so the slots are aligned with the holes in the nut and spindle.

P1-8 Install a new cotter pin and bend the ends around the retainer flange.

P1-9 Install the dust cap and be sure the hub rotates freely.

Wheel Bearing Adjustment with Two Separate Tapered Roller Bearings in the Wheel Hub

CUSTOMER CARE: Never sell a customer automotive service that is not required on his or her car. Selling preventive maintenance, however, is a sound business practice, and may save a customer some future problems. An example of preventive maintenance is selling a cooling system flush when the cooling system is not leaking but the manufacturer's recommended service interval has elapsed. If customers find out they were sold some unnecessary service, they will probably never return to the shop. They will likely tell their friends about their experience, and that kind of advertising the shop can do without.

Wheel bearing endplay is the amount of horizontal wheel bearing hub movement. If a bearing has a preload condition, a slight tension is placed on the bearing. Loose front wheel bearing adjustment results in lateral front wheel movement and reduced directional stability. If the wheel bearing adjusting nut is tightened excessively, the bearings may overheat, resulting in premature bearing failure. The bearing adjustment procedure may vary depending on the make of vehicle. Always follow the procedure in the vehicle manufacturer's service manual. A typical bearing adjustment procedure follows:

1. With the hub and bearings assembled on the spindle, tighten the adjusting nut to 17 to 25 ft-lbs (23 to 34 Nm) while the hub is rotated in the forward direction.

2. Loosen the adjusting nut 1/2 turn and retighten it to 10 to 15 in-lbs (1.0 to 1.7 Nm). This specification varies depending on the make of vehicle. Always use the manufacturer's specifications.

3. Position the adjusting nut retainer over the nut so the retainer slots are aligned with the cotter pin hole in the spindle.

4. Install a new cotter pin, and bend the ends around the retainer flange.

5. Install the grease cap, and make sure the hub and drum rotate freely.

Wheel Hub Unit Diagnosis

Special Tools

Dial indicator

When wheel bearings and hubs are an integral assembly, the bearing endplay should be measured with a dial indicator stem mounted against the hub. If the endplay exceeds 0.005 in (0.127 mm) as the hub is moved in and out, the hub and bearing assembly should be replaced. This specification is typical, but the vehicle manufacturer's specifications must be used. Hub and bearing replacement is also necessary if the bearing is rough or noisy. Integral-type bearing and hub assemblies are used on the front and rear wheels on some front-wheel-drive cars.

When the front wheel bearings are mounted in the steering knuckle, the wheel bearings may be checked with the vehicle raised on the hoist and a dial indicator positioned against the outer wheel rim lip as shown in Figure 3-11.

When the wheel is moved in and out, the maximum bearing movement on the dial indicator should be as follows:

0.020 in (0.508 mm) for 13-in (33-cm) wheels

0.023 in (0.584 mm) for 14-in (35.5-cm) wheels

0.025 in (0.635 mm) for 15-in (38-cm) wheels

If the bearing movement is excessive, check the hub nut torque before replacing the bearing. When this torque is correct and bearing movement is excessive, the bearing should be replaced.

Figure 3-11 Wheel bearing diagnosis on vehicle. (Courtesy of Chrysler Corp.)

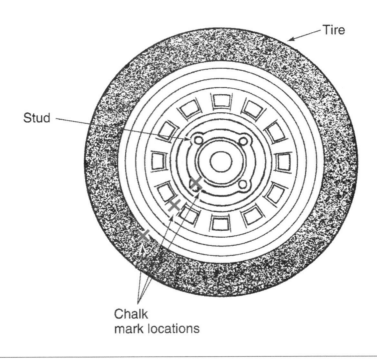

Figure 3-12 Chalk marking on wheel, tire, and stud. (Courtesy of Chrysler Corp.)

When a wheel is removed to service the wheel bearings, proper balance must be maintained between the wheel and tire and the hub. Therefore, the tire, wheel, and hub stud should be chalk marked prior to removal (Figure 3-12).

Front Drive Axle Diagnosis

On many front-wheel-drive vehicles, the front drive axles must be removed before the wheel hub unit or steering knuckle and bearing can be detached. Therefore, we will discuss front drive axle diagnosis and removal. Since drive axle noises may be confused with front wheel bearing noise, a brief discussion of drive axle noises and problems may be helpful. A defective inner drive axle joint usually causes a vibration when the vehicle is decelerating at 35 to 45 miles per hour (mph), or 56 to 72 kilometers per hour (kmph). When an outer drive axle joint is worn, a clicking noise is heard during a hard turn below 20 mph (32 kph). To determine which drive axle has the defective joint, lift the vehicle on a hoist, and allow the front wheels to drop down. This action will position the axle joints at a different angle than when the car is driven on the road. Lift the lower control arms one at a time with a floor jack, and place the transmission drive to simulate the driving conditions that provided the vibration or noise. If the vibration or noise occurs with one lower control arm lifted, that side has the defective drive axle joint.

Drive Axle Removal

Many drive axles have a **circlip** on the inner joint extension that holds the inner joint into the differential side gear. Drive axle systems vary depending on the vehicle. Follow the drive axle removal procedure in the vehicle manufacturer's service manual. A general front drive axle removal procedure follows:

1. Loosen the front wheel nuts and hub nuts.
2. Lift the vehicle on a hoist and be sure the hoist safety mechanism is in place; then remove the front wheels and tires.
3. Remove the brake calipers and rotors. Connect a piece of wire from the calipers to a suspension or chassis component. Do not allow the calipers to hang on the end of the brake line.
4. Install protective drive axle boots if these are supplied by the car manufacturer.
5. Remove the ball joint to steering knuckle clamp bolt (Figure 3-13).
6. Pry the ball joint stud from the steering knuckle.
7. Pull the inner axle from the transaxle (Figure 3-14); do not allow the axle to drop down at a severe angle.
8. Remove the hub nut and washer, and separate the outer axle joint from the wheel hub. Some outer axle joint splines, such as those used on early model Ford Escorts or Lynx,

Special Tools

Front wheel bearing hub pulling and installing tools

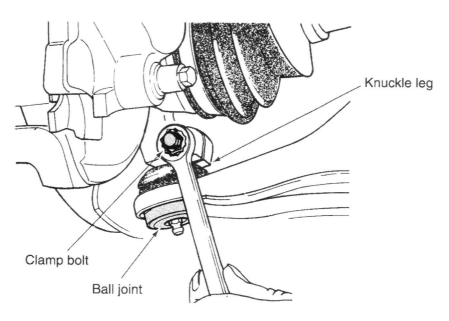

Figure 3-13 Removal of ball joint to steering knuckle clamp bolt. (Courtesy of Chrysler Corp.)

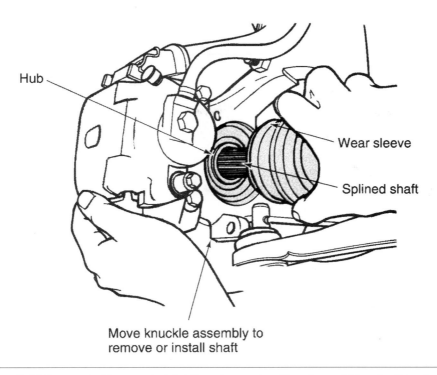

Hub

Wear sleeve

Splined shaft

Move knuckle assembly to
remove or install shaft

Figure 3-14 Removal of inner axle joint from transaxle. (Courtesy of Chrysler Corp.)

are slightly spiraled. On this type of outer axle joint, a special puller is required to separate the axle joint from the wheel (Figure 3-15).

9. Remove the drive axle from the chassis.

Reverse the drive axle removal procedure for front drive axle installation.

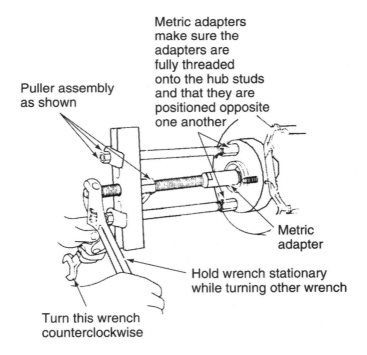

Metric adapters
make sure the
adapters are
fully threaded
onto the hub studs
and that they are
positioned opposite
one another

Puller assembly
as shown

Metric
adapter

Hold wrench stationary
while turning other wrench

Turn this wrench
counterclockwise

Figure 3-15 Removal of outer axle joint from front wheel hub. (Courtesy of Ford Motor Company)

Special Procedures for Drive Axle Removal

Some car manufacturers recommend removal of the strut from the steering knuckle rather than removal of the ball joint, when the drive axles are removed. This type of suspension system has a threaded nut on the ball joint stem to hold the ball joint into the steering knuckle. If an eccentric camber adjustment bolt is positioned in the strut, the bolt head position should be marked in relation to the strut before the bolt is removed (Figure 3-16).

On these suspension systems, the brake calipers and brake line clamps should be removed before the drive axles. Some car manufacturers supply a slide-hammer-type puller to remove the inner axle joints from the transaxle.

On Chrysler manual and automatic transaxles, the speedometer gear must be removed from the right differential extension housing before the right drive axle is removed (Figure 3-17).

Early model Chrysler transaxles had circlip retainers, which held the inner drive axle joints into the differential side gears. The differential cover had to be removed and these circlips had to be compressed with needle nose pliers before the drive axles could be removed from the differential side gears. The drive axles had to be rotated until a flat area on the axle and the circlip ends were visible (Figure 3-18).

The inner joint housings did not have tripod springs on early models. Later-model Chrysler drive axles may be pulled from the differential side gears without collapsing the circlips. To determine which type of circlip is used, grasp the inner joint housing and try to pull it out of the transaxle. If the inner joint housing moves outward and springs back into the transaxle, the joint has a later model circlip. When the inner joint is solid in the transaxle, the older model circlip, which requires compressing before removal, is used.

On Ford automatic transaxles (ATX), the left inner axle joint is inset into the transaxle housing. In this location, it is impossible to pry the axle joint from the transaxle. Therefore, the right axle joint must be removed first. A special tool is then used to drive the left inner axle joint from the differential side gear (Figure 3-19).

✔ **SERVICE TIP:** Do not remove the tool until the left inner axle joint is reinstalled in the Ford ATX. If the tool is removed without installing the left inner axle joint, the side gears may fall out of place. If this action occurs, the Ford ATX may have to be disassembled to install the side gears properly.

If both left and right drive axles are removed at the same time on Ford transaxles, the differential side gears may become dislocated. When this occurs, the differential must be removed to realign the side gears. The special tool for driving out the left inner axle joint may be left in place to support the side gears on ATX models. On manual transaxles (MTX), shipping plugs T81P-1177-B should be installed in each side gear when the drive axles are removed.

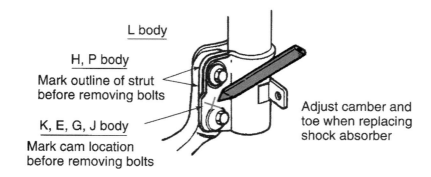

Figure 3-16 Marking eccentric strut bolt before removal. (Courtesy of Chrysler Corp.)

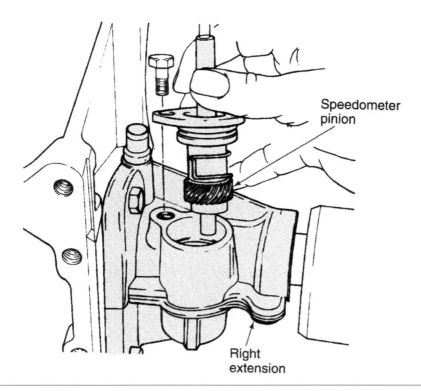

Figure 3-17 Speedometer drive gear removal before right axle removal. (Courtesy of Chrysler Corp.)

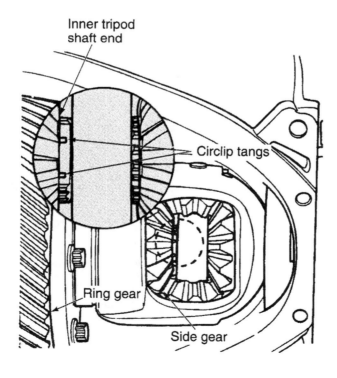

Figure 3-18 Compressible circlips in early model Chrysler transaxles. (Courtesy of Chrysler Corp.)

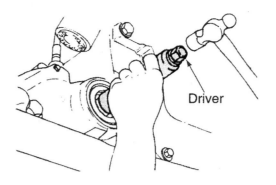

Figure 3-19 Removal of left inner drive axle joint on Ford automatic transaxles. (Courtesy of Ford Motor Company)

Front Wheel Bearing Hub Unit Removal and Replacement

The front wheel bearing removal and replacement procedure varies depending on the vehicle and the type of front wheel bearing. Always follow the front wheel bearing removal and replacement procedure in the manufacturer's service manual. The following procedure applies to front wheel bearing units that are pressed into the steering knuckle.

When front wheel bearing replacement is necessary, the steering knuckle must be removed and the wheel hub must be pressed from the bearing with a special tool (Figure 3-20).

A special puller is used to remove and replace the wheel bearing in the knuckle (Figure 3-21 and Figure 3-22).

The wheel hub must be pulled into the wheel bearing with a special tool (Figure 3-23). The proper driving tool is used to install the seal behind the bearing in the knuckle (Figure 3-24).

When two separate roller bearings are mounted in the steering knuckle, the bearing races must be driven from the knuckle with a hammer and punch. These bearings must be lubricated with

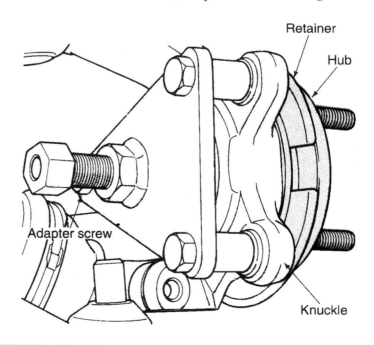

Figure 3-20 Wheel hub removal from bearing. (Courtesy of Chrysler Corp.)

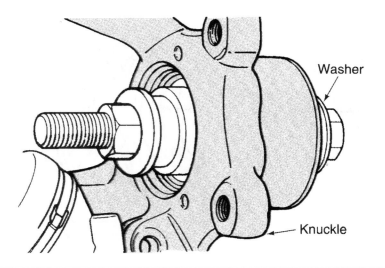

Figure 3-21 Wheel bearing removal from knuckle. (Courtesy of Chrysler Corp.)

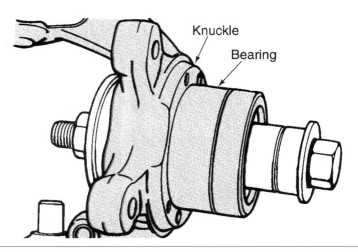

Figure 3-22 Wheel bearing installation in knuckle. (Courtesy of Chrysler Corp.)

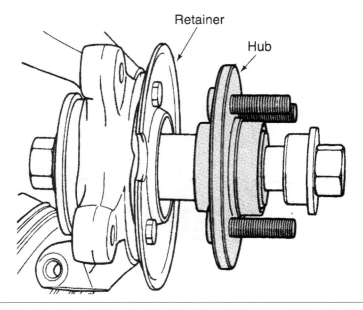

Figure 3-23 Wheel hub installation in wheel bearing. (Courtesy of Chrysler Corp.)

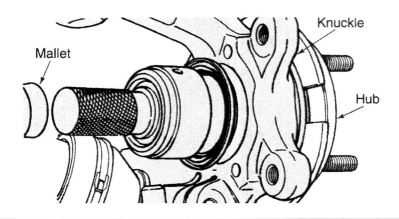

Figure 3-24 Seal installation in steering knuckle. (Courtesy of Chrysler Corp.)

wheel bearing grease prior to installation as described earlier in this chapter. When the wheel bearings are removed, all wheel bearing seals must be replaced. A staked-type hub nut must be replaced if it is removed.

On these front-wheel-drive cars, the hub nut torque applies the correct adjustment on the front wheel bearings. Therefore, this torque is extremely important. With the brakes applied, the hub nut should be tightened to the specified torque (Figure 3-25). When the hub nut is torqued to specifications, the nut lock and cotter pin should be installed (Figure 3-26).

 WARNING: Never use an impact wrench to tighten a hub nut. This action may cause wheel bearing damage.

 WARNING: Never reuse a cotter pin.

After the wheel is installed, the wheel nuts should be tightened in sequence to the specified torque (Figure 3-27). On cars with the front wheel bearings mounted in the steering knuckles, never move a car unless the front hub nuts are torqued to specifications. Lack of bearing preload could damage the bearings if the hub nuts are not tightened to specifications. If the car must be moved when the drive axles are removed, place a large bolt and nut with suitable washers through the front wheel bearing and tighten the nut to specifications.

Figure 3-25 Hub nut torquing. (Courtesy of Chrysler Corp.)

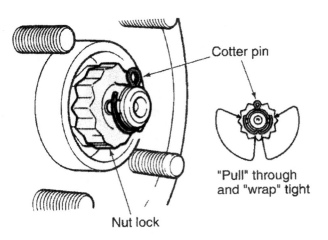

Figure 3-26 Nut lock and cotter pin installation. (Courtesy of Chrysler Corp.)

Figure 3-27 Wheel nut tightening sequence. (Courtesy of Chrysler Corp.)

Rear Axle Bearing and Seal Service, Rear-Wheel-Drive Cars

CAUTION: Use extreme caution when diagnosing problems with a vehicle raised on a hoist and the engine running with the transmission in drive. Keep away from rotating wheels, drive shafts, or drive axles.

Rear axle bearing noise may be diagnosed with the vehicle raised on a hoist. Be sure the hoist safety mechanism is engaged after the vehicle is raised on the hoist. With the engine running and the transmission in drive, operate the vehicle at moderate speed (35 to 45 mph) (56 to 72 kmph) and listen with a **stethoscope** placed on the rear axle housing directly over the axle bearings. If grinding or clicking noises are heard, bearing replacement is necessary.

Many axle shafts in rear-wheel-drive cars have a roller bearing and seal at the outer end (Figure 3-28). These axle shafts are often retained in the differential with **"C" locks** that must be removed before the axles.

Special Tools

Technician's stethoscope

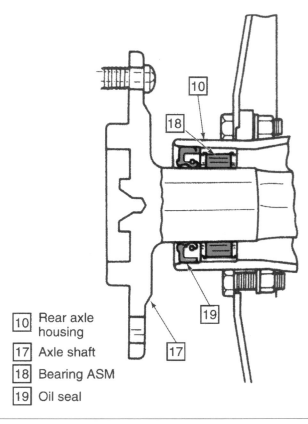

10 Rear axle housing
17 Axle shaft
18 Bearing ASM
19 Oil seal

Figure 3-28 Rear axle roller bearing and seal, rear-wheel-drive car. (Courtesy of Chevrolet Motor Division, General Motors Corp.)

The rear axle bearing removal and replacement procedure varies depending on the vehicle make and model year. Always follow the rear axle bearing removal and replacement procedure in the manufacturer's service manual. A typical rear axle shaft removal and replacement procedure on a rear-wheel-drive car with "C" lock axle retainers is as follows:

Special Tools

Seal puller
Bearing puller

1. Loosen the rear wheel nuts and chalk mark the rear wheel position in relation to the rear axle studs.
2. Raise the vehicle on a hoist and make sure the hoist safety mechanism is in place.
3. Remove the rear wheels and brake drums, or calipers and rotors.
4. Place a drain pan under the differential and remove the differential cover. Discard the old lubricant.
5. Remove the differential lock bolt, pin, pinion gears, and shaft (Figure 3-29).
6. Push the axle shaft inward and remove the axle "C" lock.
7. Pull the axle from the differential housing.

Reverse the axle removal procedure to reinstall the axle. Always use a new differential cover gasket, and fill the differential to the bottom of the filler plug opening with the manufacturer's recommended lubricant. Be sure all fasteners, including the wheel nuts, are tightened to the specified torque. A typical axle bearing and seal removal procedure:

1. Remove the axle seal with a seal puller.
2. Use the proper bearing puller to remove the axle bearing (Figure 3-30).
3. Clean the axle housing seal and bearing mounting area with solvent and a brush. Clean this area with compressed air.

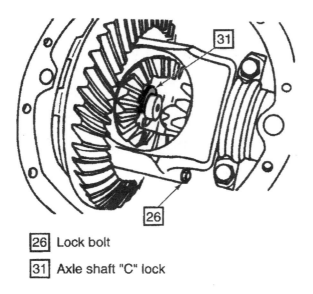

26 Lock bolt

31 Axle shaft "C" lock

Figure 3-29 Rear axle "C" lock, lock bolt, and pinion gears. (Courtesy of Chevrolet Motor Division, General Motors Corp.)

4. Check the seal and bearing mounting area in the housing for metal burrs and scratches. Remove any burrs or irregularities with a fine-toothed round file.

5. Wash the axle shaft with solvent and blow it dry with compressed air.

6. Check the bearing contact area on the axle for roughness, pits, and scratches. If any of these conditions are present, axle replacement is necessary.

7. Be sure the new bearing fits properly on the axle and in the housing. Install the new bearing with the proper bearing driver (Figure 3-31). The bearing driver must apply pressure to the outer race that is pressed into the housing.

8. Be sure the new seal fits properly on the axle shaft and in the housing. Make sure the garter spring on the seal faces toward the differential. Use the proper seal driver to install the new seal in the housing (Figure 3-32).

9. Lubricate the bearing, seal, and bearing surface on the axle with the manufacturer's specified differential lubricant.

10. Reverse the rear axle removal procedure to reinstall the rear axle.

11. Be sure all fasteners are tightened to the specified torque.

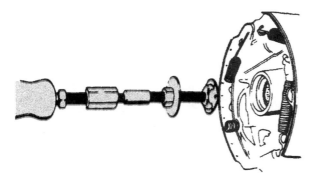

Figure 3-30 Rear axle bearing puller. (Courtesy of Chevrolet Motor Division, General Motors Corp.)

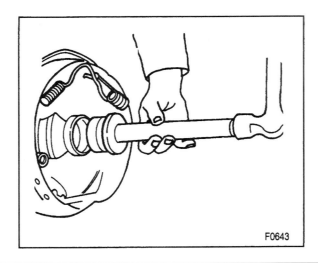

Figure 3-31 Rear axle bearing driver. (Courtesy of Chevrolet Motor Division, General Motors Corp.)

WARNING: Never use an acetylene torch to heat axle bearings or adaptor rings during the removal and replacement procedure. The heat may cause fatigue in the steel axle and the axle may break suddenly, causing the rear wheel to fall off. This action will likely result in severe vehicle damage and personal injury.

Special Tools

Axle puller

Classroom Manual
Chapter 3, page 43

Some rear axles have a sealed bearing that is pressed onto the axle shaft and held in place with an adaptor ring. These rear axles usually do not have "C" locks in the differential. A retainer plate is mounted on the axle between the bearing and the outer end of the axle. This plate is bolted to the outer end of the differential housing. After the axle retainer plate bolts are removed, a slide-hammer-type puller is attached to the axle studs to remove this type of axle. When this type of axle bearing is removed, the adaptor ring must be split with a hammer and chisel while the axle is held in a vise. Do not heat the adaptor ring or the bearing with an acetylene torch during the removal or installation process. After the adaptor ring is removed, the bearing must be pressed from the axle shaft, and the bearing must not be reused. A new bearing and adaptor ring must be pressed onto the axle shaft. The bearing removal and replacement procedure is shown in Figure 3-33.

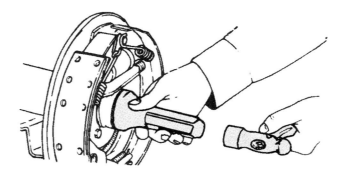

Figure 3-32 Installing rear axle seal. (Courtesy of Chevrolet Motor Division, General Motors Corp.)

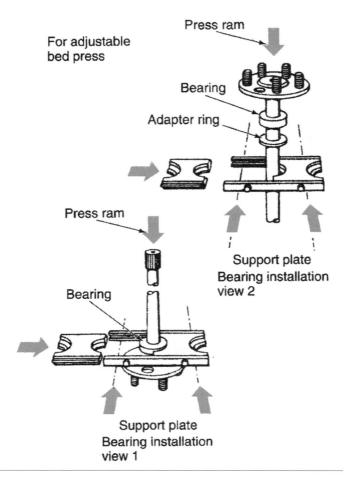

Figure 3-33 Axle bearing and adaptor ring removal and replacement. (Courtesy of Ford Motor Company)

Guidelines for Servicing Wheel Bearings and Seals

1. When a wheel bearing fails prematurely, determine the cause of the failure and correct this cause to prevent a second bearing failure.

2. Before a tire and wheel assembly is removed from a car, chalk mark the wheel and tire assembly in relation to the hub or axle to maintain wheel balance when the wheel is reinstalled.

3. Replace tapered roller bearings and races as a matched set.

4. When cleaning bearings, never spin the bearings with compressed air.

5. Do not wash sealed bearings or bearings that are shielded on both sides.

6. When drying bearings with compressed air, be sure the shop air supply is free from moisture.

7. Do not wipe bearings with paper towels or shop towels with lint on them.

8. Always inspect bearing and seal mounting areas for metal burrs and scratches. Burrs may be removed with a fine-toothed file.

9. While servicing wheel bearings, always keep hands, tools, and work area clean.

10. When two separate tapered roller bearings are mounted in the wheel hub, follow the car manufacturer's recommended bearing adjustment procedure.

11. Check wheel bearing hub units for endplay with a dial indicator.

12. On many front-wheel-drive vehicles the front drive axle must be removed to service the wheel bearing hub unit.

13. When front wheel bearings are mounted in the steering knuckle, the front hub nuts must be tightened to the specified torque before the vehicle is moved.

14. Do not use an acetylene torch to heat rear axle bearings and adaptors that are pressed into the axle shaft.

15. On some rear wheel drive cars, the rear axle "C" locks inside the differential must be removed prior to rear axle removal.

CASE STUDY

A customer complains of a bearing noise in the right front wheel of his full-size Ford wagon. The customer says the right front outer wheel bearing has been replaced twice in the last year, and this is the third failure. The technician asks the customer about the mileage intervals between bearing replacements, and the customer indicates that the wheel bearing has lasted about 8,000 miles (12,800 kilometers) each time it has been replaced. The technician finds out from the customer that no other work was done on the car each time the bearing was replaced.

When the technician removes the right front wheel and hub, the outer bearing rollers and races are badly scored. After cleaning both bearings, races, and hub, a closer examination of the outer bearing race shows an uneven wear pattern, which indicates misalignment. The technician removes the outer bearing race and finds a small metal burr behind the bearing race. This burr caused race misalignment and excessive wear on the race and rollers. The burr is removed with a fine-toothed file. The inner bearing and race have indentation wear because metal particles from the outer bearing contaminated the lubricant in the hub. The inner bearing race is removed, and the hub and spindle are thoroughly cleaned. Both bearings and the inner seal are replaced and the bearings and hub are repacked with grease. After reinstalling the hub, the bearings are carefully adjusted to the manufacturer's specifications, and the wheel nuts are tightened to the specified torque. A road test indicates that the bearing noise has been eliminated.

One small metal burr caused this customer a considerable amount of unnecessary expense. This experience proves that a technician's diagnostic capability is extremely important!

Terms to Know

Bearing abrasive step wear	Bearing frettage	Circlip
Bearing brinelling	Bearing galling	"C" locks
Bearing etching	Bearing indentations	Stethoscope
Bearing fatigue spalling	Bearing smears	

ASE Style Review Questions

1. While discussing defective bearings:
   Technician A says brinelling appears as indentations across the bearing races.
   Technician B says brinelling occurs while the bearing is rotating.
   Who is correct?
   - **A.** A only
   - **B.** B only
   - **C.** Both A and B
   - **D.** Neither A nor B

2. While discussing bearing defects:
   Technician A says misalignment wear on a front wheel bearing could be caused by a metal burr behind one of the bearing races. *T echnician B* says misalignment wear on a front wheel bearing could be caused by a bent front spindle.
   Who is correct?
   - **A.** A only
   - **B.** B only
   - **C.** Both A and B
   - **D.** Neither A nor B

3. While discussing front wheel bearing service on a vehicle with the bearing hub unit pressed into the steering knuckle:
   Technician A says the front wheel bearings may be damaged if this front-wheel-drive vehicle is moved without the hub nuts torqued to specifications.
   Technician B says the hub nut torque supplies the correct wheel bearing adjustment.
   Who is correct?
   - **A.** A only
   - **B.** B only
   - **C.** Both A and B
   - **D.** Neither A nor B

4. While discussing front wheel bearing service on a front-wheel-drive car with two separate tapered roller bearings mounted in the front steering knuckles:
   Technician A says the hub nut torque supplies the correct bearing adjustment.
   Technician B says that the brake should be applied while the front hub nuts are torqued.
   Who is correct?
   - **A.** A only
   - **B.** B only
   - **C.** Both A and B
   - **D.** Neither A nor B

5. A front-wheel-drive vehicle has 14-in (35.5-cm) rims and front wheel bearings mounted in the steering knuckles. The vehicle is lifted on a hoist and a dial indicator is positioned against the outer rim lip. The total inward and outward rim movement is 0.035 in (0.889 mm).
   Technician A says that the wheel bearing may require replacement. *Technician B* says that the hub nut torque should be checked prior to bearing replacement.
   Who is correct?
   - **A.** A only
   - **B.** B only
   - **C.** Both A and B
   - **D.** Neither A nor B

6. When cleaning and inspecting wheel bearings:
   - **A.** sealed bearings should be washed in solvent.
   - **B.** high-pressure air may be used to spin the bearings.
   - **C.** a bent bearing cage may be caused by improper tool use.
   - **D.** bearing overload may cause bearing cage wear.

7. All these statements about hub seal service are true EXCEPT:
   - **A.** The garter spring must face toward the flow of lubricant.
   - **B.** A ball-peen hammer should be used to install the seal.
   - **C.** Seal contact area on the spindle must be clean and free from metal burrs.
   - **D.** The outer edge of the seal case should be coated with sealant.

8. When servicing press-in rear axle bearings on a rear-wheel-drive car:
   - **A.** these bearing may be reused after they are removed from the axle shaft.
   - **B.** the bearing adaptor ring should be removed by splitting it with a hammer and chisel.
   - **C.** a cutting torch may be used to cut the bearing off the axle shaft.
   - **D.** an acetylene torch may be used to heat the new adaptor ring prior to installation.

9. A front-wheel-drive vehicle has two tapered roller bearings in each rear wheel hub. When adjusting these wheel bearings:
   A. the adjusting nut should be tightened to 17 to 25 ft-lbs (23 to 24 Nm), backed off 1/2 turn, and then tightened to 10 to 15 in-lbs (1.0 to 1.7 Nm).
   B. the adjusting nut should be tightened to 40 ft-lbs (54 Nm), backed off 1 turn, and then tightened to 10 ft-lbs (13.5 Nm).
   C. the adjusting nut should be tightened to 50 ft-lbs (67.5 Nm), backed off 3/4 turn, and then tightened to 10 to 15 in-lbs (1.0 to 1.7 Nm).
   D. the wheel and hub should not be rotated while adjusting the wheel bearings.

10. A unitized front wheel bearing hub that is bolted to the steering knuckle has 0.010 in (0.254 mm) of hub endplay. The proper repair procedure is to:
    A. repack and readjust the wheel bearings in the hub.
    B. tighten the hub nut to the specified torque.
    C. inspect the drive axle and hub splines, and replace the worn components.
    D. replace the wheel bearing hub assembly.

ASE Challenge Questions

1. The customer says her front-wheel-drive car makes "a moaning noise" in a turn. Which of the following could cause this problem?
   A. Outer front wheel bearing
   B. Rear axle bearing
   C. Differential gear noise
   D. Transaxle output shaft bearing

2. The customer complains of a "whining noise" in the back of her rear-wheel-drive car when driving between 30 and 40 miles per hour.
   Technician A says a good way to diagnose this problem is on a hoist with a stethoscope.
   Technician B says a good way to diagnose this problem is on the road with a microphone.
   Who is correct?
   A. A only
   B. B only
   C. Both A and B
   D. Neither A nor B

3. Upon inspecting a noisy front wheel bearing from a 4WD sport utility vehicle, brinelling damage to the outer bearing race was noticed.
   Technician A says the damage was probably caused by tightening the bearing nut with an impact wrench.
   Technician B says the damage was probably caused by driving the vehicle through deep, muddy water.
   Who is correct?
   A. A only
   B. B only
   C. Both A and B
   D. Neither A nor B

4. Discussing the cause of overheating damage of a front wheel bearing:
   Technician A says overheating may be caused by insufficient or incorrect bearing lubricant.
   Technician B says overheating may be caused by overtightening the wheel bearing nut on installation.
   Who is correct?
   A. A only
   B. B only
   C. Both A and B
   D. Neither A nor B

5. The customer says her front-wheel-drive car has a noise and vibration when decelerating and is most noticeable under 50 miles per hour. Which would be the most probable cause of this problem?
   A. Defective outer front wheel bearing
   B. Defective inner front wheel bearing
   C. Defective outer drive axle joint
   D. Defective inner drive axle joint

Job Sheet 7

Name _____ Date _____

Service Integral Wheel Bearing Hubs

Upon completion of this job sheet, you should be able to measure integral wheel bearing hub end-play and remove and replace integral wheel bearing hubs.

ASE Correlation

This job sheet is related to the ASE Suspension and Steering Task List content area: B. Suspension Systems Diagnosis and Repair, 3. Miscellaneous Service, Task 3: Diagnose and service front and/or rear wheel bearings.

Tools and Materials

A front-wheel-drive car with integral-type front wheel bearing hubs
A floor jack and safety stands, or a lift for raising the vehicle
Dial indicator
Integral wheel bearing hub puller
Torque wrench

Vehicle Make and Model Year _____
Vehicle VIN _____

Procedure

1. Road test the vehicle and listen for any abnormal wheel bearing noises; check for steering looseness or wander that may be caused by loose wheel bearings.

   Type of abnormal noise   Possible causes

   a. _____   a. _____

   b. _____   b. _____

   c. _____   c. _____

   Steering wander and looseness

   yes _____ no _____

2. Use a dial indicator to measure the endplay on the front wheel bearing hubs with the wheel installed on the hub.

   Endplay left side _____ Endplay right side _____

   Specified endplay _____

   Necessary repairs _____

3. Loosen and remove the cotter pin, lock nut, and drive axle nut before raising the vehicle.

4. Raise the vehicle on a lift and remove the front wheels.

5. Remove the brake calipers and brake rotors. Use a length of wire to tie the calipers to the chassis.

6. Use a dial indicator to measure the endplay directly on the front wheel bearing hubs.

Endplay left side _____ Endplay right side _____

Specified endplay _____

Necessary repairs _____

7. If a wheel speed sensor for the antilock brake system (ABS) is integral in the front wheel hubs, disconnect the wheel speed sensor connectors.

8. Remove the hub-to-knuckle bolts, and place the transaxle in park.

9. Use the proper puller to remove the wheel bearing hub from the drive axle.

10. Install the new hub and bearing assembly on the drive axle splines, and install a new drive axle nut. Tighten the drive axle nut to pull the hub onto the splines. Do not tighten this nut to the specified torque at this time.

11. Place the transaxle in neutral and install the wheel bearing hub bolts. Tighten these bolts to the specified torque.

Specified wheel bearing hub bolt torque _____

Actual wheel bearing hub bolt torque _____

12. Install the wheel sensor connectors, rotors, and calipers. Tighten the caliper mounting bolts to the specified torque. Place a large punch between the rotor fins and the caliper, and tighten the drive axle nut to the specified torque.

Specified caliper bolt torque _____

Actual caliper bolt torque _____

Specified drive axle nut torque _____

Actual drive axle nut torque _____

13. Install the drive axle lock nut and tighten to 50 ft-lbs. Lock nut properly installed: yes _____ no _____

14. Install the wheels in their original positions, and tighten the wheel nuts to the specified torque.

Specified wheel nut torque _____

Actual wheel nut torque _____

15. Lower the vehicle onto the shop floor, and tighten the drive axle nut to the specified torque. Install the lock nut and a new cotter pin.

Specified drive axle nut torque _____

Actual drive axle nut torque _____

Are the lock nut and cotter pin properly installed and tightened?

Instructor check _____

16. Road test the car and listen for abnormal wheel bearing noises.

Wheel bearing noise: satisfactory _____ unsatisfactory _____

✓ **Instructor Check** _____

Job Sheet 8

8

Name _____ Date _____

Diagnose Wheel Bearings

Upon completion of this job sheet, you should be able to diagnose wheel bearings.

ASE Correlation

This job sheet is related to the ASE Suspension and Steering Task List content area: B Suspension Systems Diagnosis and Service, 3. Miscellaneous Service, Task 3: Diagnose and service front and/or rear wheel bearings.

Tools and Materials

A vehicle with wheel bearing noise or bearing noise from another source

Vehicle Make and Model Year _____

Vehicle VIN _____

Procedure

1. Road test the vehicle and listen for any abnormal wheel noises, and check for steering looseness and wander that may be caused by loose wheel bearings.

2. Wheel bearing noise: satisfactory _____ unsatisfactory _____

3. Is the bearing noise coming from the front of rear of the vehicle?

   front _____ rear _____

4. Is this noise more noticeable when turning a corner at low speeds?

   yes _____ no _____

5. Is this noise more noticeable during acceleration? yes _____ no _____

6. Is this noise more noticeable during deceleration? yes _____ no _____

7. Is this noise more noticeable when driving at a steady speed? yes _____ no _____

   If the answer to this question is yes, state the speed when the bearing noise is most noticeable. _____

8. Is this noise most noticeable when driving down a narrow street at a steady low speed?

   yes _____ no _____

9. State the exact cause of the bearing noise, and explain the reason for your diagnosis.

   _____

   _____

   _____

☑ **Instructor Check** _____

Job Sheet 9

Name _____ Date _____

Clean Lubricate, Install, and Adjust Nonsealed Wheel Bearings

Upon completion of this job sheet, you should be able to clean, lubricate, install, and adjust non-sealed wheel bearings.

ASE Correlation

This job sheet is related to the ASE Suspension and Steering Task List content area: B Suspension Systems Diagnosis and Service, 3. Miscellaneous Service, Task 3: Diagnose and service front and/or rear wheel bearings.

Tools and Materials

A vehicle with tapered roller-type front wheel bearings, or a front-wheel-drive car with tapered roller-type rear wheel bearings.
A floor jack and safety stands, or a lift for raising the vehicle
Wheel bearing grease
Cleaning solution
Seal drivers
Torque wrench

Vehicle Make and Model Year _____

Vehicle VIN _____

Procedure

1. Raise the vehicle on a lift with the chassis supported on the lift.

2. Chalk mark the wheels in relation to the studs; remove the wheels, wheel hubs, and brake drums or rotors.

3. Remove the inner hub seal and wheel bearings from the hub.

4. Clean the hub and bearings with an approved cleaning solvent.

   Are the hub and bearings properly cleaned? yes _____ no _____

   Instructor check _____

5. Inspect the wheel bearings, bearing cones, and hubs.

Type of defects	Necessary replacement parts
a. _____	a. _____
b. _____	b. _____
c. _____	c. _____

6. Inspect all bearing-related conditions such as lubrication, alignment, spindle condition, hub condition. State the cause(s) of the bearing defects listed in step 5.

a _____

b _____

c _____

7. Repack the wheel bearings and hub with the car manufacturer's specified wheel bearing grease.

Are the wheel bearings and hub repacked with manufacturer's recommended grease?

yes _____ no _____

Instructor check _____

8. Install a new inner hub seal with the proper seal driver. Lubricate the seal lips with a light coating of the car manufacturer's recommended wheel bearing grease.

Is the new inner hub seal properly installed? yes _____ no _____

Instructor check _____

9. Inspect, clean, and lubricate the spindles and seal contact area.

Spindle and seal contact condition: satisfactory _____ unsatisfactory _____

If unsatisfactory, state necessary repairs. _____

_____

Is the spindle and seal contact area properly cleaned and lubricated? yes _____ no _____

Instructor check _____

10. Install the wheel hubs, wheel bearings, washer, and retaining nut. Adjust the wheel bearing retaining nut to the specified initial torque.

Specified initial wheel bearing torque _____

Actual initial wheel bearing torque _____

11. Back off the wheel bearing retaining nut the specified amount.

Specified portion of a turn to back off the wheel bearing retaining nut _____

Actual portion of a turn the wheel bearing retaining nut is backed off _____

12. Tighten the wheel bearing retaining nut to the final specified torque.

Specified final wheel bearing retaining nut torque _____

Actual final wheel bearing retaining nut torque _____

13. Install a new cotter pin through the wheel bearing retaining nut and spindle opening.

Is the new cotter pin properly installed? yes _____ no _____

Instructor check _____

14. Install the wheel bearing dust covers. Install the wheels in their original position, and tighten the wheel nuts to the specified torque.

Specified wheel nut torque _____

Actual wheel nut torque _____

15. Road test the car and listen for abnormal wheel bearing noises.

Wheel bearing noise: satisfactory _____ unsatisfactory _____

◥ **Instructor Check** _____

Table 3-1   ASE Task

Diagnose and service front and/or rear wheel bearings.

Problem Area	Symptoms	Possible Causes	Classroom Manual	Shop Manual
NOISES	Growling noise at low speed, more noticeable while cornering	Defective front wheel bearings	43	57
	Growling noise at low speed	Defective rear wheel bearings	56	57
STEERING QUALITY	Reduced directional stability	Loose front wheel bearings	50	66

Wheel and Tire Servicing and Balancing

Upon completion and review of this chapter, you should be able to:

❏ Diagnose tire thump and vibration problems.

❏ Diagnose steering pull problems related to tire condition.

❏ Rotate tires according to the vehicle manufacturer's recommended procedure.

❏ Remove and replace wheel-and-tire assemblies.

❏ Demount, inspect, repair, and remount tires.

❏ Inspect wheel rims.

❏ Measure tire and wheel radial and lateral runout.

❏ Diagnose problems caused by excessive radial or lateral wheel and tire runout.

❏ Measure tire tread wear.

❏ Perform off-car static wheel balance procedures.

❏ Perform off-car dynamic wheel balance procedures.

❏ Diagnose tire wear problems caused by tire and wheel imbalance.

❏ Perform on-car balance procedures.

Basic Tools

Basic technician's tool set

Service manual

Tire repair kit

Tread depth gauge

Tire thump may be defined as a pounding noise caused by tire and wheel rotation.

Tire vibration is a fast shaking of the tire that is transferred to the chassis and passenger compartment.

Steering pull is the tendency of the steering to gradually pull to the right or left when the vehicle is driven straight ahead on a reasonably smooth, straight road.

Tire conicity occurs when the tire belt is wound off-center in the manufacturing process creating a cone-shaped belt that results in steering pull.

Tire Noises and Steering Problems

Diagnosis of Tire Noises

 SERVICE TIP: Tire noise varies with road surface conditions, whereas differential noise is not affected when various road surfaces are encountered.

Uneven tread surfaces may cause tire noises that seem to originate elsewhere in the vehicle. These noises may be confused with differential noise. Differential noise usually varies with acceleration and deceleration, whereas tire noise remains more constant in relation to these forces. Tire noise is most pronounced on smooth asphalt road surfaces at speeds of 15 to 45 mph (24 to 72 kmph).

Tire Thump and Vibration

When **tire thump** and **tire vibration** are present, check these items:

1. Cupped tire treads

2. Excessive tire **radial runout**

3. Manufacturing defects such as heavy spots, weak spots, or tread chunking

4. Incorrect wheel balance

Steering Pull

A vehicle should maintain the straight-ahead forward direction on smooth, straight road surfaces without excessive steering wheel correction by the driver. If the steering gradually pulls to one side on a smooth, straight road surface, a tire, steering, or suspension defect is present. Tires of different types, sizes, designs, or inflation pressures on opposite sides of a vehicle cause **steering pull.** Sometimes a tire manufacturing defect occurs in which the belts are wound off-center on the tire. This condition is referred to as **tire conicity.** A cone-shaped object rolls in the direction of its smaller diameter. Similarly, a tire with conicity tends to lead, or pull, to one side, which causes the vehicle to follow the action of the tire (Figure 4-1).

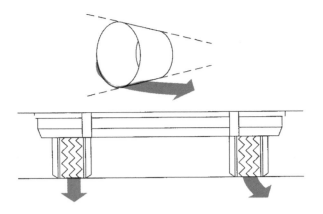

Figure 4-1 Tire conicity. (Courtesy of Chrysler Corp.)

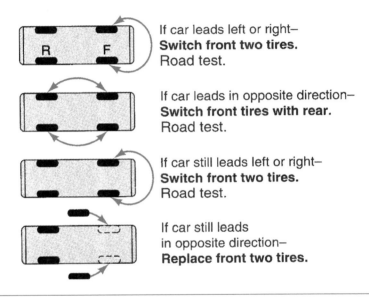

If car leads left or right—
Switch front two tires.
Road test.

If car leads in opposite direction—
Switch front tires with rear.
Road test.

If car still leads left or right—
Switch front two tires.
Road test.

If car still leads
in opposite direction—
Replace front two tires.

Figure 4-2 Tire conicity diagnosis. (Courtesy of Chrysler Corp.)

 SERVICE TIP: Tire conicity is not visible. It can only be diagnosed by changing the tire and wheel position.

Since tire conicity cannot be diagnosed by a visual inspection, it must be diagnosed by switching the two front tires, and reversing the front and rear tires (Figure 4-2). Incorrect front suspension alignment angles also cause steering pull.

Tire Rotation

Driving habits determine tire life to a large extent. Severe brake applications, high-speed driving, turning at high speeds, rapid acceleration and deceleration, and striking curbs are just a few driving habits that shorten tire life. Most car manufacturers recommend **tire rotation** at specified intervals to obtain maximum tire life. The exact tire rotation procedure depends on the model year, the type of tires, and whether the vehicle has a conventional spare or a compact spare (Figure 4-3). Tire rotation procedures do not include the compact spare. The vehicle manufacturer provides tire rotation information in the owner's manual and service manual. Vehicle manufacturers usually recommend different tire rotation procedures for bias ply tires than for radial tires (Figure 4-4).

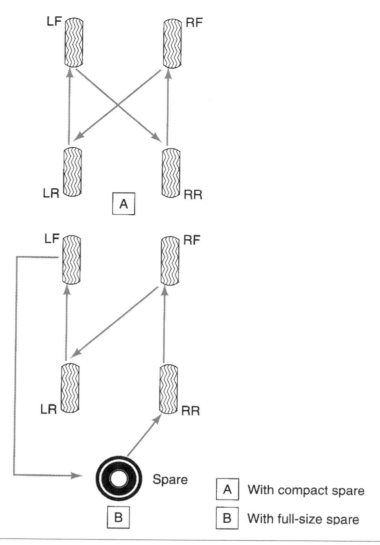

Figure 4-3 Radial tire rotation procedure. (Courtesy of Chevrolet Motor Division, General Motors Corp.)

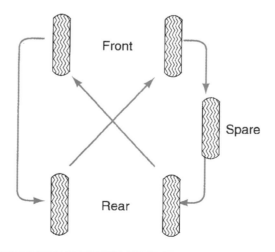

Figure 4-4 Bias ply tire rotation procedure. (Courtesy of Pontiac Motor Division, General Motors Corp.)

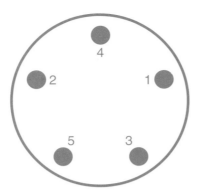

Figure 4-5 Wheel nut tightening sequence. (Courtesy of Chevrolet Motor Division, General Motors Corp.)

When tires and wheels are installed on a vehicle, it is very important that the wheel nuts are torqued to manufacturer's specifications in the proper sequence (Figure 4-5). Do not use an impact wrench when tightening wheel nuts to the specified torque.

Wheel and Tire Service

Wheel and Tire Removal

 CUSTOMER CARE: During many automotive service operations, including tire and wheel service, the technician literally has the customer's life in his or her hands! Always perform tire and wheel service carefully and thoroughly. Always watch for unsafe tire or wheel conditions, and report these problems to the customer. When you prove to the customer that you are concerned about vehicle safety, you will probably have a steady customer.

SERVICE TIP: Some wheel covers have fake plastic lug nuts that must be removed to access the lug nuts. Be careful not to break the fake lug nuts.

When it is necessary to remove a wheel and tire assembly, follow these steps:

1. Remove the wheel cover. If the vehicle is equipped with **antitheft locking wheel covers,** the lock bolt for each wheel cover is located behind the ornament in the center of the wheel cover. A special key wrench is supplied to the owner for ornament and lock bolt removal. If the customer's key wrench has been lost, a master key is available from the vehicle dealer.

2. Loosen the wheel lug nuts about one-half turn, but do not remove the wheel nuts. Some vehicles are equipped with **antitheft wheel nuts.** A special lug nut key is supplied to the vehicle owner. This lug nut key has a hex nut on the outer end and a special internal projection that fits in the wheel nut opening. Install the lug nut key on the lug nuts and connect the lug nut wrench on the key hex nut to loosen the lug nuts.

CAUTION: Before the vehicle is raised on a hoist, be sure that the hoist is lifting on the car manufacturer's recommended lifting points.

CAUTION: If the vehicle is lifted with a floor jack, place safety stands under the suspension or frame and lower the vehicle onto the safety stands. Then remove the floor jack from under the vehicle.

3. Raise the vehicle on a hoist or with a floor jack to a convenient working level.

4. Chalk mark the tire, wheel, and one of the lug nuts, so the wheel and tire can be reinstalled in the same position.

 WARNING: If heat is used to loosen a rusted wheel, the wheel and/or wheel bearings may be damaged.

5. Remove the lug nuts and the wheel-and-tire assembly. If the wheel is rusted and will not come off, hit the inside of the wheel with a large rubber mallet. Do not hit the wheel with a steel hammer, because this action could damage the wheel. Do not heat the wheel.

Tire and Wheel Service Precautions

There are many different types of tire changing equipment in the automotive service industry. However, specific precautions apply to the use of any tire changing equipment. These precautions include the following:

1. Before you operate any tire changing equipment, always be absolutely certain that you are familiar with the operation of the equipment.

2. When operating tire changing equipment, always follow the equipment manufacturer's recommended procedure.

3. Always deflate a tire completely before attempting to demount the tire.

4. Clean the bead seats on the wheel rim before mounting the tire on the wheel rim.

5. Lubricate the outer surface of the tire beads with rubber lubricant before mounting the tire on the wheel rim.

6. When the tire is mounted on the wheel rim, be sure the tire is positioned evenly on the wheel rim.

7. While inflating a tire, do not stand directly over the tire. An air hose extension allows the technician to stand back from the tire during the inflation process.

8. Do not overinflate tires.

9. When mounting tires on cast aluminum alloy wheel rims or cast magnesium alloy wheel rims, always use the tire changing equipment manufacturer's recommended tools and procedures.

Tire Demounting

 WARNING: If hand tools or tire irons are used to demount tires, tire bead and wheel rim damage may occur.

Always use a tire changer to demount tires (Figure 4-6). Do not use hand tools or tire irons for this purpose. A typical tire demounting procedure follows.

1. Remove the valve core and be sure the tire is completely deflated.

2. Place the wheel and tire on the tire changer with the narrow bead ledge facing upward.

3. Follow the operating procedure recommended by the manufacturer of the tire changer to force the tire bead inward and separate it from the rim on both sides.

4. Push one edge of the top bead into the drop center of the rim.

Figure 4-6 Tire changer.

5. Place the tire changer's bar or lever between the bead and the rim on the opposite side of the rim from where the bead is in the drop center.

6. Operate the tire changer to rotate the bar or lever and move the bead over the top of the rim.

7. Repeat steps 4, 5, and 6 to move the lower bead over the top of the rim.

Tire Inspection and Repair

To find a leak in a tire and wheel, inflate the tire to the pressure marked on the sidewall, and then submerge the tire and wheel in a tank of water. An alternate method of leak detection is to sponge soapy water on the tire and wheel. Bubbles will appear wherever the leak is located in the tire or wheel. Mark the leak location in the tire or wheel rim with a crayon, and mark the tire at the valve stem location so the tire can be reinstalled in the same position on the wheel to maintain proper balance.

A puncture is the most common cause of a tire leak, and many punctures can be repaired satisfactorily. Do not attempt to repair punctures over 1/4 inch in diameter. Punctures in the sidewalls or on the tire shoulders should not be repaired. The repairable area in belted bias ply tires is approximately the width of the belts (Figure 4-7). The belts in radial tires are wider than those in bias ply tires. The repairable area in radial tires is also the width of the belts. Since compact spare tires have thin treads, do not attempt to repair these tires.

Inspect the tire; do not repair a tire with any of the following defects, signs of damage, or excessive wear:

1. Tires with the wear indicators showing

2. Tires worn until the fabric or belts are exposed

3. Bulges or blisters

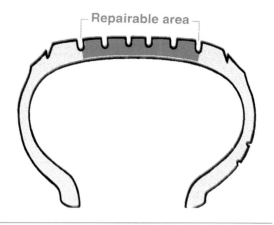

Figure 4-7 Repairable area on bias ply and belted bias ply tires. (Courtesy of Cadillac Motor Car Division, General Motors Corp.)

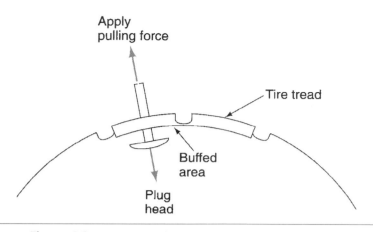

Figure 4-8 Plug installation procedure.

4. Ply separation

5. Broken or cracked beads

6. Cuts or cracks anywhere in the tire

Since most vehicles are equipped with tubeless tires, we will discuss this type of tire repair. If the cause of the puncture, such as a nail, is still in the tire, remove it from the tire. Most punctures can be repaired from inside the tire with a service plug or vulcanized patch service kit. The instructions of the tire service kit manufacturer should be followed, but we will discuss three common tire repair procedures.

Plug Installation Procedure

1. Buff the area around the puncture with a wire brush or wire buffing wheel.

2. Select a plug slightly larger than the puncture opening, and insert the plug in the eye of the insertion tool.

3. Wet the plug and the insertion tool with vulcanizing fluid.

4. While holding and stretching the plug, pull the plug into the puncture from the inside of the tire (Figure 4-8). The head of the plug should contact the inside of the tire. If the plug pulls through the tire, repeat the procedure.

5. Cut the plug off 1/32 inch from the tread surface. Do not stretch the plug while cutting.

Cold Patch Installation Procedure

 WARNING: Radial tire patches should have arrows that must be positioned parallel to the radial plies.

1. Buff the area around the puncture with a wire brush or buffing wheel.

2. Apply vulcanizing fluid to the buffed area and allow it to dry until it is tacky.

3. Peel the backing from the patch, and apply the patch over the puncture. Center the patch over the puncture.

4. Run a stitching tool back and forth over the patch to improve bonding.

Hot Patch Installation Procedure

1. Buff the area around the puncture with a wire brush or buffing wheel.
2. Apply vulcanizing fluid to the buffed area, if required.
3. Peel the backing from the patch and install the patch so it is centered over the puncture on the inside of the tire. Many hot patches are heated with an electric heating element clamped over the patch. This element should be clamped in place for the amount of time recommended by the equipment or patch manufacturer.
4. After the heating element is removed, allow the patch to cool for a few minutes and be sure the patch is properly bonded to the tire.

Wheel Rim Service

 WARNING: The use of abrasive cleaners, alkaline-base detergents, or caustic agents on aluminum or magnesium wheel rims may cause discoloration or damage to the protective coating.

Steel rims should be spray cleaned with a water hose. Aluminum or magnesium wheel rims should be cleaned with a mild soap and water solution, and rinsed with clean water. The use of abrasive cleaners, alkaline-base detergents, or caustic agents may damage aluminum or magnesium wheel rims. Clean the rim bead seats on these wheel rims thoroughly with the mild soap and water solution. The rim bead seats on steel wheel rims should be cleaned with a wire brush or coarse steel wool.

 WARNING: Steel wheel rims must not be welded, heated, or peened with a ball peen hammer. These procedures may weaken the rim and create a safety hazard.

WARNING: Installing an inner tube to correct leaks in a tubeless tire or wheel rim is not an approved procedure.

Steel wheel rims should be inspected for excessive rust and corrosion, cracks, loose rivets or welds, bent or damaged bead seats, and elongated lug nut holes. Aluminum or magnesium wheel rims should be inspected for damaged bead seats, elongated lug nut holes, cracks, and porosity. If any of these conditions are present on either type of wheel rim, replace the wheel rim.

Many shops always replace the tire valve assembly when a tire is repaired or replaced. This policy helps prevent future problems with tire valve leaks. The inner end of the valve may be cut off with a pair of diagonal pliers, and then the outer end may be pulled from the rim. Coat the new valve with rubber tire lubricant and pull it into the rim opening with a special puller screwed onto the valve threads.

Wheel Rim Leak Repair

A wheel rim leak may be repaired if the leak is not caused by excessive rust on a steel rim, and the rim is in satisfactory condition. Follow these steps for wheel rim leak repair:

1. Use #80-grit sandpaper to thoroughly clean the area around the leak on the tire side of the rim.
2. Use a shop towel to remove any grit from the leak area.
3. Be sure the wheel rim is at room temperature, and apply a heavy coating of silicone rubber sealer over the leak area.
4. Spread the sealer over the entire sanded area with a putty knife.
5. Allow the sealer to cure for 6 hours before remounting the tire.

Tire Remounting Procedure

1. Be sure the wheel rim bead seats are thoroughly cleaned.

2. Coat the tire beads and the wheel rim bead seats with rubber tire lubricant.

3. Secure the wheel rim on the tire changer with the narrow bead ledge facing upward, and place the tire on top of the wheel rim with the bead on the lower side of the tire in the drop center of the wheel rim.

4. Use the tire changer bar or lever under the tire bead to install the tire bead over the wheel rim. Always operate the tire changer with the manufacturer's recommended procedure.

5. Repeat the procedure in steps 3 and 4 to install the upper bead over the wheel rim.

6. Rotate the tire on the wheel rim until the crayon mark is aligned with the valve stem. This mark was placed on the tire prior to demounting.

    ▲ **WARNING:** When a bead expander is installed around the tire, never exceed 10 psi (69 kPa) pressure in the tire. A higher pressure may cause the expander to break and fly off the tire, causing serious personal injury or property damage.

    ▲ **WARNING:** When a bead expander is not used, never exceed 40 psi (276 kPa) tire pressure to move the tire beads out tightly against the wheel rim. A higher pressure may blow the tire bead against the rim with excessive force, and this action could burst the rim or tire, resulting in serious personal injury or property damage.

    ▲ **WARNING:** While inflating a tire, do not stand directly over a tire. In this position, serious injury could occur if the tire or wheel rim flies apart.

7. Follow the recommended procedure supplied by the manufacturer of the tire changer to inflate the tire. This procedure may involve the use of a bead expander installed around the center of the tire tread to expand the tire beads against the wheel rim. If a bead expander is used, inflate the tire to 10 psi (69 kPa) to move the beads out tightly against the wheel rim. Never exceed this pressure with a bead expander installed on the tire. Always observe the circular marking around the tire bead as the tire is inflated. This mark should be centered around the wheel rim (Figure 4-9). Always observe both beads while a tire is inflated. If the circular mark around the tire bead is not centered on the rim, deflate the tire and center it on the wheel rim. If a bead expander is not used, never exceed 40 psi (276 kPa) to try and move tire beads out tightly against the rim. When either tire bead will not move out tightly against the wheel rim with 40 psi (276 kPa) tire pressure, deflate the tire and center it on the wheel rim again.

Photo Sequence 2 shows a typical procedure for demounting and mounting a tire on a wheel assembly.

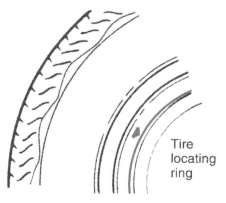

Tire locating ring

Figure 4-9 The circular ring around the tire bead must be centered on the wheel rim. (Courtesy of Oldsmobile Motor Division, General Motors Corp.)

Photo Sequence 2
Typical Procedure for Demounting and Mounting a Tire on a Wheel Assembly

P2-1 A typical tire changer.

P2-2 Demounting the tire from the wheel begins with releasing the air, removing the air valve core, and unseating the tire from the rim. The machine does the unseating. The technician merely guides the operating lever.

P2-3 Once both sides of the tire are unseated, place the assembly onto the machine.

P2-4 Depress the pedal that clamps the wheel to the tire machine.

P2-5 Lower the machine's arm into position on the tire-and-wheel assembly.

P2-6 Insert the tire iron between the tire and wheel. Depress the pedal that causes the wheel to rotate. This will free the tire from the wheel.

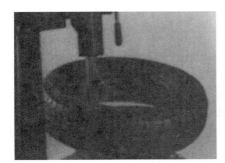

P2-7 After the tire is totally free from the rim, remove the tire.

P2-8 To prepare the wheel for the mounting of a new tire, use a wire brush to remove all of the dirt and rust from the sealing surface.

P2-9 Rubber compound should be liberally applied to the bead area of the new tire.

Photo Sequence 2
Typical Procedure for Demounting and Mounting a Tire on a Wheel Assembly (continued)

P2-10 Place the tire onto the wheel assembly and lower the arm onto the assembly. As the wheel rotates, the tire will be forced over the rim.

P2-11 After the tire is completely over the rim, install the air ring over the tire. Activate this to seat the tire against the wheel.

P2-12 Reinstall the air valve core and inflate the tire to the recommended pressure.

Tire and Wheel Runout Measurement

Radial tire runout refers to excessive variations in the tread circumference.

Special Tools

Tire runout gauge

Ideally, a tire-and-wheel assembly should be perfectly round. However, this condition is rarely achieved. A tire-and-wheel assembly that is out-of-round is said to have radial runout. If the radial runout exceeds manufacturer's specifications, a vibration may occur because the radial runout causes the spindle to move up and down (Figure 4-10). A defective tire with a variation in stiffness may also cause this up-and-down spindle action.

A dial indicator gauge may be positioned against the center of the tire tread as the tire is rotated slowly to measure radial runout (Figure 4-11). Radial runout of more than 0.060 in (1.5 mm) will cause vehicle shake. If the radial runout is between 0.045 in to 0.060 in (1.1 mm to 1.5 mm), vehicle shake may occur. These are typical radial runout specifications. Always consult the vehicle manufacturer's specifications. Mark the highest point of radial runout on the tire with a crayon, and mark the valve stem position on the tire.

If the radial tire runout is excessive, demount the tire and check the runout of the wheel rim with a dial indicator positioned against the lip of the rim while the rim is rotated (Figure 4-12). Use a crayon to mark the highest point of radial runout on the wheel rim. Radial wheel runout should not exceed 0.035 in (0.9 mm), whereas the maximum lateral wheel runout is 0.045 in (1.1 mm). If the highest point of wheel radial runout coincides with the chalk mark from the highest point of maximum tire radial runout, the tire may be rotated 180° on the wheel to reduce radial runout. Tires or wheels with excessive runout are usually replaced.

Lateral tire runout may be defined as excessive variations in the sidewalls of the tire.

Classroom Manual
Chapter 4, page 62

Lateral tire runout may be measured with a dial indicator located against the sidewall of the tire. Excessive lateral runout causes the tire to waddle as it turns, and this waddling sensation may be transmitted to the passenger compartment (Figure 4-13). A chassis waddling action may also be caused by a defective tire in which the belt is not straight. If the lateral runout exceeds 0.080 in (2.0 mm), wheel shake problems will occur on the vehicle. Chalk mark the tire and wheel at the highest point of radial runout. When the tire runout is excessive, the tire should be removed from the wheel, and the wheel lateral runout should be measured with a dial indicator positioned against the edge of the wheel as the wheel is rotated. Wheels or tires with excessive lateral runout should be replaced.

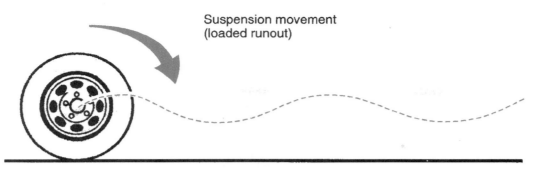

Suspension movement
(loaded runout)

Caused by

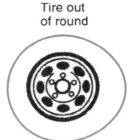

Tire out
of round

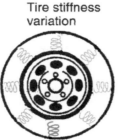

Tire stiffness
variation

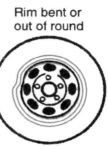

Rim bent or
out of round

Figure 4-10 Vertical tire and wheel vibrations caused by radial tire or wheel runout, or variation in tire stiffness. (Courtesy of Oldsmobile Motor Division, General Motors Corp.)

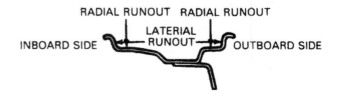

MEASURING WHEEL RUNOUT

INBOARD SIDE

TIRE MOUNTED
ON WHEELS

OUTBOARD SIDE

LATERAL RUNOUT

LATERAL RUNOUT

RADIAL RUNOUT

RADIAL RUNOUT*

* IF WHEEL DESIGN MAKES THIS OUTBOARD
MEASUREMENT IMPOSSIBLE, THE INBOARD
SIDE ONLY MAY BE USED.

TIRE REMOVED FROM WHEEL

RADIAL RUNOUT   RADIAL RUNOUT

LATERAL
RUNOUT

INBOARD SIDE

OUTBOARD SIDE

Figure 4-11 Measuring tire radial runout. (Courtesy of Chrysler Corp.)

Figure 4-12 Measuring wheel radial runout. (Courtesy of Oldsmobile Motor Division, General Motors Corp.)

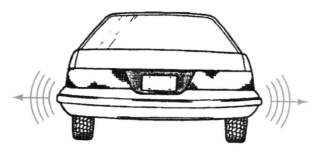

Tire waddle often caused by
- Steel belt not straight within tire
- Excessive lateral runout

Figure 4-13 Chassis waddling action caused by lateral tire or wheel runout, or a defective tire with a belt that is not straight. (Courtesy of Oldsmobile Motor Division, General Motors Corp.)

Tread Wear Measurement

On most tires, the **tread wear indicators** appear as wide bands across the tread when tread depth is worn to 1/16 in (1.6 mm). Most tire manufacturers recommend tire replacement when the tread wear indicators appear across two or more tread grooves at three locations around the tire (Figure 4-14). If tires do not have wear indicators, a tread depth gauge may be used to measure the tread depth (Figure 4-15). The tread depth gauge reads in 32nds of an inch. Tires with 2/32 inch of tread depth or less should be replaced.

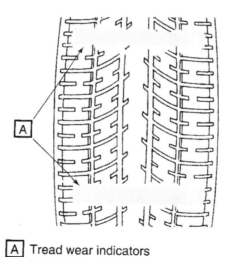

A Tread wear indicators

Figure 4-14 Tire tread wear indicators. (Courtesy of Oldsmobile Motor Division, General Motors Corp.)

Figure 4-15 Tread depth gauge. (Courtesy of Snap-on Tools Corp.)

Preliminary Wheel Balancing Checks

These preliminary checks should be completed before a wheel and tire are balanced:

1. Check for objects in tire tread.

2. Check for objects inside tire.

3. Inspect the tread and sidewall.

4. Check inflation pressure.

5. Measure tire and wheel runout.

6. Check wheel bearing adjustment.

7. Check for mud collected on the inside of the wheel.

CAUTION: On many wheel balancers, the tire and wheel are spun at high speed during the dynamic balance procedure. Be sure that all wheel weights are attached securely, and check for other loose objects on the tire and wheel, such as stones in the tread. If loose objects are detached from the tire or wheel at high speed, they may cause serious personal injury or property damage.

CAUTION: On the type of wheel balancer that spins the tire and wheel at high speed during the dynamic balance procedure, always attach the tire-and-wheel assembly securely to the balancer. Follow the equipment manufacturer's recommended wheel-mounting procedure. If the tire and wheel assembly becomes loose on the balancer at high speed, serious personal injury or property damage may result.

CAUTION: Prior to spinning a tire and wheel at high speed on a wheel balancer, always lower the protection shield over the tire. This shield provides protection in case anything flies off the tire or wheel.

All of the items on the preliminary check list influence wheel balance or safety. Therefore, it is extremely important that the preliminary checks be completed. Since a tire-and-wheel assembly is rotated at high speed during the dynamic balance procedure, it is very important that objects such as stones be removed from the treads. Centrifugal force may dislodge objects from the treads and cause serious personal injury. For this reason, it is also extremely important that the old wheel weights be removed from the wheel prior to balancing, and the new weights attached securely to the wheel during the balance procedure. Follow these preliminary steps before attempting to balance a wheel and tire:

1. When most types of wheel balancers are used, the wheel and tire must be removed from the vehicle and installed on the balancer. All mud, dust, and debris must be washed from the wheel after it is removed.

2. Objects inside a tire, such as balls of rubber, make balance impossible. When the wheel-and-tire assmbly is mounted on the balancer, be absolutely sure that the wheel is securely tightened on the balancer. As the tire is rotated slowly, listen for objects rolling inside the tire. If such objects are present, they must be removed prior to wheel balancing.

3. The tire should be inspected for tread and sidewall defects before the balance procedure. These defects create safety hazards and they may influence wheel balance. For example, tread chunking makes the wheel balancing difficult.

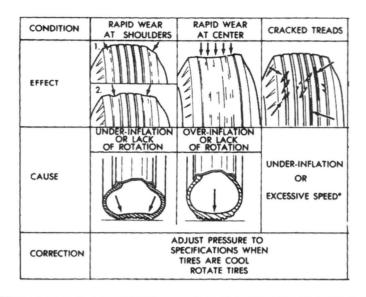

CONDITION	RAPID WEAR AT SHOULDERS	RAPID WEAR AT CENTER	CRACKED TREADS
EFFECT			
CAUSE	UNDER-INFLATION OR LACK OF ROTATION	OVER-INFLATION OR LACK OF ROTATION	UNDER-INFLATION OR EXCESSIVE SPEED*
CORRECTION	ADJUST PRESSURE TO SPECIFICATIONS WHEN TIRES ARE COOL ROTATE TIRES		

Figure 4-16 Tire tread wear caused by underinflation and overinflation. (Courtesy of Chrysler Corp.)

Tire Inflation Pressure

A tire depends on correct inflation pressure to maintain its correct shape and support the vehicle weight. Excessive inflation pressure causes the following problems:

1. Excessive center tread wear (Figure 4-16)
2. Hard ride
3. Damage to tire carcass

When tires are underinflated, the following problems will be evident:

1. Excessive wear on each side of the tread (Figure 4-16)
2. Hard steering
3. Wheel damage
4. Excessive heat buildup in the tire, and possible severe tire damage with resultant hazardous driving

The tires should be inflated to the car manufacturer's recommended pressure prior to the balance procedure. Loose wheel bearings allow lateral wheel shaking, and simulate an imbalanced wheel condition when the vehicle is in motion. Therefore, wheel bearing adjustments should be checked when wheel balance conditions are diagnosed.

Static Wheel Balance Procedure

Static wheel balance refers to proper balance of the tire when it is at rest.

Many different types of wheel balancers are available in the automotive service industry. The exact wheel balance procedure may vary depending on the type of wheel balancer. One of the simplest types of wheel balancers for static balancing is the bubble balancer (Figure 4-17).

The bubble balancer will only perform a static balance procedure. This type of balancer does not have dynamic balance capabilities. Therefore, the required static weight should be divided, and

Figure 4-17 Bubble-type wheel balancer. (Courtesy of Mac Tools, Inc.)

half of this weight should be placed on each side of the wheel directly opposite each other. When this type of wheel balancer is used to static balance a wheel, use the following procedure.

1. Complete the preliminary checks outlined previously.

2. Adjust the legs on the balancer to center the bubble.

3. Position the wheel securely on the balancer, and align the triangular mark with the valve stem on the wheel.

4. Observe the spirit level on the balancer to determine the heavy spot and the light spot on the wheel. The light spot will be directly opposite the heavy spot (Figure 4-18).

5. Place enough weight at the wheel light spot until the spirit level indicates a balanced wheel.

6. Chalk mark from the weight position on the sidewall radially around the tire to the inside bead.

7. Remove the wheel from the balancer and divide the weight by 2.

8. Install the required weights securely on the rim at the chalk-marked positions. The combined weight of the two wheel weights must equal the original weight installed in step 5 (Figure 4-19).

9. Reinstall the wheel on the balancer and check for proper balance.

Special Tools

Off-car wheel balancer

Many wheel weights have a spring clamp, which holds the weight on the edge of the rim. A special pair of wheel weight pliers is used to tap the wheel weight onto the rim and remove the weight from the rim (Figure 4-20). Magnesium or aluminum wheels may require the use of stick-on wheel weights. This type of weight must also be used on some Cadillac wheels on which the wheel covers interfere with the conventional weights. Regardless of the wheel weight type, they must be attached securely to the wheel.

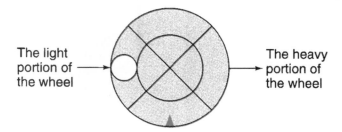

The light portion of the wheel → ← The heavy portion of the wheel

Figure 4-18 Bubble and triangular mark on a bubble balancer. (Courtesy of the Bada Company.)

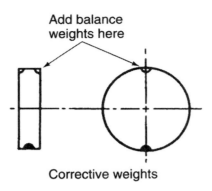

Add balance
weights here

Corrective weights

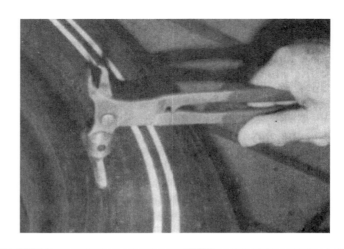

Figure 4-19 Dynamic wheel balance procedure. (Courtesy of Oldsmobile Motor Division, General Motors Corp.)

Figure 4-20 Wheel weights are removed and installed with special wheel weight pliers.

On some types of wheel balancers during the **static balance** procedure, the wheel is allowed to rotate by gravity. A **heavy spot** rotates the tire until this spot is at the bottom. The necessary static balance weights are then added at the top of the wheel 180° from the heavy spot. When the wheel-and-tire assembly is balanced statically, gravity does not rotate the wheel from the at-rest position. Rotate the tire by hand and check the static balance at 120° intervals.

The electronic wheel balancer is the most common type in use at the present time (Figure 4-21). There are many different types of electronic wheel balancers in the automotive service industry. With an electronic balancer, the operator must enter the wheel diameter, width, and offset. The balancer must have this information to perform its calculations.

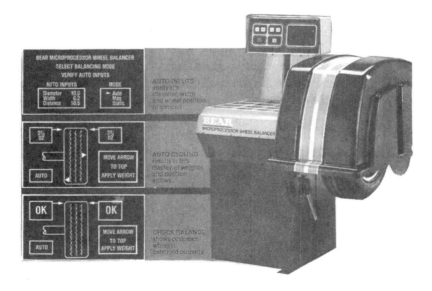

Figure 4-21 Electronic wheel balancer. (Courtesy of Bear Automotive Service Equipment Company)

Dynamic Wheel Balance Procedure

Dynamic imbalance is corrected with lead weights. Many electronic wheel balancers have an electric drive motor that spins the tire at high speed. On some electronic wheel balancers, the tire is spun by hand during the balance procedure. The electronic-type balancer performs static and dynamic balance calculations simultaneously, and indicates the correct weight size and location to the operator. The tire-and-wheel assembly must be fastened securely on the balancer with the correct size adapters. Off-car electronic high-speed wheel balancers have a safety hood that must be positioned over the tire prior to dynamic balancing. Hand-spun electronic balancers do not require a safety hood. The preliminary checks mentioned previously in this chapter also apply to dynamic balancing. Wheel balancers must always be operated according to the manufacturer's instructions.

When the heavy spot is located on the inside edge of the tread, the correct size weight installed on the inside of the wheel 180° from the heavy spot provides proper dynamic balance (Figure 4-22). When the heavy spot is on the outside edge of the tread, the correct size wheel weight installed at location C 180° from the heavy spot on the outside of the wheel provides proper dynamic balance (Figure 4-23).

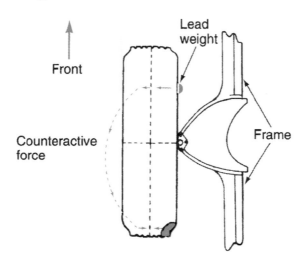

Figure 4-22 Dynamic wheel balance with heavy spot on the inside edge of the tread.

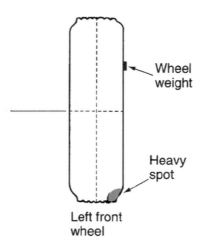

Figure 4-23 Dynamic wheel balance with heavy spot on the outside edge of the tread.

Dynamic wheel balance refers to proper balance of the tire during tire and wheel rotation.

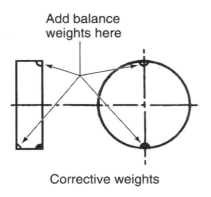

Add balance weights here

Corrective weights

Figure 4-24 Dynamic wheel balance procedure. (Courtesy of Oldsmobile Motor Division, General Motors Corp.)

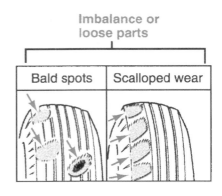

Imbalance or loose parts

Bald spots | Scalloped wear

Figure 4-25 Cupped tire wear and bald spots on tire tread caused by static or dynamic wheel imbalance. (Courtesy of Chrysler Corp.)

Some dynamic wheel balance procedures require the installation of the correct size wheel weight 180° from the heavy spot, and an equal weight on the opposite side of the wheel 180° from the first weight. This second weight maintains static balance (Figure 4-24). Static or dynamic wheel imbalance causes cupped tire wear and bald spots on the tire tread (Figure 4-25).

On-Car Wheel Balancing

Special Tools

On-car wheel balancer

On-car wheel balancers contain a drum driven by an electric motor. This drum is positioned against the tire on the vehicle, which allows the electric motor to rotate the wheel. Many on-car balancers have a strobe light with a meter, and an electronic vibration sensor. Off-car wheel balancing and on-car wheel balancing are a complementary combination for fine-tuning wheel balance. For example, a wheel vibration problem may still exist after an off-car balance procedure. If this problem occurs, the on-car balancer may be used to correct the problem. The on-car balance procedure corrects imbalance problems in all rotating components, including brake drums or rotors. When a front wheel on a rear-wheel-drive car is balanced, use this procedure:

1. Perform the preliminary checks listed previously in this chapter.
2. Raise the wheel being balanced 5 in (12 cm) off the floor and be sure the chassis is supported on safety stands so the wheel drops downward.
3. Install the electronic vibration sensor between the lower control arm and the floor.
4. Chalk mark a reference mark on the outer sidewall of the tire.
5. Spin the wheel just fast enough to produce vibration on the front bumper.
6. When vibration causes strobe light flashing, move the balancer drum away from the tire and allow the tire to spin freely.
7. Shine the strobe light around the tire sidewall and note the chalk-mark position.
8. Note the pointer position on the meter.
9. Use the balancer's brake plate to slow and stop the wheel.
10. Rotate the wheel until the chalk mark is in the exact position where it appeared under the strobe light. The heavy spot is now at the bottom of the wheel and the balancing weight should be attached 180° from the heavy spot. Install the amount of weight indicated on the meter.

Figure 4-26 On-car wheel balancer. (Courtesy of Hunter Engineering Company)

11. Spin the wheel again (Figure 4-26). If the wheel balance is satisfactory, the meter pointer will read in the balanced position. When the pointer does not indicate a balanced wheel, shine the strobe light on the tire sidewall. If the installed wheel weight is at the 12 o'clock position, additional weight is required, whereas the 6 o'clock weight position indicates excessive weight. A 3 o'clock or 9 o'clock weight position may be corrected by moving the weight 1 in (2.5 cm) toward the 12 o'clock position.

Additional on-car wheel balancer precautions are these:

1. Do not spin the wheel at excessive speeds.

2. Do not spin the front wheels on a front-wheel-drive vehicle with the floor jack under the chassis and the suspension dropped downward. Under this condition, severe angles exist in the front drive axle joints, and these joints may be damaged if the wheels are rotated with the balancer. Place the floor jack under the lower control arm to raise the wheel.

Rear Wheel Balancing

If an on-car balancer is used on the rear wheels of a rear-wheel-drive car, the technician must determine if the vehicle has a conventional or a limited slip differential. With the transmission in park, or in gear with a manual transmission, rotate one rear wheel by hand. If the vehicle has a limited slip differential, the rear wheels will not rotate, whereas a free-turning wheel indicates a conventional differential.

 WARNING: When a rear wheel is rotated with an on-car balancer, do not allow the speed indicated on the speedometer to exceed 35 mph (56 kmph).

When the vehicle has a conventional differential, use the same front wheel balance procedure on the rear wheels. Raise only the wheel to be balanced from the floor, and do not allow the speed indicated on the speedometer to exceed 35 mph (56 kmph). When the other rear tire is resting on the floor, the speed of the wheel being balanced is 70 mph (112 kmph). If the vehicle has a limited slip differential, proceed as follows for on-car rear wheel balancing:

1. Raise the vehicle and place safety stands under the frame to support the vehicle weight.

2. Lift the rear axle housing with a floor jack to reduce the universal joint angles, but do not take the vehicle weight off the safety stands.

Classroom Manual
Chapter 4, page 83

3. Remove the wheel that is not being balanced. Install and torque the wheel nuts on the brake drum to hold the drum in place.

4. Balance the opposite wheel using the same procedure as for front wheels.

5. Leave the balanced wheel on the vehicle and install and balance the other wheel. Be sure to torque the wheel nuts properly. Do not allow rear wheel speed to exceed 55 mph (90 kmph) on the speedometer.

Guidelines for Servicing Wheels and Tires

1. Tire noises vary with road conditions, but differential noise is not affected by road conditions.

2. Steering pull may be caused by defects in the suspension or steering systems, and by tire conicity.

3. Tire conicity occurs when the belt in a tire is wound off-center during the manufacturing process.

4. Tire conicity cannot be diagnosed by a visual inspection.

5. The tire rotation procedure varies depending on the model year, type of tires, and whether the vehicle has a compact spare or a conventional spare.

6. Hand tools and tire irons should not be used to demount tires.

7. Tire punctures over 1/4 inch in diameter should not be repaired.

8. Compact spare tires should not be repaired.

9. Do not use caustic agents, alkaline-base detergents, or abrasive cleaners to clean aluminum or magnesium wheels.

10. When a bead expander is used to mount a tire, do not increase tire pressure above 10 psi (69 kPa).

11. When a bead expander is not used to mount a tire, do not increase tire pressure above 40 psi (276 kPa) to move the tire beads out against the wheel rim.

12. While mounting a tire, be sure that the tire beads and wheel rim bead seats are coated with rubber tire lubricant.

13. Radial tire and wheel runout causes tire thump.

14. Lateral tire and wheel runout causes tire and vehicle waddling.

15. Balls of rubber or other objects inside a tire make proper balance impossible.

16. Aluminum or magnesium wheels require the use of stick-on wheel weights.

17. Electronic wheel balancers perform the necessary wheel weight calculations and indicate the exact locations where these weights should be installed.

CASE STUDY

A customer complains about severe front end waddling on a 1998 Cadillac Brougham equipped with steel-belted radial tires. The technician questions the customer further regarding this problem and learns that the problem only occurs after the car has sat over night, and it only lasts for approximately six blocks when the customer starts driving the car.

The technician carefully checks all steering and suspension components, wheel bearing adjustment, and tire runout. No problems are found during these checks, and the technician asks the customer to leave the car in the shop over night. The technician installs the front tires on the rear of the car and rotates the rear tires to the front, and the car is left in the shop over night. A road test the next morning indicates rear-end waddling, which proves the problem must be caused by a tire problem such as an improperly installed steel belt, since the tires were the only thing changed on the car. The technician installs and balances two new tires on the rear wheel rims, and installs these wheels and tires on the front of the car. The front tires and wheels were moved to the rear wheels. The customer reports later that the problem has been completely corrected.

Terms to Know

Antitheft lug nuts	Ply separation	Tire thump
Antitheft wheel covers	Radial runout	Tire vibration
Dynamic imbalance	Static balance	Tread wear indicators
Heavy spot	Tire conicity	
Lateral tire runout	Tire rotation	

ASE Style Review Questions

1. While discussing a tire thumping problem:
   Technician A says this problem may be caused by cupped tire treads.
   Technician B says a heavy spot in the tire may cause this complaint.
   Who is correct?
   A. A only **C.** Both A and B
   B. B only **D.** Neither A nor B

2. While discussing a vehicle that pulls to one side:
   Technician A says that excessive radial runout on the right front tire may cause this problem.
   Technician B says that tire conicity may be the cause of this complaint.
   Who is correct?
   A. A only **C.** Both A and B
   B. B only **D.** Neither A nor B

3. While discussing tire noise:
   Technician A says that tire noise varies with road surface conditions.
   Technician B says that tire noise remains constant when the vehicle is accelerated and decelerated.
   Who is correct?
   A. A only **C.** Both A and B
   B. B only **D.** Neither A nor B

4. While discussing tire wear:
   Technician A says that static imbalance causes feathered tread wear.
   Technician B says that dynamic imbalance causes cupped wear and bald spots on the tire tread.
   Who is correct?
   A. A only **C.** Both A and B
   B. B only **D.** Neither A nor B

5. While discussing on-car wheel balancing:
   Technician A says that during an on-car wheel balancing procedure on a rear wheel of a rear-wheel-drive car, the speed indicated on the speedometer should not exceed 65 mph (105 kmph).
   Technician B says that the speed indicated on the speedometer on this car must not exceed 35 mph (56 kmph).
   Who is correct?
   A. A only **C.** Both A and B
   B. B only **D.** Neither A nor B

6. A front tire has excessive wear on both edges of the tire tread. The most likely cause of this problem is:
   A. overinflation.
   B. underinflation.
   C. improper static balance.
   D. improper dynamic balance.

7. When measuring radial tire and wheel runout, the maximum runout on most automotive tire-and-wheel assemblies should be:
   A. 0.015 in (0.038 mm).   C. 0.045 in (1.143 mm).
   B. 0.025 in (0.635 mm).   D. 0.070 in (1.77 mm).

8. When measuring lateral wheel runout with the tire demounted from the wheel, the maximum runout on most automotive wheels is:
   A. 0.020 in (0.508 mm).   C. 0.045 in (1.143 mm).
   B. 0.030 in (0.762 mm).   D. 0.055 in (1.397 mm).

9. All of these statements about improper wheel balance are true EXCEPT:
   A. Dynamic imbalance may cause wheel shimmy.
   B. Dynamic imbalance may cause steering pull in either direction.
   C. Static imbalance causes wheel tramp.
   D. Static imbalance causes rapid wear on suspension components.

10. When diagnosing wheel balance problems:
    A. balls of rubber inside the tire have no effect on wheel balance.
    B. loose wheel bearing adjustment may simulate improper static wheel balance.
    C. improper dynamic wheel balance may be caused by a heavy spot in the center of the tire tread.
    D. after a tire patch is installed, the tire and wheel may be improperly balanced.

ASE Challenge Questions

1. The owner of a rear-wheel-drive car with aftermarket alloy wheels says he has replaced the wheel bearings three times in the past two years. He wants to know why the bearings fail.
   Technician A says excessive radial runout of the wheel may be the cause of the problem.
   Technician B says excessive offset of the wheel may be the cause of the problem.
   Who is correct?
   A. A only   C. Both A and B
   B. B only   D. Neither A nor B

2. A customer returns with recently purchased radial tires saying that the rear of the car feels like "it's riding on Jello™." All of the follow could cause this problem EXCEPT:
   A. the radial belt of a rear tire not straight.
   B. the wheel improperly mounted.
   C. excessive lateral wheel runout.
   D. excessive radial tire runout.

3. A customer says the new tires he just purchased vibrate. The installer says he balanced the wheels and tires with a bubble balancer before placing them on the car.
   Technician A says correcting imbalance requires the balancing weights to be placed on both sides of the rim.

   Technician B says a tire that is in perfect static balance may exhibit imbalance when it is in motion.
   Who is correct?
   A. A only   C. Both A and B
   B. B only   D. Neither A nor B

4. After a set of radial tires is rotated, the customer returns saying he feels vibration and steering shimmy. To correct this problem, you should:
   A. balance the wheels and tires with a bubble balancer.
   B. return the wheels and tires to their original positions.
   C. check the wheel bearings.
   D. balance the wheels and tires with an on-car balancer.

5. A customer says there is a "thumping" vibration in the wheels and an inspection of the tires shows the two front wheels have flat spots on the tire treads.
   Technician A says heavy spots in the tires may have caused this condition.
   Technician B says locking the wheels and skidding on pavement caused this condition.
   Who is correct?
   A. A only   C. Both A and B
   B. B only   D. Neither A nor B

Job Sheet 10

Name _____ Date _____

Tire Demounting and Mounting

Upon completion of this job sheet, you should be able to demount and mount tires.

Tools and Materials

Tire changer
Tire and wheel assembly

Procedure

1. Remove the valve core to release all the air pressure from the tire. Chalk mark the tire at the valve stem opening in the wheel so the tire may be re-installed in the same position to maintain proper wheel balance.

   Is all the air pressure released from the tire? yes _____ no _____

   Is the tire chalk marked at the valve stem location in the wheel?

   Instructor check _____

   ✓ **SERVICE TIP:** The following is a generic tire demounting and mounting procedure.

2. Position the wheel and tire on the tire changer with the narrow bead ledge facing upward. Follow the tire changer manufacturer's recommended procedure to attach the wheel to the tire changer.

   Is the wheel and tire securely attached to the tire changer according to the tire changer manufacturer's recommended procedure? yes _____ no _____

   Instructor check _____

   CAUTION: Do not proceed to demount the tire unless the wheel and tire are securely attached to the tire changer. This action may cause personal injury.

3. Position the top bead breaker on the tire changer properly to loosen the top tire bead from the wheel. Is the top bead breaker properly positioned to loosen the top bead?
   yes _____ no _____

4. Operate the upper bead breaker on the tire changer to loosen the top tire bead. Is the top tire bead loosened from the wheel? yes _____ no _____

5. Position the lower bead breaker on the tire changer to loosen the lower bead from the wheel. Is the lower bead breaker properly positioned to loosen the lower bead?
   yes _____ no _____

6. Operate the lower bead breaker on the tire changer to loosen the lower tire bead from the wheel. Is the lower tire bead loosened from the wheel? yes _____ no _____

   Instructor check _____

    SERVICE TIP: Some tire changers rotate a bar or lever to move the tire beads over the wheel. Other tire changers rotate the tire and wheel to perform this function.

7. Push one part of the upper tire bead into the drop center in the wheel. Position the tire changer bar, lever, or tool under the opposite side of the upper bead from the part of the bead in the drop center of the wheel.

   Is the tire changer bar, lever, or tool properly positioned to remove the upper tire bead over the top edge of the wheel? yes _____ no _____

8. Operate the tire changer bar, lever, or tool to remove the upper tire bead over the top edge of the wheel. Is the upper tire bead over the top edge of the wheel? yes _____ no _____

9. Push one part of the lower tire bead into the drop center in the wheel. Position the tire changer bar, lever, or tool under the opposite side of the lower bead from the part of the bead in the drop center of the wheel.

   Is the tire changer bar, lever, or tool properly positioned to remove the lower tire bead over the top edge of the wheel? yes _____ no _____

10. Operate the tire changer bar, lever, or tool to remove the lower tire bead over the top edge of the wheel. Is the lower tire bead over the top edge of the wheel? yes _____ no _____

    Instructor check _____

11. Remove the tire from the wheel and remove the wheel from the tire changer.

12. Thoroughly clean the wheel to tire bead sealing surfaces with a wire brush. Inspect the wheel for rust or damage.

    Wheel condition: satisfactory _____ unsatisfactory _____

    If the wheel condition is unsatisfactory, state the defective condition(s) and the necessary corrective action. _____

    _____

    Wheel to tire bead sealing surfaces cleaned: yes _____ no _____

13. Coat the tire beads with tire lubricant.

14. Position the wheel on the tire changer with the narrow bead ledge facing upward. Follow the tire changer manufacturer's recommended procedure to attach the wheel to the tire changer.

    Is the wheel and tire securely attached to the tire changer according to the tire changer manufacturer's recommended procedure? yes _____ no _____

    Instructor check _____

    CAUTION: Do not proceed to mount the tire unless the wheel and tire are securely attached to the tire changer. This action may cause personal injury.

15. Push one part of the lower tire bead into the drop center in the wheel. Position the tire changer bar, lever, or tool under the opposite side of the upper bead from the part of the bead in the drop center of the wheel.

    Is the tire changer bar, lever, or tool properly positioned to install the lower tire bead over the top edge of the wheel? yes _____ no _____

16. Operate the tire changer bar, lever, or tool to install the lower tire bead over the top edge of the wheel. Is the lower tire bead over the top edge of the wheel? yes _____ no _____

17. Push one part of the upper tire bead into the drop center in the wheel. Position the tire changer bar, lever, or tool under the opposite side of the upper bead from the part of the bead in the drop center of the wheel.

    Is the tire changer bar, lever, or tool properly positioned to install the upper tire bead over the top edge of the wheel? yes _____ no _____

18. Operate the tire changer bar, lever, or tool to install the upper tire bead over the top edge of the wheel. Is the upper tire bead over the top edge of the wheel? yes _____ no _____

Instructor check _____

19. Rotate the tire on the wheel so the chalk mark placed on the tire in step 8 is aligned with the valve stem opening in the wheel.

Is the chalk mark on the tire aligned with the valve stem position? yes _____ no _____

Instructor check _____

20. Install and tighten the valve core in the valve stem.

> ✓ **SERVICE TIP:** Different makes of tire changers have various ways of tire bead seating on the wheel. Some tire changers have an air inflation ring that is positioned on top of the wheel and tire. Other tire changers have air jets in each clamping jaw that hold the wheel on the tire changer. Air pressure from these jets helps force the tire beads against the wheel sealing surfaces.

> ✓ **SERVICE TIP:** Some tire changers have an air hose and air pressure gauge designed into the tire changer. On other tire changers a shop air hose must be used to inflate the tire.

> ■ **CAUTION:** When inflating a tire, do not stand or lean over the tire and wheel. If the tire bead comes off the rim or the rim cracks, personal injury may occur. When using a shop air hose to inflate the tire, install an extension on the air hose.

21. Supply air to the air ring or air jets on the tire changer to move the tire beads outward against the wheel. Are the tire beads moved outward against the wheel?

yes _____ no _____

Instructor check _____

22. Connect an air supply hose to the tire valve stem. When using a shop air hose, connect an extension to the air hose. Inflate the tire to the specified pressure.

Specified tire pressure _____

Actual tire pressure _____

✓ **Instructor Check** _____

Job Sheet 11

Name _____ Date _____

Tire and Wheel Runout Measurement

Upon completion of this job sheet, you should be able to measure radial and lateral tire and wheel runout.

ASE Correlation

This job sheet is related to the ASE Suspension and Steering Task List content area: D. Wheel and Tire Diagnosis and Repair, Task: 5. Measure wheel, tire, axle flange, and hub runout (radial and lateral); determine needed repairs.

Tools and Materials

Dial indicator
Tire and wheel assembly

Wheel type and diameter _____

Tire type and manufacturer _____

Procedure

SERVICE TIP: The tire and wheel may be installed on an off-car wheel balancer to measure tire and wheel runout.

1. Position a dial indicator against the center of the tire tread as the tire is rotated slowly to measure radial runout.

   Radial runout _____

   Specified radial runout _____

2. Mark the highest point of radial runout on the tire with a crayon, and mark the valve stem position on the tire.

3. If the radial tire runout is excessive, demount the tire and check the runout of the wheel rim with a dial indicator positioned against the lip of the rim while the rim is rotated.

   Wheel radial runout _____

   Specified wheel radial runout _____

4. Use a crayon to mark the highest point of radial runout on the wheel rim.

5. If the highest point of wheel radial runout coincides with the chalk mark from the highest point of maximum tire radial runout, the tire may be rotated 180° on the wheel to reduce radial runout. Tires or wheels with excessive runout are usually replaced.

   Does the highest point of tire radial runout coincide with the highest point of wheel radial runout? yes _____ no _____

   Instructor check _____

   State the required action to correct excessive radial runout and the reason for this action.

   _____

   _____

6. Position a dial indicator located against the sidewall of the tire to measure lateral runout.

   Tire lateral runout _____

   Specified tire lateral runout _____

7. Chalk mark the tire and wheel at the highest point of radial runout.

8. If the tire runout is excessive, the tire should be demounted from the wheel and the wheel lateral runout measured.

9. If the tire runout was excessive, measure the wheel lateral runout with a dial indicator positioned against the edge of the wheel as the wheel is rotated.

   Wheel lateral runout _____

   Specified wheel lateral runout _____

   State the required action to correct excessive wheel lateral runout and the reason for this action. _____

   _____

✔ **Instructor Check** _____

Job Sheet 12

Name _____ Date _____

Off-Car Wheel Balancing

Upon completion of this job sheet, you should be able to perform off-car wheel balancing procedures.

ASE Correlation

This job sheet is related to the ASE Suspension and Steering Task List content area: D. Wheel and Tire Diagnosis and Repair, Task: 7. Balance wheel and tire assembly (static and/or dynamic).

Tools and Materials

Electronic off-car wheel balancer
Tire and wheel
Wheel weights

Wheel diameter _____

Manufacturer and type of tire _____

Procedure

1. Wash mud and debris from the wheel and tire.

2. Mount the tire and wheel assembly on the wheel balancer with the wheel balancer manufacturer's recommended mounting components.

3. Tighten the tire and wheel retaining nut on the wheel balancer to the specified torque.

   Tire and wheel retaining nut on the wheel balancer tightened to the specified torque?
   yes _____ no _____

   Instructor check _____

4. Remove stones and other objects from the tire tread.

   Are stones and other objects removed from the tire tread? yes _____ no _____

   Instructor check _____

5. Check for objects inside tire.

   Are there objects inside the tire? yes _____ no _____

   Instructor check _____

6. Remove all the old wheel weights from the wheel.

   Are the old wheel weights removed? yes _____ no _____

   Instructor check _____

7. Inspect the tread and sidewall.

   List defective tread and sidewall conditions. _____

   _____

   List the types of tire tread wear and state the causes of this wear.

   _____

   _____

   Based on your inspection of the tire, list the required tire service.

   _____

   _____

8. Check tire inflation pressure.

   Tire inflation pressure _____

   Specified tire inflation pressure _____

9. Measure tire and wheel runout.

   Radial tire and wheel runout _____

   Specified tire and wheel radial runout _____

   Lateral tire and wheel runout _____

   Specified lateral tire and lateral wheel runout _____

   Recommended tire and wheel service. _____

   _____

10. Lower the safety hood on the wheel balancer.

    Is the safety hood lowered? yes _____ no _____

11. Enter the required information on wheel balancer keypad, such as wheel diameter, wheel
    width, and wheel offset or distance.

    Is the required wheel information entered in the wheel balancer? yes _____ no _____

    Instructor check _____

12. If the wheel balancer has a mode selector, select the desired balance mode.

    Balance mode selected _____

13. Spin the wheel on the balancer, then apply the balancer brake to stop the wheel.

14. Observe the balancer display indicating the necessary amount of wheel weights and the
    required location of these weights. Install the required wheel weights in the locations
    indicated on the balancer. Be sure the wheel weights are securely attached to the rim.

    Are the wheel weight(s) securely attached to the rim at the locations indicated by the wheel
    balancer? yes _____ no _____

    Instructor check _____

15. Spin the tire-and-wheel assembly on the the balancer again to be sure this assembly is
    properly balanced.

    Does the wheel balancer indicate proper wheel balance? yes _____ no _____

    If the answer to this question is no, repeat the balance procedure.

✔ **Instructor Check** _____

Job Sheet 13

Name _____ Date _____

On-Car Wheel Balancing

Upon completion of this job sheet, you should be able to perform on-car wheel balancing procedures.

ASE Correlation

This job sheet is related to the ASE Suspension and Steering Task List content area: D. Wheel and Tire Diagnosis and Repair, Task: 7. Balance wheel-and-tire assembly (static and/or dynamic).

Tools and Materials

On-car wheel balancer
Car
Wheel weights

Vehicle make and model year _____

Vehicle VIN _____

Procedure

1. Wash mud and debris from the wheel and tire.

2. Raise the wheel being balanced 5 in (12 cm) off the floor; be sure the chassis is securely supported on safety stands.

   ⚠️ **WARNING:** Do not spin the front wheels on a front-wheel-drive vehicle with the floor jack under the chassis and the suspension dropped downward. Under this condition severe angles exist in the front drive axle joints, and these joints may be damaged if the wheels are rotated with the balancer. Place the floor jack under the lower control arm to raise the wheel.

3. Remove stones and other objects from the tire tread.

   Are all objects removed from the tire tread? yes _____ no _____

   Instructor check _____

4. Inspect the tire tread and sidewall condition.

   Tread and sidewall condition _____

   _____

   Types of tire tread wear _____

   _____

   Causes of tire tread wear _____

   _____

   Based on your inspection of the tire tread and sidewall state the necessary tire service.

   _____

   _____

5. Check the wheel bearing adjustment and adjust to specifications if necessary.

Wheel bearing adjustment: satisfactory _____ unsatisfactory _____

6. Install the electronic vibration sensor between the lower control arm and the floor.

Is the vibration sensor properly installed? yes _____ no _____

Instructor check _____

7. Chalk mark a reference mark on the outer sidewall of the tire.

▲ **WARNING:** To obtain proper wheel balance, spin the wheel only to the speed where the vibration occurs. Do not spin the wheel at excessive speeds.

8. Spin the wheel just fast enough to produce vibration on the front bumper.

9. When vibration causes strobe light flashing, move the balancer drum away from the tire and allow the tire to spin freely.

10. Shine the strobe light around the tire sidewall and note the chalk-mark position.

Chalk mark clock position _____

11. Note the pointer position on the meter.

Meter pointer position _____

12. Use the balancer's brake plate to slow and stop the wheel.

13. Rotate the wheel until the chalk mark is in the exact position where it appeared under the strobe light. The heavy spot is now at the bottom of the wheel and the balancing weight should be attached 180° from the heavy spot. Install the amount of weight indicated on the meter.

Amount of wheel weight installed _____

14. Spin the wheel again. If the wheel balance is satisfactory, the meter pointer will read in the balanced position. If the pointer does not indicate a balanced wheel, shine the strobe light on the tire sidewall. If the installed wheel weight is at the 12 o'clock position, additional weight is required. A 6 o'clock weight position indicates excessive weight. A 3 o'clock or 9 o'clock weight position may be corrected by moving the weight 1 in (2.5 cm) toward the 12 o'clock position.

Wheel and tire balance: satisfactory _____ unsatisfactory _____

If the wheel balance is unsatisfactory, state the action required to balance tire and wheel.

_____

_____

✔️ **Instructor Check** _____

Table 4-1 ASE TASK

Diagnose tire wear patterns; determine needed repairs.

Problem Area	Symptoms	Possible Causes	Classroom Manual	Shop Manual
TIRE TREAD WEAR	Wear on center or edges of tread	Overinflation or underinflation	76	108
	Cupped wear on tread	Improper wheel balance	79	108

Table 4-2 ASE TASK

Inspect tires; check and adjust air pressure.

Problem Area	Symptoms	Possible Causes	Classroom Manual	Shop Manual
NOISE	Tire thumping	Tire defects, weak spots, bulges	76	104

Table 4-3 ASE TASK

Diagnose wheel/tire vibration, shimmy, and noise problems; determine needed repairs.

Problem Area	Symptoms	Possible Causes	Classroom Manual	Shop Manual
RIDE QUALITY	Vibration above a certain speed	Improper static wheel balance	79	104
	Shimmy above a certain speed	1. Improper dynamic wheel balance	81	107
		2. Excessive lateral runout	62	100
TIRE NOISE	Thumping	1. Excessive tire radial runout	66	100
		2. Cupped tire treads	80	107
		3. Improper wheel balance	79	104
		4. Defects, heavy spots, bulges	79	104

Table 4-4 ASE TASK

Rotate tires/wheels according to manufacturer's recommendations.

Problem Area	Symptoms	Possible Causes	Classroom Manual	Shop Manual
TIRE TREAD WEAR	Premature tread wear	Lack of tire rotation	65	91

Table 4-5 ASE TASK

Measure wheel, tire, axle flange, and hub runout (radial and lateral); determine needed repairs.

Problem Area	Symptoms	Possible Causes	Classroom Manual	Shop Manual
NOISE	Tire thumping	Excessive radial wheel or tire runout	78	100
VIBRATION	Chassis waddle	Excessive lateral wheel or tire runout	78	100

Table 4-6 ASE TASK

Diagnose tire pull (lead) problems; determine corrective actions.

Problem Area	Symptoms	Possible Causes	Classroom Manual	Shop Manual
DIRECTIONAL CONTROL	Steering pull	Tire conicity	65	90

Table 4-7 ASE TASK

Balance wheel and tire assembly (static and/or dynamic).

Problem Area	Symptoms	Possible Causes	Classroom Manual	Shop Manual
RIDE QUALITY	Chassis vibration above a certain speed	1. Improper static wheel balance	79	104
	Chassis shimmy above a certain speed	2. Improper dynamic wheel balance	81	107

Shock Absorber and Strut Diagnosis and Service

Upon completion and review of this chapter, you should be able to:

❏ Perform a visual shock absorber inspection.

❏ Perform a shock absorber bounce test, and determine shock absorber condition.

❏ Determine shock absorber condition from a manual shock absorber test.

❏ Remove and replace shock absorbers.

❏ Diagnose shock absorber and strut noise complaints.

❏ Remove and replace front and rear struts.

❏ Remove struts from coil springs.

❏ Install coil springs on struts.

❏ Follow the vehicle manufacturer's recommended strut disposal procedure.

❏ Perform off-car strut cartridge replacement procedures.

❏ Perform on-car strut cartridge replacement procedures.

Shock Absorber Visual Inspection

> ✓ **SERVICE TIP:** During a visual shock absorber inspection, check the rebound, or strikeout, bumpers on the control arms, or chassis. If the rebound bumpers are severely worn, the shock absorbers may be worn out.

Bolt Mounting

Shock absorbers should be inspected for loose mounting bolts and worn mounting bushings. If these components are loose, rattling noise is evident, and replacement of the bushings and bolts is necessary.

Bushing Condition

In some shock absorbers, the bushing is permanently mounted in the shock, and the complete unit must be replaced if the bushing is worn. When the mounting bushings are worn, the shock absorber will not provide proper spring control.

Oil Leakage

Shock absorbers and struts should be inspected for oil leakage. A slight oil film on the lower oil chamber is acceptable. Any indication of oil dripping is not acceptable, and unit replacement is necessary (Figure 5-1).

Struts and shock absorbers should be inspected visually for physical damage such as a bent condition and severe dents or punctures. When any of these conditions are present, unit replacement is required (Figure 5-2).

Basic Tools

Basic technician's tool set

Service manual

Hydraulic floor jack

Jack stands

Center punch

A strut is similar internally to a shock absorber, but a strut also supports the steering knuckle. In many applications, the coil spring is mounted on the strut.

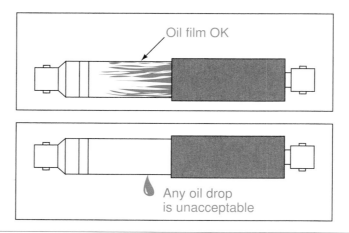

Figure 5-1 Shock absorber and strut oil leak diagnosis. (Courtesy of Delco Products Division of General Motors Corp.)

Shock Absorber or Strut Bounce Test

CUSTOMER CARE: Always be willing to spend a few minutes explaining problems, including safety concerns, regarding the customer's vehicle. When customers understand why certain repairs are necessary, they feel better about spending the money. For example, if you explain that worn-out shock absorbers cause excessive chassis oscillations and may result in loss of steering control on irregular road surfaces, the customer is more receptive to spending the money for shock absorber replacement.

When the **bounce test** is performed, the bumper is pushed downward with considerable weight applied on each corner of the vehicle. The bumper is released after this action, and one free upward bounce should stop the vertical chassis movement if the shock absorber or strut provides proper spring control. Shock absorber replacement is required if more than one free upward bounce occurs.

Shock Absorber Manual Test

CAUTION: Gas-filled shock absorbers will extend when disconnected.

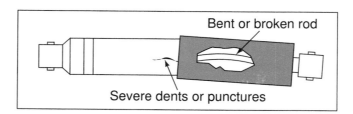

Figure 5-2 Damaged shock absorber inspection. (Courtesy of Delco Products Division of General Motors Corp.)

A **manual test** may be performed on shock absorbers. When this test is performed, disconnect the lower end of the shock and move the shock up and down as rapidly as possible. A satisfactory shock absorber should offer a strong, steady resistance to movement on the entire compression and rebound strokes. The amount of resistance may be different on the compression stroke compared to the rebound stroke. If a loss of resistance is experienced during either stroke, shock replacement is essential.

Upward wheel movement is referred to as wheel jounce, and downward wheel movement is called rebound.

Some defective shock absorbers or struts may have internal clunking, clicking, and squawking noises, or binding conditions. When these shock absorber noises or conditions are experienced, shock absorber or strut replacement is necessary.

Air Shock Absorber Diagnosis and Replacement

Air shock absorbers are similar to conventional shocks except they have air pressure applied to them to control chassis curb height. Some air shock absorbers must be pressurized with the shop air hose. This type of unit contains a valve for inflation purposes. Other air shock absorbers are inflated by an onboard compressor with interconnecting plastic lines between the compressor and the shocks. Shock absorber lines must be inspected for breaks, cracks, and sharp bends. If any of these defects are present, the line must be replaced. The shock absorber lines must be secured to the chassis, and they must not rub against other components.

When air shock absorbers slowly lose their air pressure and reduce the curb riding height, shock replacement is required. Before removing an air shock, relieve the air pressure in the shock.

Shock Absorber Replacement

CAUTION: Never apply heat to the lower shock absorber or strut chamber with an acetylene torch. Excessive heat may cause a shock absorber or strut explosion, which could result in personal injury.

When shock absorber replacement is necessary, follow this procedure:

1. Before replacing rear shock absorbers, lift the vehicle on a hoist and support the rear axle on safety stands so the shock absorbers are not fully extended.
2. When a front shock absorber must be changed, lift the front end on the vehicle with a floor jack, then place safety stands under the lower control arms. Lower the vehicle onto the safety stands and remove the floor jack.
3. Disconnect the upper shock mounting nut and grommet.
4. Remove the lower shock mounting nut or bolts, and remove the shock absorber.
5. Reverse steps 1 through 4 to install the new shock absorber and grommets.
6. With the full vehicle weight supported on the suspension, tighten the shock mounting nuts to the specified torque.

Classroom Manual
Chapter 5, page 87

Diagnosis of Front Spring and Strut Noise

Strut chatter may be heard when the steering wheel is turned with the vehicle not moving, or moving at low speed. To verify the location of this chattering noise, place one hand on a front coil

Strut chatter is heard as a clicking noise when the front wheels are turned.

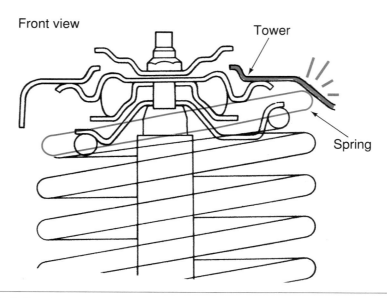

Front view

Tower

Spring

Figure 5-3 Coil spring to upper tower interference. (Courtesy of Chrysler Corp.)

spring while someone turns the steering wheel. If strut chatter is present, the spring binds and releases as it turns. This condition is caused by the upper spring seat binding against the strut bearing mount. A revised spring seat is available to correct this problem on some models.

A noise that occurs on sharp turns or during front suspension jounce may be caused by one of the following problems:

A strut tower is a raised, circular, reinforced area inboard of the front fenders, that supports the upper end of the coil spring and strut assembly.

1. Interference between the upper strut rebound stop and the upper mount or **strut tower.**

2. Interference between the coil spring and tower (Figure 5-3).

3. Interference between the coil spring and the upper mount (Figure 5-4).

On some models, these coil-spring interference problems may be corrected by installing upper coil spring spacers on top of the coil spring. Spring removal from the strut is required to install these spacers.

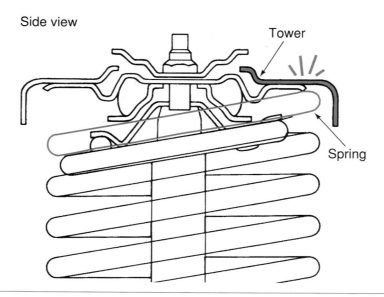

Side view

Tower

Spring

Figure 5-4 Coil spring to upper mount interference. (Courtesy of Chrysler Corp.)

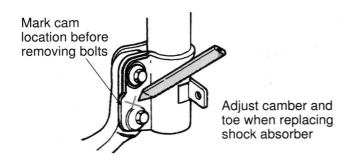

Mark cam location before removing bolts

Adjust camber and toe when replacing shock absorber

Figure 5-5 Camber bolt marking for strut removal. (Courtesy of Chrysler Corp.)

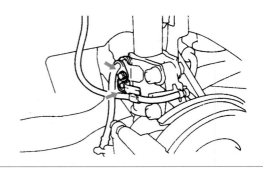

Figure 5-6 Brake line and ABS wheel-speed sensor wire removal from strut. (Reprinted with permission)

If the lateral movement of the strut rod and nut above the strut tower exceeds 3/16 in (4.76 mm), the upper strut bearing and mount assembly should be replaced.

Worn **spring insulators** or broken coil springs cause a rattling noise on road irregularities. Broken coil springs result in reduced curb riding height and harsh riding.

Strut Removal and Replacement

Before a front strut and spring assembly is removed, the strut must be removed from the steering knuckle, and top strut mount bolts must be removed from the strut tower. If an **eccentric camber bolt** is used to attach the strut to the knuckle, always mark the bolt head in relation to the strut and reinstall the bolt in the same position (Figure 5-5).

An eccentric camber bolt has an oblong head, which provides inward or outward steering knuckle movement in relation to the strut as this bolt is rotated.

Always follow the vehicle manufacturer's recommended procedure in the service manual for removal of the strut and spring assembly. A typical procedure for strut and spring assembly removal follows:

1. Raise the vehicle on a hoist or floor jack. If a floor jack is used to raise the vehicle, lower the vehicle onto safety stands placed under the chassis so the lower control arms and front wheels drop downward. Remove the floor jack from under the vehicle.

2. Remove the brake line and antilock brake system (ABS) wheel-speed sensor wire from clamps on the strut (Figure 5-6). In some cases, the clamps may also have to be removed from the strut.

3. Remove the strut to steering knuckle retaining bolts, and remove the strut from the knuckle (Figure 5-7).

4. Remove the upper strut mounting bolts on top of the strut tower, and remove the strut and spring assembly (Figure 5-8).

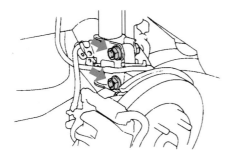

Figure 5-7 Removing strut to steering knuckle retaining bolts. (Reprinted with permission)

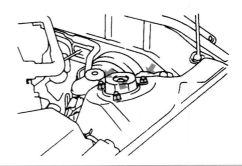

Figure 5-8 Removing upper strut mounting bolts on top of strut tower. (Reprinted with permission)

Removal of Strut from Coil Spring

CAUTION: Always use a coil spring compressing tool according to the tool or vehicle manufacturer's recommended service procedure. Be sure the tool is properly installed on the spring. If a coil spring slips off the tool when the spring is compressed, severe personal injury or property damage may occur.

CAUTION: Never loosen the upper strut mount retaining nut on the end of the strut rod unless the spring is compressed enough to remove all spring tension from the upper strut mount. If this nut is loosened with spring tension on the upper mount, this mount becomes a very dangerous projectile that may cause serious personal injury or property damage.

WARNING: Never clamp the lower strut or shock absorber chamber in a vise with excessive force. This action may distort the lower chamber and affect piston movement in the strut or shock absorber.

The coil spring must be compressed with a special tool before the strut can be removed. All the tension must be removed from the upper spring seat before the upper strut piston rod nut is loosened. Many different **spring-compressing tools** are available, and they must always be used according to the manufacturer's recommended procedure. *If the coil spring has an enamel-type coating, tape the spring where the compressing tool contacts the spring.* The spring may break prematurely if this coating is chipped. A typical procedure for removing a strut from a coil spring follows:

Special Tools

Coil spring
compressing tool

1. Install the spring compressing tool on the coil spring according to the tool or vehicle manufacturer's recommended procedure.

2. Turn the nut on top of the compressing tool until all the spring tension is removed from the upper strut mount (Figure 5-9).

3. Install a bolt and two nuts in the upper strut to knuckle mounting bolt holes. Install a nut on each side of the strut flange (Figure 5-10). Clamp this bolt securely in a vise to hold the strut and coil assembly and the compressing tool.

4. Use the bar on the spring compressing tool to hold the strut and spring assembly from turning, and loosen the nut on the upper strut mount (Figure 5-11). Be sure all the spring tension is removed from the upper strut mount before loosening this nut.

5. Remove the upper strut mount, **upper insulator,** coil spring, spring bumper, and **lower insulator** (Figure 5-12).

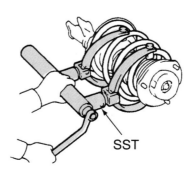

SST

Figure 5-9 The spring compressing tool is tightened until all the spring tension is removed from the upper strut mount. (Reprinted with permission)

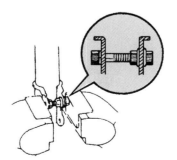

Figure 5-10 Bolt and two nuts installed in upper strut to knuckle mounting hole to hold strut and spring assembly in a vise. (Reprinted with permission)

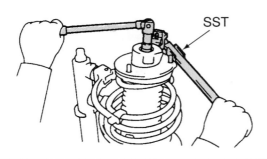

Figure 5-11 Removal of nut from strut piston rod. (Reprinted with permission)

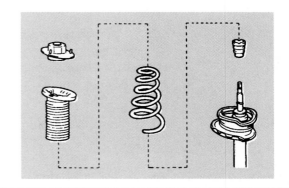

Figure 5-12 Removal of upper strut mount, upper insulator, spring, spring bumper, and lower insulator. (Reprinted with permission)

Strut Disposal Procedure

⚠️ **WARNING:** Always follow the vehicle manufacturer's recommended procedure for strut disposal. Do not throw gas-filled struts or shock absorbers in a fire of any kind or in a dumpster. If the vehicle manufacturer recommends drilling the strut to release the gas charge, drill the strut at the manufacturer's recommended location.

The following is a typical strut drilling procedure that is performed prior to strut disposal:

1. Fully extend the strut rod.
2. Center punch the strut at the manufacturer's recommended drilling location (Figure 5-13).
3. Drill a small hole at the center-punched position.

Installation of Coil Spring on Strut

A typical procedure for installing a coil spring on a strut follows:

1. Install the lower insulator on the lower strut spring seat, and be sure the insulator is properly seated (Figure 5-14).
2. Install the spring bumper on the strut rod (Figure 5-15).
3. With the coil spring compressed in the spring compressing tool, install the spring on the strut (Figure 5-16). Be sure the spring is properly seated on the lower insulator spring seat.

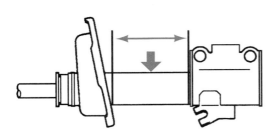

Figure 5-13 Strut drilling location. (Reprinted with permission)

Figure 5-14 Insulator installation on lower spring seat. (Reprinted with permission)

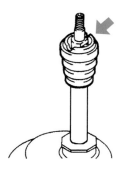

Figure 5-15 Spring bumper installation on strut piston rod. (Reprinted with permission)

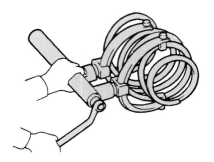

Figure 5-16 Installing compressed spring on strut with compressing tool. (Reprinted with permission)

4. Be sure the strut piston rod is fully extended and install the upper insulator on top of the coil spring.

5. Install the upper strut mount on the upper insulator (Figure 5-17).

6. Be sure the spring, upper insulator, and upper strut mount are properly positioned and seated on the coil spring and strut piston rod (Figure 5-18).

7. Install a bolt and nuts in the upper strut to knuckle retaining bolt, hole and clamp this bolt in a vise to hold the strut, spring, and compressing tool as in the disassembly procedure.

8. Use the compressing tool bar to hold the strut and spring from turning, and tighten the strut piston rod nut to the specified torque (Figure 5-19).

9. Rotate the upper strut mount until the lowest bolt in this mount is aligned with the tab on the lower spring seat (Figure 5-20).

10. Gradually loosen the nut on the compressing tool until all the spring tension is released from the tool, and remove the tool from the spring.

Installation of Spring and Strut Assembly in Vehicle

A typical installation procedure for a spring and strut assembly follows:

1. Install the strut and spring assembly with the upper strut mounting bolts extending through the bolt holes in the strut tower. Tighten the nuts on the upper strut mounting bolts to the specified torque (Figure 5-21).

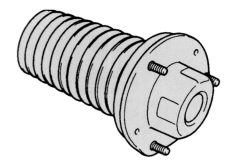

Figure 5-17 Upper insulator and upper strut mount installed on coil spring. (Reprinted with permission)

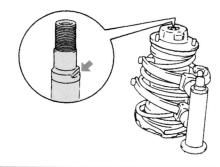

Figure 5-18 Upper strut mount properly positioned on strut piston rod. (Reprinted with permission)

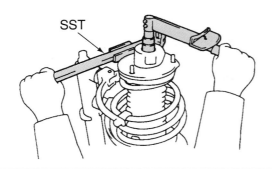

Figure 5-19 Tightening nut on strut piston rod. (Reprinted with permission)

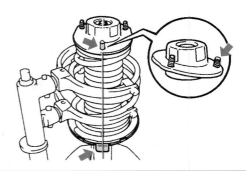

Figure 5-20 Aligning lowest bolt on upper strut mount with tab on lower spring seat. (Reprinted with permission)

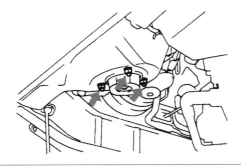

Figure 5-21 Nuts installed on upper strut mounting bolts. (Reprinted with permission)

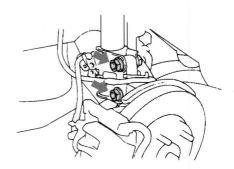

Figure 5-22 Lower end of strut installed in steering knuckle. (Reprinted with permission)

2. Install the lower end of the strut in the steering knuckle to the proper depth, and tighten the strut to knuckle retaining bolts to the specified torque (Figure 5-22). If one of the strut to knuckle bolts is an eccentric camber bolt, be sure the eccentric is aligned with the mark placed on the strut during the removal procedure.

3. Install the brake hose in the clamp on the strut. Place the ABS wheel-speed sensor wire in the strut clamp if the vehicle is equipped with ABS (Figure 5-23).

Photo Sequence 3 shows a typical procedure for removing and replacing a MacPherson strut.

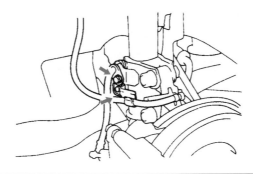

Figure 5-23 Brake hose and ABS wire installed in strut clamps. (Reprinted with permission)

Photo Sequence 3
Typical Procedure for Removing and Replacing a MacPherson Strut

P3-1 The top of the strut assembly is mounted directly to the chassis of the car. Prior to loosening the strut to chassis bolts, scribe alignment marks on the strut bolts and the chassis.

P3-2 With the top stut bolts or nuts removed, raise the car to a working height. It is important that the car be supported on its frame and not on its suspension components.

P3-3 Remove the wheel assembly. The strut is accessible from the wheel well after the wheel is removed.

P3-4 Remove the bolt that fastens the brake line or hose to the strut assembly.

P3-5 Remove the strut to steering knuckle bolts.

P3-6 Support the steering knuckle with wire and remove the strut assembly from the car.

P3-7 Install the strut assembly into the proper type spring compressor. Then compress the spring until it is possible to safely loosen the retaining bolts.

P3-8 Remove the old strut assembly from the spring and install the new strut. Compress the spring to allow for reassembly and tighten the retaining bolts.

P3-9 Reinstall the strut assembly into the car. Make sure all bolts are properly tightened and in the correct locations.

Rear Strut Replacement

The rear strut replacement procedure varies depending on the type of rear suspension. Always follow the vehicle manufacturer's recommended procedure outlined in the appropriate service manual. A typical rear strut replacement procedure follows:

1. Lift the vehicle with a floor jack and lower the vehicle onto safety stands placed under the chassis to support the vehicle weight.

2. Place the floor jack under the lower control arm and operate the jack to support some of the spring tension.

3. Remove the tire-and-wheel assembly.

4. Remove the strut to spindle bolts.

5. Pull upward on the strut to remove the strut from the spindle. If necessary, lower the floor jack slightly to remove the strut from the spindle.

6. Disconnect the upper strut mount from the chassis, and remove the strut.

7. When the new strut and/or mount is installed, reverse steps 1 through 6. Tighten all the bolts to the specified torque.

8. Check vehicle alignment.

The rear coil springs may be removed from the rear struts using the same basic procedure for spring removal from the front struts.

Installing Strut Cartridge, Off-Car

 SERVICE TIP: Check the cost of the strut cartridges versus the price of new struts. Give customers the best value for their repair dollar!

Many struts are a sealed unit, and thus rebuilding is impossible. However, some manufacturers supply a replacement cartridge that may be installed in the strut housing after the strut has been removed from the vehicle. Always follow the **strut cartridge** manufacturer's recommended installation procedure. The following is a typical off-car strut cartridge installation procedure:

1. Install a bolt and two nuts in the upper strut to knuckle mounting bolt hole. Place a nut on the inside and outside of the strut flange.

2. Clamp this bolt in a vise to hold the strut.

3. Locate the line groove near the top of the strut body, and use a pipe cutter installed in this groove to cut the top of the strut body.

4. After the cutting procedure remove the strut piston assembly from the strut (Figure 5-24).

5. Remove the strut from the vise and dump the oil from the strut.

6. Place the special tool supplied by the vehicle manufacturer or cartridge manufacturer on top of the strut body. Strike the tool with a plastic hammer until the tool shoulder contacts the top of the strut body. This action removes burrs from the strut body, and places a slight flare on the body.

7. Remove the tool from the strut body.

8. Place the new cartridge in the strut body, turn the cartridge until it settles into indentations in the bottom of the strut body.

9. Place the new nut over the cartridge.

A strut cartridge contains the inner working part of the strut, which may be installed in the outer housing of the old strut.

Special Tools

Pipe cutter

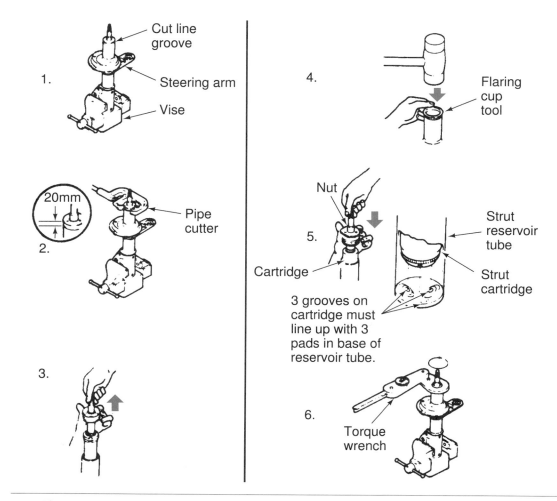

Figure 5-24 Installation of strut cartridge. (Courtesy of Pontiac Motor Division, General Motors Corp.)

10. Using a special tool supplied by the vehicle or cartridge manufacturer, tighten the nut to the specified torque.

11. Move the strut piston rod in and out several times to check for proper strut operation.

Installing Strut Cartridge, On-Car

WARNING: If a vehicle is hoisted or lifted in any way during an on-car strut cartridge replacement, the coil spring may fly off the strut, causing vehicle damage and personal injury.

On some vehicles, the front strut cartridge may be removed and replaced with the strut installed in the vehicle. Always consult the vehicle manufacturer's service manual for the proper strut service procedure. A typical on-car strut cartridge installation procedure follows:

1. Remove the nuts from the upper strut mount and remove the strut mount cover (Figure 5-25).

2. Use the special tool supplied by the vehicle manufacturer and a number 50 torx bit to remove the nut from the strut piston rod (Figure 5-26). Never lift or hoist the vehicle once this nut is removed.

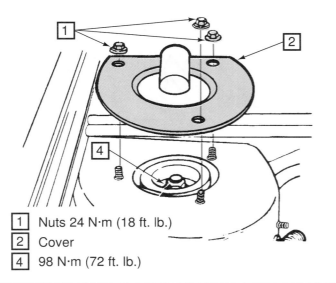

1   Nuts 24 N·m (18 ft. lb.)
2   Cover
4   98 N·m (72 ft. lb.)

Figure 5-25 Removing upper strut mount nuts and cover. (Courtesy of Delco Products Division of General Motors Corp.)

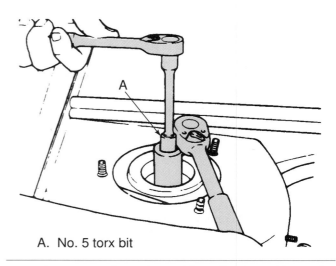

A.   No. 5 torx bit

Figure 5-26 Removing nut from the strut piston rod. (Courtesy of Delco Products Division of General Motors Corp.)

3. Remove the bushing from the upper strut mount (Figure 5-27).

4. Thread the special tool supplied by the vehicle manufacturer onto the top of the strut piston rod, and push this rod downward into the strut (Figure 5-28). Remove this tool and then remove the jounce bumper retainer (Figure 5-29).

5. Remove jounce bumper (Figure 5-30).

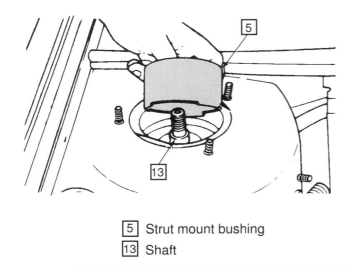

5   Strut mount bushing
13   Shaft

Figure 5-27 Removing upper strut mount bushing. (Courtesy of Delco Products Division of General Motors Corp.)

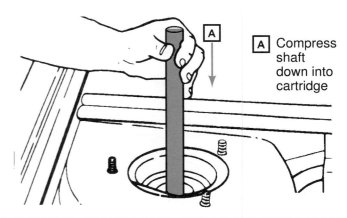

A   Compress shaft down into cartridge

Figure 5-28 Pushing strut piston rod downward. (Courtesy of Delco Products Division of General Motors Corp.)

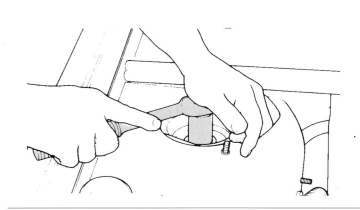

Figure 5-29 Removing jounce bumper retainer. (Courtesy of Delco Products Division of General Motors Corp.)

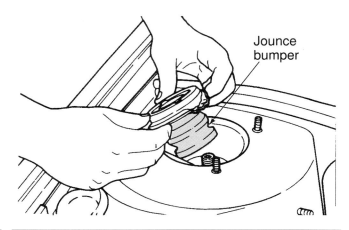

Figure 5-30 Removing jounce bumper. (Courtesy of Delco Products Division of General Motors Corp.)

6. Thread the special tool onto the strut piston rod, and re-extend this rod (Figure 5-31). Use the special tool supplied by the vehicle manufacturer to remove the closure nut on top of the strut (Figure 5-32).

7. Grasp the top of the strut piston rod and remove the strut valve mechanism (Figure 5-33).

8. Remove the oil from the strut tube with a hand-operated suction pump.

9. Install the new strut cartridge in the strut, and tighten the closure nut to the specified torque (Figure 5-34). Reverse steps 1 through 6 to complete the strut cartridge replacement. Place a light coating of the vehicle manufacturer's recommended engine oil on the upper strut mount bushing prior to bushing installation.

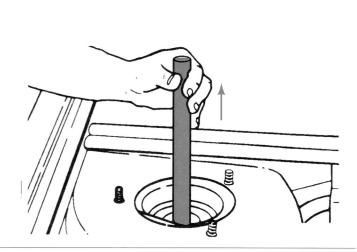

Figure 5-31 Re-extending strut piston rod. (Courtesy of Delco Products Division of General Motors Corp.)

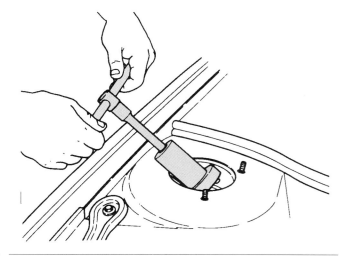

Figure 5-32 Removing strut closure nut. (Courtesy of Delco Products Division of General Motors Corp.)

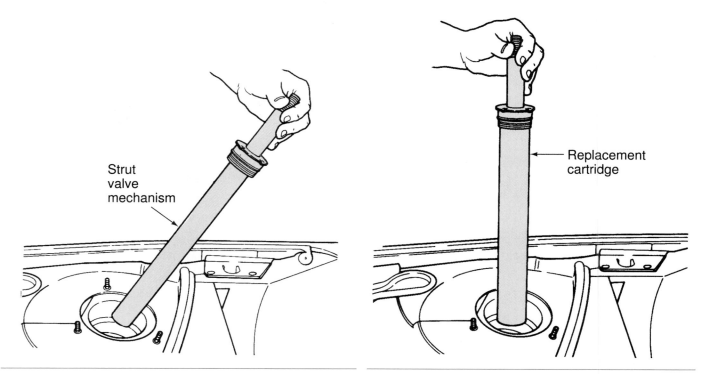

Figure 5-33 Removing strut valve mechanism. (Courtesy of Delco Products Division of General Motors Corp.)	Figure 5-34 Installing new strut cartridge. (Courtesy of Delco Products Division of General Motors Corp.)

Diagnosis of Electronically Controlled Shock Absorbers

The actuators on electronically controlled shock absorbers may be removed by pushing inward simultaneously on the two actuator retaining tabs and lifting the actuator off the top of the strut (Figure 5-35).

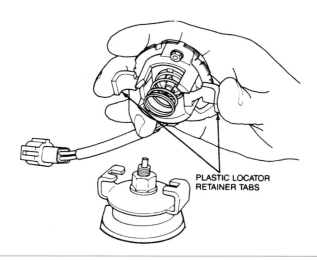

Figure 5-35 Electronically controlled strut actuator. (Courtesy of Ford Motor Company)

WARNING: If the strut piston rod nut and actuator mounting bracket are removed, do not move or raise the vehicle. This action releases the coil spring tension and may result in personal injury and vehicle damage.

Follow these steps to diagnose the electronic actuator:

1. With the actuator removed from the strut and the actuator wiring harness connected, turn the ignition switch on. Move the ride control switch to the auto position and wait 5 seconds.

2. Move the ride control switch to the firm position, and wait 5 seconds. If the actuator control tube on the bottom of the actuator rotated, the actuator is operating. If the actuator control tube did not rotate, proceed with the actuator tests.

3. With the ride control switch in the firm position, place matching H's beside the control tube and the actuator.

4. With the ride control switch in the auto position, place matching S's beside the control tube and the actuator.

5. Turn the ignition switch off and disconnect the actuator electrical connector. The actuator control tube may be rotated with a small screwdriver.

SERVICE TIP: Wiring harness colors vary depending on the vehicle make and model year. The wiring colors in the following steps are based on Ford vehicles. If you are working on a different make of vehicle, refer to the wire colors in the vehicle manufacturer's service information.

6. Connect a pair of ohmmeter leads to the position sense wire and the signal return wire. The position sense wire is white or white with colored tracer, and the signal return wire is yellow with a black tracer. With the actuator in the S position, the position sense should be closed and the ohmmeter should indicate less than 10 ohms.

7. Rotate the actuator to the H position. Under this condition, the position sense switch should be open and the ohmmeter should indicate over 1,000 ohms. If the ohmmeter readings are not within specifications, replace the actuator.

8. Connect the ohmmeter leads to the position sense and soft power terminals in the actuator electrical connector. The position sense wire is white or white with a colored tracer, and the soft power wire is tan with a red tracer. If the ohmmeter indicates over 1,000 ohms, the actuator is satisfactory. An ohmmeter reading below 10 ohms, replace the actuator.

9. Connect the ohmmeter leads from the signal return terminal in the wiring harness side of the actuator connector to a chassis ground. The ohmmeter should indicate less than 10 ohms. If the ohmmeter reading is higher than specified, check the signal return wire and the programmed ride control (PRC) module.

Guidelines for Servicing Shock Absorbers and Struts

1. Shock absorber condition may be determined by performing a visual inspection and a bounce test or manual test.

2. A slight oil film on the lower shock absorber chamber is acceptable, but any sign of excessive oil leaks from a shock absorber or strut indicates that replacement is necessary.

3. Shock absorbers or struts that are bent, dented, or punctured should be replaced.

4. During the bounce test, if the vehicle bumper does more than one free upward bounce, the shock absorber or strut is defective.

5. During a manual test, the shock absorber should offer a strong, steady resistance to movement on the entire compression and rebound strokes. The amount of resistance may be greater on the rebound stroke than on the compression stroke, depending on shock absorber ratio. If there is a reduction or loss of resistance on either stroke, shock absorber replacement is necessary.

6. Front strut chatter may be caused by upper strut mount binding.

7. Front strut noise on sharp turns or during suspension jounce may be caused by interference between the coil spring and the strut tower, or between the coil spring and the upper mount.

8. If one of the front strut to steering knuckle bolts has an eccentric cam, the cam position should be marked on the strut prior to bolt removal.

9. When a coil spring compressing tool is used, it is extremely important to follow the tool manufacturer's and vehicle manufacturer's recommended spring compressing procedure. Never loosen the strut piston rod nut until the spring is compressed so all the tension is removed from the upper strut mount.

10. If the vehicle manufacturer recommends drilling a gas-filled strut or shock absorber prior to strut disposal, always drill the hole at the manufacturer's specified location.

11. When a strut and coil spring are reassembled, always make sure the spring and insulators are properly seated on the spring seats.

12. Before assembling the strut and coil spring in the vehicle, always make sure the upper strut mount is properly aligned with the lower spring seat.

13. Off-car strut cartridge replacements are possible on some vehicles, whereas on other cars on-car strut cartridge replacement is possible. On some struts, cartridge replacement is not possible, and struts must be replaced if they are defective.

CASE STUDY

A customer complained about a squeaking noise in the rear suspension of a 1993 Dodge Aries. The customer said the noise occurred during normal driving at lower speeds. The technician lifted the car on a hoist and made a check of all rear suspension bushings in shock absorbers, track bar, and trailing arms. The spring insulators and all suspension bolts were checked visually. All of these items were in good condition, and there was no evidence of a squeaking noise as the chassis was bounced gently. Since the exact source of chassis noise can sometimes be difficult to locate, the technician performed a visual check of bushings, insulators, and fasteners in the front suspension. No problems were found in the front suspension.

One of the first requirements for successful automotive diagnosis is to obtain as much information as possible from the customer. The customer with the squeaking rear suspension in a Dodge Aries lived in a part of the country where cold temperatures occur in the winter. This complaint occurred in January. The technician questioned the customer further about the conditions when the squeaking suspension noise was heard, and the customer revealed that the noise occurred when the temperature was severely cold. The customer also indicated that the noise disappeared at warmer temperatures.

The technician informed the customer that the only way to find the exact cause of this annoying squeak was to leave the car on the lot at the shop all night and check it first thing in the morning when it was colder. The customer complied with this suggestion, and the technician drove the car into the shop immediately the next morning. The squeaking noise occurred as the car was driven across the parking lot. The technician lifted the car on a hoist and listened with a stethoscope at each rear suspension bushing as a coworker gently bounced the rear suspension. No squeaking noise was heard at any of the rear suspension bushings. However, when the stethoscope pickup was placed on the left rear shock absorber, the squeaking noise was loud and clear. The shock absorber was quickly removed, and the squeaking noise was gone when the rear chassis was bounced gently. Replacement of the left rear shock absorber corrected this complaint, and made a customer happy.

Terms to Know

Bounce test

Eccentric camber bolt

Lower insulator

Manual test

Spring-compressing tools

Spring insulators

Strut cartridge

Strut chatter

Strut tower

Upper insulator

ASE Style Review Questions

1. A slight oil film appears on the lower shock absorber oil chamber, and the shock absorber performs satisfactorily in a bounce test:
   Technician A says the shock absorber is satisfactory.
   Technician B says the shock absorber may contain excessive pressure.
   Who is correct?
   A. A only
   B. B only
   C. Both A and B
   D. Neither A nor B

2. While discussing a shock absorber and strut bounce test:
   Technician A says the shock absorber is satisfactory if the bumper makes two free upward bounces.
   Technician B says the bumper must be pushed downward with considerable force.
   Who is correct?
   A. A only
   B. B only
   C. Both A and B
   D. Neither A nor B

3. While discussing a shock absorber manual test:
   Technician A says the shock absorber's resistance to movement should be jerky and erratic.
   Technician B says the shock absorber's resistance to movement may be greater on the rebound stroke compared to the compression stroke.
   Who is correct?
   A. A only
   B. B only
   C. Both A and B
   D. Neither A nor B

4. When the front wheels are turned on a vehicle equipped with front struts, the left front coil spring provides a chattering action and noise:
   Technician A says the strut has internal defects and strut replacement is necessary.
   Technician B says the upper strut bearing and mount is defective.
   Who is correct?
   A. A only
   B. B only
   C. Both A and B
   D. Neither A nor B

5. While discussing enamel-coated coil springs:
   Technician A says if the enamel-type coating on a coil spring is chipped, the spring may break prematurely.
   Technician B says coil springs should be taped in the compressing tool contact areas to prevent chipping of the enamel-type coating.
   Who is correct?
   A. A only
   B. B only
   C. Both A and B
   D. Neither A nor B

6. During service procedures, gas-filled shock absorbers may:
   A. be heated with an acetylene torch.
   B. be thrown in an incinerator.
   C. extend when disconnected.
   D. retract when disconnected and extended.

7. During wheel jounce, a thumping noise is heard in the right front suspension, but the strut bounce test is satisfactory. All of these defects could cause the problem EXCEPT:
   A. a worn upper strut mount.
   B. worn spring insulators.
   C. a broken coil spring.
   D. a fluid leak in the strut.

8. When removing and replacing a front strut on a MacPherson strut suspension:
   A. the cam-type strut to knuckle bolt should be marked in relation to the strut.
   B. the strut rod nut may be loosened with spring tension applied to the upper strut mount.
   C. the lower strut chamber may be clamped in a vise to support the strut.
   D. the spring compressor must be installed before the strut is removed from the vehicle.

9. When assembling and installing a strut and coil spring in a MacPherson strut front suspension:
   A. the upper spring insulator should be installed between the upper strut mount and the strut tower.
   B. the lower spring insulator should be installed between the coil spring and the spring bumper.
   C. on some vehicles, the upper strut mount must be properly aligned in relation to the lower spring seat on the strut.
   D. final compressing of the coil spring may be done by tightening the nut on the strut rod.

10. During an on-car strut cartridge replacement:
    A. oil must be removed from the strut with a hand-operated suction pump.
    B. the front suspension may be lifted with a floor jack.
    C. the strut rod must be pulled upward before removing the jounce bumper.
    D. the new strut cartridge must be filled with the manufacturer's specified fluid.

ASE Challenge Questions

1. A customer says his MacPherson strut front suspension is making a chattering noise when the steering wheel is turned hard to the left. He says he also feels a "kind of vibration" in the steering wheel when the chatter is heard. To correct this problem, you should:
   A. replace the front struts.
   B. check the lower strut spring seating.
   C. check the stabilizer links.
   D. check the upper bearing and strut mounting.

2. When performing a bounce test on a twin I-beam front suspension the front tires noticeably flex as the front end of the vehicle is vigorously exercised. When the vehicle is released, the front end rebounds once then settles, but the front wheel now has a slight, but obvious negative camber.
   Technician A says the shock absorbers are worn and should be replaced.

Technician B says the twin I-beam suspension can change camber during a bounce and rebound.
Who is correct?

A. A only
C. Both A and B
B. B only
D. Neither A nor B

3. While a vehicle with a torsion bar front suspension is one the hoist the technician notices that the front rebound bumpers are damaged and worn. All of the following could cause this problem EXCEPT:

A. loss of torsion bar tension.
B. worn shock absorbers.
C. trailer towing.
D. bed overloading.

4. A manual test of a shock absorber shows a stronger resistance to rebound than compression.
Technician A says the shock is defective.
Technician B says the shock has a 50/50 ratio.
Who is correct?

A. A only
C. Both A and B
B. B only
D. Neither A nor B

5. *Technician A* says any visible presence of oil on a strut or shock requires replacement.
Technician B says a slight film of oil on a strut or shock is OK if it performs satisfactorily in a bounce test.
Who is correct?

A. A only
C. Both A and B
B. B only
D. Neither A nor B

Job Sheet 14

Name _____ Date _____

Remove Strut and Spring Assembly and Disassemble Spring and Strut

Upon completion of this job sheet, you should be able to remove a strut and spring assembly from a car and disassemble the spring and strut.

ASE Correlation

This job sheet is related to the ASE Suspension and Steering Task List content area: B. Suspension Systems Diagnosis and Repair, 1. Front Suspensions, Task: 6. Inspect and replace front suspension system coil springs and spring insulators (silencers).

Tools and Materials

Front-wheel-drive car
Floor jack
Safety stands
Coil spring compressor

Vehicle make and model year _____

Vehicle VIN _____

Procedure

1. With the vehicle parked on the shop floor, perform a strut bounce test.

   Based on the bounce test results, state the strut condition and give the reason for your diagnosis. _____

   _____

   _____

2. Raise the vehicle on a hoist or with a floor jack. If a floor jack is used to raise the vehicle, lower the vehicle onto safety stands placed under the chassis so the lower control arms and front wheels drop downward. Remove the floor jack from under the vehicle.

   Is the vehicle securely supported on safety stands? yes _____ no _____

   Instructor check _____

3. Remove the brake line and antilock brake system (ABS) wheel-speed sensor wire from clamps on the strut. In some cases, these clamps may have to be removed from the strut.

   Is the brake line and ABS wheel-speed sensor wire disconnected? yes _____ no _____

   Instructor check _____

4. Punch mark the cam bolt in relation to the strut, remove the strut to steering knuckle retaining bolts, and remove the strut from the knuckle.

Is the cam bolt marked in relation to the strut? yes _____ no _____

Is the strut removed from the knuckle? yes _____ no _____

Instructor check _____

State the reason for marking the cam bolt in relation to the strut.

_____

_____

5. Remove the upper strut mounting bolts on top of the strut tower; remove the strut and spring assembly.

> **CAUTION:** Always use a coil spring compressing tool according to the tool or vehicle manufacturer's recommended service procedure. Be sure the tool is properly installed on the spring. If a coil spring slips off the tool when the spring is compressed, severe personal injury or property damage may occur.

> **CAUTION:** Never loosen the upper strut mount retaining nut on the end of the strut rod unless the spring is compressed enough to remove all spring tension from the upper strut mount. If this nut is loosened with spring tension on the upper mount, this mount becomes a very dangerous projectile that may cause serious personal injury or property damage.

> **WARNING:** Never clamp the lower strut or shock absorber chamber in a vise with excessive force. This action may distort the lower chamber and affect piston movement in the strut or shock absorber.

6. Install the spring compressing tool on the coil spring according to the tool or vehicle manufacturer's recommended procedure.

Is the compressing tool properly installed on the spring and strut assembly?
yes _____ no _____

Instructor check _____

> **WARNING:** If the coil spring has an enamel-type coating and the compressing tool contacts the coil spring, tape the spring where the compressing tool contacts the spring.

7. Turn the nut on top of the compressing tool until all the spring tension is removed from the upper strut mount.

Is all the spring tension removed from the upper strut mount? yes _____ no _____

Instructor check _____

8. Install a bolt and two nuts in the upper strut to knuckle mounting bolt holes. Install a nut on each side of the strut flange. Clamp this bolt securely in a vise to hold the strut and coil assembly and the compressing tool.

9. Use the bar on the spring compressing tool to hold the strut and spring assembly from turning, and loosen the nut on the upper strut mount. Be sure all the spring tension is removed from the upper strut mount before loosening this nut.

Is all the spring tension removed from the upper strut mount? yes _____ no _____

Is the upper strut mount retaining nut loosened? yes _____ no _____

Instructor check _____

10. Remove the nut, upper strut mount, upper insulator, coil spring, spring bumper and lower insulator.

11. Inspect the strut, upper strut mount, coil spring, spring insulators, and spring bumper. On the basis of this inspection list the necessary strut and spring service, and give the reasons for your diagnosis. _____

_____

_____

_____

Job Sheet 15

Name _____ Date _____

Assemble Strut and Spring Assembly and Install Strut and Spring Assembly in the Car

Upon completion of this job sheet, you should be able to assemble a spring and strut and install a spring and strut assembly in a car.

ASE Correlation

This job sheet is related to the ASE Suspension and Steering Task List content area: B. Suspension Systems Diagnosis and Repair, 1. Front Suspensions, Task: 6. Inspect and replace front suspension system coil springs and spring insulators (silencers).

Tools and Materials

Front-wheel-drive car
Floor jack
Safety stands
Coil spring compressor

Vehicle make and model year _____

Vehicle VIN _____

Procedure

1. Install a bolt in the upper strut to knuckle retaining bolt, and clamp this bolt in a vise to hold the strut, spring, and compressing tool as in the disassembly procedure.

2. Install the lower insulator on the lower strut spring seat and be sure the insulator is properly seated.

   Is the lower insulator properly seated? yes _____ no _____

   Instructor check _____

3. Install the spring bumper on the strut rod.

   Is the spring bumper properly installed? yes _____ no _____

   Instructor check _____

4. With the coil spring compressed in the spring compressing tool, install the spring on the strut. Be sure the spring is properly seated on the lower insulator spring seat.

   Is the coil spring properly seated on the lower insulator? yes _____ no _____

   Instructor check _____

5. Be sure the strut piston rod is fully extended and install the upper insulator on top of the coil spring.

   Is the strut rod fully extended? yes _____ no _____

   Is the upper insulator properly installed? yes _____ no _____

   Instructor check _____

6. Install the upper strut mount on the upper insulator.

7. Be sure the spring, upper insulator, and upper strut mount are properly positioned and seated on the coil spring and strut piston rod.

    Are the spring, upper insulator, and upper strut mounted properly positioned?

    yes _____ no _____

    Instructor check _____

8. Use the compressing tool bar to hold the strut and spring from turning, then tighten the strut piston rod nut to the specified torque.

    Specified strut piston rod nut torque _____

    Actual strut piston rod nut torque _____

9. Rotate the upper strut mount until the lowest bolt in this mount is aligned with the tab on the lower spring seat.

    Is the upper strut mount properly aligned? yes _____ no _____

    Instructor check _____

10. Gradually loosen the nut on the compressing tool until all the spring tension is released from the tool, and remove the tool from the spring.

11. Install the strut and spring assembly with the upper strut mounting bolts extending through the bolt holes in the strut tower. Tighten the nuts on the upper strut mounting bolts to the specified torque.

    Specified upper strut mount nut torque _____

    Actual upper strut mount nut torque _____

12. Install the lower end of the strut in the steering knuckle to the proper depth. Align the punch marks on the cam bolt and strut that were placed on these components during disassembly, and tighten the strut to knuckle retaining bolts to the specified torque.

    Is the cam bolt properly positioned? yes _____ no _____

    Specified strut to knuckle bolt torque _____

    Actual strut to knuckle bolt torque _____

    Instructor check _____

13. Install the brake hose in the clamp on the strut. Place the ABS wheel-speed sensor wire in the strut clamp if the vehicle is equipped with ABS.

    Is the brake hose properly installed and tightened? yes _____ no _____

    Is the ABS wheel-speed sensor properly installed and tightened? yes _____ no _____

    Instructor check _____

14. Raise the vehicle with a floor jack, remove the safety stands, and lower the vehicle onto the shop floor.

☑ **Instructor Check** _____

Job Sheet 16

Name _____ Date _____

Install Strut Cartridge Off-Car

Upon completion of this job sheet, you should be able to remove and replace a strut cartridge with the strut removed from the car.

ASE Correlation

This job sheet is related to the ASE Suspension and Steering Task List content area: B. Suspension Systems Diagnosis and Repair, 1. Front Suspensions, Task: 10. Inspect and replace MacPherson strut cartridge or assembly.

Tools and Materials

Front strut removed from car
Pipe cutter
Torque wrench

Vehicle make and model year _____

Vehicle VIN _____

Procedure

1. List the strut conditions and suspension operating conditions that indicate a strut should be serviced or replaced.

   _____
   _____
   _____
   _____

2. Install a bolt and two nuts in the upper strut to knuckle mounting bolt hole. Place a nut on the inside and outside of the strut flange.

3. Clamp this bolt in a vise to hold the strut.

4. Locate the line groove near the top of the strut body, and use a pipe cutter installed in this groove to cut the top of the strut body.

5. After the cutting procedure, remove the strut piston assembly from the strut.

   Is the strut piston assembly removed from the strut? yes _____ no _____

   Instructor check _____

6. Remove the strut from the vise and dump the oil from the strut.

7. Place the special tool supplied by the vehicle manufacturer or cartridge manufacturer on top of the strut body and strike the tool with a plastic hammer until the tool shoulder contacts the top of the strut body. This action removes burrs from the strut body and places a slight flare on the body.

   Is the strut body properly flared? yes _____ no _____

   Instructor check _____

8. Remove the tool from the strut body.

9. Place the new cartridge in the strut body and turn the cartridge until it settles into indentations in the bottom of the strut body.

   Is the new strut cartridge properly seated in strut body indentations? yes _____ no _____

   Instructor check _____

10. Place the new nut over the cartridge.

11. Using a special tool supplied by the vehicle or cartridge manufacturer, tighten the nut to the specified torque.

    Specified strut nut torque _____

    Actual strut nut torque _____

12. Move the strut piston rod in and out several times to check for proper strut operation.

    Strut operation: satisfactory _____ unsatisfactory _____

✔ **Instructor Check** _____

Table 5-1   ASE TASK

Inspect and replace shock absorbers.

Problem Area	Symptoms	Possible Causes	Classroom Manual	Shop Manual
RIDE QUALITY	Excessive chassis oscillations	Worn-out shock absorbers	87	126
NOISE	Rattling on road irregularities	Worn shock absorber bushings, grommets	88	125

Table 5-2   ASE TASK

Inspect and replace air shock absorbers, lines, and fittings.

Problem Area	Symptoms	Possible Causes	Classroom Manual	Shop Manual
CURB RIDING HEIGHT	Low curb riding height	Worn-out air shocks, lines, fittings	99	127

Table 5-3   ASE TASK

Inspect and replace front suspension system coil springs and spring insulators (silencers).

Problem Area	Symptoms	Possible Causes	Classroom Manual	Shop Manual
NOISE	Rattling on road irregularities	1. Worn spring insulators	93	127
		2. Broken coil spring	93	128
RIDE HARSHNESS	Low curb riding height	Weak coil springs	93	128

Table 5-4   ASE TASK

Inspect and replace MacPherson strut upper bearing and mount.

Problem Area	Symptoms	Possible Causes	Classroom Manual	Shop Manual
NOISE	Chatter while cornering	Worn upper strut mount	94	128

Table 5-5 ASE TASK

Inspect and replace MacPherson strut cartridge or assembly.

Problem Area	Symptoms	Possible Causes	Classroom Manual	Shop Manual
RIDE QUALITY	Excessive chassis oscillations	Worn-out strut cartridge	92	135

Front Suspension System Service

Upon completion and review of this chapter, you should be able to:

❏ Measure curb riding height.

❏ Diagnose and correct curb riding height problems.

❏ Adjust torsion bars.

❏ Diagnose front suspension noise and body sway.

❏ Remove and replace ball joints, and check ball joint condition.

❏ Remove and replace steering knuckles, and check knuckle condition.

❏ Remove and replace lower control arms, and check control arm and bushing condition.

❏ Remove and replace coil springs, and check spring and insulator condition.

❏ Remove and replace upper control arms, and check control arm and bushing condition.

❏ Remove and replace control arm bushings.

❏ Inspect and replace rebound bumpers.

❏ Diagnose, remove, and replace stabilizer bars.

❏ Diagnose, remove, and replace strut rods.

❏ Diagnose, remove, and replace leaf springs.

❏ Replace torsion bars, and check torsion bar condition.

Curb Riding Height Measurement

Regular inspection and proper maintenance of suspension systems is extremely important for maintaining vehicle safety. The **curb riding height** is determined mainly by spring condition. Other suspension components, such as control arm bushings, affect curb riding height if they are worn. Since incorrect curb riding height affects most of the other suspension angles, this measurement is critical.

Reduced curb riding height on the front suspension may cause decreased **directional stability.** If the curb riding height is reduced on one side of the front suspension, the steering may pull to one side. Reduced rear suspension height increases **steering effort** and causes rapid steering wheel return after turning a corner. Harsh riding occurs when the curb riding height is less than specified. The curb riding height must be measured at the vehicle manufacturer's specified location, which varies depending on the type of suspension system.

When the vehicle is on a level floor or on an alignment rack, measure the curb riding height from the floor to the center of the lower control arm mounting bolt on both sides of the front suspension (Figure 6-1). On the rear suspension system, measure the curb riding height from the floor to the center of the strut rod mounting bolt (Figure 6-2).

If the curb riding height is less than specified, the control arms and bushings should be inspected and replaced as necessary. When the control arms and bushings are in normal condition, the reduced curb riding height may be caused by sagged springs that require replacement.

Front Suspension Diagnosis and Service

Torsion Bar Adjustment

● **CUSTOMER CARE:** Each time undercar service is performed, make a quick inspection of suspension and steering components. Advise the customer regarding any necessary repairs. This procedure often obtains additional work for the shop, and the customer will be impressed that you are interested in his or her car and personal safety.

Basic Tools

Basic technician's tool set

Service manual

Floor jack

Safety stands

Pry bar

3/8 electric drill and drill bits

Directional stability refers to the tendency of the vehicle steering to remain in the straight-ahead position when driven straight ahead on a reasonably smooth, level road surface.

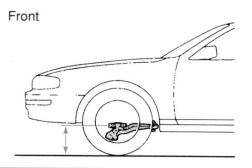

Front

Rear

Figure 6-1 Curb riding height measurement, front suspension. (Reprinted with permission)

Figure 6-2 Curb riding height measurement, rear suspension. (Reprinted with permission)

Special Tools

Machinist's rule

On torsion bar front suspension systems, the torsion bars may be adjusted to correct the curb riding height. The curb riding height must be measured at the location specified by the vehicle manufacturer. On some torsion bar front suspension systems, the curb riding height is measured with the vehicle on a lift and the tires supported on the lift. Measure the distance from the center of the front lower control arm bushing to the lift. Then measure the distance from the lower end of the front spindle to the lift (Figure 6-3). The difference between these two readings is the curb riding height.

If the curb riding height is not correct on a torsion bar front suspension, the torsion bar anchor adjusting bolts must be rotated until the curb riding height equals the vehicle manufacturer's specifications (Figure 6-4).

Checking Ball Joints

✓ **SERVICE TIP:** Regular chassis lubrication at the vehicle manufacturer's recommended service interval is one of the keys to long ball joint life. Always advise the customer of this fact.

Wear Indicators. Some ball joints have a grease fitting installed in a floating retainer. The grease fitting and retainer may be used as a **ball joint wear indicator.** With the vehicle weight resting on the wheels, grasp the grease fitting and check for movement (Figure 6-5).

Some car manufacturers recommend ball joint replacement if any grease fitting movement is present. In some other ball joints, the grease fitting retainer extends a short distance through the ball joint surface (Figure 6-6). On this type of joint, replacement is necessary if the grease fitting shoulder is flush with or inside the ball joint cover.

Ball Joint Unloading

On many suspension systems, ball joint looseness is not apparent until the weight has been removed from the joint. When the coil spring is positioned between the lower control arm and the

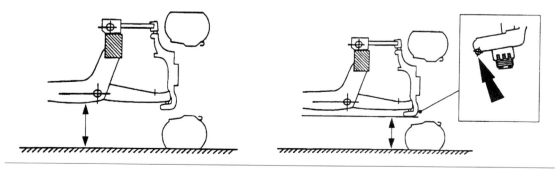

Figure 6-3 Curb riding height measurement, torsion bar front suspension. (Courtesy of Ford Motor Company)

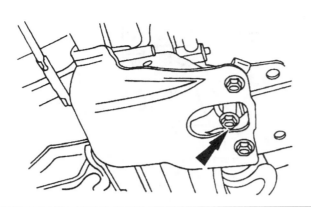

Figure 6-4 Curb riding height adjustment, torsion bar front suspension. (Courtesy of Ford Motor Company)

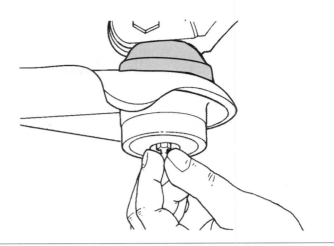

Figure 6-5 Ball joint grease fitting wear indicator. (Courtesy of Chrysler Corp.)

chassis, place a floor jack near the outer end of the lower control arm and raise the tire off the floor (Figure 6-7). Be sure the rebound bumper is not in contact with the control arm or frame.

When the coil springs are positioned between the upper control arm and the chassis, place a steel block between the upper control arm and the frame. With this block in place, raise the tire off the floor with a floor jack under the front crossmember (Figure 6-8).

Ball Joint Vertical Measurement

The vehicle manufacturer may provide ball joint vertical and radial (horizontal) tolerances. A dial indicator is one of the most accurate ball joint measuring devices (Figure 6-9). Always install the dial indicator at the vehicle manufacturer's recommended location for ball joint measurement. When measuring the **ball joint vertical movement** in a compression-loaded ball joint, attach the dial indicator to the lower control arm and position the dial indicator stem on the lower side of the

Special Tools

Dial indicator for ball joint measurement

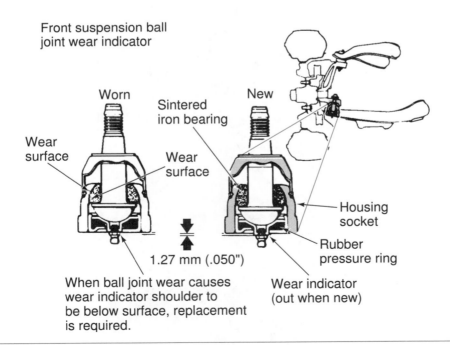

Figure 6-6 Ball joint wear indicator with grease fitting extending from ball joint surface. (Courtesy of Chevrolet Motor Division, General Motors Corp.)

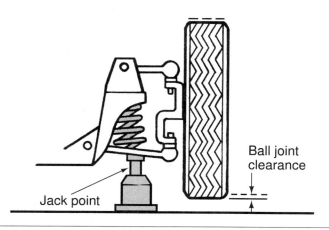

Figure 6-7 Floor jack position to check ball joint wear with spring between the lower control arm and the chassis. (Courtesy of Sealed Power Corp.)

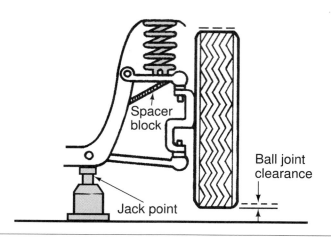

Figure 6-8 Floor jack position to check ball joint wear with spring between the upper control arm and the chassis. (Courtesy of Sealed Power Corp.)

steering knuckle beside the ball joint stud (Figure 6-10). Depress the dial indicator stem approximately 0.250 in (6.35 mm), and zero the dial indicator. Place a pry bar under the tire and pry straight upward while observing the vertical ball joint movement on the dial indicator. If this movement is more than specified, ball joint replacement is necessary.

On a tension-loaded ball joint, clean the top end of the lower ball joint stud, and position the dial indicator stem against the top end of this stud. Depress the dial indicator plunger approximately 0.250 in (6.35 mm) and zero the dial indicator.

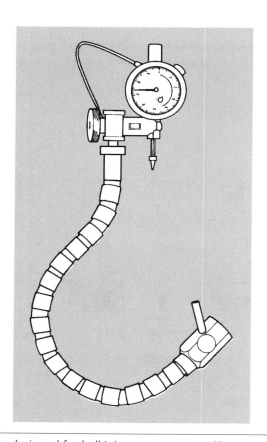

Figure 6-9 Dial indicator designed for ball joint measurement. (Courtesy of Moog Automotive)

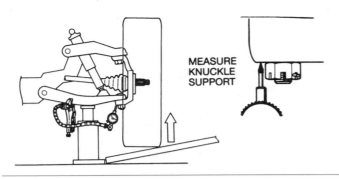 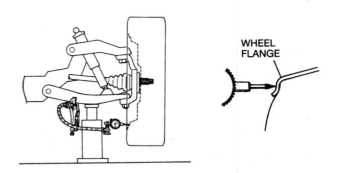

Figure 6-10 Dial indicator installed to measure vertical ball joint movement on a compression-loaded ball joint. (Courtesy of Moog Automotive)

Figure 6-11 Dial indicator installed to measure radial ball joint movement. (Courtesy of Moog Automotive)

Lift upward with a pry bar under the tire and observe the dial indicator reading. If the vertical ball joint movement exceeds the manufacturer's specifications, ball joint replacement is required.

Ball Joint Radial Measurement

Worn ball joints cause improper **camber angles** and **caster angles,** which result in reduced directional stability and tire tread wear. Connect the dial indicator to the lower control arm of the ball joint being checked and position the dial indicator stem against the edge of the wheel rim (Figure 6-11).

Be sure the front wheel bearings are adjusted properly prior to the **ball joint radial measurement.** While an assistant grasps the top and bottom of the raised tire and attempts to move the tire and wheel horizontally in and out, observe the reading on the dial indicator (Figure 6-12).

The lower ball joint on a MacPherson strut front suspension should be checked for radial movement with a dial indicator when the tire is lifted off the floor (Figure 6-13). Since the spring load is carried by the upper and lower spring seats when the tire is lifted off the floor, it is not necessary to unload this type of ball joint. Photo Sequence 4 shows a typical procedure for measuring the lower ball joint radial movement on a MacPherson strut front suspension.

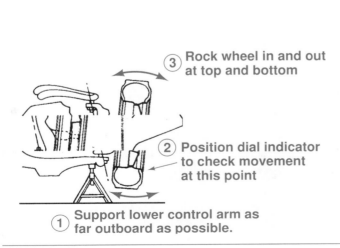

③ Rock wheel in and out at top and bottom

② Position dial indicator to check movement at this point

① Support lower control arm as far outboard as possible.

Figure 6-12 Measuring radial ball joint movement. (Courtesy of Chevrolet Motor Division, General Motors Corp.)

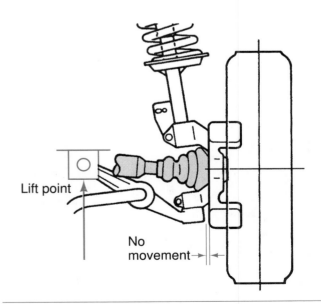

Lift point

No movement

Figure 6-13 Ball joint wear measurement on MacPherson strut front suspension. (Courtesy of Ford Motor Company)

Photo Sequence 4
Typical Procedure for Measuring the Lower Ball Joint Radial Movement on a MacPherson Strut Front Suspension

P4-1 Raise the front suspension with a floor jack and place safety stands under the chassis at the vehicle manufacturer's recommended lifting points.

P4-2 Grasp the front tire at the top and bottom and rock the tire inward and outward while a coworker visually checks for movement in the front wheel bearing. If there is movement in the front wheel bearing, adjust or replace the bearing.

P4-3 Position a dial indicator against the inner edge of the rim at the bottom. Preload and zero the dial indicator.

P4-4 Grasp the bottom of the tire and pull outward.

P4-5 With the tire held outward, read the dial indicator.

P4-6 Push the bottom of the tire inward and be sure the dial indicator reading is zero. Adjust the dial indicator as required.

P4-7 Grasp the bottom of the tire and pull outward.

P4-8 With the tire held in this position, read the dial indicator.

P4-9 If the dial indicator reading is more than specified, replace the lower ball joint.

Checking Ball Joints, Twin I-Beam Axles

Follow these steps to check the ball joints on a twin I-beam axle:

1. Lift the front end of the vehicle with a floor jack and place safety stands near the outer end of the I-beams.

2. Lower the vehicle onto the safety stands.

3. While an assistant grasps the wheel at the bottom and moves the wheel in and out, watch for movement between the lower part of the axle jaw and the spindle lower arm. If this movement exceeds 1/32 inch, replace the lower ball joint.

4. As an assistant grasps the top of the wheel and moves the wheel in and out, watch for movement between the upper axle jaw and the spindle upper arm. If this movement exceeds 1/32 inch, replace the upper ball joint.

Classroom Manual
Chapter 6, page 110

Front Steering Knuckle Diagnosis and Service, Front-Wheel-Drive Car

Front Steering Knuckle Diagnosis. Steering looseness may be caused by a worn tie-rod end opening or ball joint opening in the steering knuckle. If these knuckle openings are worn, reduced directional stability may be experienced. Since a bent steering knuckle affects front suspension alignment angles, this problem may cause reduced directional stability and tire tread wear. Many steering arm and knuckle assemblies must be replaced as a unit. If the steering arms are bent, knuckle replacement is required.

Steering Knuckle Removal and Replacement. The steering knuckle replacement procedure varies depending on the type of front suspension. Always follow the vehicle manufacturer's steering knuckle replacement procedure in the service manual. Follow these steps to replace the steering knuckle on a front-wheel-drive vehicle with a MacPherson strut front suspension system.

1. Remove the wheel cover and loosen the front wheel nuts and the drive axle nut.

2. Lift the vehicle chassis on a hoist and allow the front suspension to drop downward. Remove the front wheel, brake caliper, brake rotor, and drive axle nut. Tie the brake caliper to a suspension component. Do not allow the caliper to hang on the end of the brake hose.

3. Remove the inner end of the drive axle from the transaxle with a pulling or prying action. On some Chrysler products, the differential cover must be removed and axle circlips compressed before the drive axle is removed from the transaxle.

4. Remove the outer end of the drive axle from the steering knuckle. On some early model Ford Escorts and Lynx, a puller is required for this operation.

5. Be sure the vehicle weight is supported on the hoist with the front suspension dropped downward. Disconnect the lower ball joint nuts from the mounting bolts in the lower control arm (Figure 6-14).

6. Remove the cotter pin from the outer tie-rod nut (Figure 6-15). Remove the outer tie-rod nut, and then use a puller to disconnect the tie-rod end from the steering knuckle (Figure 6-16).

7. If an eccentric cam is used on one of the strut-to-knuckle bolts, mark the cam and bolt position in relation to the strut, and remove the strut-to-knuckle bolts.

8. Remove the knuckle from the strut, and lift the knuckle out of the chassis (Figure 6-17).

9. Pry the dust deflector from the steering knuckle with a large flat-blade screwdriver (Figure 6-18).

Figure 6-14 Removal of nuts from lower ball joint mounting bolts under the lower control arm. (Reprinted with permission)

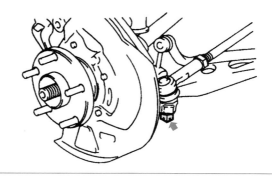

Figure 6-15 Removing cotter pin and nut from outer tie-rod end. (Reprinted with permission)

Special Tools

Ball joint puller

10. Use a puller to remove the ball joint from the steering knuckle (Figure 6-19). Check the ball joint and tie-rod end openings in the knuckle for wear and out-of-round. Replace the knuckle if these openings are worn or out-of-round.

11. Use the proper driving tool to reinstall the dust deflector in the steering knuckle (Figure 6-20).

12. Reverse steps 1 through 10 to reinstall the knuckle. Service or replace the wheel bearing as required. (Refer to Chapter 3 for wheel bearing service.) Torque all nuts to specifications and install cotter pins as required.

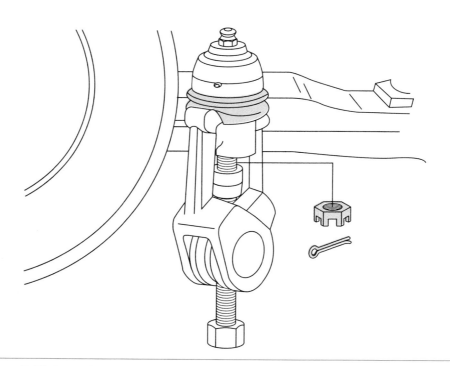

Figure 6-16 Removing outer tie-rod end from steering knuckle. (Courtesy of Chrysler Corp.)

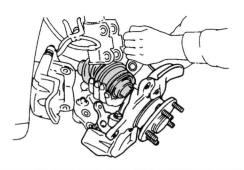

Figure 6-17 Removing the steering knuckle from the lower control arm. (Reprinted with permission)

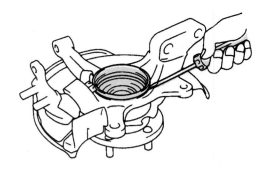

Figure 6-18 Removing dust deflector from steering knuckle. (Reprinted with permission)

Front Steering Knuckle Replacement Rear-Wheel-Drive Cars

Proceed as follows for steering knuckle replacement on short-and-long arm front suspension systems with the coil spring positioned between the lower control arm and the frame.

1. Remove the wheel cover and loosen the wheel nuts.

2. Lift the vehicle with a floor jack and place safety stands under the chassis so the front suspension drops downward. Lower the vehicle onto the safety stands and remove the floor jack.

3. Remove the front wheel, brake caliper, and the brake rotor and hub with the wheel bearings. Attach the brake caliper to a chassis component with a piece of wire. Do not allow the caliper to hang on the end of the brake hose.

4. Remove the outer tie-rod cotter pin and nut, and remove the outer tie-rod end from the steering arm.

5. Remove the cotter pins from the upper and lower ball joint nuts and loosen, but do not remove, the nuts.

CAUTION: During the knuckle replacement procedure, the floor jack supports the spring tension when the ball joints are disconnected from the knuckle. Do not lower the floor jack. Chain the coil spring to the lower control arm as a safety precaution.

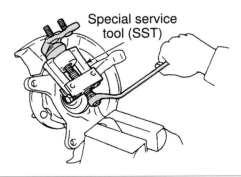

Figure 6-19 Removing ball joint from steering knuckle. (Reprinted with permission)

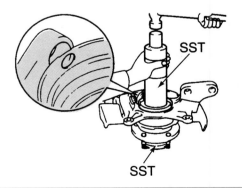

Figure 6-20 Installing dust deflector in steering knuckle. (Reprinted with permission)

6. Support the lower control arm with a floor jack, and raise the floor jack until the spring tension is supported by the jack.

7. Remove both ball joint nuts, and remove the ball joint studs from the steering knuckle. A threaded spreader tool may be used to push the upper and lower ball joints from the knuckle (Figure 6-21 and Figure 6-22).

8. Pull upward on the upper control arm to remove the upper ball joint stud from the knuckle. Remove the knuckle from the lower ball joint stud. Check the ball joint and tie-rod end openings in the knuckle for wear and out-of-round. Replace the knuckle if these conditions are present.

9. Reverse steps 1 through 8 to install the steering knuckle. Tighten the ball joint nuts and the outer tie-rod nut to the specified torque, and install the cotter pins in these nuts. Adjust the front wheel bearings, and tighten the wheel nuts to the specified torque.

Ball Joint Replacement

Results of Ball Joint Wear. Worn ball joints affect steering angles and cause reduced directional stability and excessive tire tread wear.

Ball Joint Removal and Replacement. Ball joints may be threaded, pressed, bolted, or riveted into the control arms (Figure 6-23). The ball joint replacement procedure varies depending on the type of suspension and the method of ball joint attachment. Always follow the vehicle manufacturer's recommended ball joint replacement procedure in the service manual. The following are typical steps that apply to all three methods of ball joint attachment.

1. Remove the wheel cover and loosen the wheel nuts.

2. Lift the vehicle with a floor jack and place safety stands under the chassis so the front suspension is allowed to drop downward. Lower the vehicle onto the safety stands and remove the floor jack.

3. Remove the wheel and place a floor jack under the outer end of the lower control arm. Operate the floor jack and raise the lower control arm until the ball joints are unloaded.

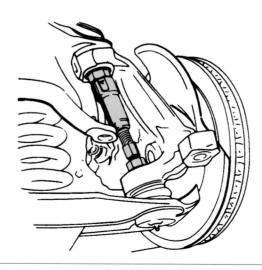

Figure 6-21 Removing the lower ball joint from the steering knuckle. (Courtesy of Chevrolet Motor Division, General Motors Corp.)

Figure 6-22 Removing the upper ball joint from the steering knuckle. (Courtesy of Chevrolet Motor Division, General Motors Corp.)

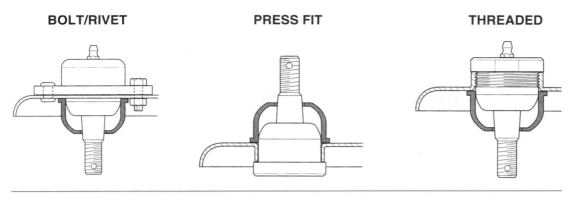

BOLT/RIVET   **PRESS FIT**   **THREADED**

Figure 6-23 Methods of ball joint attachment. (Courtesy of Cooper Industries, Inc.)

Remove other components such as the brake caliper, rotor, and drive axle as required to gain access to the ball joints.

4. Remove the cotter pin in the ball joint or joints that requires replacement, and loosen, but do not remove, the ball joint stud nuts.

☑ **SERVICE TIP:** Some ball joints have a tapered stud, and a nut is threaded onto the top of this stud to retain the ball joint in the steering knuckle. Other ball joints have a straight stud, and a pinch bolt extends through the steering knuckle and a notch in the side of the ball joint stud to hold the ball joint in the steering knuckle.

5. Loosen the ball joint stud tapers in the steering knuckle. A threaded expansion tool is available for this purpose.

6. Remove the ball joint nut and lift the knuckle off the ball joint stud. Block or tie up the knuckle and hub assembly to access the ball joint.

7. If the ball joint is riveted to the control arm, drill and punch out the rivets, and bolt the new ball joint to the control arm (Figure 6-24).

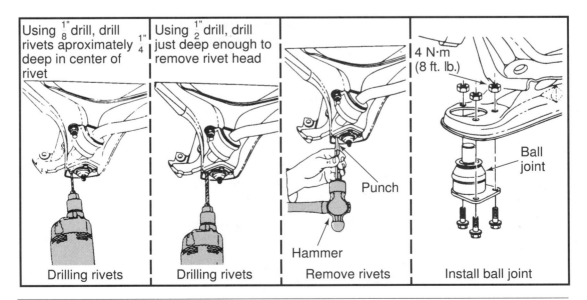

Using $\frac{1}{8}$" drill, drill rivets aproximately $\frac{1}{4}$" deep in center of rivet

Using $\frac{1}{2}$" drill, drill just deep enough to remove rivet head

4 N·m (8 ft. lb.)

Ball joint

Punch

Hammer

Drilling rivets | Drilling rivets | Remove rivets | Install ball joint

Figure 6-24. Replacement of riveted ball joint. (Courtesy of Pontiac Motor Division, General Motors Corp.)

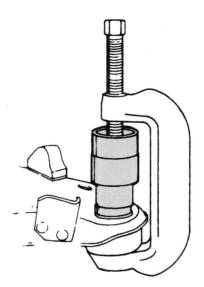

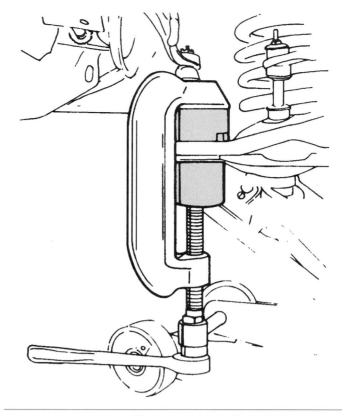

Figure 6-25 Removing pressed-in ball joint from lower control arm. (Courtesy of Chevrolet Motor Division, General Motors Corp.)

Figure 6-26 Installing pressed-in ball joint in lower control arm. (Courtesy of Chevrolet Motor Division, General Motors Corp.)

Special Tools

Ball joint pressing tools

8. If the ball joint is pressed into the lower control arm, remove the ball joint dust boot, and use a pressing tool to remove and replace the ball joint (Figure 6-25 and Figure 6-26).

9. If the ball joint housing is threaded into the control arm, use the proper size socket to remove and install the ball joint. The replacement ball joint must be torqued to the manufacturer's specifications. If a minimum of 125 foot-pounds of torque cannot be obtained, the control arm threads are damaged and control arm replacement is necessary.

10. If the ball joint is bolted to the lower control arm, install the new ball joint and tighten the bolt and nuts to the specified torque (Figure 6-27).

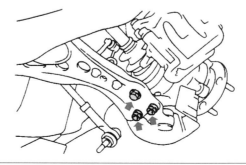

Figure 6-27 Installing ball joint retaining bolts and nuts in lower control arm. (Reprinted with permission)

11. Clean and inspect the ball joint stud tapered opening in the steering knuckle. If this opening is out-of-round or damaged, the knuckle must be replaced.

12. Check the fit of the ball joint stud in the steering knuckle opening. This stud should fit snugly in the opening and only the threads on the stud should extend through the knuckle. If the ball joint stud fits loosely in the knuckle tapered opening, either this opening is worn or the wrong ball joint has been supplied.

CAUTION: Never back off a ball joint stud nut to align the cotter pin openings in the nut and stud. Always tighten the nut to the next hole to install the cotter pin.

13. Install the ball joint stud in the steering knuckle opening, making sure the stud is straight and centered. Install the stud nut and tighten this nut to the specified torque. Install a new cotter pin through the stud and nut. Do not loosen the nut to align the nut and stud openings.

14. Reassemble the components that were removed in step 3. Make sure the wheel nuts are tightened to the specified torque.

15. After ball joint replacement, the front suspension alignment should be checked.

Control Arm Diagnosis and Service

Control Arm Diagnosis and Replacement

Upper and lower control arms should be inspected for cracks, bent conditions, and worn bushings. If the control arm bushings are worn, steering is erratic, especially on irregular road surfaces. Worn control arm bushings may cause a rattling noise while driving on irregular road surfaces. Dry or worn control arm bushings may cause a squeaking noise on irregular road surfaces. Caster and camber angles on the front suspension are altered by worn upper and lower control arm bushings. Incorrect caster or camber angles may cause the vehicle to pull to one side. Tire wear may be excessive when the camber angle is not within specifications.

Lower Control Arm Replacement, MacPherson Strut Suspension

The upper or lower control arm removal and replacement procedure varies depending on the type of suspension. Always follow the recommended procedure in the vehicle manufacturer's service manual. The following is a lower control arm replacement procedure for a MacPherson strut front suspension:

1. Remove the wheel cover and loosen the wheel nuts and drive axle nut.

2. Lift the front of the vehicle with a floor safety, and place jack stands under the chassis. Lower the vehicle onto the safety stands and allow the front suspension to drop downward.

3. Remove the front wheel and the front fender apron seal (Figure 6-28). Remove the drive axle nut (Figure 6-29).

4. Remove the cotter pin and nut from the outer tie-rod end, and use a puller to remove this tie-rod end from the steering arm (Figure 6-30).

5. Remove the stabilizer brackets from the lower control arm (Figure 6-31).

6. Remove the lower ball joint nuts and bolt from the lower control arm (Figure 6-32).

7. Remove the drive axle from the axle hub (Figure 6-33), and secure the drive axle to a suspension component with a piece of wire (Figure 6-34). Do not allow the drive axle to hang downward.

Caster angle on the front suspension is the tilt of a line that intersects the center of the lower ball joint and the center of the upper strut mount in relation to a vertical line through the center of the wheel and spindle viewed from the side of the vehicle.

Camber angle on the front or rear suspension is the tilt of a line inward or outward in relation to the vertical line through the center of the wheel and tire as viewed from the front.

Special Tools

Tie-rod end puller

Figure 6-28 Removal of front fender apron seal. (Reprinted with permission)

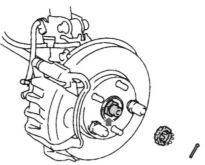

Figure 6-29 Removal of drive axle cotter pin, lock cap, and nut. (Reprinted with permission)

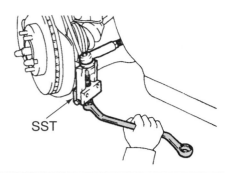

Figure 6-30 Removing outer tie-rod end. (Reprinted with permission)

8. Remove the two attaching bolts at the front side of the lower control arm (Figure 6-35).

9. Remove the bolt and nut on the rear side of the lower control arm (Figure 6-36), and remove the lower arm bushing stopper from the lower arm shaft.

Reverse steps 1 through 9 to install the replacement control arm. Tighten all bolts and nuts to the specified torque, and install cotter pins as required.

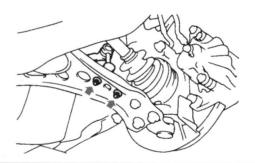

Figure 6-31 Removing stabilizer brackets from lower control arm. (Reprinted with permission)

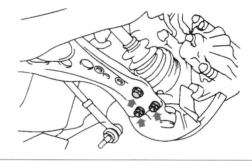

Figure 6-32 Removing lower ball joint bolts and nuts from lower control arm. (Reprinted with permission)

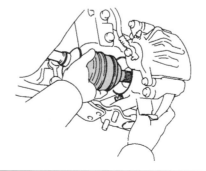

Figure 6-33 Removing drive axle from axle hub. (Reprinted with permission)

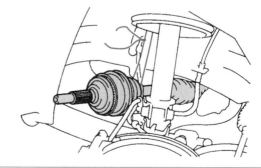

Figure 6-34 Securing drive axle with a piece of wire. (Reprinted with permission)

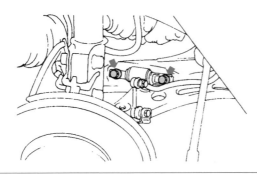

Figure 6-35 Removing bolts at the front side of the lower control arm. (Reprinted with permission)

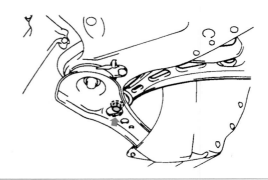

Figure 6-36 Removing bolt and nut from rear side of the lower control arm. (Reprinted with permission)

Lower Control Arm and Spring Replacement, Short-and-Long Arm Suspension

⚠️ **WARNING:** During control arm and spring replacement, the coil spring tension is supported by a compressing tool or floor jack. Always follow the vehicle manufacturer's recommended control arm and spring replacement procedures very carefully. Serious personal injury or property damage may occur if the spring tension is released suddenly.

Broken coil springs may cause a rattling noise while driving on irregular road surfaces. Weak or broken coil springs also reduce curb riding height. A rattling noise while driving on road irregularities may also be caused by worn or broken spring insulators. Since weak or broken coil springs affect front suspension alignment angles, this problem may cause reduced directional stability, excessive tire wear, and harsh riding.

Follow this procedure when a lower control arm and/or spring is replaced on a vehicle with a short-and-long arm suspension system with the coil springs positioned between the lower control arm and the frame:

1. Lift the vehicle on a hoist until the tires are a short distance off the floor, and allow the front suspension to drop downward. An alternate method is to lift the vehicle with a floor jack, and then support the chassis securely on safety stands so the front suspension drops downward.

2. Disconnect the lower end of the shock absorber. On some applications, the shock absorber must be removed.

3. Disconnect the stabilizer bar from the lower control arm.

4. Install a spring compressor and turn the spring compressor bolt until the spring is compressed (Figure 6-37). Make sure all the spring tension is supported by the compressing tool.

5. Place a floor jack under the lower control arm, and raise the jack until the control arm is raised and the rebound bumper is not making contact.

6. Remove the lower ball joint cotter pin and nut, and use a threaded expansion tool to loosen the lower ball joint stud.

7. Lower the floor jack very slowly to lower the control arm and coil spring.

8. Disconnect the lower control arm inner mounting bolts, and remove the lower control arm.

9. Rotate the compressing tool bolt to release the spring tension, and remove the spring from the control arm.

Special Tools

Spring compressing tool, short-and-long arm suspension

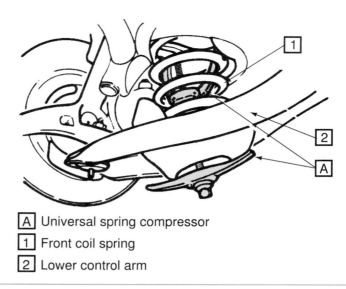

A Universal spring compressor
1 Front coil spring
2 Lower control arm

Figure 6-37 Spring compressing tool installed on spring in short-and-long-arm suspension system. (Courtesy of Chevrolet Motor Division, General Motors Corp.)

10. Inspect the lower control arm for a bent condition or cracks. If either of these conditions is present, replace the control arm. Visually inspect all control arm bushings. Loose or worn bushings must be replaced. Visually inspect upper and lower spring insulators for cracks and wear, and inspect the spring seat areas in the chassis and lower control arm. Worn or cracked spring insulators must be replaced.

11. Reverse steps 1 through 9 to install the lower control arm. Be sure the coil spring and insulators are properly seated in the lower control arm and in the upper spring seat.

Note: See Chapter 5 for Spring and strut service.

Upper Control Arm Removal and Replacement, Short-and-Long Arm Suspension System

Proceed as follows when replacing an upper control arm on a short-and-long arm suspension system with the front coil springs located between the lower control arm and the frame.

1. Remove the wheel cover and loosen the wheel nuts.

2. Lift the vehicle with a floor jack and install safety stands under the chassis. Lower the vehicle onto the safety stands so the front suspension drops downward.

3. Remove the front wheel and tire.

4. Place a floor jack under the lower control arm and raise this arm. Make sure the rebound bumper is not making contact.

5. Remove the cotter pin and loosen, but do not remove, the upper ball joint nut.

6. Use a threaded expansion tool to loosen the ball joint stud in the control arm. Make sure the floor jack is supporting the spring tension.

CAUTION: Since the floor jack is supporting the spring tension, do not lower the floor jack until the ball joint is reconnected and the ball joint nut is tightened to the specified torque.

7. Remove the ball joint nut and the inner control arm mounting bolts. If shims are located on the inner control arm mounting bolts, note the shim position.

8. Remove the control arm from the chassis.

9. Visually inspect the upper control arm for a bent or cracked condition. Inspect the control arm bushing openings for wear, and replace the control arm if it is bent, cracked, or worn in the bushing areas. Inspect the control arm shaft and bushings for wear, and replace all worn parts (Figure 6-38).

10. Install the control arm and place the original number of shims on the inner mounting bolts. Tighten these bolts to the manufacturer's specifications.

11. Install the ball joint stud in the steering knuckle, and tighten the stud nut to the specified torque.

12. Install the upper ball joint cotter pin and the front wheel. Tighten the wheel nuts to the specified torque.

The front suspension alignment angles should be checked after upper control arm replacement.

Rebound Bumpers

 SERVICE TIP: Excessively worn rebound bumpers indicate worn-out shock absorbers, reduced curb riding height, or driving continually on irregular road surfaces.

Rebound bumpers are usually bolted to the lower control arm or to the chassis. Inspect the rebound bumpers for cracks, wear, and flattened conditions (Figure 6-39). Damaged rebound bumpers may be caused by sagged springs and insufficient curb riding height, or worn-out shock

Rebound bumpers may be called strikeout bumpers.

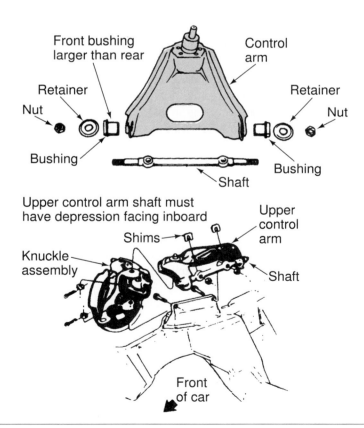

Figure 6-38 Upper control arm components. (Courtesy of Chevrolet Motor Division, General Motors Corp.)

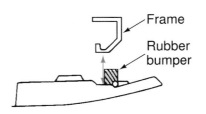

Figure 6-39 Rebound bumper. (Courtesy of Sealed Power Corp.)

absorbers and struts. If the rebound bumpers must be replaced, remove the mounting bolts and the bumper. Install the new bumper and tighten the mounting bolts to the specified torque.

Front Lower Control Arm Bushing Removal and Replacement

Special Tools

Control arm bushing removal, replacement, and flaring tools

All suspension bushings should be checked periodically for wear, looseness, and deterioration. These bushings are important for providing quiet suspension operation and preventing the transmission of suspension vibration to the chassis and passenger compartment. When a bushing is contained in a steel sleeve, press only on the outer sleeve during the removal and replacement procedure. The control arm bushing removal and replacement procedure varies depending on the type of suspension system. Always follow the vehicle manufacturer's recommended procedure in the service manual. Special bushing removal and replacement tools are required for control arm bushing replacement. A front control arm bushing removal tool and spacer is used to remove the control arm bushing (Figure 6-40). The spacer is installed between the bushing support lugs to prevent distorting these lugs during the bushing removal process.

A different tool is required to install the new front control arm bushing (Figure 6-41). The same spacer that is used during the bushing removal is installed between the bushing lugs during bushing installation.

After the front control arm bushing is installed, a bushing flaring tool is used to flare the bushing (Figure 6-42). Notice that a spacer is installed between the bushing lugs during the flaring process. Flaring retains the bushing securely (Figure 6-43).

The rear control arm bushing is replaced with the same procedure as the front control arm bushing. Some of the same special tools are used for rear control arm bushing replacement (Figure 6-44 and Figure 6-45).

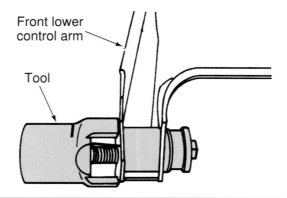

Figure 6-40 Removal of front control arm bushing. (Courtesy of Chevrolet Motor Division, General Motors Corp.)

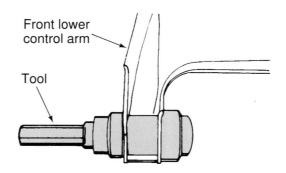

Figure 6-41 Installation of front control arm bushing. (Courtesy of Chevrolet Motor Division, General Motors Corp.)

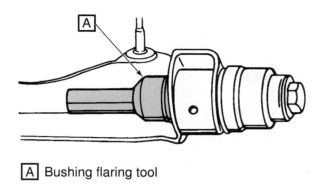

[A] Bushing flaring tool

Figure 6-42 Flaring the front control arm bushing. (Courtesy of Chevrolet Motor Division, General Motors Corp.)

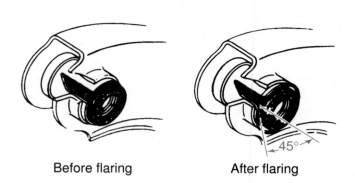

Before flaring   After flaring

Figure 6-43 Front control arm bushing after flaring. (Courtesy of Chevrolet Motor Division, General Motors Corp.)

Stabilizer Bar Diagnosis and Replacement

Worn **stabilizer bar** mounting bushings, grommets, or mounting bolts cause a rattling noise as the vehicle is driven on irregular road surfaces. A weak stabilizer bar or worn bushings and grommets cause harsh riding and excessive body sway while driving on irregular road surfaces. Worn or very dry stabilizer bar bushings may cause a squeaking noise on irregular road surfaces. All stabilizer bar components should be visually inspected for wear. Stabilizer bar removal and replacement procedures vary depending on the vehicle. Always follow the vehicle manufacturer's recommended procedure in the service manual. The following is a typical stabilizer bar removal and replacement procedure:

A stabilizer bar may be referred to as a sway bar.

1. Lift the vehicle on a hoist and allow both sides of the front suspension to drop downward as the vehicle chassis is supported on the hoist.

2. Remove the mounting bolts at the outer ends of the stabilizer bar and remove the bushings, grommets, brackets, or spacers (Figure 6-46).

3. Remove the mounting bolts in the center area of the stabilizer bar.

4. Remove the stabilizer bar from the chassis.

5. Visually inspect all stabilizer bar components such as bushings, bolts, and spacer sleeves. Replace the stabilizer bar, grommets, bushings, brackets, or spacers as

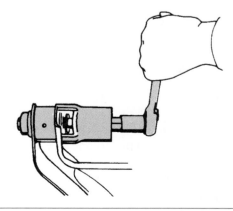

Figure 6-44 Rear control arm bushing removal. (Courtesy of Chevrolet Motor Division, General Motors Corp.)

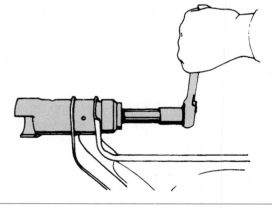

Figure 6-45 Rear control arm bushing installation. (Courtesy of Chevrolet Motor Division, General Motors Corp.)

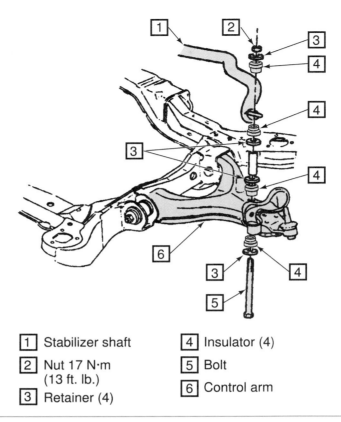

1	Stabilizer shaft	4	Insulator (4)
2	Nut 17 N·m (13 ft. lb.)	5	Bolt
3	Retainer (4)	6	Control arm

Figure 6-46 Stabilizer bar components. (Courtesy of Oldsmobile Motor Division, General Motors Corp.)

required. Split bushings may be removed over the stabilizer bar. Bushings that are not split must be pulled from the bar.

6. Reverse steps 2 through 4 to install the stabilizer bar. Make sure all stabilizer bar components are installed in the original position, and tighten all fasteners to the specified torque.

Some vehicle manufacturers specify that stabilizer bars must have equal distances between the outer bar ends and the lower control arms. Always refer to the manufacturer's recommended measurement procedure. If this measurement is required, adjust the nut on the outer stabilizer bar mounting bolt until equal distances are obtained between the outer bar ends and the lower control arms. Worn grommets cause these distances to be unequal.

Strut Rod Diagnosis and Replacement

Some front suspension systems have a **strut rod** connected from the lower control arm to the frame. Rubber grommets isolate the rod from the chassis components. Worn strut rod grommets may allow the lower control arm to move rearward or forward. This movement changes the caster angle, which may affect steering quality and cause the vehicle to pull to one side. A bent strut rod also causes steering pull. Worn strut rod grommets or loose mounting bolts cause a rattling noise while driving on irregular road surfaces. Inspect the strut rod grommets visually for wear and deterioration. With the vehicle lifted on a hoist, grasp the strut rod firmly and apply vertical and horizontal force to check the rod and grommets for movement. Worn grommets must be replaced. If a strut rod is bent, replace the rod.

Follow these steps for strut rod replacement:

1. Lift the vehicle on a hoist.

2. Remove the strut rod nut from the front end of the rod.

3. Remove the strut rod bolts from the lower control arm.

4. Pull the strut rod rearward to remove the rod (Figure 6-47).

5. Remove the bushings from the opening in the chassis.

6. Visually inspect the strut rod, bushings, washers, and retaining bolts. Replace all worn parts. Reverse steps 1 through 5 to reinstall the strut rod. Tighten the strut rod nut and bolts to the specified torque.

Since strut rod and bushing conditions affect front suspension alignment, check front suspension alignment after strut rod service.

Front Leaf Spring Inspection and Replacement

Front leaf springs are used in some truck suspension systems. These springs are mounted longitudinally with a spring on each side of the suspension. Many leaf springs have plastic silencers between the spring leafs. If these silencers are worn out, creaking and squawking noises will be heard when the vehicle is driven over road irregularities at low speeds.

When the silencers require checking or replacement, lift the vehicle with a floor jack and support the frame on safety stands so the suspension hangs downward. With the vehicle weight no longer applied to the springs, the spring leafs may be pried apart with a pry bar to remove and replace the silencers. On some leaf springs, the clamps must be removed to replace the silencers.

Leaf springs should be inspected for a sagged condition, which causes the curb riding height to be less than specified. Leaf springs should also be visually inspected for broken leafs, broken center bolts, and worn shackles or bushings. Weak or broken leaf springs affect front suspension alignment angles and cause excessive tire tread wear, reduced directional stability, and harsh riding. A rattling noise while driving on irregular road surfaces may be caused by worn shackles or bushings. A broken center bolt usually allows one side of the front suspension to move rearward in relation to the other side. This action may cause steering pull to one side. Worn shackles and bushings lower

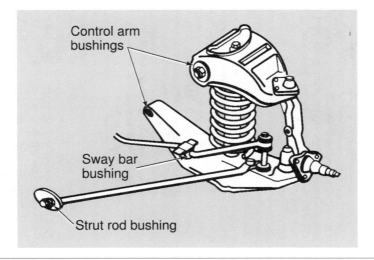

Figure 6-47 Strut rod. (Courtesy of Sealed Power Corp.)

curb riding height and cause a rattling noise on road irregularities. The following is a typical front leaf spring replacement procedure:

1. Lift the front end of the vehicle with a floor jack under the front axle and place safety stands under the frame. Lower the vehicle weight onto the safety stands, but leave the floor jack under the front axle to support some of the front suspension weight.

2. Remove the nuts from the spring U-bolts, and remove the U-bolts and lower spring plate. If the spring plate is attached to the shock absorber, this plate may be left on the shock absorber and moved out of the way. If shims are positioned between the spring plate and the front axle, be sure to note the position and the number of shims. These shims must be reinstalled in their original position because they set the front wheel caster angle.

3. Be sure the floor jack is lowered enough to relieve the vehicle weight from the springs.

4. Remove the front shackle assembly.

5. Remove the rear spring mounting bolt, and remove the spring from the chassis (Figure 6-48).

6. Check all the spring hangers, bolts, bushings, and shackle plates for wear, and replace as required. Springs with broken or sagged leafs are usually sent to a spring rebuilding shop for repair.

7. Check the spring center bolt to be sure it is not broken.

8. Reverse steps 1 through 5 to install the front springs. Tighten all bolts to the manufacturer's specifications.

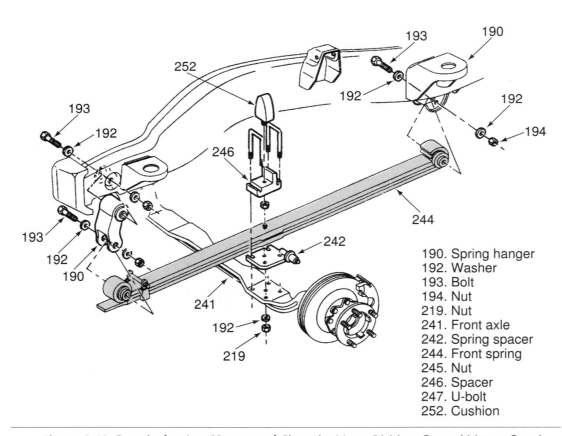

190. Spring hanger
192. Washer
193. Bolt
194. Nut
219. Nut
241. Front axle
242. Spring spacer
244. Front spring
245. Nut
246. Spacer
247. U-bolt
252. Cushion

Figure 6-48 Front leaf spring. (Courtesy of Chevrolet Motor Division, General Motors Corp.)

Removing and Replacing Longitudinally Mounted Torsion Bars

Worn torsion bar components such as pivot cushion bushings and control arm bushings result in harsh riding and suspension noise when driving on road irregularities. A worn torsion bar hex and anchor may cause a rattling noise while driving on irregular road surfaces. Weak torsion bars or those with worn bushings or anchors cause reduced curb riding height, which may result in reduced directional stability and excessive tire tread wear. Always follow the vehicle manufacturer's recommended procedure in the service manual for torsion bar removal and replacement. The following is a typical torsion bar removal and replacement procedure:

CAUTION: Some torsion bar front suspension systems on sport utility vehicles (SUVs) are combined with an air suspension system. When servicing these systems, the air suspension switch in the rear jack storage area must be shut off before performing any suspension service or hoisting, jacking, or towing the vehicle. If this procedure is not followed, personal injury and vehicle damage may occur.

1. Raise the vehicle on a lift with the tires supported on the lift.
2. Remove the torsion bar cover plate bolts and the cover plate (Figure 6-49).
3. Measure and record the distance from the lip on the head of the torsion bar adjusting bolt to the casting surface this bolt is threaded into (Figure 6-50). When reinstalled, this bolt must be adjusted to this same measurement.
4. Install the torsion bar tool and adapters (Figure 6-51). Tighten this tool until the torsion bar adjuster lifts off the adjustment bolt.

WARNING: The torsion bar adjusting bolt is coated with a dry adhesive. This bolt must be replaced if it is backed off or removed. Failure to follow this procedure may cause the bolt to loosen during operation, resulting in improper wheel alignment, reduced steering control, increased tire wear, and reduced ride quality.

5. Remove the torsion bar adjustment bolt and nut.
6. Loosen the torsion bar tool to remove all the tension from the torsion bar (Figure 6-52).
7. Use the end of a screwdriver to place matching alignment marks on the torsion bar and the adjuster so these components may be reassembled in the same position. Remove the torsion bar insulator. Pull the torsion bar to the rear to remove this bar from the lower control arm.

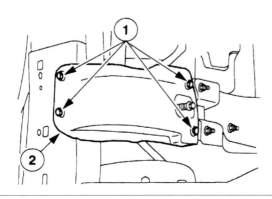

Figure 6-49 Torsion bar cover plate. (Courtesy of Ford Motor Company)

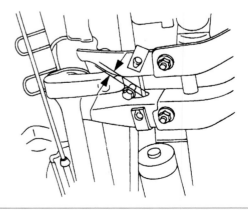

Figure 6-50 Measuring the torsion bar adjusting bolt position. (Courtesy of Ford Motor Company)

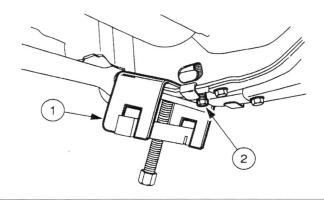

Figure 6-51 Torsion bar tool and adapters. (Courtesy of Ford Motor Company)

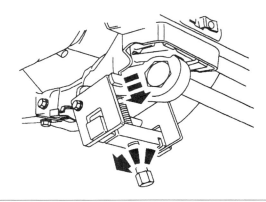

Figure 6-52 Loosening the torsion bar tool. (Courtesy of Ford Motor Company)

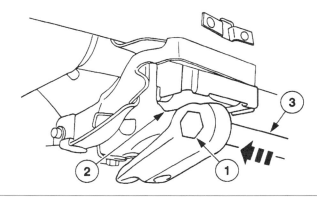

Figure 6-53 Alignment marks placed on the torsion bar and adjuster. (Courtesy of Ford Motor Company)

8. Position the torsion bar in the lower control arm.

9. Install the torsion bar adjuster on the torsion bar with the alignment marks properly aligned (Figure 6-53).

10. Install the torsion bar tool and adapters. Tighten the torsion bar tool until a new adjustment bolt can be installed. Tighten this bolt until the measurement obtained in step 3 is obtained between the head of the bolt and the casting the bolt is threaded into.

11. Install the torsion bar cover plate, and tighten the plate retaining bolts to the specified torque.

12. Measure the suspension riding height as explained previously this chapter. If the suspension riding height is not within specifications, adjust the torsion bar adjusting bolt to obtain the specified riding height.

Guidelines for Servicing Front Suspension Systems

1. The curb riding height measurement location varies depending on the vehicle. Always measure this height at the manufacturer's recommended location shown in the service manual.

2. The curb riding height is critical because it affects most other front suspension alignment angles.

3. On a torsion bar front suspension, the torsion bars may be adjusted to correct improper curb riding height.

4. Some ball joints have wear indicators, and on these applications the grease fitting position indicates ball joint condition.

5. The spring tension must be unloaded from a ball joint before the joint wear is measured.

6. Steering knuckles should always be checked for wear and out-of-round in the ball joint and tie-rod end openings.

7. Ball joints may be pressed, threaded, bolted, or riveted to the control arm.

8. When installing a cotter pin in a ball joint or tie-rod end nut, never back off the nut to install the cotter pin, and always use a new cotter pin.

9. After replacement of front suspension components such as control arms, control arm bushings, springs, and strut rods, front suspension alignment should be checked.

10. Bent control arms or worn control arm bushings affect curb riding height and front suspension alignment angles.

11. A weak stabilizer bar or worn stabilizer bar bushings causes harsh riding, excessive body sway, and suspension noise.

12. Excessive wear on rebound bumpers may be caused by improper curb riding height or worn-out shock absorbers.

13. Some control arm bushings require flaring after installation to secure the bushing.

14. Bent strut rods or worn strut rod bushings cause improper front suspension alignment and reduced directional stability.

15. Worn strut rod bushings or loose mounting bolts cause suspension noise.

16. Squeaking front leaf springs may be caused by worn-out silencers between the spring leafs.

CASE STUDY

A customer complained about steering pull to the left while braking on a 1993 Oldsmobile Toronado. The technician questioned the customer about other symptoms, but the customer stated that the steering did not pull while driving, and the car had no other problems. Further questioning of the customer revealed that extensive brake work had been done in an attempt to correct the problem. The front brake pads had been replaced and the front brake rotors had been turned.

While performing a road test, the technician discovered the car had a definite pull to the left while braking, but the steering was normal while driving. During the road test, no other problems were evident. The technician removed the front wheels and brake rotors. A careful examination of the brake linings and rotor surfaces indicated these components were in excellent condition. The technician checked the rotor surfaces to be sure they were both in the same condition. The pistons in both front calipers moved freely. A check of the brake lines and hoses did not reveal any visible problems, and a pressure check at each front brake caliper indicated equal pressures at both front wheels during a brake application.

Next, the technician made a visual check of the steering and suspension components, and discovered a worn, loose left front strut rod bushing. This bushing was replaced, and the strut rod nut tightened to the specified torque. The strut rod was in satisfactory condition. A road test of the car indicated no steering pull during brake application.

Terms to Know

Ball joint radial movement

Ball joint vertical movement

Ball joint wear indicator

Camber angles

Caster angles

Curb riding height

Directional stability

Rebound bumper

Stabilizer bar

Steering effort

Strut rod

ASE Style Review Questions

1. While discussing curb riding height:
   Technician A says worn control arm bushings reduce curb riding height.
   Technician B says incorrect curb riding height affects most other front suspension angles.
   Who is correct?
   A. A only
   B. B only
   C. Both A and B
   D. Neither A nor B

2. The shoulder on the ball joint grease fitting is inside the ball joint cover:
   Technician A says the ball joint should be replaced.
   Technician B says a longer grease fitting should be installed.
   Who is correct?
   A. A only
   B. B only
   C. Both A and B
   D. Neither A nor B

3. While discussing ball joint unloading on a short-and-long arm suspension system with the coil springs between the lower control arm and the chassis:
   Technician A says a steel spacer should be installed under the upper control arm.
   Technician B says a floor jack should be placed under the lower control arm to unload the ball joints.
   Who is correct?
   A. A only
   B. B only
   C. Both A and B
   D. Neither A nor B

4. While discussing ball joint radial measurement:
   Technician A says the dial indicator should be positioned against the top of the ball joint stud.
   Technician B says the front wheel bearing adjustment does not affect the ball joint radial measurement.
   Who is correct?
   A. A only
   B. B only
   C. Both A and B
   D. Neither A nor B

5. While discussing ball joint installation:
   Technician A says the ball joint nut may be backed off to install the cotter pin.
   Technician B says only the ball joint threads should extend above the opening in the steering knuckle.
   Who is correct?
   A. A only
   B. B only
   C. Both A and B
   D. Neither A nor B

6. When installing a new threaded ball joint in a lower control arm, the technician can only torque the ball joint to 90 ft-lbs. The necessary repair for this problem is to:
   A. weld the ball joint into the control arm.
   B. place Loctite on the ball joint threads.
   C. replace the lower control arm.
   D. install a larger diameter ball joint.

7. All of these defects could result in worn-out rebound bumpers EXCEPT:
   A. sagged springs.
   B. worn-out shock absorbers.
   C. continual driving on rough roads.
   D. curb riding height more than specified.

8. When removing and replacing control arm bushings:
   A. if the bushing is contained in a steel sleeve, press on the rubber bushing.
   B. a spacer should be installed between the control arm support lugs during bushing removal and installation.
   C. the same tool is used for bushing removal and replacement.
   D. bushing flaring is necessary to expand the bushing.

9. A car experiences excessive body sway when cornering, but there is no abnormal noise in the suspension. The most likely cause of this problem is:
   A. stabilizer bar bushing is missing.
   B. a weak stabilizer bar.
   C. stabilizer bar grommets worn out.
   D. a broken stabilizer bar.

10. The steering on a vehicle pulls to the left while driving straight ahead. The most likely cause of this problem is:
    A. the left front strut rod is bent.
    B. the left lower ball joint is worn.
    C. worn front stabilizer bushings on the left side.
    D. reduced curb riding height on both sides of the front suspension.

ASE Challenge Questions

1. The customer says her 1987 Blazer pulls to the left. A cursory check of the vehicle shows that the right front tire height is too low. The next step to diagnose the condition would be:
   A. check the ball joints.
   B. check the torsion bar.
   C. check strut rod bushings.
   D. check stabilizer bar bushings.

2. A customer says she drives her front-wheel-drive car has a steering wander problem on irregular road surfaces. All of the following could cause this problem EXCEPT:
   A. worn stabilizer bar bushings.
   B. worn ball joints.
   C. worn strut rod bushings.
   D. worn tie-rod ends.

3. While diagnosing a customer's complaint of steering instability on a MacPherson strut front suspension with wear indicator type ball joints, an inspection of the ball joints shows the grease fittings to be solid.
   Technician A says no movement of the fitting means the ball joint is good.
   Technician B says ball joint wear may not be apparent with the weight on the joint.
   Who is correct?
   A. A only
   B. B only
   C. Both A nad B
   D. Neither A nor B

4. A customer says her rear-wheel-drive car has excessive body sway while cornering. In diagnosing this problem, which of the following components should you check first?
   A. Control arm bushings
   B. Strut rod bushings
   C. Stabilizer bar bushings
   D. Shock absorber bushings

5. *Technician A* says a bent strut rod can cause a car to pull to one side.
   Technician B says a deteriorated strut rod bushing can cause steering and braking problems.
   Who is correct?
   A. A only
   B. B only
   C. Both A nad B
   D. Neither A nor B

Job Sheet 17

Name _____ Date _____

Measure Lower Ball Joint Vertical and Radial Movement, Short-and-Long Arm Suspension Systems

NATEF and ASE Correlation

This job sheet is related to NATEF and ASE Automotive Suspension and Steering Task List content area: B. Suspension System Diagnosis and Repair, 1. Front Suspensions, Task 4: Inspect and replace upper and lower ball joints (with or without wear indicators).

Tools and Materials

Ball joint dial indicator
Floor jack
Safety stands

Vehicle make and model year _____

Vehicle VIN _____

Procedure

1. Raise the front of the vehicle with a floor jack lift pad is positioned on the specified lifting point.

2. Install safety stands near the outer ends of the lower control arms. Lower the floor jack so the control arms are supported on the safety stands. Remove the floor jack.

3. Attach a dial indicator for ball joint measurement to the lower control arm, and position the dial indicator stem against the lower end of the steering knuckle next to the ball joint retaining nut if the suspension has a lower compression-loaded ball joint. When the suspension has a tension-loaded ball joint, place the indicator stem against the top of the ball joint stud.

   Type of ball joint _____

   Dial indicator stem position _____

4. Preload the dial indicator stem 0.250 in (6.35 mm), and zero the dial indicator.

   Is the dial indicator preloaded and placed in the zero position? yes _____ no _____

   Instructor check _____

5. Place a pry bar under the front tire and lift straight upward on the pry bar while a coworker observes the dial indicator.

6. Compare the reading on the dial indicator to the vehicle manufacturer's specifications. If the vertical ball joint movement exceeds specifications, replace the ball joint.

   Specified ball joint vertical movement_____

   Actual ball joint vertical movement _____

   Necessary ball joint service_____

7. Be sure the front wheel bearings are properly adjusted.

8. Attach the dial indicator to the lower control arm, and position the dial indicator against the inner edge of the wheel rim. Preload the dial indicator 0.250 in (6.35 mm) and place the dial in the zero position.

   Is the dial indicator preloaded and placed in the zero position? yes _____ no _____

   Instructor check _____

9. Grasp the tire at the top and bottom, and try to rock the tire inward and outward while a coworker observes the dial indicator.

10. Compare the reading on the dial indicator to the vehicle manufacturer's specifications. If the radial ball joint movement exceeds specifications, replace the ball joint.

    Specified ball joint radial movement _____

    Actual ball joint radial movement _____

11. Repeat the measurements in steps 3 through 10 on the opposite side of the front suspension.

12. Based on your ball joint measurements, state all the necessary ball joint service and explain the reasons for your diagnosis. _____

_____

_____

_____

☑ Instructor Check _____

Job Sheet 18

18

Name _____ Date _____

Ball Joint Replacement

NATEF and ASE Correlation

This job sheet is related to NATEF and ASE Automotive Suspension and Steering Task List content area: B. Suspension System Diagnosis and Repair, 1. Front Suspensions, Task 4: Inspect and replace upper and lower ball joints (with or without wear indicators).

Tools and Materials

Floor jack
Safety stands
Ball joint loosening tool
Torque wrench

Vehicle make and model year_____
Vehicle VIN_____

Procedure

1. Remove the wheel cover and loosen the wheel nuts.

2. Lift the vehicle with a floor jack and place safety stands under the chassis so the front suspension is allowed to drop downward. Lower the vehicle onto the safety stands and remove the floor jack.

3. Remove the wheel and place a floor jack under the outer end of the lower control arm. Operate the floor jack and raise the lower control arm until the ball joints are unloaded.

   Are the ball joints properly unloaded? yes _____ no _____

   Instructor check _____

4. Remove other components, such as the brake caliper, rotor, and drive axle as required to gain access to the ball joints.

5. Remove the cotter pin in the ball joint or joints that require replacement, and loosen, but do not remove, the ball joint stud nuts.

   Is the ball joint stud nut loosened? yes _____ no _____

   Instructor check _____

6. Loosen the ball joint stud tapers in the steering knuckle. A threaded expansion tool is available for this purpose.

   Are the ball joint stud tapers loosened? yes _____ no _____

   Instructor check _____

7. Remove the ball joint nut and lift the knuckle off the ball joint stud. Block or tie up the knuckle and hub assembly to access the ball joint.

8. If the ball joint is riveted to the control arm, drill and punch out the rivets and bolt the new ball joint to the control arm.

9. If the ball joint is pressed into the lower control arm, remove the ball joint dust boot and use a pressing tool to remove and replace the ball joint.

10. If the ball joint housing is threaded into the control arm, use the proper size socket to remove and install the ball joint. The replacement ball joint must be torqued to the manufacturer's specifications. If a minimum of 125 foot-pounds of torque cannot be obtained, the control arm threads are damaged and control arm replacement is necessary.

Ball joint torque, threaded ball joint _____

Control arm condition, threaded ball joint:
satisfactory _____ unsatisfactory _____

11. If the ball joint is bolted to the lower control arm, install the new ball joint and tighten the bolt and nuts to the specified torque.

Ball joint bolt torque _____

12. Clean and inspect the ball joint stud tapered opening in the steering knuckle. If this opening is out-of-round or damaged, the knuckle must be replaced.

Condition of ball joint stud opening in the steering knuckle:

satisfactory _____ unsatisfactory _____

13. Check the fit of the ball joint stud in the steering knuckle opening. This stud should fit snugly in the opening and only the threads on the stud should extend through the knuckle. If the ball joint stud fits loosely in the knuckle tapered opening, either this opening is worn or the wrong ball joint has been supplied.

Ball joint stud smooth tapered area appearing above steering knuckle surface:

yes _____ no _____

If the answer to this question is yes, state the necessary repairs and explain the reason for your diagnosis.

_____

_____

Ball joint stud fit in the steering knuckle opening:

satisfactory _____ unsatisfactory _____

If the ball joint stud fit in the steering knuckle opening is unsatisfactory, state the necessary repairs and explain the reason for your diagnosis.

_____

_____

CAUTION: Never back off a ball joint stud to align the cotter pin openings in the nut and stud. This may cause the ball joint stud to loosen, resulting in a suspension failure. Always tighten the nut to the next hole to install the cotter pin.

14. Install the ball joint stud in the steering knuckle opening, making sure the stud is straight and centered. Install the stud nut and tighten this nut to the specified torque. Install a new cotter pin through the stud and nut. Do not loosen the nut to align the nut and stud openings.

Specified ball joint nut torque _____

Actual ball joint nut torque _____

Is a new cotter pin properly installed? yes _____ no _____

Instructor check _____

15. Reassemble the components that were removed in step 4. Make sure the wheel nuts are tightened to the specified torque.

Specified wheel nut torque _____

Actual wheel nut torque _____

✔ **Instructor Check** _____

Job Sheet 19

19

Name _____ Date _____

Steering Knuckle Removal, MacPherson Strut Front Suspension

NATEF and ASE Correlation

This job sheet is related to NATEF and ASE Automotive Suspension and Steering Task List content area: B. Suspension System Diagnosis and Repair, 1. Front Suspensions, Task 5: Inspect and replace steering knuckle/spindle assemblies and steering arms.

Tools and Materials

Floor jack
Safety stands
Tire-rod end puller
Torque wrench
Driving tools

Vehicle make and model year_____

Vehicle VIN_____

Procedure

1. Remove the wheel cover and loosen the front wheel nuts and the drive axle nut.

2. Lift the vehicle chassis on a hoist and allow the front suspension to drop downward. Remove the front wheel, brake caliper, brake rotor, and drive axle nut. Tie the brake caliper to a suspension component; do not allow the caliper to hang on the end of the brake hose.

   Is the vehicle weight supported on hoist with front suspension dropped downward?

   yes _____ no _____

   Instructor check _____

3. Remove the inner end of the drive axle from the transaxle with a pulling or prying action. On some Chrysler products, the differential cover must be removed and axle circlips compressed prior to drive axle removal from the transaxle.

   Is the inner end of drive axle removed from transaxle? yes _____ no _____

   Instructor check _____

4. Remove the outer end of the drive axle from the steering knuckle and hub. On some early model Ford Escorts and Lynx, a puller is required for this operation.

   Is the outer end of drive axle removed from steering knuckle and hub? yes _____ no _____

   Instructor check _____

5. Be sure the vehicle weight is supported on the hoist with the front suspension dropped downward. the mounting bolts in the lower control arm.

   Are the lower ball joint retaining nuts removed? yes _____ no _____

   Instructor check _____

6. Remove the cotter pin from the tie-rod nut. Remove the outer tie-rod nut and use a puller to disconnect the tie-rod end from the steering knuckle.

Is the outer tie-rod end removed? yes _____ no _____

Instructor check _____

7. If an eccentric cam is used on one of the strut to knuckle bolts, mark the cam and bolt position in relation to the strut and remove the strut to knuckle bolts.

Is the eccentric camber marked in relation to strut? yes _____ no _____

Instructor check _____

8. Remove the knuckle from the strut and lift the knuckle out of the chassis.

9. Pry the dust deflector from the steering knuckle with a large flat-blade screwdriver.

10. Use a puller to remove the ball joint from the steering knuckle. Check the ball joint and tie-rod end openings in the knuckle for wear and out-of-round. Replace the knuckle if these openings are worn or out-of-round.

Condition of ball joint opening in knuckle: satisfactory _____

unsatisfactory _____

Condition of tie-rod end opening in the knuckle:

satisfactory _____ unsatisfactory _____

State the necessary steering knuckle and related repairs, and explain the reason for your diagnosis. _____

_____

_____

11. Use the proper driving tool to reinstall the dust deflector in the steering knuckle.

✓ **Instructor Check** _____

Table 6-1 NATEF and ASE TASK

Diagnose front suspension system noises, body sway/roll, and ride height problems; determine needed repairs.

Problem Area	Symptoms	Possible Causes	Classroom Manual	Shop Manual
NOISES	Rattles on road irregularities	**1.** Worn control arm bushings	105	155
		2. Worn stabilizer bar bushings	116	173
		3. Worn strut rod grommets	114	174
		4. Worn leaf spring shackles and bushings	135	177
		5. Worn torsion bars, anchors, and bushings	131	167
		6. Broken springs or spring insulators	116	173
BODY SWAY	Excessive sway on road irregularities	Weak stabilizer bar, worn bushings	116	155
RIDE HEIGHT PROBLEMS	Steering pulls to one side	Low ride height, one side of front suspension	138	155
	Reduced directional stability	Low ride height on front suspension	138	155
	Increased steering effort	Low ride height on rear suspension	139	155
	Rapid steering wheel return	Low ride height on rear suspension	139	155

Table 6-2 NATEF and ASE TASK

Measure vehicle ride height; determine needed repairs.

Problem Area	Symptoms	Possible Causes	Classroom Manual	Shop Manual
RIDE QUALITY	Harsh riding	Low ride height	138	155
DIRECTIONAL STABILITY	Steering pull	Low ride height one side of front suspension	138	155
STEERING QUALITY	Reduced directional stability	Low ride height on front suspension	139	155
	Increased steering effort	Low ride height on rear suspension	140	155
	Rapid steering wheel return	Low ride height on rear suspension	140	155

Table 6-3 NATEF and ASE TASK

Inspect and replace upper and lower control arms, bushings, shafts, and rebound bumpers.

Problem Area	Symptoms	Possible Causes	Classroom Manual	Shop Manual
NOISES	Rattles on road irregularities	1. Worn control arm bushings or shafts	116	167
		2. Worn strut rod grommets	116	167
		3. Worn stabilizer bar bushings	117	173
	Squeaks on road irregularities	1. Dry or worn control arm bushings	118	169
		2. Dry or worn stabilizer bushings	119	173
RIDE QUALITY	Harsh riding	Reduced curb riding height	116	155
STEERING QUALITY	Reduced directional stability	1. Worn control arm bushings	116	169
		2. Reduced curb riding height	117	155
TIRE LIFE	Tire tread wear	Worn control arm bushings	119	169

Table 6-4 NATEF and ASE TASK

Inspect, adjust, and replace strut rods/radius arm (compression/tension), and bushings.

Problem Area	Symptoms	Possible Causes	Classroom Manual	Shop Manual
NOISE	Rattles on road irregularities	Worn strut rod grommets	114	174
STEERING QUALITY	Steering pull	1. Worn strut rod bushings	114	174
		2. Bent strut rod	114	174

Table 6-5 NATEF and ASE TASK

Inspect and replace upper and lower ball joints (with or without wear indicators).

Problem Area	Symptoms	Possible Causes	Classroom Manual	Shop Manual
STEERING QUALITY	Reduced directional stability	Worn ball joints	111	156
TIRE LIFE	Tire tread wear	Worn ball joints	111	156

Table 6-6 NATEF and ASE TASK

Inspect and replace steering knuckle/spindle assemblies and steering arms.

Problem Area	Symptoms	Possible Causes	Classroom Manual	Shop Manual
STEERING QUALITY	Steering looseness	Worn tie-rod end opening in knuckle	117	161
	Reduced directional stability	**1.** Worn ball joint stud opening in knuckle	117	161
		2. Bent steering knuckle	117	161
TIRE LIFE	Tire tread wear	Bent steering knuckle	117	161

Table 6-7 NATEF and ASE TASK

Inspect and replace front suspension system coil springs and spring insulators (silencers).

Problem Area	Symptoms	Possible Causes	Classroom Manual	Shop Manual
NOISE	Rattling on road irregularities	**1.** Broken coil springs **2.** Worn or broken spring insulators	104 105	167 168
STEERING QUALITY	Reduced directional stability	Weak or broken coil springs	105	169
TIRE LIFE	Tire tread wear	Weak or broken coil springs	105	170
RIDE QUALITY	Harsh riding	Weak or broken spring	106	171

Table 6-8 NATEF and ASE TASK

Inspect and replace front suspension system leaf spring(s), leaf spring insulators (silencers), shackles, brackets, bushings, and mounts.

Problem Area	Symptoms	Possible Causes	Classroom Manual	Shop Manual
NOISE	Squeaking on road irregularities	Worn leaf spring silencers	108	175
	Rattling on road irregularities	**1.** Broken spring leafs **2.** Worn shackles, bushings, mounts	108 108	176 176
STEERING QUALITY	Reduced directional stability	**1.** Weak or broken spring **2.** Broken spring center bolt	108 108	176 176
TIRE LIFE	Tire tread wear	Weak or broken spring	108	176
RIDE QUALITY	Harsh riding	Weak or broken spring	108	176

Table 6-9 NATEF and ASE TASK

Inspect, replace, and adjust front suspension system torsion bars; inspect mounts.

Problem Area	Symptoms	Possible Causes	Classroom Manual	Shop Manual
RIDE QUALITY	Harsh riding	Reduced curb riding height	121	177
NOISE	Rattling on road irregularities	1. Worn torsion bar pivot bushings	121	178
		2. Worn torsion bar hex and anchor	121	178

Table 6-10 NATEF and ASE TASK

Inspect and replace stabilizer bar (sway bar) bushings, brackets, and links.

Problem Area	Symptoms	Possible Causes	Classroom Manual	Shop Manual
RIDE QUALITY	Excessive body sway	1. Weak or broken torsion bar	117	173
		2. Worn torsion bar bushings	117	173
NOISE	Rattling on road irregularities	Worn torsion bar bushings, links	117	173

Rear Suspension Service

Upon completion and review of this chapter, you should be able to:

- ❏ Diagnose rear suspension noises.
- ❏ Diagnose rear suspension sway and lateral movement.
- ❏ Measure and correct rear suspension curb riding height.
- ❏ Remove and replace rear coil springs.
- ❏ Inspect rear springs, insulators, and seats.
- ❏ Inspect strut or shock absorber bushings and upper mount, replace strut cartridge.
- ❏ Remove, inspect, and replace lower control arms.
- ❏ Inspect, remove, and replace rear ball joints.
- ❏ Inspect, remove, and replace suspension adjustment links.
- ❏ Diagnose, remove, and replace rear leaf springs.
- ❏ Diagnose, remove, and replace stabilizer bars.
- ❏ Diagnose, remove, and replace track bars.
- ❏ Inspect, remove, and replace tie rods.

Rear Suspension Diagnosis and Service

Noise Diagnosis

A squeaking noise in the rear suspension may be caused by a suspension bushing or a defective strut or shock absorber. If a rattling noise occurs in the rear suspension, check these components:

1. Worn or missing suspension bushings, such as control arm bushings, track bar bushings, stabilizer bar bushings, trailing arm bushings, and strut rod bushings
2. Worn strut or shock absorber bushings or mounts
3. Defective struts or shock absorbers
4. Broken springs or worn spring insulators

Sway and Lateral Movement Diagnosis

Excessive **body sway** or roll on road irregularities may be caused by a weak stabilizer bar or loose stabilizer bar bushings. If **lateral movement** is experienced on the rear of the chassis, the track bar or track bar bushings may be defective. Lateral movement is sideways movement.

Curb Riding Height Measurement

 SERVICE TIP: Proper curb riding height must be maintained to provide normal steering quality and tire wear.

Regular inspection and proper maintenance of suspension systems are extremely important to maintaining vehicle safety. The **curb riding height** is determined mainly by spring condition. Other suspension components, such as control arm bushings, affect curb riding height if they are worn. *Since incorrect curb riding height affects most of the other suspension angles, this measurement is critical.* Reduced rear suspension height increases steering effort and causes rapid steering wheel return after turning a corner. Harsh riding occurs when the curb riding height is less than specified. The curb riding height must be measured at the vehicle manufacturer's specified location, which varies depending on the type of vehicle and suspension system. On some vehicles, the curb riding height on the rear suspension is measured from the floor to the center of the strut rod mounting bolt when the vehicle is on a level floor or an alignment rack (Figure 7-1). Photo Sequence 5 shows a typical procedure for measuring front and rear curb riding height.

Basic Tools

Basic technician's tool set

Service manual

Machinist's rule

Floor jack

Safety stands

Transmission jack

Foot-pound torque wrench

Pry bar

Photo Sequence 5
Typical Procedure for Measuring Front and Rear Curb Riding Height

P5-1 Check the trunk for extra weight.

P5-2 Check the tires for normal inflation pressure.

P5-3 Park the car on a level shop floor or alignment track.

P5-4 Find the vehicle manufacturer's specified curb riding height measurement locations in the service manual.

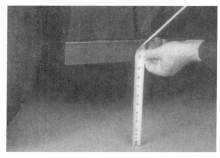

P5-5 Measure and record the right front curb riding height.

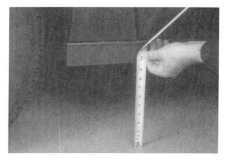

P5-6 Meaure and record the left front curb riding height.

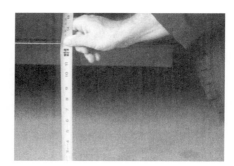

P5-7 Measure and record the right rear curb riding height.

P5-8 Measure and record the left rear curb riding height.

P5-9 Compare the measurement results to the specified curb riding height in the service manual.

Rear

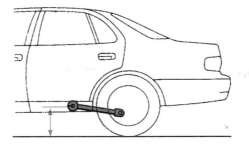

Figure 7-1 Curb riding height measurement, rear suspension. (Reprinted with permission)

Rear Strut, Coil Spring, and Upper Mount Service and Diagnosis

> **CAUTION:** When the rear coil springs are mounted on the struts, never loosen the upper strut nut until all the spring tension is removed from the upper support with a spring compressor. If this nut is removed with spring tension on the upper support, the spring tension turns the spring and the upper support into a dangerous projectile, which may cause serious personal injury and/or property damage.

Weak springs cause harsh riding and reduced curb riding height. Broken springs or spring insulators cause a rattling noise while driving on road irregularities. Worn-out struts or shock absorbers cause excessive chassis oscillations and harsh riding. (Refer to Chapter 5 for strut and shock absorber service.) Loose or worn strut or shock absorber bushings cause a rattling noise on road irregularities. The rear coil spring removal and replacement procedure varies depending on the type of rear suspension system. Always follow the vehicle manufacturer's recommended procedure in the service manual. The following is a typical rear strut and spring removal and replacement procedure on a MacPherson strut independent rear suspension system with the coil springs mounted on the struts:

1. Remove the rear seat and the package trim tray (Figure 7-2).

2. Remove the wheel cover and loosen the wheel nuts.

3. Lift the vehicle with a floor jack, and lower the chassis onto safety stands so the rear suspension is allowed to drop downward.

4. Place a wooden block between the floor jack and the rear spindle on the side where the strut and spring removal is taking place. Raise the floor jack to support some of the suspension system weight (Figure 7-3).

5. Remove the rear wheel, disconnect the nut from the small spring in the lower arm (Figure 7-4), and remove the brake hose and antilock brake system (ABS) wire from the strut (Figure 7-5).

6. Remove the stabilizer bar link from the strut (Figure 7-6), and loosen the strut-to-spindle mounting bolts (Figure 7-7).

7. Remove three upper support nuts under the package tray trim (Figure 7-8), and lower the floor jack to remove the strut from the knuckle (Figure 7-9). Remove the strut from the chassis.

> **SERVICE TIP:** If the plastic coating on a coil spring is chipped, the spring may break prematurely. The spring may be taped in the compressing tool contact areas to prevent chipping.

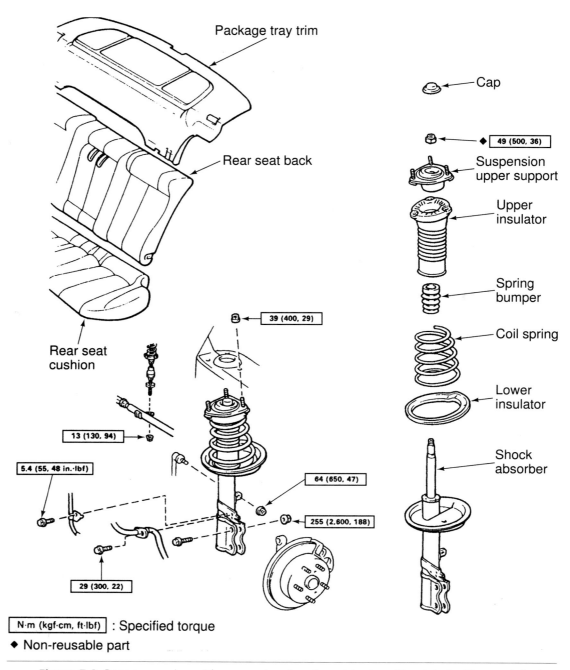

Package tray trim

Rear seat back

Rear seat cushion

Cap

49 (500, 36)

Suspension upper support

Upper insulator

Spring bumper

Coil spring

Lower insulator

Shock absorber

39 (400, 29)

13 (130, 94)

5.4 (55, 48 in.·lbf)

64 (650, 47)

255 (2,600, 188)

29 (300, 22)

N·m (kgf·cm, ft·lbf) : Specified torque

◆ Non-reusable part

Figure 7-2 Rear suspension with rear seat and package trim tray. (Reprinted with permission)

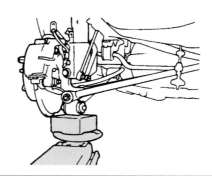

Figure 7-3 Floor jack supporting some of the rear suspension weight. (Reprinted with permission)

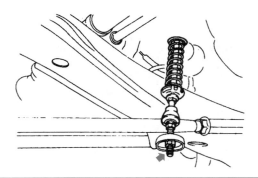

Figure 7-4 Disconnecting nut from small spring in lower arm. (Reprinted with permission)

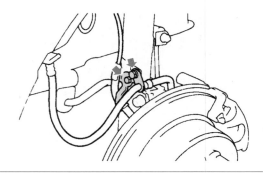

Figure 7-5 Disconnecting brake hose and ABS wire from rear strut. (Reprinted with permission)

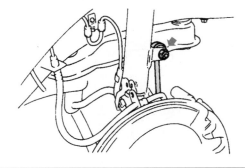

Figure 7-6 Removing stabilizer bar from rear strut. (Reprinted with permission)

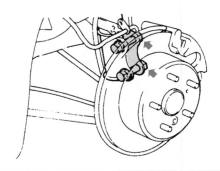

Figure 7-7 Loosening strut-to-rear spindle bolts. (Reprinted with permission)

Figure 7-8 Removing upper mount nuts. (Reprinted with permission)

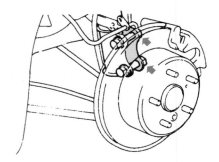

Figure 7-9 Removing strut from rear spindle. (Reprinted with permission)

8. Following the car or equipment manufacturer's recommended procedure, install a spring compressor on the coil spring, and tighten the spring compressor until all the spring tension is removed from the upper support (Figure 7-10).

9. Install a bolt in the upper strut-to-spindle bolt hole and tighten two nuts on the end of this bolt. One nut must be on each side of the strut bracket (Figure 7-11). Clamp this bolt in a vise to hold the strut, coil spring, and spring compressor.

10. Using the special tool to hold the spring and strut from turning, loosen the nut from the strut rod nut (Figure 7-12).

11. Remove the strut rod nut, upper support, and upper insulator.

12. Remove the strut from the lower end of the spring.

Special Tools

Coil-spring compressor

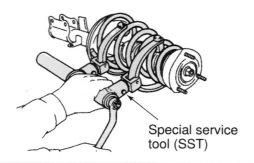

Special service
tool (SST)

Figure 7-10 Spring compressor installed on rear spring.
(Reprinted with permission)

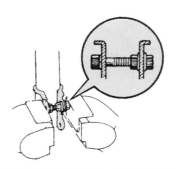

Figure 7-11 Strut, coil spring, and spring compressor
retained in a vise with a bolt and two nuts in upper
strut-to-spindle bolt hole. (Reprinted with permission)

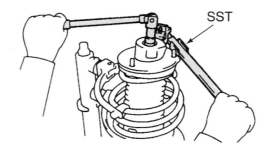

SST

Figure 7-12 Loosening strut rod nut. (Reprinted with
permission)

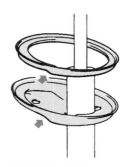

Figure 7-13 Inspecting lower spring seat and insulator.
(Reprinted with permission)

13. If the spring is to be replaced, rotate the compressor bolt until all the spring tension is removed from the compressing tool, and then remove the spring from the tool.

14. Inspect the lower insulator and spring seat on the strut (Figure 7-13). If the spring seat is warped or damaged, replace the strut. A new cartridge may be installed in some rear struts. (The strut cartridge replacement procedure is explained in Chapter 5.)

15. Visually inspect the upper mount, insulator, and spring bumper. If any of these components are damaged, worn, or distorted, replacement is necessary. Worn upper mounts and insulators or damaged spring seats cause suspension noise while driving on road irregularities. Assemble the spring bumper, spring, upper mount, and insulator (Figure 7-14), and reverse steps 1 through 13 to reinstall the coil spring and strut. Tighten all fasteners to the specified torque.

CUSTOMER CARE: When talking to customers, always remember the two Ps, pleasant and polite. There may be many days when we do not feel like being pleasant and polite. Perhaps we have several problem vehicles in the shop with symptoms that are difficult to diagnose and correct. Some service work may be behind schedule, and customers may be irate because their vehicles are not ready on time. However, we should always remain pleasant and polite with customers. Our attitude does much to make the customer feel better and realize their business is appreciated. A customer may not feel very happy about an expensive repair bill, but a pleasant attitude on our part may help improve the customer's feelings. When the two Ps are remembered by service personnel, customer relations are enhanced, and the customer will return to the shop. Conversely, if service personnel have a grouchy, indifferent attitude, the customer may be turned off and take his or her business to another shop.

Classroom Manual
Chapter 7, page 145

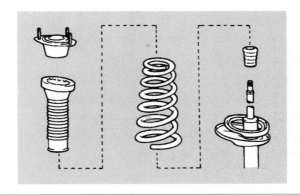

Figure 7-14 Assembly of rear strut, spring bumper, spring, upper mount, and insulator. (Reprinted with permission)

Lower Control Arm and Ball Joint Diagnosis and Replacement

Worn bushings on the lower control arms may cause incorrect rear wheel camber or toe, which results in rear tire wear and steering pull. Bent lower control arms must be replaced. When ball joints with **wear indicators** are in normal condition, there is 0.050 in (1.27 mm) between the grease nipple shoulder and the cover. If the ball joint is worn, the grease nipple shoulder is flush with or inside the cover (Figure 7-15). A worn ball joint causes improper rear wheel toe and/or camber, which may result in tire tread wear or steering pull.

The lower control arm removal and replacement procedure varies depending on the vehicle and the type of suspension. Always follow the vehicle manufacturer's recommended procedure in the service manual. The following is a typical lower control arm removal procedure:

1. Lift the vehicle on a hoist with the chassis supported on the hoist and the control arms dropped downward. The vehicle may be lifted with a floor jack, and the chassis supported on safety stands.

2. Remove the wheel and tire assembly.

3. Remove the stabilizer bar from the knuckle bracket.

Directional stability refers to the tendency of the vehicle steering to remain in the straight-ahead position when driven straight ahead on a reasonably smooth, level road surface.

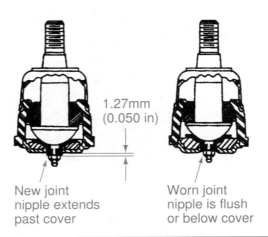

1.27mm (0.050 in)

New joint nipple extends past cover

Worn joint nipple is flush or below cover

Figure 7-15 Ball joint wear indicator. (Courtesy Oldsmobile Motor Division, General Motors Corp.)

4. Remove the parking brake cable retaining clip from the lower control arm.

5. If the car has electronic level control (ELC), disconnect the height sensor link from the control arm.

Special Tools

Lower control arm support tool

6. Install a special tool to support the lower control arm in the bushing areas (Figure 7-16).

7. Place a transmission jack under the special tool and raise the jack enough to remove the tension from the control arm bushing retaining bolts. If the car was lifted with a floor jack and supported on safety stands, place a floor jack under the special tool.

8. Place a safety chain through the coil spring and around the lower control arm.

9. Remove the bolt from the rear control arm bushing.

10. Be sure the jack is raised enough to relieve the tension on the front bolt in the lower control arm and remove this bolt.

11. Lower the jack slowly and allow the control arm to pivot downward. When all the tension is released from the coil spring, remove the safety chain, coil spring, and insulators.

Check the coil spring for distortion and proper free length. If the spring has a vinyl coating, check this coating for scratches and nicks. Check the spring insulators for cracks and wear. After the coil spring is removed, follow these steps to remove the lower control arm:

1. Remove the nut on the inner end of the **suspension adjustment link,** and disconnect this link from the lower control arm (Figure 7-17).

2. Remove the cotter pin from the ball joint nut, and loosen, but do not remove, the nut from the ball joint stud.

Special Tools

Ball joint stud removal tool

Ball joint pressing tool

3. Use a special ball joint removal tool to loosen the ball joint in the knuckle.

4. Remove the ball joint nut and the lower control arm (Figure 7-18).

Check the lower control arm for bends, distortion, and worn bushings. A special ball joint pressing tool is used to press the ball joint from the lower control arm (Figure 7-19), and the same pressing tool with different adapters is used to press the new ball joint into the control arm (Figure 7-20).

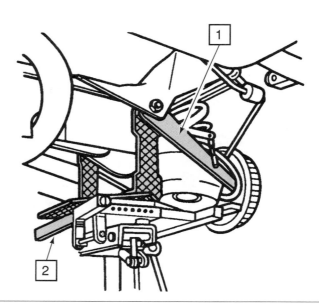

Figure 7-16 Special tool installed to support the inner end of the control arm. (Courtesy of Oldsmobile Motor Division, General Motors Corp.)

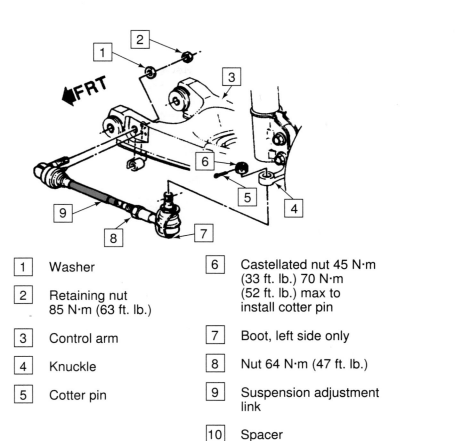

1	Washer	6	Castellated nut 45 N·m (33 ft. lb.) 70 N·m (52 ft. lb.) max to install cotter pin
2	Retaining nut 85 N·m (63 ft. lb.)	7	Boot, left side only
3	Control arm	8	Nut 64 N·m (47 ft. lb.)
4	Knuckle	9	Suspension adjustment link
5	Cotter pin	10	Spacer

Figure 7-17 Suspension adjustment link removal from the lower control arm. (Courtesy of Oldsmobile Motor Division, General Motors Corp.)

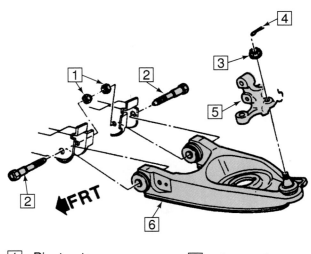

1	Pivot nuts 115 N·m (85 ft. lb.)	4	Cotter pin
2	Pivot bolts 187 N·m (138 ft. lb.)	5	Knuckle
3	Castellated nut	6	Control arm

Figure 7-18 Removing lower control arm and ball joint. (Courtesy of Oldsmobile Motor Division, General Motors Corp.)

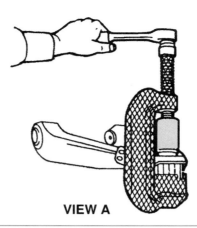

VIEW A

Figure 7-19 Ball joint removal from lower control arm. (Courtesy of Oldsmobile Motor Division, General Motors Corp.)

CAUTION: When disconnecting a linkage joint, such as the ends of the suspension adjustment link, do not use a wedge-type tool between the joint and the attached part. This method of joint removal may damage the joint seal.

Special Tools

Suspension adjustment link removal tool

Remove the nut and cotter pin on the outer end of the suspension adjustment link, and use a special puller to remove this link from the knuckle (Figure 7-21). Remove the suspension adjustment link and inspect the joints. If the joints are loose or the seals are damaged, joint replacement is necessary. The joint studs must fit snugly in the knuckle and lower control arm openings. When the joint studs are worn, joint replacement is necessary, and worn stud openings in the knuckle or lower control arm require component replacement.

Follow this procedure to install the lower control arm and spring:

CAUTION: Never loosen a ball joint nut to install a cotter pin, because this action causes improper torquing of the nut.

1. Install the ball joint stud in the knuckle and install the nut on the ball joint stud. Tighten the ball joint nut to the specified torque, and then tighten the nut an additional 2/3 turn. If necessary, tighten the nut slightly to align the nut castellations with the cotter pin hole in the ball joint stud, and install the cotter pin.

VIEW B

Figure 7-20 Ball joint installation in lower control arm. (Courtesy of Oldsmobile Motor Division, General Motors Corp.)

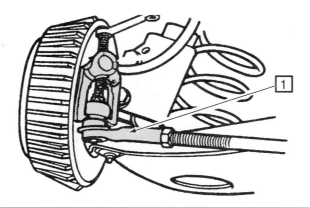

Figure 7-21 Suspension adjustment link removal from the knuckle. (Courtesy of Oldsmobile Motor Division, General Motors Corp.)

2. Snap the upper insulator on the coil spring. Install the lower spring insulator and the spring in the lower control arm (Figure 7-22).

3. Be sure the top of the coil spring is properly positioned in relation to the front of the vehicle (Figure 7-23).

4. Install the special tool on the inner ends of the control arm, and place the transmission jack under the special tool.

5. Slowly raise the transmission jack until the control arm bushing openings are aligned with the openings in the chassis.

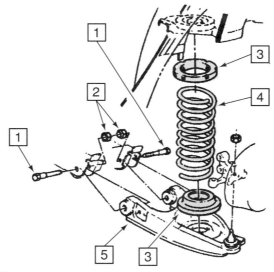

1	Pivot bolt 187 N·m (138 ft. lb.)
2	Pivot nut 115 N·m (85 ft. lb.)
3	Insulator
4	Coil spring
5	Control arm

Figure 7-22 Installation of spring and insulators. (Courtesy of Oldsmobile Motor Division, General Motors Corp.)

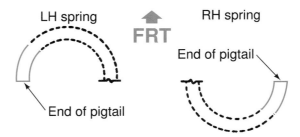

Top views of upper ends of springs

LH spring   **FRT**   RH spring

End of pigtail

End of pigtail

Figure 7-23 Proper location of upper coil spring ends in relation to the front of the vehicle. (Courtesy of Oldsmobile Motor Division, General Motors Corp.)

CAUTION: The pivot bolts and nuts in the inner ends of the lower control arm must be tightened to the specified torque with the vehicle weight supported on the wheels, and the suspension at normal curb height. Failure to follow this procedure may adversely affect ride quality and steering characteristics.

6. Install the bolts and nuts in the inner ends of the control arm. Do not torque these bolts and nuts at this time.

7. Install the stabilizer-bar-to-knuckle bracket and tighten the fasteners to the specified torque.

8. Install the parking brake retaining clip.

9. If the vehicle has ELC, install the height sensor link, and tighten the fastener to the specified torque.

10. Install the suspension adjustment link and tighten the retaining nuts to the specified torque. Install cotter pins as required.

11. Remove the transmission jack and install the wheel-and-tire assembly.

12. Lower the vehicle onto the floor and tighten the wheel hub nuts and lower control arm bolts and nuts to the specified torque.

Rear Leaf-Spring Diagnosis and Replacement

This leaf-spring discussion applies to multiple-leaf springs on rear suspension systems that have two springs mounted longitudinally in relation to the chassis. Many leaf springs have plastic **spring silencers** between the spring leafs. If these silencers are worn out, creaking and squawking noises are heard when the vehicle is driven over road irregularities at low speeds.

When the silencers require checking or replacement, lift the vehicle with a floor jack and support the frame on safety stands so the rear suspension moves downward. With the vehicle weight no longer applied to the springs, the spring leafs may be pried apart with a pry bar to remove and replace the silencers.

Worn shackle bushings, brackets, and mounts cause excessive chassis lateral movement and rattling noises. With the normal vehicle weight resting on the springs, insert a pry bar between the rear outer end of the spring and the frame. Apply downward pressure on the bar and observe the

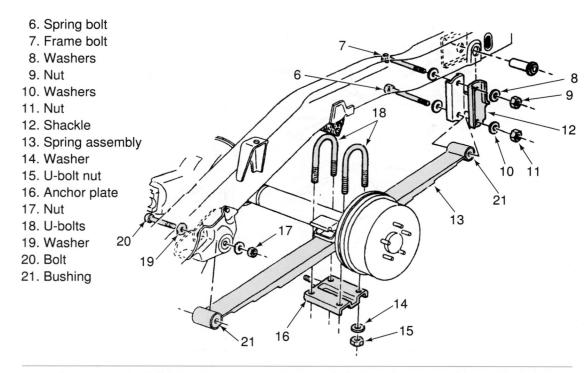

6. Spring bolt
7. Frame bolt
8. Washers
9. Nut
10. Washers
11. Nut
12. Shackle
13. Spring assembly
14. Washer
15. U-bolt nut
16. Anchor plate
17. Nut
18. U-bolts
19. Washer
20. Bolt
21. Bushing

Figure 7-24 Leaf-spring rear suspension. (Courtesy of Chevrolet Motor Division, General Motors Corp.)

rear shackle for movement. Shackle bushings, brackets, or mounts must be replaced if there is movement in the shackle. The same procedure may be followed to check the front bushing in the main leaf. A broken spring center bolt may allow the rear axle assembly to move rearward on one side. This movement changes rear wheel tracking, which results in handling problems, tire wear, and reduced directional stability. Sagged rear springs reduce the curb riding height. Spring replacement is necessary if the springs are sagged. When rear leaf-spring replacement is necessary, proceed as follows:

1. Lift the vehicle with a floor jack and place safety stands under the frame. Lower the vehicle weight onto the safety stands, and leave the floor jack under the differential housing to support the rear suspension weight.

2. Remove the nuts from the spring U-bolts, and remove the U-bolts and lower spring plate (Figure 7-24). The spring plate may be left on the rear shock absorber and moved out of the way.

3. Be sure the floor jack is lowered sufficiently to relieve the vehicle weight from the rear springs.

4. Remove the rear shackle nuts, plate, shackle, and bushings.

5. Remove the front spring mounting bolt and remove the spring from the chassis. Check the spring center bolt to be sure it is not broken.

6. Check the front hanger, bushing, and bolt, and replace as necessary.

7. Check the rear shackle, bushings, plate, and mount, and replace the worn components.

8. Reverse steps 1 through 5 to install the spring. Tighten all bolts and nuts to the specified torque.

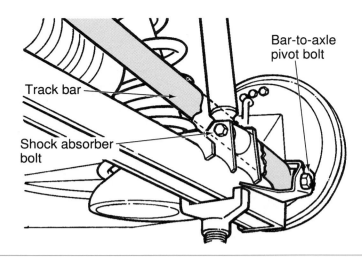

Figure 7-25 Checking track bar bushings. (Courtesy of Chrysler Corp.)

Track Bar Diagnosis and Replacement

Some rear suspension systems have a track bar to control lateral chassis movement. Rubber mounting bushings insulate the track bar from the chassis components. Worn track bar mounts and bushings may cause rattling and excessive lateral chassis movement.

When the track bar is checked, lift the vehicle on a hoist or floor jack with the rear suspension in the normal riding height position. If the vehicle is lifted with a floor jack, place safety stands under the rear axle to support the vehicle weight. Grasp the track bar firmly and apply vertical and horizontal force. If there is movement in the track bar mountings, track bar, or bushing, replacement is essential (Figure 7-25).

Another track bar checking method is to leave the vehicle on the shop floor and observe the track bar mounts as an assistant applies side force to the chassis or rear bumper. If there is lateral movement in the track bar bushings or brackets, replace the bushings and check the bracket bolts. Bent track bars must be replaced.

When the track bar is replaced, remove the mounting bolts, bushings, grommets, and track bar. Inspect the mounting bolt holes in the chassis for wear. After the track bar is installed with the proper grommets and bushings, tighten the mounting bolts to the specified torque.

☑ **SERVICE TIP:** On rear suspension systems with an inverted U-channel, the stabilizer bar inside the U-channel sometimes breaks away where it is welded to the end plate in the U-channel. This results in a rattling, scraping noise when the car is driven over road irregularities.

Stabilizer Bar Diagnosis and Service

A stabilizer bar may be referred to as a sway bar.

Worn stabilizer bar mounting bushings, grommets, or mounting bolts cause a rattling noise as the vehicle is driven on irregular road surfaces. A weak stabilizer bar, or worn bushings and grommets cause harsh riding and excessive body sway while driving on irregular road surfaces. Worn or very dry stabilizer bar bushings may cause a squeaking noise on irregular road surfaces. All stabilizer bar components should be visually inspected for wear. Stabilizer bar removal and replacement procedures vary depending on the vehicle. Always follow the vehicle manufacturer's recommended procedure in the service manual. Following is a typical rear stabilizer bar removal and replacement procedure:

1. Lift the vehicle on a hoist and allow both sides of the rear suspension to drop downward as the vehicle chassis is supported on the hoist.

2. Remove the mounting bolts at the outer ends of the stabilizer bar and remove the bushings, grommets, brackets, or spacers (Figure 7-26).

3. Remove the mounting bolts in the center area of the stabilizer bar.

4. Remove the stabilizer bar from the chassis.

5. Visually inspect all stabilizer bar components such as bushings, bolts, and spacer sleeves. Replace the stabilizer bar, grommets, bushings, brackets, or spacers as required. Split bushings may be removed over the stabilizer bar. Bushings that are not split must be pulled from the bar.

6. Reverse steps 2 through 4 to install the stabilizer bar. Make sure all stabilizer bar components are installed in the original position, and tighten all fasteners to the specified torque.

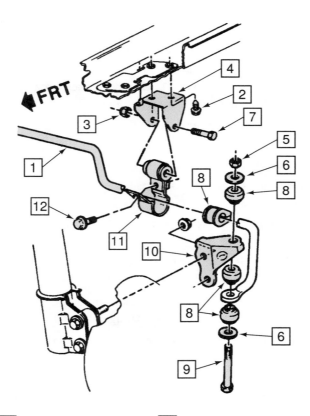

1	Stabilizer shaft	7	Bolt 50 N·m (37 ft. lb.)
2	Bolt 18 N·m (14 ft. lb.)	8	Insulator
3	Nut 47 N·m (35 ft. lb.)	9	Bolt
4	Bracket	10	Stabilizer shaft bracket
5	Nut 18 N·m (14 ft. lb.)	11	Link
6	Retainer	12	Bolt 50 N·m (37 ft. lb.)

Figure 7-26 Stabilizer bar, bushings, grommets, spacers, and brackets. (Courtesy of Oldsmobile Motor Division, General Motors Corp.)

Rear Suspension Tie Rod Inspection and Replacement

Classroom Manual
Chapter 7, page 156

Rear tie rods should be inspected for worn grommets, loose mountings, and bent conditions. Loose tie-rod bushings or a bent tie rod will change the rear wheel tracking and result in reduced directional stability. Worn tie-rod bushings also cause a rattling noise on road irregularities. When the rear tie rod is replaced, remove the front and rear rod mounting nuts. The lower control arm or rear axle may have to be pried rearward to remove the tie rod. Check the tie-rod grommets and mountings for wear, and replace parts as required. When the tie rod is reinstalled, tighten the mounting bolts to specifications, and check the rear wheel toe.

Guidelines for Servicing Rear Suspension Systems

1. A squeaking noise may be caused by a suspension bushing, or a defective strut or shock absorber.

2. A rattling noise may be caused by worn suspension bushings, loose shock absorber or strut bushings or mounts, broken springs or spring insulators, or defective struts or shock absorbers.

3. Excessive body sway or roll, when one wheel strikes a road irregularity, may be caused by a defective stabilizer bar or bushings.

4. Excessive lateral movement of the rear chassis may be caused by a defective track bar or bushings.

5. Reduced curb riding height causes harsh riding.

6. Reduced curb riding height may be caused by sagged springs, worn control arm bushings, or bent control arms.

7. Sagged rear springs cause excessive steering effort and rapid steering wheel return after a turn.

8. When the coil spring is mounted on the rear strut, never loosen the upper strut nut until the spring is compressed with a spring compressing tool.

9. If a coil spring has a vinyl coating, the spring should be taped in the compressing tool contact areas to prevent chipping the coating.

10. On some independent rear suspension systems, a ball joint connects the lower control arm to the knuckle. Many of these ball joints have a conventional wear indicator and a press fit in the control arm.

11. When some lower control arms are installed, the control arm bolts should be tightened with the normal vehicle weight supported on the tires.

12. Suspension adjustment link studs should not be loosened with a wedge-type tool.

13. A nut should never be loosened to install a cotter pin.

14. A broken leaf-spring center bolt may allow the rear axle to move, which results in improper tracking and reduced directional stability.

CASE STUDY

A customer complained about steering pull to the left on a Chevrolet Celebrity. The customer also said the problem had just occurred in the last few days. The technician road tested the car and found the customer's description of the complaint to be accurate. A careful inspection of the tires indicated there was no abnormal tire wear on the front or rear tires. The vehicle was inspected for recent collision damage, but there was no evidence of this type of damage. After lifting the vehicle on a hoist, the technician checked all the front suspension components, including ball joints, control arms, control arm bushings, and wheel bearings. However, none of these front suspension components indicated any sign of wear or looseness.

Realizing that improper rear wheel tracking causes steering pull, the technician inspected the rear suspension. A pry bar was used to apply downward and rearward force to the trailing arms on the rear suspension. When this action was taken on the right rear trailing arm, the technician discovered the trailing arm bushing was very loose. This defect had allowed the right side of the rear axle to move rearward a considerable amount, which explained why the steering pulled to the left.

After installing a new trailing arm bushing, the alignment was checked on all four wheels, and the wheel alignment was within specifications. A road test indicated no evidence of steering pull or other steering problems.

Terms to Know

Body sway

Curb riding height

Lateral movement

Spring silencers

Suspension adjustment link

Wear indicators

ASE Style Review Questions

1. While discussing lateral movement of the rear chassis:
   Technician A says this problem may be caused by loose stabilizer bar bushings.
   Technician B says this problem may be caused by loose track bar bushings.
   Who is correct?
   A. A only
   B. B only
   C. Both A and B
   D. Neither A nor B

2. While discussing excessive rear body sway or roll when driving on road irregularities:
   Technician A says this problem may be caused by worn spring insulators.
   Technician B says this problem may be the result of sagged rear springs.
   Who is correct?
   A. A only
   B. B only
   C. Both A and B
   D. Neither A nor B

3. While discussing rear suspension curb riding height:
   Technician A says reduced rear curb riding height causes harsh riding.
   Technician B says reduced rear curb riding height increases steering effort.

Who is correct?
   A. A only
   B. B only
   C. Both A and B
   D. Neither A nor B

4. While discussing curb riding height:
   Technician A says the curb riding height is measured at the same location on each vehicle.
   Technician B says the curb riding height has no effect on other alignment angles.
   Who is correct?
   A. A only
   B. B only
   C. Both A and B
   D. Neither A nor B

5. While discussing spring removal on a MacPherson strut rear suspension:
   Technician A says all the spring tension must be removed from the upper mount before the upper strut nut is loosened.
   Technician B says a vinyl-coated spring should be taped in the compressing tool contact area.
   Who is correct?
   A. A only
   B. B only
   C. Both A and B
   D. Neither A nor B

6. The most likely cause of excessive rear chassis lateral movement is (are):
   A. a weak stabilizer bar.
   B. worn track bar bushings.
   C. worn out struts.
   D. sagged rear coil springs.

7. All these statements about rear suspension systems with lower control arms and ball joints are true EXCEPT:
   A. Worn lower control arm bushings do not affect tire tread wear.
   B. Some rear suspension ball joints have wear indicators.
   C. On some rear suspensions the ball joint is pressed into the lower control arm.
   D. On some vehicles the vehicle weight must be resting on the tires when the inner, lower control arm bolts are tightened.

8. Rear suspension adjustment links:
   A. may be lengthened or shortened to adjust rear wheel camber.
   B. are connected from the lower control arm to the knuckle.

   C. may be loosened on each end with a wedge-type tool.
   D. prevent fore-and-aft control arm and wheel movement.

9. On a rear suspension system with longitudinally mounted leaf springs:
   A. improper vehicle tracking may be caused by a broken spring center bolt.
   B. worn spring silencers may cause reduced curb riding height.
   C. a shackle is mounted between the front of the spring and the chassis.
   D. longitudinal rear leaf springs are usually installed on independent rear suspension systems.

10. While diagnosing improper rear wheel tracking:
    A. improper rear wheel tracking may be caused when both rear springs are sagged the same amount.
    B. a bent rear suspension tie rod does not affect rear wheel tracking.
    C. improper rear wheel tracking may result in steering pull when driving straight ahead.
    D. improper rear wheel tracking may cause front wheel shimmy.

ASE Challenge Questions

1. A squeak in the rear suspension could be caused by all of the following EXCEPT:
   A. the suspension bushing.
   B. weak spring leaves.
   C. worn spring anti-friction pads.
   D. a defective shock absorber.

2. A vehicle with a live-axle coil spring rear suspension has become hard to steer with harsh ride quality. Which of the following could be the cause of this problem?
   A. Worn lateral link bushings
   B. Weak rear coil springs
   C. Bent rear shock rod
   D. Worn stabilizer bar bushings

3. A customer says his 1993 Isuzu Trooper has become "real antsy" on a curve. An initial check of the rear suspension by pushing sideways against the bumper with the truck on the floor indicates wear.
   Technician A says worn sway bar bushings could be the problem.
   Technician B says a worn shock mount could be the problem.

Who is correct?
A. A only   C. Both A nad B
B. B only   D. Neither A nor B

4. A car with independent rear suspension has excessive rear tire wear. An inspection of the rear tires shows they are worn on the inside edge and the tread is feathered.
   Technician A says the problem could be the tires are toeing out during acceleration.
   Technician B says worn control arm bushings could be the cause of the problem.
   Who is correct?
   A. A only   C. Both A nad B
   B. B only   D. Neither A nor B

5. The steering on a front-wheel-drive car pulls to the right.
   Technician A says the strut on the right rear suspension assembly could be the problem.
   Technician B says worn bushings of the left rear lower arm assembly could be the problem.
   Who is correct?
   A. A only   C. Both A nad B
   B. B only   D. Neither A nor B

Job Sheet 20

Name _____ Date _____

Remove and Service Rear Suspension Strut and Coil Spring

NATEF and ASE Correlation

This job sheet is related to NATEF and ASE Automotive Suspension and Steering Task List content area: B. Suspension System Diagnosis and Repair, 2. Rear Suspensions, Task 2: Inspect and replace rear suspension system coil springs and spring insulators (silencers). Task 5: Inspect and replace rear MacPherson strut cartridge or assembly, and upper mount assembly.

Tools and Materials

Floor jack
Safety stands
Hoist
Coil-spring compressing tool

Vehicle make and model year_____

Vehicle VIN _____

Procedure

1. Remove the rear seat and the package trim tray.

2. Remove the wheel cover and loosen the wheel nuts.

3. Lift the vehicle with a floor jack, and lower the chassis onto safety stands so the rear suspension is allowed to drop downward.

    Is the rear suspension properly supported on the safety stands? yes _____ no _____

    Instructor check _____

4. Place a wooden block between the floor jack and the rear spindle on the side where the strut and spring removal is taking place. Raise the floor jack to support some of the suspension system weight.

    Is the floor jack placed under the rear spindle and supprting some of the vehicle weight?
    yes _____ no _____

    Instructor check _____

5. Remove the rear wheel, disconnect the nut from the small spring in the lower arm, and remove the brake hose and antilock brake system (ABS) wire from the strut.

6. Remove the stabilizer bar link from the strut, and loosen the strut-to-spindle mounting bolts.

7. Remove three upper support nuts under the package tray trim, and lower the floor jack to remove the strut from the knuckle. Remove the strut from the chassis.

    ✔ **SERVICE TIP:** If the plastic coating on a coil spring is chipped, the spring may break prematurely. The spring may be taped in the compressing tool contact areas to prevent chipping.

8. Following the car and equipment manufacturers' recommended procedures, install a spring compressor on the coil spring, and tighten the spring compressor until all the spring tension is removed from the upper support.

Is the spring compressor properly installed on the coil spring? yes _____ no _____

Is the spring compressor tightened so all tension is removed from the upper support?

Instructor check _____

9. Install a bolt in the upper strut-to-spindle bolt hole and tighten two nuts on the end of this bolt. One nut must be on each side of the strut bracket.

Clamp this bolt in a vise to hold the strut, coil spring, and spring compressor.

10. Using the special tool to hold the spring and strut from turning, loosen the nut from the strut rod nut.

Is all the coil-spring tension removed from the upper strut mount by the spring compressor? yes _____ no _____

Is the strut rod nut loosened? yes _____ no _____

Instructor check _____

11. Remove the strut rod nut, upper support, and upper insulator.

12. Remove the strut from the lower end of the spring.

13. If the spring is to be replaced, rotate the compressor bolt until all the spring tension is removed from the compressing tool, then remove the spring from the tool.

14. Inspect the lower insulator and spring seat on the strut. Is the spring seat is warped or damaged, replace the strut. A new cartridge may be installed in some rear struts.

Lower insulator condition: satisfactory_____ unsatisfactory_____

Spring seat condition: satisfactory_____ unsatisfactory_____

Strut condition: satisfactory_____ unsatisfactory_____

List all the components that require replacement and explain the reasons for your diagnosis.

_____

_____

_____

15. Visually inspect the coil spring, upper mount, insulator , and spring bumper. If any of these components are damaged, worn, or distorted, replacement is necessary.

Coil spring condition: satisfactory_____ unsatisfactory_____

Upper mount condition: satisfactory_____ unsatisfactory_____

Insulator condition: satisfactory_____ unsatisfactory_____

Spring bumper condition: satisfactory_____ unsatisfactory_____

List all the components that require replacement and explain the reasons for your diagnosis.

_____

_____

_____

✔ **Instructor Check** _____

Job Sheet 21

21

Name _____ Date _____

Remove and Rear Suspension Lower Control Arm and Ball Joint

NATEF and ASE Correlation

This job sheet is related to NATEF and ASE Automotive Suspension and Steering Task List content area: B. Suspension System Diagnosis and Repair, 2. Rear Suspensions, Task 3: Inspect and replace rear suspension system transverse links (track bars), control arms, stabilizer bars (sway bars), bushings and mounts. Task 7: Inspect and replace rear ball joints and tie-rod assemblies.

Tools and Materials

Floor jack
Safety stands
Hoist
Control arm removing tool
Transmission jack
Ball joint removal and replacement tools

Vehicle make and model year _____

Vehicle VIN _____

Procedure

1. Lift the vehicle on a hoist with the chassis supported in the hoist, and control arms dropped downward. The vehicle may be lifted with a floor jack and the chassis supported on safety stands.

2. Remove the wheel and tire assembly.

3. Remove the stabilizer bar from the knuckle bracket

4. Remove the parking brake cable retaining clip from the lower control arm.

5. If the car has electronic level control (ELC) disconnect the height sensor link from the control arm.

6. Install a special tool to support the lower control arm in the bushing areas.

7. Place a transmission jack under the special tool and raise the jack enough to remove the tension from the control arm bushing retaining bolts. If the car was lifted with a floor jack and supported on safety stands, place a floor jack under the special tool.

   Is the special control arm support tool properly installed and supported?

   yes _____ no _____

   Instructor check _____

8. Place a safety chain through the coil spring and around the lower control arm.

   Is the saftey chain properly installed? yes _____ no _____

   Instructor check _____

9. Remove the bolt from the rear control arm bushing.

10. Be sure the jack is raised enough to relieve the tension on the front bolt in the lower control arm and remove this bolt.

11. Lower the jack slowly and allow the control arm to pivot downward. When all the tension is released from the coil spring, remove the safety chain, coil spring, and insulators.

12. Inspect the coil spring for distortion and proper free length. If the spring has a vinyl coating, check this coating for scratches or nicks. Check the spring insulators for cracks and wear.

    Coil spring condition: satisfactory_____ unsatisfactory_____

    List all the components that require replacement and explain the reasons for your diagnosis.

    _____

    _____

    _____

13. Remove the nut on the inner end of the suspension adjustment link and disconnect this link from the lower control arm.

14. Remove the cotter pin from the ball joint nut, and loosen, but do not remove, the nut from the ball joint stud.

    Is the ball joint nut loosened? yes _____ no _____

    Instructor check _____

15. Use a special ball joint removal tool to loosen the ball joint in the knuckle.

16. Remove the ball joint nut and the control arm.

17. Inspect the lower control arm for a bends, distortion, and worn bushings.

    Lower control arm condition: satisfactory_____ unsatisfactory_____

    List the control arm and related parts that require replacement and explain the reasons for your diagnosis. _____

    _____

    _____

    Instructor check _____

18. Use a special ball joint pressing tool to press the ball joint from the lower control arm.

19. Use the same pressing tool with different adapters to press the new ball joint into the control arm.

Job Sheet 22

22

Name _____ Date _____

Install Rear Suspension Lower Control Arm and Ball Joint Assembly

NATEF and ASE Correlation

This job sheet is related to NATEF and ASE Automotive Suspension and Steering Task List content area: B. Suspension System Diagnosis and Repair, 2. Rear Suspensions, Task 3: Inspect and replace rear suspension system transverse links (track bars), control arms, stabilizer bars (sway bars), bushings and mounts.

Tools and Materials

Floor jack
Safety stands
Hoist
Control arm removing tool
Transmission jack

Vehicle make and model year _____

Vehicle VIN _____

Procedure

1. Install the ball joint stud in the knuckle and install the nut on the ball joint stud. Tighten the ball joint nut to the specified torque, and then tighten the nut an additional 2/3 turn. If necessary, tighten the nut slightly to align the nut castellations with the cotter pin hole in the ball joint stud, and install the cotter pin.

   Specified ball joint stud nut torque_____

   Actual ball joint stud nut torque_____

2. Snap the upper insulator on the coil spring. Install the lower spring insulator and the spring in the lower control arm.

   Is the coil spring and insulator properly installed in lower control arm?

   yes _____ no _____

   Instructor check _____

3. Be sure the top of the coil spring is properly positioned in relation to the front of the vehicle.

   Is the top of the spring properly positioned? yes _____ no _____

   Instructor check _____

4. Install the special tool on the inner ends of the control arm, and place the transmission jack or floor jack under the special tool.

   Is the special tool properly supported on the control arm? yes _____ no _____

   Instructor check _____

   Is the special tool properly supported on the transmission or floor jack?

   Instructor check _____

5. Slowly raise the transmission jack until the control arm bushing openings are aligned with the openings in the chassis.

   > **CAUTION:** The pivot belts and nuts in the inner ends of the lower control arm must be tightened to the specified torque with the vehicle supported on the wheels and the suspension at normal curb height. Failure to follow this procedure may adversely affect ride quality and steering characteristics.

6. Install the bolts and nuts in the inner ends of the control arm. Do not torque these bolts and nuts at this time.

7. Install the stabilizer-bar-to-knuckle bracket fasteners to the specified torque.

   Specified torque on stabilizer-bar-to-knuckle bracket fasteners_____

   Actual torque on stabilizer-bar-to-knuckle bracket fasteners_____

8. Install the parking brake retaining clip.

9. If the vehicle has ELC, install the height sensor link, and tighten the fastener to the specified torque.

10. Install the suspension adjustment link and tighten the fastener to the specified torque. Install cotter pins as required.

    Specified adjustment link retaining nut torque_____

    Actual adjustment link retaining nut torque_____

    Are the cotter pins properly installed in adjustment link retaining nuts?

    yes _____ no _____

    Instructor check _____

11. Remove the transmission jack or floor jack, and install the wheel-and-tire assembly.

12. Lower the vehicle onto the floor. Tighten the wheel hub nuts and lower control arm bolts and nuts to the specified torque.

    Specified wheel nut torque_____

    Actual wheel nut torque_____

    Specified control arm retaining nut torque_____

    Actual control arm retaining nut torque_____

☑ **Instructor Check** _____

Table 7-1 NATEF and ASE TASK

Diagnose suspension system noises, body sway, and ride height problems; determine needed repairs.

Problem Area	Symptoms	Possible Causes	Classroom Manual	Shop Manual
NOISE	Rattling on road irregularities	**1.** Worn suspension bushings, grommets	150	193
		2. Defective struts, shock absorbers	150	193
	Squeaking on road irregularities	Worn, dry suspension bushings, grommets	152	195
RIDE QUALITY	Excessive body sway	Defective stabilizer or bushings	145	206
	Harsh riding	Insufficient ride height	145	193
STEERING QUALITY	Excessive steering effort	Insufficient rear ride height	145	193

Table 7-2 NATEF and ASE TASK

Inspect and replace rear suspension system coil springs and spring insulators (silencers).

Problem Area	Symptoms	Possible Causes	Classroom Manual	Shop Manual
NOISE	Rattling on road irregularities	Broken springs or spring insulators	145	195
STEERING QUALITY	Excessive steering effort	Sagged rear springs	146	195

Table 7-3 NATEF and ASE TASK

Inspect and replace rear suspension system transverse links (track bars), control arms, stabilizer bars (sway bars), bushings, and mounts.

Problem Area	Symptoms	Possible Causes	Classroom Manual	Shop Manual
NOISE	Rattling on road irregularities	Worn bushings in track bar, stabilizer bar, or control arms	152	206
RIDE QUALITY	Excessive body sway	Worn stabilizer bushings, loose brackets	152	206
	Excessive lateral movement, rear chassis	Worn track bar bushings, loose brackets	152	206
STEERING QUALITY	Steering pulls to one side	Bent lower control arm, worn bushings	152	199
TIRE WEAR	Excessive tread wear	Bent lower control arm, worn bushings	152	199

Table 7-4 NATEF and ASE TASK

Inspect and replace rear suspension system leaf spring(s), leaf-spring insulators (silencers), shackles, brackets, bushings, and mounts.

Problem Area	Symptoms	Possible Causes	Classroom Manual	Shop Manual
NOISE	Squeaking on road irregularities	Worn silencers	146	204
	Rattling on road irregularities	Worn bushings, shackles, brackets	146	205
RIDE QUALITY	Excessive body sway	Worn bushings, shackles, brackets	146	205
	Harsh riding	Reduced ride height	146	205

Table 7-5 NATEF and ASE TASK

Inspect and replace rear MacPherson strut cartridge or assembly, and upper mount assembly.

Problem Area	Symptoms	Possible Causes	Classroom Manual	Shop Manual
NOISE	Rattling on road irregularities	**1.** Loose strut bushings, upper mount	152	199
		2. Defective struts	153	199
RIDE QUALITY	Harsh riding, excessive chassis oscillations	Defective struts	153	200

Table 7-6 NATEF and ASE TASK

Inspect and replace rear ball joints and tie rod assemblies.

Problem Area	Symptoms	Possible Causes	Classroom Manual	Shop Manual
NOISE	Rattling on road irregularities	Worn tie rod bushings	155	199
	Steering pull	**1.** Worn tie rod bushings	155	200
		2. Bent tie rod	155	200
TIRE LIFE	Tread wear	Worn ball joint	155	200

CHAPTER 8

Computer-Controlled Suspension System Service

Upon completion and review of this chapter, you should be able to:

❏ Diagnose programmed ride control (PRC) systems.

❏ Diagnose computer command ride (CCR) systems.

❏ Diagnose electronic air suspension systems.

❏ Remove, replace, and inflate air springs.

❏ Adjust front and rear trim height on electronic air suspension systems.

❏ Service and repair nylon air lines.

❏ Adjust trim height on a rear load-levelling air suspension system.

❏ Diagnose rear load-levelling air suspension systems.

❏ Diagnose automatic air suspension systems.

❏ Diagnose air suspension systems with speed levelling capabilities.

❏ Diagnose automatic ride control (ARC) systems.

❏ Diagnose road sensing suspension systems.

Preliminary Inspection of Computer-Controlled Suspension Systems

Prior to diagnosing a computer-controlled suspension system, a preliminary inspection should be performed. The preliminary inspection may locate a minor defect that is the cause of the problem. If the preliminary inspection is not performed, a lot of time may be wasted performing advanced diagnosis when the problem is a minor defect. When the preliminary diagnosis does not locate any minor problems, further diagnosis is required. Follow these steps to complete the preliminary diagnosis:

1. Talk to the customer and find out the exact complaint regarding vehicle operation.

2. If necessary, road test the vehicle to determine the exact symptoms.

3. Inspect the vehicle for any work that was completed recently, including body work. Wires or air lines may have been damaged by a collision.

4. Check all electrical connections and wiring in the computer-controlled suspension system.

5. If the vehicle has a computer-controlled air suspension system, listen for air leaks in the system.

6. If the vehicle has a computer-controlled air suspension system, check all the air lines for cracks, breaks, damage, and kinks.

7. Measure the vehicle curb (trim) height.

Programmed Ride Control System Diagnosis

General Diagnosis

The **programmed ride control (PRC)** control module monitors the complete system while the vehicle is driven. If a defect occurs in the PRC system, the mode indicator light starts flashing.

When the mode selector switch is moved to the opposite position and returned to the same position, erroneous codes are cleared from the control module memory. If the mode indicator light continues flashing, the self-test should be performed.

Self-Test

The self-test connector is located under the ash tray (Figure 8-1). This connector has two terminals that must be connected during the self-test. A jumper tool should be fabricated to connect these terminals, and this tool should be soldered to the end of a 7-inch slotted screwdriver blade to access the self-test terminals (Figure 8-2).

Follow this procedure to perform the self-test:

1. Turn the ignition switch off and be sure the headlights and parking lights are off. These lights must remain off during the self-test.

2. Position the mode select switch in the Auto position.

3. Remove the ash tray and insert the fabricated tool in the self-test connector terminals.

4. Start the engine and leave the mode select switch in the Auto position.

5. When the engine has been running for 20 seconds or more, remove the tool from the self-test connector.

6. Count the mode indicator light flashes to obtain the trouble codes. For example, if the light flashes six times code 6 has been provided. The light flashes each code four times at nine-second intervals. The mode indicator light is in the tachometer (Figure 8-3).

Diagnostic Trouble Codes (DTCs)

✓ **SERVICE TIP: Diagnostic trouble codes (DTCs)** together with the vehicle manufacturer's service manual diagnostic procedures will usually locate the problem in a PRC system.

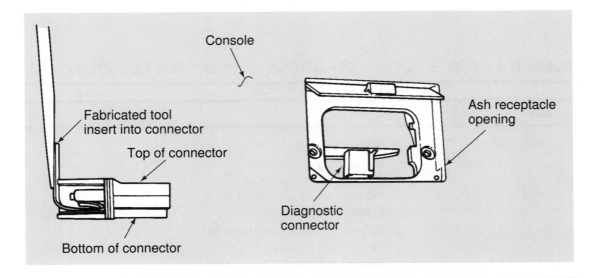

Figure 8-1 Self-test terminals, programmed ride control system. (Courtesy of Ford Motor Company)

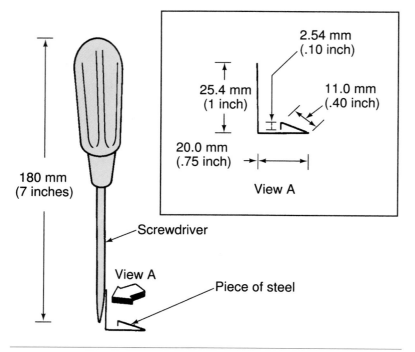

Figure 8-2 Fabricated tool to connect self-test terminals. (Courtesy of Ford Motor Company)

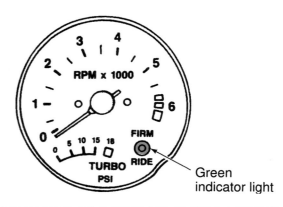

Figure 8-3 Mode indicator light in tachometer flashes trouble codes. (Courtesy of Ford Motor Company)

Several DTCs are available on the PRC system (Table 8-1).

The DTCs vary depending on the vehicle make and year. Always use the DTC list in the manufacturer's service manual. A trouble code indicates a defect in a specific area. For example, if DTC 2 is received the defect may be in the right rear actuator or in the connecting wires.

After a PRC system defect has been corrected, the DTCs should be cleared. The system should be checked for DTCs again to make sure there are no other faults in the system.

Classroom Manual
Chapter 8, page 172

Table 8-1 PRC SYSTEM DIAGNOSTIC TROUBLE CODES (DTCS)

Code	Defect
6	No problem
1	Left rear actuator circuit
2	Right rear actuator circuit
3	Right front actuator circuit
4	Left front actuator circuit
5	Soft relay control circuit shorted
7	PRC control module
13	Firm relay control circuit shorted
14	Relay control circuit

(Courtesy of Ford Motor Company)

Diagnosis of Computer Command Ride System

SPECIAL TOOLS

Jumper wires

Diagnostic Trouble Code (DTC) Diagnosis

If a defect occurs in the **computer command ride (CCR)** system, both **light-emitting diodes (LEDs)** in the driver select switch are illuminated. Should the defect correct itself, the LEDs go out, but a DTC should be retained in the control module nonvolatile **electronically erasable programmable read only memory (EEPROM).** When a DTC is set in the control module memory, the control module performs a 1-second retest at 3-minute intervals, or when the driver select switch position is changed. Both switch LEDs go out if the defect is corrected. If the defect is still present, the LEDs remain illuminated.

A jumper wire may be connected between terminals A and C in the **data link connector (DLC)** to obtain the DTCs from the control module in the computer command ride system (Figure 8-4).

After the jumper wire is installed across terminals A and C in the DLC, the ignition switch is turned on, and both the driver select switch LEDs blink at the same time to provide the fault codes. If these LEDs flash twice followed by a pause and five more flashes, DTC 25 is indicated. DTC 12 is always flashed first. This code indicates the beginning of the code sequence. Each DTC flashes three times, and the DTCs are displayed in numerical order. The DTC sequence keeps repeating until the ignition switch is turned off or the jumper wire is removed. A DTC indicates a problem in a specific area, but voltmeter and ohmmeter tests may be necessary to locate the exact cause of the problem. These tests are explained in detail in the manufacturer's service manual.

After a defect is corrected, the DTC may be cleared by alternately connecting and disconnecting the jumper wire between DLC terminals A and C three times with the ignition switch on. Each jumper wire connection must be held for 1 second, and there should be a 1-second pause between each connection. When the DTC erasing procedure is completed, turn the ignition switch off and connect a jumper wire between terminals A and C in the DLC. Turn the ignition switch on and check the driver select switch LED flashes. If the defect is corrected and the DTCs are erased, only code 12 flashes. A CCR system DTC list is provided in Table 8-2.

Continual Strut Cycling

When diagnosing a CCR system always follow the service procedures in the vehicle manufacturer's service manual (Photo Sequence 6). If a 6.6Ω resistor is connected between terminals A and C in the DLC, the CCR control module enters a **continual strut cycling** mode. In this mode, the control module cycles all the struts from the Normal position to the Firm position every 2 seconds. When a defect is present in a strut, that strut does not cycle until a retest is performed. If the defect has corrected itself before the 3-minute retest, the strut cycles when the retest is completed.

A light-emitting diode (LED) is a special type of diode from which light is emitted when the diode conducts current.

A nonvolatile electronically erasable programmable read only memory (EEPROM) is an electronic chip that retains information in memory when ignition voltage is removed from the chip.

A test terminal called the assembly line diagnostic link (ALDL) is located under the center or left side of the instrument panel on many General Motors products. Test connections are completed at this terminal to diagnose many of the electronic systems on the vehicle.

The Society of Automotive Engineers (SAE) J1930 terminology is an attempt to standardize electronics terminology in the automotive industry.

In the SAE J1930 terminology the term data link connector (DLC) replaces older terms such as ALDL.

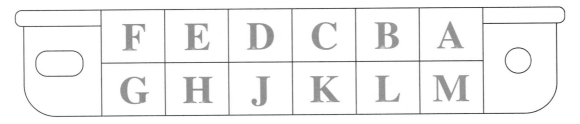

Figure 8-4 Data link connector (DLC)

Table 8-2 COMPUTER COMMAND RIDE (CCR) SYSTEM DIAG-
NOSTIC TROUBLE CODES (DTCs)

Code	Defect
12	Initiation
13	Left front actuator over current
14	Right front actuator over current
15	Left rear actuator over current
16	Right rear actuator over current
23	Left front actuator position error
24	Right front actuator position error
25	Left rear actuator position error
26	Right rear actuator position error
32	Mercury switch accelerometer error
33	Driver select switch input error
34	Vehicle speed signal error

(Courtesy of Oldsmobile Motor Division, General Motors Corp.)

Strut Connector Precautions

Classroom Manual
Chapter 8, page 178

When a strut connector is removed, the connector cavity must be clean and dry before the connector is reconnected. The connector should be inspected for corroded terminals and damaged wires. The strut connector alignment groove must align properly with the strut alignment pin as the connector is installed (Figure 8-5). Marking the alignment pin position on the outside of the strut with a wax pencil provides easier groove to pin alignment.

CCR strut connector

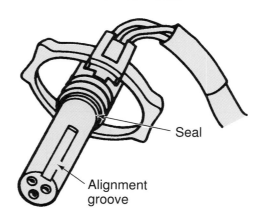

Figure 8-5 Strut connector. (Courtesy of Oldsmobile Motor Division, General Motors Corp.)

Photo Sequence 6
Typical Procedure for Diagnosing a Computer Command Ride (CCR) System

P6-1 Road test the car to check CCR operation and warm up the CCR components.

P6-2 Park vehicle in shop service bay.

P6-3 Visually inspect wires and terminals on strut actuators.

P6-4 Look up diagnostic procedure and trouble codes in the service manual.

P6-5 Locate the data link connector (DLC) under the instrument panel.

P6-6 Connect a jumper tool or wire between terminals A and C in the DLC.

P6-7 Turn the ignition switch on.

P6-8 Observe the CCR indicator light flashes to read the trouble codes.

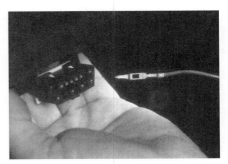

P6-9 Turn off the ignition switch and remove the jumper tool in the DLC.

Electronic Air Suspension Service and Diagnosis

Diagnostic Procedure

If the air suspension warning lamp is illuminated with the engine running, the control module has detected a defect in the electronic air suspension system. The electronic air suspension diagnostic and service procedures vary depending on the vehicle. Always follow the vehicle manufacturer's recommended procedures in the service manual. The following diagnostic and service procedures apply to electronic air suspension systems on some Lincoln Continentals and Mark VIIs. When the air suspension warning lamp indicates a system defect, the diagnostic procedure may be entered as follows:

1. Be sure the air suspension system switch is turned on.

2. Turn the ignition switch on for 5 seconds and then turn it off. Leave the driver's door open and the other doors closed.

3. Ground the diagnostic lead located near the control module and close the driver's door with the window down.

4. Turn the ignition switch on. The warning lamp should blink continuously at 1.8 times per second to indicate that the system is in the diagnostic mode.

There are 10 tests in the diagnostic procedure. The control module switches from one test to the next when the driver's door is opened and closed. The first three tests in the diagnostic procedure are the following:

1. Rear suspension

2. Right front suspension

3. Left front suspension

During these three tests, each suspension location should be raised for 30 seconds, lowered for 30 seconds, and raised for 30 seconds. For example, in test 2 this procedure should be followed on the right front suspension. If the expected signal or an illegal signal is received during the test procedure, the test stops and the air suspension warning lamp is illuminated. If all the signals and commands are normal during the first three tests, the warning lamp continues to flash at 1.8 times per second.

While tests 4 through 10 are being performed, the air suspension warning lamp flashes the test number. For example, during test 4 the warning lamp flashes 4 times followed by a pause and 4 more flashes. This flash sequence continues while test 4 is completed. The driver's door must be opened and closed to move to the next test. During tests 4 through 10, the technician must listen to or observe various components to detect abnormal operation. The warning lamp only indicates the test number during these tests. Actions performed by the control module during test 4 through 10 are as follows:

4. The compressor cycles on and off at 0.25 cycles per second. This action is limited to 50 cycles.

5. The vent solenoid opens and closes at 1 cycle per second.

6. The left front air valve opens and closes at 1 cycle per second, and the vent solenoid is opened. When this occurs, the left front corner of the vehicle should drop slowly.

7. The right front air valve opens and closes at 1 cycle per second and the vent solenoid is opened. This action causes the right front corner of the vehicle to drop slowly.

8. During this test the right rear air valve opens and closes at 1 cycle per second and the vent valve is opened. This action should cause the right rear corner of the vehicle to drop slowly.

9. The left rear solenoid opens and closes at 1 cycle per second and the vent valve is opened, which should cause the left rear corner of the vehicle to drop slowly.

10. Return the module from the diagnostic mode to normal operation by disconnecting the diagnostic lead from ground. This mode change also occurs if the ignition switch is turned off or when the brake pedal is depressed.

If defects are found during the test sequence, specific electrical tests may be performed on air valve or vent valve windings and connecting wires to locate the problem.

Air Spring Removal and Installation

CAUTION: The system control switch must be in the Off position when system components are serviced to prevent personal injury and damage to system components.

WARNING: The system control switch must be turned off prior to hoisting, jacking, or towing the vehicle. If the front of the chassis is lifted with a bumper jack, the rear suspension moves downward. The electronic air suspension system will attempt to restore the rear trim height to normal, and this action may cause the front of the chassis to fall off the bumper jack, resulting in personal injury or vehicle damage.

WARNING: When air spring valves are being removed, always rotate the valve to first stage until all the air escapes from the air spring. Never turn the valve to the second (release) stage until all the air is released from the spring.

Many components in an electronic air suspension system, such as control arms, shock absorbers, and stabilizer bars are diagnosed and serviced in the same way as the components in a conventional suspension system. However, the air spring service procedures are different compared to coil-spring service procedures on a conventional suspension system.

Follow these steps for air spring removal:

1. Turn off the electronic air suspension switch in the trunk.

2. Hoist the vehicle and allow the suspension to drop downward, or lift the vehicle with a floor jack and place safety stands under the chassis. Lower the vehicle onto the safety stands and allow the suspension to drop downward.

3. Disconnect the nylon air line from the spring solenoid valve, and rotate the valve to the first stage to allow the air to escape from the spring. Never turn the valve to the second stage until all the air is exhausted from the spring.

4. Disconnect the lower spring retainer and remove the spring from the chassis.

5. Before an air spring is installed, it must be properly folded over the piston at the bottom of the membrane (Figure 8-6).

6. Install the spring in the chassis and connect the lower spring retainer. Be sure the top of the spring is properly seated in the spring seat. When an air spring is installed in the front or rear suspension, the spring must be properly positioned to eliminate folds and creases in the membrane (Figure 8-7).

Air Spring Inflation

WARNING: Do not allow the suspension to compress an air spring until the air-spring is inflated. This action may damage the air spring.

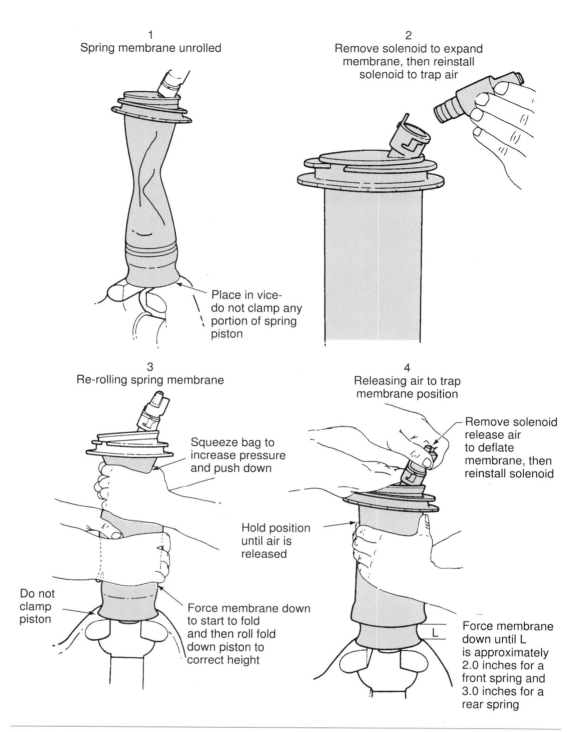

1
Spring membrane unrolled

Place in vice- do not clamp any portion of spring piston

2
Remove solenoid to expand membrane, then reinstall solenoid to trap air

3
Re-rolling spring membrane

Squeeze bag to increase pressure and push down

Do not clamp piston

Force membrane down to start to fold and then roll fold down piston to correct height

4
Releasing air to trap membrane position

Remove solenoid release air to deflate membrane, then reinstall solenoid

Hold position until air is released

Force membrane down until L is approximately 2.0 inches for a front spring and 3.0 inches for a rear spring

Figure 8-6 Air spring folding. (Courtesy of Ford Motor Company)

The weight of the vehicle must not be allowed to compress an uninflated air spring. When an air spring is being inflated use this procedure:

1. With the vehicle chassis supported on a hoist, lower the hoist until a slight load is placed on the suspension. Do not lower the hoist until the suspension is heavily loaded.

2. Turn on the air suspension system switch.

3. Turn the ignition switch from Off to Run for 5 seconds with the driver's door open and the other doors shut. Turn the ignition switch off.

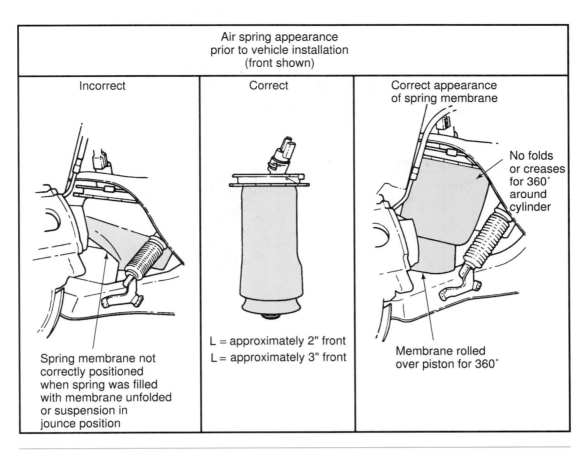

Air spring appearance prior to vehicle installation (front shown)

Incorrect	Correct	Correct appearance of spring membrane

Spring membrane not correctly positioned when spring was filled with membrane unfolded or suspension in jounce position

L = approximately 2" front
L = approximately 3" front

No folds or creases for 360° around cylinder

Membrane rolled over piston for 360°

Figure 8-7 When an air spring is installed in the front or rear suspension, the spring must be properly positioned to eliminate folds and creases in the membrane. (Courtesy of Ford Motor Company)

4. Ground the diagnostic lead.

5. Apply the brake pedal and turn the ignition to the run position. The warning lamp will flash every 2 seconds to indicate the fill mode.

6. To fill a rear spring or springs, close and open the driver's door once. After a 6-second delay, the rear spring will fill for 60 seconds.

7. To fill a front spring or springs, close and open the driver's door twice. After a 6-second delay the front spring will fill for 60 seconds.

8. When front and rear springs require filling, fill the rear springs first. Once the rear springs are filled, close and open the driver's door once to begin filling the front springs.

9. The spring fill mode is terminated if the diagnostic lead is disconnected from ground. Termination also occurs if the ignition switch is turned off or the brake pedal is applied.

Trim Height Adjustment

The **trim height** should be measured on the front and rear suspension at the locations specified by the vehicle manufacturer (Figure 8-8).

    If the rear suspension trim height is not within specifications, it may be adjusted by loosening the attaching bolt on the top height sensor bracket (Figure 8-9). When the bracket is moved one index mark up or down, the ride height is lowered or raised 0.25 in (6.35 mm).

Front and rear trim height is the distance between the vehicle chassis and the road surface measured at locations specified by the vehicle manufacturer.

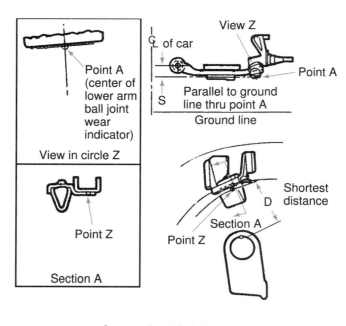

Suspension ride height

VEHICLE		S	D
MARK VII/ CONTINENTAL	INCHES	0.24	5.06
	MM	6.0	128.6

Figure 8-8 Trim height measurement locations. (Courtesy of Ford Motor Company)

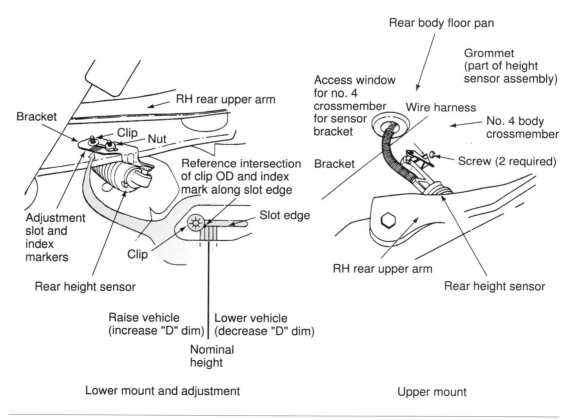

Figure 8-9 Rear trim height adjustment. (Courtesy of Ford Motor Company)

The front suspension trim height may be adjusted by loosening the lower height sensor attaching bolt. Three adjustment positions are located in the lower front height sensor bracket (Figure 8-10). If the height sensor attaching bolt is moved one position up or down, the front suspension height is lowered or raised 0.5 in (12.7 mm).

Line Service

Nylon lines on the electronic air suspension system have quick disconnect fittings. These fittings should be released by pushing downward and holding the plastic release ring, and then pulling outward on the nylon line (Figure 8-11). Simply push the nylon line into the fitting until it seats to reconnect an air line.

If a line fitting is damaged, it may be removed by looping the line around your fingers and pulling on the line without pushing on the release ring (Figure 8-11). When a new collet and release ring is installed, the O-ring under the collet must be replaced. If a leak occurs in a nylon line, a sharp knife may be used to cut the defective area out of the line. A service fitting containing a collet fitting in each end is available for line repairs. After the defective area is cut out of the line, push the two ends of the line into the service fitting (Figure 8-12).

Classroom Manual
Chapter 8, page 182

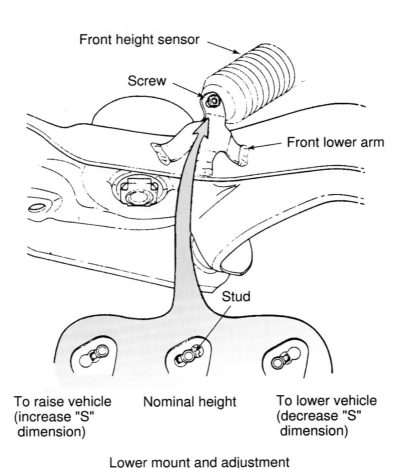

Front height sensor

Screw

Front lower arm

Stud

To raise vehicle
(increase "S"
dimension)

Nominal height

To lower vehicle
(decrease "S"
dimension)

Lower mount and adjustment

Figure 8-10 Front trim height adjustment. (Courtesy of Ford Motor Company)

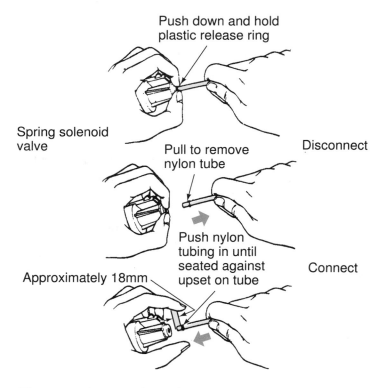

Push down and hold
plastic release ring

Spring solenoid
valve

Disconnect

Pull to remove
nylon tube

Push nylon
tubing in until
seated against
upset on tube

Connect

Approximately 18mm

Disconnect/connect from air spring soleniod
shown - same procedure for air line disconnect/
connect from air compressor dryer

Figure 8-11 Air line removal from air spring valves or compressor outlets. (Courtesy of Ford Motor Company)

Rear Load-Levelling Air Suspension System Service and Diagnosis

Trim Height Adjustment

The trim height is measured between the top of the rear axle housing and the frame (Figure 8-13). Follow this procedure to measure and adjust the rear suspension height:

1. Position the vehicle on a level floor surface over an open service pit or equivalent. Place the transmission in Park and apply the parking brake.

2. Be sure the air suspension switch is on, then open the driver's door and turn the ignition switch to the Run position.

3. Add weight to the rear bumper and wait for the suspension air compressor to run and restore the rear suspension to the trim height.

4. After the compressor stops and the rear suspension is at trim height, remove the weight from the rear bumper and close the driver's door. Allow the rear suspension drop to trim height.

5. Rock the rear suspension laterally and allow the suspension to stabilize.

6. Turn off the air suspension switch.

7. Remove the weld flash from the frame in the suspension height measurement location.

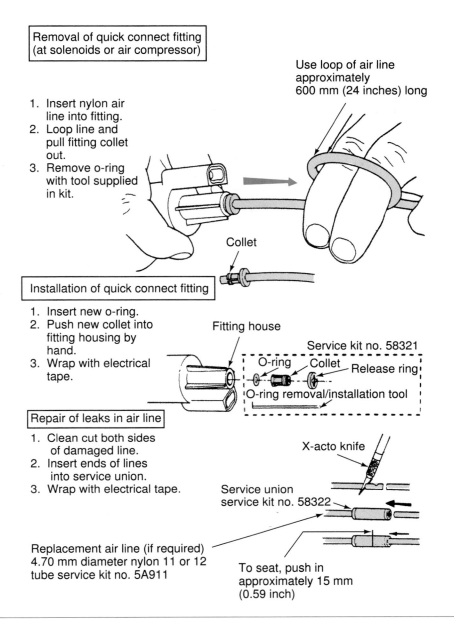

Removal of quick connect fitting (at solenoids or air compressor)

1. Insert nylon air line into fitting.
2. Loop line and pull fitting collet out.
3. Remove o-ring with tool supplied in kit.

Use loop of air line approximately 600 mm (24 inches) long

Collet

Installation of quick connect fitting

1. Insert new o-ring.
2. Push new collet into fitting housing by hand.
3. Wrap with electrical tape.

Fitting house

Service kit no. 58321

O-ring Collet Release ring

O-ring removal/installation tool

Repair of leaks in air line

1. Clean cut both sides of damaged line.
2. Insert ends of lines into service union.
3. Wrap with electrical tape.

X-acto knife

Service union service kit no. 58322

Replacement air line (if required) 4.70 mm diameter nylon 11 or 12 tube service kit no. 5A911

To seat, push in approximately 15 mm (0.59 inch)

Figure 8-12 Removal of quick disconnect fittings and air line repair procedure. (Courtesy of Ford Motor Company)

8. Install the suspension height gauge between the top of the axle housing and the lower side of the frame on the left side of the vehicle (Figure 8-14). Release the thumb screw on the gauge and slide the gauge post upward until the post contacts the frame. Tighten the thumb screw in this position.

9. Remove the suspension height gauge and read the dimension on the gauge post at the edge of the gauge body (Figure 8-14).

10. If a trim height adjustment is required, the upper height sensor mounting bolts may be loosened and the bracket moved downward to increase trim height, or upward to decrease trim height (Figure 8-15).

11. After the suspension height adjustment, repeat steps 2 through 9 to be sure the adjustment is accurate.

SPECIAL TOOLS

Suspension height gauge

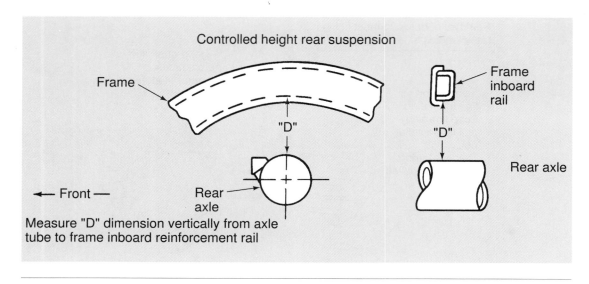

Controlled height rear suspension

Measure "D" dimension vertically from axle tube to frame inboard reinforcement rail

Figure 8-13 Rear trim height measurement location. (Courtesy of Ford Motor Company)

Drive Cycle Diagnostics

SERVICE TIP: It is difficult to diagnose a rear load-levelling suspension system without a scan tool.

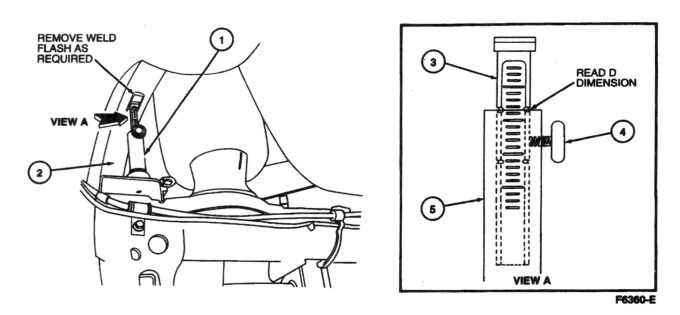

Item	Part Number	Description
1	T90P-5995-B	Height Checking Gauge
2	—	Frame Rail
3	—	Sliding Post and Scale (Part of T90P-5995-B)
4	—	Set Screw (Part of T90P-5995-B)
5	—	Gauge Housing (Part of T90P-5995-B)

Figure 8-14 Gauge installed to measure rear suspension height. (Courtesy of Ford Motor Company)

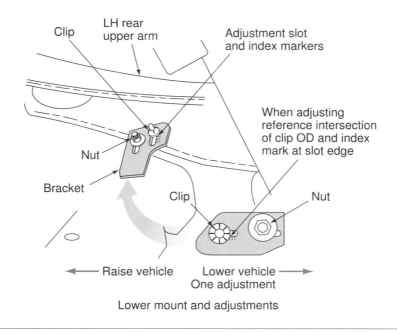

Clip
LH rear upper arm
Adjustment slot and index markers

When adjusting reference intersection of clip OD and index mark at slot edge

Nut

Bracket

Clip

Nut

← Raise vehicle

Lower vehicle →
One adjustment

Lower mount and adjustments

Figure 8-15 Trim height adjustment procedure. (Courtesy of Ford Motor Company)

When diagnosing the rear load-levelling air suspension system, these tests may be completed:

1. Drive cycle diagnostics
2. Auto test diagnostics
3. Functional tests

Drive Cycle Diagnostics

During the **drive cycle diagnostics**, the car is driven at speeds above 25 mph (40 kmph). Duplicate any driving conditions at which the customer indicated a problem with the air suspension system or any conditions that caused the check air suspension light to be illuminated. Park the car in the shop and turn off the ignition switch. Connect a **scan tool** to the air suspension data link connector (DLC) in the trunk (Figure 8-16). Press the appropriate button on the scan tool to turn it on. Operate the scan tool according to the scan tool manufacturer's instructions and retrieve any diagnostic trouble codes (DTCs) stored in the air suspension control module (Figure 8-17).

Auto Test Diagnostics

When the **auto test diagnostic mode** is entered with the scan tool, the air suspension control module performs these tests:

1. Module test (DTC 70)
2. Unstable battery voltage test (DTC 80)
3. Shorted or open conditions that could result in DTCs 39-46
4. Shorted or open conditions that could result in DTCs 68-71

If any shorted or open conditions are found, DTC 13 is displayed and the automatic part of the test is ended. When no defects are located, the module continues with the automatic part of the test. In the next test stage, the suspension control module raises and lowers the suspension height to verify the high, trim, and low suspension height signals from the rear suspension height sensor. When the air suspension system is operating normally, the rear suspension is at trim height after this test is completed. If all three height sensor signals are not satisfactory, DTC 13 is displayed on

A scan tool is a digital tester that may be used to diagnose various onboard computer systems.

SPECIAL TOOLS

Scan tool

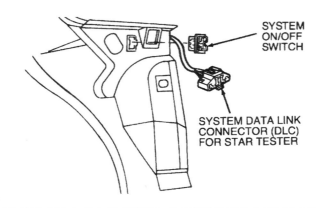

Figure 8-16 Data link connector (DLC) in the trunk for rear load-levelling air suspension system. (Courtesy of Ford Motor Company)

DTC	Pinpoint Test	Description	Service Priority
10		Diagnostics Entered, Auto Test in Progress	
11		Vehicle Passes[a]	
12		Auto Tested Passed, Perform Manual Inputs	
13		Auto Test Failed, Perform Manual Inputs	
15	—	No Drive Cycle Errors Detected	
16	—	EVO Diagnostic Trouble Code (Refer to Section 11-00)	
17	—	EVO Diagnostic Trouble Code (Refer to Section 11-00)	
18	—	EVO Diagnostic Trouble Code (Refer to Section 11-00)	
23	b	Functional Test, Vent Rear	
26	b	Functional Test, Compress Rear	
28	P	Steering Select Switch Not Detected (Town Car Only)	4th
31	—	Functional Test, Air Compressor Relay Toggle	
32	b	Functional Test, Vent Solenoid Toggle	
33	b	Functional Test, Air Spring Solenoid Toggle	
35	—	Drive Cycle Error Codes Erased OK	
39	A	Compressor Relay Control Circuit Short to Battery	2nd
40	A	Compress or Relay Control Circuit Short to Ground.	2nd
42	B	Air Spring Solenoid Valve Short to Ground	2nd
43	C	Air Spring Solenoid Valve Short to Battery	2nd
44	D	Vent Solenoid Valve Short to Battery	2nd
45	E	Short in Vent Solenoid Valve Circuit	2nd
46	F	Air Suspension Height Sensor Supply Circuit Shorted	2nd
51	G	Unable to Detect Lowering of Rear	3rd
54	H	Unable to Detect Raising of Rear	3rd
68	J	Short in Rear Air Suspension Height Sensor Circuit	2nd
70	—	Replace Control Module	
71	K	Air Suspension Height Sensor Circuit Open	3rd
72	L	Did Not Detect 4 Open and Closed Door Signals	4th
74	—	EVO Diagnostic Trouble Code (Refer to Section 11-00)	
80	M	Control Module Detects Low Battery Voltage	1st
—	N	Unable to Enter Auto Test Diagnostics or Warning Light Remains On After Vehicle Passes (Code 11)	

a   If vehicle is still low or high in rear, check ride height as outlined.
b   Functional tests after during Auto Manual Test.

Figure 8-17 Diagnostic trouble codes (DTCs) for rear load-levelling air suspension system. (Courtesy of Ford Motor Company)

the scan tool. When all height sensor signals are received properly by the control module, DTC 12 is displayed on the scan tool. When DTC 12 is displayed, the manual part of the test may be completed. In the manual portion of this test, each door is opened and closed, and the steering wheel is turned about 1/4 turn in each direction. During these actions, the control module monitors each door switch and the steering sensor. At the end of the manual test mode, the scan tool displays DTC 11 indicating a system pass or other DTCs indicating defects in the system. If several DTCs are displayed, each DTC appears on the scan tool for 15 seconds.

Functional Tests

 WARNING: Extended use of the functional tests may overheat the compressor and cause the circuit breaker in the compressor to open. This circuit breaker should cool and reset itself if the compressor is inoperative for 15 minutes.

SERVICE TIP: Various scan tools have different diagnostic capabilities. Some test functions may not be available in certain scan tools.

WARNING: Most scan tools require the installation of the proper module in the scan tool for the system being tested. If the wrong module is installed in the scan tool for the system being tested, diagnosis is impossible and scan tool or electronic system damage may occur.

The **functional tests** can only be run after auto test diagnostics if there are no faults located in the auto test diagnostics. The functional tests may be useful in diagnosing air suspension defects. Specific scan tool buttons may have to be pressed to complete this test sequence as directed on the scan tester display. During the functional tests the control module performs these functions:

1. DTC 23: During this test, the control module opens the rear air spring solenoid valves and the vent valve; the rear suspension height should become lower.

2. DTC 26: During this test, the control module starts the compressor and opens the rear air spring solenoid valves; the rear suspension height should become higher.

3. DTC 31: During this test, the control module cycles the compressor relay on and off at 1-second intervals.

4. DTC 32: During this test, the control module cycles the vent solenoid on and off at 1-second intervals.

5. DTC 33: During this test, the control module cycles both air spring solenoid valves on and off at 1-second intervals.

WARNING: If the scan tool is disconnected while it is turned on or while the ignition switch is turned on, the module may be damaged in the system being tested. Be sure the scan tool is turned off and the ignition switch is off before disconnecting the scan tool from the DLC.

Automatic Air Suspension Service and Diagnosis

Drive Cycle Diagnostics

The drive cycle diagnostic procedure displays any diagnostic trouble codes (DTCs) that have occurred during the last drive cycle. This test procedure detects intermittent faults that may not appear during the other test procedures.

The car should be driven for a minimum of 5 minutes at different speeds and road conditions, or until a ride control message is illuminated in the instrument panel. After this driving period, the

The same control module operates the rear load-levelling air suspension system and the electronic variable orifice (EVO) steering system. In the EVO steering system, the control module operates a solenoid in the power steering pump so that less effort is required to turn the steering wheel at low speeds, and more steering effort is required at higher speeds

Classroom Manual
Chapter 8, page 190

car should be driven directly to the service area and parked with the ignition switch off. The DTCs are now stored in the air suspension control module for 1 hour of ignition off time, or until the ignition switch is turned on. If the ignition switch is turned on during this 1-hour period, the DTCs are erased in the air suspension module memory. The automatic air suspension on/off switch in the trunk must be on during the test procedure.

A scan tool with air suspension test capabilities should be connected to the diagnostic connector near the air suspension module in the trunk (Figure 8-18). Always follow the instructions in the vehicle manufacturer's service manual, and in the scan tester manufacturer's manual. The scan tool displays a pass DTC, or one or more of 32 possible DTCs. The DTCs are displayed continuously for 1 hour or until the ignition switch is turned on. The DTCs should be written down for later reference.

Service Bay Diagnostics

Auto/Manual Diagnostic Check.   During the **auto/manual test** the air suspension control module verifies itself and checks other system components. After this check the scan tool displays "12" (okay do manual checks), or "13" (faults detected do manual checks). The manual tests allow the technician to test the system inputs to the air suspension module.

A hard fault code indicates a fault that is present all the time including the time of testing.

Diagnostic Trouble Code (DTC) Display.   After the auto/manual test procedure the DTCs may be displayed by pressing the appropriate scan tool button (Figure 8-19). Each DTC is displayed for approximately 15 seconds. The DTC display continues until the next test mode is entered. The DTCs obtained during the drive cycle diagnosis should be compared to DTCs that appear during the **service bay diagnostics.** When a DTC appears during both test modes, the DTC represents a **hard fault**. If the DTC is only present during the drive cycle diagnosis, the code indicates an **intermittent fault.**

An intermittent fault code indicates fault that occurs and then disappears.

Pinpoint Tests.   Many of the DTCs have a **pin point test** procedure represented by two letters. Detailed pinpoint test procedures must be followed in the car manufacturer's service manual. The

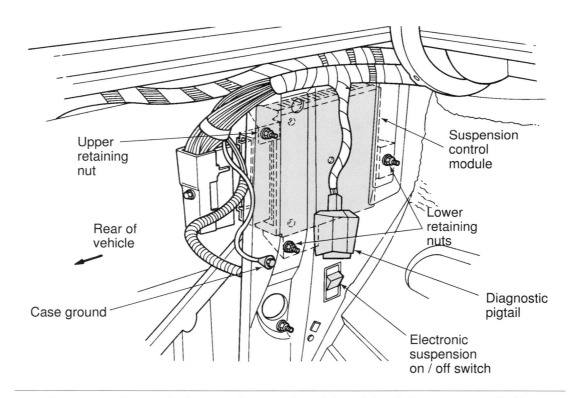

Figure 8-18 Automatic air suspension control module and data link connector (DLC). (Courtesy of Ford Motor Company)

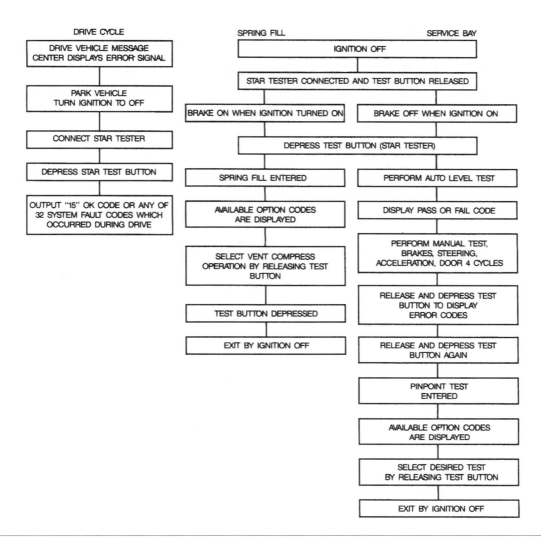

Figure 8-19 Drive cycle, service bay, and spring fill diagnostic procedures. (Courtesy of Ford Motor Company)

pinpoint tests are designed to locate the exact cause of a DTC. For example, DTC 40 may be obtained indicating a problem in the left front air spring solenoid valve circuit. Pinpoint test EA will inform the technician to perform specific tests to prove if the trouble is in the solenoid valve, connecting wires, or module (Figure 8-20).

The pinpoint tests are prioritized, with 1 being the highest priority and 7 the lowest. One fault may result in the display of a second fault. The pinpoint tests must be performed in the order of priority starting with the highest priority (lowest numbers).

 SERVICE TIP: When fault codes are provided on a priority basis, always recheck the system for codes after a fault is corrected. Some computers are only capable of storing a specific number of codes in order of priority. If only one code is displayed, there could be another fault of lesser priority in the system.

Classroom Manual
Chapter 8, page 194

Spring Fill Diagnostics

The **spring fill diagnostics** allow the technician to force the automatic air suspension control module to fill and vent each air spring. This procedure indicates inoperative components such as air spring solenoids or the module. During the spring fill procedure, the technician can check for air line leaks.

STAR CODE	PINPOINT PROCEDURE	DESCRIPTION	SERVICE PRIORITY
10		Service Bay Diagnostics Entered	
11		System Checked Out Okay	
12		Automatic Test Completed — No Faults Detected — Perform Manual Inputs	
13		Automatic Test Completed — Faults Detected — Perform Manual Inputs	
15		No Faults Detected	
21		Vent Right Front Air Spring	
22		Vent Left Front Air Spring	
23		Vent Right Rear Air Spring	
24		Inflate Right Front Air Spring	
25		Inflate Left Front Air Spring	
26		Inflate Right Rear Air Spring	
27		Vent Left Rear Air Spring	
28		Inflate Left Rear Air Spring	
31		Air Compressor Toggle	
32		Vent Solenoid Valve Toggle	
33		Air Spring Solenoid Valve Toggle	
34		Shock Actuator Toggle (Firm/Soft)	
35		Door Open & Door Closed Detection	
40	EA	Short — Left Front Air Spring Solenoid Valve Circuit	2nd
41	EB	Short — Right Front Air Spring Solenoid Valve Circuit	2nd
42	EC	Short — Left Rear Air Spring Solenoid Valve Circuit	2nd
43	ED	Short — Right Rear Air Spring Solenoid Valve Circuit	2nd
44	EE	Short — Vent Solenoid Valve Circuit	2nd
45	EF	Short — Air Compressor Relay Circuit	2nd
46	EG	Short — Height Sensor Power Supply Circuit	2nd
47	EH	Short — Soft Shock Actuator Relay Circuit	2nd
48	EI	Short — Firm Shock Actuator Relay Circuit	2nd
49	HA	Unable to Detect Lowering of Right Front Corner	5th
50	HB	Unable to Detect Lowering of Left Front Corner	5th
51*	HC	Unable to Detect Lowering of Right Rear Corner	5th
51*		Unable to Detect Lowering of Rear of Vehicle	5th
52	IA	Unable to Detect Raising of Right Front Corner	6th
53	IB	Unable to Detect Raising of Left Front Corner	6th
54*	IC	Unable to Detect Raising of Right Rear Corner	6th
54*		Unable to Detect Raising of Rear of Vehicle	6th
55	JA	Speed Greater Than 15 mph Not Detected	7th
56	GA	Soft Not Detected — Left Rear Shock Actuator Circuit	4th
57	GB	Soft Not Detected — Right Front Shock Actuator Circuit	4th
58	GC	Soft Not Detected — Left Front Shock Actuator Circuit	4th
59	GD	Soft Not Detected — Right Rear Shock Actuator Circuit	4th
60	GA	Firm Not Detected — Left Rear Shock Actuator Circuit	4th
61	GB	Firm Not Detected — Right Front Shock Actuator Circuit	4th
62	GC	Firm Not Detected — Left Frt Shock Actuator Circuit	4th
63	GD	Firm Not Detected — Right Rear Shock Actuator Circuit	4th
64	GE	Soft Not Detected — All Shock Actuator Circuits	
65	GE	Firm Not Detected — All Shock Actuator Circuits	4th
66	EJ	Short — Right Front Height Sensor Circuit	2nd
67	EK	Short — Left Front Height Sensor Circuit	2nd
68	EL	Short — Rear Height Sensor Circuit	2nd
69	FA	Open — Rilght Front Height Sensor Circuit	3rd
70	FB	Open — Left Front Height Sensor Circuit	3rd
71	FC	Open — Rear Height Sensor Circuit	3rd
72	JB	At Least Four Open & Closed Door Signals Not Detected	7th
73	JC	Brake Pressure Switch Activation Not Detected	7th
74	JD	Steering Wheel Rotations Not Detected	7th
75	JE	Acceleration Signal Not Detected	7th
78	HD	Unable to Detect Lowering of Left Rear Corner	5th
79	ID	Unable to Detect Raising of Left Rear Corner	6th
80	DA	Insufficient Battery Voltage to Run Diagnostics	1st

NOTE: System faults have been prioritized for repair. Start with those codes identified with a service priority of: 1st, then 2nd, then 3rd, . . . and finally 7th.

Figure 8-20 Diagnostic trouble codes (DTCs) and pinpoint tests. (Courtesy of Ford Motor Company)

Diagnosis of Air Suspension with Speed Levelling Capabilities

Drive Cycle Diagnostics

The first step in diagnosing the air suspension system is to verify any customer complaints. Road test the vehicle if these complaints cannot be verified with the car in the shop. Before performing an electronic diagnosis of the air suspension system, visually inspect the system for mechanical or electronic defects (Figure 8-21). Repair any defects found during the visual inspection before proceeding with further diagnosis.

The drive cycle diagnostic test on this air suspension system is similar to the same test procedure on other air suspension systems. If the car has not been driven for over one hour or the air suspension switch has been turned off, turn this switch on and drive the vehicle. During the road test, try to duplicate the conditions the customer described. If the air suspension module detects an electrical defect in the air suspension system, the air suspension warning light will illuminate. Return the car to the shop, turn the ignition switch off, and leave the air suspension switch on. The air suspension diagnostic trouble codes (DTCs) remain stored in the air suspension module for one hour after the ignition switch is turned off. Connect a scan tool to the air suspension data link connector (DLC) located on the right front shock absorber tower. Be sure the proper cartridge for the air suspension diagnosis is installed in the scan tool. Record any DTCs displayed on the scan tool (Figure 8-22 and Figure 8-23). Some DTCs may be displayed during the drive cycle or auto test modes, whereas other DTCs are only displayed during the auto test mode.

Auto Test Diagnostics

During the auto test sequence, the scan tool displays certain DTCs representing electrical defects in the air suspension system. In the auto test mode the technician must activate certain air suspension components, and the air suspension module detects any problems in these components and displays appropriate DTCs. Follow these steps to complete the auto test diagnostics:

1. Connect a battery charger to the vehicle battery terminals with the correct polarity. Select a low charging rate on the charger.

2. Open the trunk so the air suspension switch is accessible.

3. Connect the scan tool to the air suspension data link connector (DLC) positioned on the right front shock absorber tower. Be sure the proper cartridge for air suspension diagnosis is installed in the scan tool.

4. Be sure all the vehicle doors are closed; leave the driver's window down so the ignition switch is accessible.

5. Turn the air suspension switch off and then turn this switch on.

6. Turn the ignition switch from the Off to the Run position.

7. Select air suspension diagnosis and auto test mode on the scan tool.

Mechanical	Electrical
• Restricted suspension movement • Overloaded luggage compartment	• Open fuses: — Power Distribution Box-Fuse 4 (15A Maxi), Fuse 13 (60A Maxi) — Fuse Junction Panel-Fuse 28 (10A) • Loose connectors • Corroded connectors • Air suspension service switch OFF • Corroded battery terminals

Figure 8-21 Visual inspection of air suspension system. (Courtesy of Ford Motor Company)

AIR SUSPENSION/EVO CONTROL MODULE DIAGNOSTIC TROUBLE CODE (DTC) INDEX

DTC	PPT	Description	Error Handling	Display	Generated	Validation
10		Auto test in progress			AT	
11		Auto test passed			AT	
12		Perform manual tests			AT	
15		No faults stored in memory			AT	
18	DTC 18	Control Module Detects Low Battery Voltage	Disable AS and EVO until control module senses more than 13 volts for 1 second continuous		AT	Less than 11 volts for 1 second continuous
19	DTC 19	Control Module Detects High Battery Voltage	Disable AS and EVO until control module senses less than 17.5 volts for 1 second continuous		AT	More than 19 volts for 1 second continuous
20		Control Module Memory Error 2	Disable AS	On after 1 second	DC AT	Replace control module
25	DTC 25	Air Suspension Height Sensor Supply Not 5 Volts	Disable AS	On after 1 second	DC AT	
29	DTC 29	Two Door Cycles Not Detected During Test			AT	
35	DTC 35	EVO Power Steering Control Valve Actuator Concern	AS normal, EVO full assist		DC AT	Fault after 1 second continuous
45	DTC 45	Steering Rotation Not Detected			AT	
50	DTC 50	LH Front Air Suspension Height Sensor Signal Out Of Range	Disable AS	On after 1 second	DC AT	Fault after 1 second continuous
55	DTC 55	RH Front Air Suspension Height Sensor Signal Out Of Range	Disable AS	On after 1 second	DC AT	Fault after 1 second continuous
60	DTC 60	LH Rear Air Suspension Height Sensor Signal Out Of Range	Disable AS	On after 1 second	DC AT	Fault after 1 second continuous
70	DTC 70	Vent Solenoid Valve	Disable AS	On after 1 second	DC AT	Fault after 1 second continuous
75	DTC 75	Compressor Relay Control Circuit	Disable AS	On after 1 second	DC AT	Fault after 1 second continuous
80	DTC 80	LH Front Spring Solenoid Valve Circuit Failure	Disable AS	On after 1 second	DC AT	Fault after 1 second continuous

Figure 8-22 Diagnostic trouble codes (DTCs) for air suspension system with speed levelling capabilities. (Courtesy of Ford Motor Company)

 SERVICE TIP: Do not lean on the car or open the car doors while DTC 10 is displayed in the auto test. This action causes errors in the test results.

8. When DTC 10 is displayed on the scan tool the auto test is running. If the air suspension module detects any faults, the auto test is interrupted and the DTCs representing the fault or faults are displayed on the scan tool.

DTC	PPT	Description	Error Handling	Display	Generated	Validation
85	DTC 85	RH Front Spring Solenoid Valve Circuit Failure	Disable AS	On after 1 second	DC AT	Fault after 1 second continuous
90	DTC 90	LH Rear Spring Solenoid Valve Circuit Failure	Disable AS	On after 1 second	DC AT	Fault after 1 second continuous
95	DTC 95	RH Rear Spring Solenoid Valve Circuit Failure	Disable AS	On after 1 second	DC AT	Fault after 1 second continuous
98	DTC 98	Compressor Run Time Exceeded	Disable AS	On after ON-REST-ON cycle. 11.5 minutes	DC AT	Fault after ON-REST-ON cycle. 11.5 minutes
99	DTC 99	Unable To Detect Raising or Lowering Of One Or More Corners	Disable AS	On after either maximum vent or maximum pump time-out (90-second vent, 90-second pump)	DC AT	Fault after either max. vent or max. pump time-out (90-second vent, 90-second pump)
	A	Unable to enter Auto Test				
	B	Speed Greater Than 8 km/h (5 mph) Not Detected In Last 16 Ignition Cycles—Speed Sensor Circuit Diagnosis				
	C	No Codes Displayed, Auto Test Is Running	No Codes Displayed, Auto Test Is Running			

Abbreviations Used:

- DC = Drive Cycle
- AT = Auto Test
- AS = Air Suspension
- PPT = Pinpoint Test

Figure 8-23 (continued) Diagnostic trouble codes (DTCs) for air suspension system with speed levelling capabilities. (Courtesy of Ford Motor Company)

9. When DTC 12 is displayed on the scan tool, open the driver's door, turn the steering wheel 1/4 turn in each direction, and close the driver's door. After this procedure, open and close the passenger's door. Press the appropriate button on the scan tool to continue the auto test.

10. Record the DTCs displayed on the scan tool. If DTC 11 is displayed, the air suspension module did not detect any electrical defects during the auto test.

11. Turn off the ignition switch and disconnect the scan tool.

Functional Tests

During the functional test procedure, the scan tool commands the air suspension module to operate specific air suspension components. The technician can determine if the appropriate components are functional during each test. Complete these steps to perform the functional tests:

1. Connect a battery charger to the vehicle battery terminals with the correct polarity. Select a low charging rate on the charger.

2. Open the trunk and shut off the air suspension switch.

3. Connect a scan tool to the air suspension DLC located on the right front shock absorber tower. Select air suspension and functional tests on the scan tool.

4. Be sure all the car doors are shut; leave the driver's window down so the ignition switch is accessible.

5. Turn on the ignition switch.

6. Select functional test 211 and and record any DTCs displayed on the scan tool.

7. Select functional test 228 to clear any DTCs from the air suspension module memory.

▲ **WARNING:** If the compressor runs for more than 2 minutes in functional test 221, the internal circuit breaker in the compressor may open the compressor circuit and cause the compressor to stop. If this action occurs, turn off the ignition switch for 10 minutes to allow this circuit breaker to reset.

8. Select each functional test on the scan tool display, and check the appropriate air suspension component for the desired operation during each functional test (Figure 8-24).

Classroom Manual
Chapter 8, page 200

9. At the conclusion of the functional tests, turn off the ignition switch and disconnect the scan tool.

Test Code	Description
211	Display All Diagnostic Trouble Codes In Memory
212	LH Front Pump with Audible Air Suspension Height Sensor Check
213	LH Front Vent with Audible Air Suspension Height Sensor Check
214	RH Front Pump with Audible Air Suspension Height Sensor Check

Test Code	Description
219	RH Rear Vent with Audible Air Suspension Height Sensor Check
221	Compressor Run
222	Actuator Output Test (Cycles All Solenoids and EVO Power Steering Control Valve Actuator)
223	LH Front Air Suspension Height Sensor Trim Detection, Audible Output
224	RH Front Air Suspension Height Sensor Trim Detection, Audible Output
225	LH Rear Air Suspension Height Sensor Trim Detection, Audible Output

Test Code	Description
215	RH Front Vent with Audible Air Suspension Height Sensor Check
216	LH Rear Pump with Audible Air Suspension Height Sensor Check
217	LH Rear Vent with Audible Air Suspension Height Sensor Check
218	RH Rear Pump with Audible Air Suspension Height Sensor Check

Test Code	Description
226	Vehicle Speed Sensor Detection
227	Pulse EVO Power Steering Control Valve Actuator through Duty Cycle
228	Erase All Diagnostic Trouble Codes Stored in Control Module Memory

Figure 8-24 Functional tests for air suspension system with speed levelling capabilities. (Courtesy of Ford Motor Company)

Diagnosis of Automatic Ride Control (ARC) Systems

Diagnostic Trouble Code (DTC) Retrieval and Module Parameter Identification (PID)

The first step in diagnosing the **automatic ride control (ARC)** system is to verify any customer complaints. Road test the vehicle if these complaints cannot be verified with the vehicle in the shop. Before performing an electronic diagnosis of the ARC system, visually inspect the system for mechanical or electronic defects (Figure 8-25). Repair any defects found during the visual inspection before proceeding with further diagnosis.

The scan tool is connected to the ARC system DLC positioned below the steering column. Be sure the ignition switch is off when connecting or disconnecting the scan tool. All doors including the rear tailgate must be closed, and the transfer case selector switch must be in two-wheel drive. The throttle should be closed. With the scan tool connected, turn the ignition switch on, and obtain and record the **continuous DTCs** on the scan tool display (Figure 8-26). Continuous DTCs represent faults in the ARC system that are present at the time of testing.

The **parameter identification (PID) screen** in the scan tool may be used to determine if ARC system components are operating electrically. When the PID screen is displayed on the scan tool, the ARC system component operation is monitored on the screen (Figure 8-27). The scan tool only displays a certain number of PIDs on the screen at one time. Use the appropriate buttons on the scan tool to scroll through the PIDs. The operation of some PIDs may be checked with the vehicle in the shop. For example, the DR-OPEN PID may be checked by opening and closing a door. When this action is taken, this PID should change from open to closed. If the brake pedal is depressed and released the brake on/off (BOO)-ARC PID should change from on to off if the brake switch signal is received by the ARC module. The vehicle has to be driven on the road to check other PIDs such as the vehicle speed sensor (VSS-ARC). If any ARC component does not operate properly, the technician may perform voltmeter and ohmmeter tests to determine if the problem is in the fuses, wiring harness, or the inoperative component. These tests may be called pinpoint tests.

Module Diagnostics

The module diagnostics tests may be used to obtain the DTCs in the ARC system. During this test mode, active commands may be used to operate specific system components and determine if the

Mechanical	Electrical

• Restricted suspension movement
• Excessive vehicle load
• Crimped, kinked, or leaking air line
	— Air suspension switch is OFF
	— Blown fuse:
	  — Power distribution box — maxi-fuse 9 (40A), maxi-fuse 8 (20A)
	  — Fuse junction panel — fuse 10 (7.5A), fuse 9 (7.5A)
	— Loose connectors
	— Corroded connectors
	— Damaged solenoid valve

Figure 8-25 Visual inspection of automatic ride control (ARC) system. (Courtesy of Ford Motor Company)

ARC Module Diagnostic Trouble Code (DTC) Index

DTC	Caused By	Description	Action
B1318	Battery Voltage Low	ARC Module	GO to Pinpoint Test U.
B1342	ECU Is Defective	ARC Module	GO to Pinpoint Test V.
B1485	Brake Pedal Position Switch Input Short to Battery	ARC Module	GO to Pinpoint Test W.
B1565	Door Ajar Input Short to Power	ARC Module	GO to Pinpoint Test X.
C1439	Acceleration Input Signal Circuit Failure	ARC Module	GO to Pinpoint Test F.
C1724	Height Sensor Power Circuit Failure	ARC Module	GO to Pinpoint Test G.
C1725	Front Pneumatic Failure	ARC Module	GO to Pinpoint Test H.
C1726	Rear Pneumatic Failure	ARC Module	GO to Pinpoint Test J.
C1756	Front Height Sensor Circuit Failure	ARC Module	GO to Pinpoint Test K.
C1760	Rear Height Sensor Circuit Failure	ARC Module	GO to Pinpoint Test K.
C1770	Vent Solenoid Circuit Failure	ARC Module	GO to Pinpoint Test L.
C1830	Air Compressor Relay Circuit Failure	ARC Module	GO to Pinpoint Test M.
C1845	Front Fill Solenoid Circuit Failure	ARC Module	GO to Pinpoint Test N.
C1865	Rear Fill Solenoid Circuit Failure	ARC Module	GO to Pinpoint Test N.
C1869	Rear Gate Solenoid	ARC Module	GO to Pinpoint Test P.
C1901	Right Rear Shock Absorber Circuit Failure	ARC Module	GO to Pinpoint Test Q.
C1905	Left Rear Shock Absorber Circuit Failure	ARC Module	GO to Pinpoint Test R.
C1909	Right Front Shock Absorber Circuit Failure	ARC Module	GO to Pinpoint Test S.
C1913	Left Front Shock Absorber Circuit Failure	ARC Module	GO to Pinpoint Test T.
P1807	4X4 High Range Input Short to Ground	ARC Module	GO to Pinpoint Test D.
P1808	4X4 Low Range Input Circuit Failure (4.0L), Ride Control Switch Input Circuit Failure (5.0L)	ARC Module	GO to Pinpoint Test E.

Figure 8-26 ARC system diagnostic trouble codes (DTCs). (Courtesy of Ford Motor Company)

system responds properly. If there are any defects when these components are operated, DTCs will be set in the ARC module to help the technician locate the defective component. Follow this procedure to complete the module diagnostics.

PID	Description	Expected Values
4X4HIGH	4X4 High Input	IN, OUT
4X4__LOW	4X4 Low Input	IN, OUT
AS__COMP	Compressor Relay Status	ON---, ONO--, ON-B-, ON--G, OFF---, OFFO--, OFF-B-, OFF--G
AS__GATE	Front Gate Solenoid Status	ON---, ONO--, ON-B-, ON--G, OFF---, OFFO--, OFF-B-, OFF--G
AS__VENT	Vent Solenoid Status	ON---, ONO--, ON-B-, ON--G, OFF---, OFFO--, OFF-B-, OFF--G
BOO__ARC	Brake Pedal Position Switch Input	ON, OFF
CCNTARC	Number Of Continuous DTCs Counted by the ARC Module	one count per bit
DR__OPEN	Door Ajar Input	OPEN, CLOSED
F__FILL	Front Fill Solenoid Status	ON---, ONO--, ON-B-, ON--G, OFF---, OFFO--, OFF-B-, OFF--G
FHGTSEN	Front Height Sensor	#.## VDC
HGTSENS	Height Sensor	ON, OFF
IGN__RUN	Detection of Ignition Switch in the RUN Position	RUN, notRUN
LFSHK__E	Left Front Shock Encoder Status	SOFT, FIRM
LRSHK__E	Left Rear Shock Encoder Status	SOFT, FIRM
OFF ROAD	Vehicle Off Road Status	ON, OFF
OPSTRAT	Operational Strategy	ARC
PCM__ACC	Acceleration Signal From the Powertrain Control Module (PCM)	YES, NO
R__FILL	Rear Fill Solenoid Status	ON---, ONO--, ON-B-, ON--G, OFF---, OFFO--, OFF-B-, OFF--G
RASGATE	Rear Gate Solenoid Status	ON---, ONO--, ON-B-, ON--G, OFF---, OFFO--, OFF-B-, OFF--G
RFSHK__E	Right Front Shock Encoder Status	SOFT, FIRM
RHGTSEN	Rear Height Sensor	#.## VDC
RRSHK__E	Right Rear Shock Encoder Status	SOFT, FIRM
STEER__A	Steering Rotation Sensor A	LOW, HIGH

Figure 8-27 ARC system parameter identification (PID). (Courtesy of Ford Motor Company)

SERVICE TIP: During the module diagnostics test procedure, compressor operation may lower the battery voltage and cause inaccurate test results. A battery charger connected to the vehicle battery prevents this problem.

1. Connect a battery charger to the vehicle battery terminals with the correct polarity. Select a low charging rate on the battery charger.

2. Open any vehicle door.

3. Place the transfer case selector switch in 4X4 LOW if the vehicle has a 4.0L engine. This switch position is identified as 4X4 OFF ROAD if the vehicle is equipped with a 5.0L engine.

4. Connect the scan tool to the ARC DLC.

5. Select the on-demand self-test on the scan tool display.

6. Observe and record all the on-demand DTCs on the scan tool display. **On-demand DTCs** represent intermittent ARC system faults.

7. Use the clear codes function on the scan tool to clear the DTCs.

⚠️ **WARNING:** If the compressor runs for more than 2 minutes in functional test 221, the internal circuit breaker in the compressor may open the compressor circuit and cause the compressor to stop. If this action occurs, turn off the ignition switch for 10 minutes to allow this circuit breaker to reset.

8. Select these active commands on the scan tool to raise the vehicle: front fill, rear fill, gate valve, and compressor. Operate the compressor for 30 seconds.

9. Select these active commands on the scan tool to lower the vehicle: front fill, rear fill, gate valve, and vent. All the possible active commands are shown in Figure 8-28.

10. Observe and record all DTCs displayed on the scan tool.

11. Use the clear codes function on the scan tool to clear the DTCs.

Intermittent Fault Test

The intermittent fault test may be used to find intermittent problems such as loose wiring connectors. All DTCs must be cleared from the ARC module before performing the intermittent fault test. Follow this procedure for the intermittent fault test:

1. Open any door on the vehicle.

2. Turn on the ignition switch.

3. Place the transmission selector in the Park position.

Active Command	Display	Action
RIDE CONTROL MODULE	COMPRESSR	ON, OFF
	FRNT FILL	ON, OFF
	GATEVALVE	ON, OFF
	LF SMOTOR	ON, OFF
	LFSH FIRM	ON, OFF
	LR SMOTOR	ON, OFF
	LRSH FIRM	ON, OFF
	MC NORMAL	ON, OFF
	PWR CTRL	ON, OFF
	REAR FILL	ON, OFF
	RF SMOTOR	ON, OFF
	RFSH FIRM	ON, OFF
	RR SMOTOR	ON, OFF
	RRSH FIRM	ON, OFF
	VENT	ON, OFF

Figure 8-28 ARC system active commands. (Courtesy of Ford Motor Company)

PID	State
IGN__RUN	RUN
RRSHK__E	FIRM
LRSHK__E	FIRM
RFSHK__E	FIRM
LFSHK__E	FIRM
STEER__A	HIGH
STEER__B	HIGH
BOO__ARC	ON
DR__OPEN	AJAR
4X4__LOW (4.0L)	IN
OFFROAD (5.0L)	IN
4X4 HIGH	OUT
PCM__ACC	YES

Figure 8-29 PID monitor displayed prior to the intermittent fault test. (Courtesy of Ford Motor Company)

4. Place the transfer case selector switch in 4X4 low if the vehicle has a 4.0L engine or in the OFF ROAD position when the vehicle has a 5.0L engine. Be sure the 4X4 low indicator is illuminated.

5. Fully depress the brake pedal and the accelerator pedal.

6. Connect the scan tool to the DLC.

7. Select the PID monitor in the scan tool, and be sure the PIDs shown in Figure 8-29 are displayed.

8. Select I/O Circuit Check on the scan tool.

9. All displayed inputs should indicate NO CHANGE. Wiggle any suspected wiring connectors or harness. If any input indicates CHANGED, an intermittent open circuit has occurred at the connection or harness that was wiggled.

Classroom Manual
Chapter 8, page 202

Diagnosis of Road Sensing Suspension System

Diagnostic Trouble Code (DTC) Display

WARNING: Never remove or install the wiring connector on a computer or computer system component with the ignition switch on. This action may result in computer damage.

WARNING: Do not supply voltage to or ground any circuit or component in a computer system unless instructed to do so in the vehicle manufacturer's service manual. This action may damage computer system components.

WARNING: During computer system diagnosis, use only the test equipment recommended in the vehicle manufacturer's service manual to prevent damage to computer system components.

The powertrain control module (PCM) controls such output functions as electronic fuel injection, spark advance, and emission devices.

The instrument panel cluster (IPC) module controls most instrument panel displays.

The air conditioning programmer (ACP) controls the air conditioning output functions.

The supplemental inflatable restraint (SIR) module controls air bag inflation.

The traction control system (TCS) module is combined with the antilock brake system (ABS) module, and this module controls wheel lockup during hard braking and wheel spinning during acceleration.

The real time damping (RTD) module controls the road sensing suspension (RSS) system. As mentioned in the Classroom Manual, the term RTD is used in place of RSS in the system diagnosis.

☑ **SERVICE TIP:** When removing, replacing, or servicing an electronic component on a vehicle, always disconnect the negative battery cable before starting the service procedure. If the vehicle is equipped with an air bag or bags, wait 1 minute after the battery negative cable is removed to prevent accidental air bag deployment. Many air bag computers have a backup power supply capable of deploying the air bag for a specific length of time after the battery is disconnected.

To enter the diagnostic mode, turn the ignition switch on and press the off and warmer buttons simultaneously on the **climate control center (CCC).** Hold these buttons until the segment display occurs in the instrument panel cluster (IPC). The segment display verifies all the segments are operational in the IPC. The turn signal indicators are not illuminated during this check. Do not proceed with the diagnostics unless all IPC segments are illuminated during the segment display. If any of the IPC segments are not illuminated, erroneous diagnosis may occur. The IPC must be replaced if any segments do not illuminate.

Once the diagnostics are entered, the HI and LO buttons in the CCC may be used to select or reject test displays. Pressing the HI button may be compared to a "yes" input, the LO button may be considered a "no" input. After the diagnostics are entered, the technician may select diagnostic code displays from these computers:

1. Powertrain control module (PCM)
2. Instrument panel cluster (IPC)
3. Air conditioning programmer (ACP)
4. Supplemental inflatable restraint (SIR)
5. Traction control system (TCS)
6. Real time damping (RTD)

The abbreviation for each computer appears in the instrument panel display. The technician presses the HI button to select fault codes from the computer displayed. If the technician does not want fault codes from a computer, the LO button is pressed and the display moves to the next computer abbreviation (Figure 8-30).

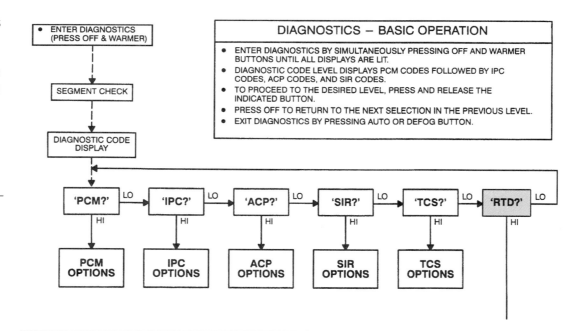

Figure 8-30 Selecting fault code displays from various computers during the diagnostic procedure. (Courtesy of Cadillac Motor Car Division, General Motors Corp.)

The fault codes from any computer are a three-digit number with a one-letter prefix and a one-letter suffix. The fault codes have these prefixes;

1. PCM - P
2. IPC - I
3. APC - A
4. SIR - R
5. TCS - T
6. RTD - S

Therefore, any fault codes in the RTD module are prefixed by the letter S. The suffix is a C or an H. When a code has a C suffix, it is a **current code** that is present at the time of diagnosis. An H suffix indicates the fault code is a **history code** that represents an intermittent fault. The fault codes are displayed in numerical order.

If there are no fault codes in the RTD module, NO S CODES appears in the IPC display. If NO S DATA is displayed, the RTD module cannot communicate with the IPC display. When the AUTO button is pressed any time during the fault code display, the system changes from the diagnostic mode to normal operation.

Optional Diagnostic Modes

After the fault code display, the technician may select these options:

1. RTD data
2. RTD inputs
3. RTD outputs
4. RTD clear codes

When one of these options is displayed in the IPC, the technician selects the option by pressing the HI button. The technician rejects the displayed option by pressing the LO button, and the display moves to the next option.

If RTD data are selected, the data from a number of inputs may be displayed in numerical order beginning with number SD01, R.F. accelerometer (Figure 8-31). This means the car must be driven to obtain a variable input from this sensor. This sensor input appears as a numerical range from -100 to 100%. As the vehicle is accelerated and decelerated, this sensor input should change within the specified range. Always consult the car manufacturer's service manual for more detailed diagnosis and specifications.

The next test in the data mode is selected by pressing the HI button, and the display goes back to the previous test parameter when the LO button is pressed. If the technician selects the RTD inputs, the lift/dive discrete input may be checked. When the technician selects RTD outputs, the RTD module cycles the displayed output. The next output in the test sequence is selected by pressing the HI button.

When CLEAR CODES is selected, the codes are erased and RTD CODES CLEAR is displayed. The AUTO or DEFOG button may be pressed to exit the diagnostics.

Classroom Manual
Chapter 8, page 209

CUSTOMER CARE: While discussing computer-controlled suspension systems with customers, remember that the average customer is not familiar with automotive electronics terminology. Always use basic terms that customers can understand when explaining electronic suspension problems. Most customers appreciate a few minutes spent by service personnel to explain their automotive electronic problems. It is not necessary to provide customers with a lesson in electronics, but it is important that customers understand the basic cause of the problem with their car so they feel satisfied the repair expenditures are necessary. A satisfied customer is usually a repeat customer!

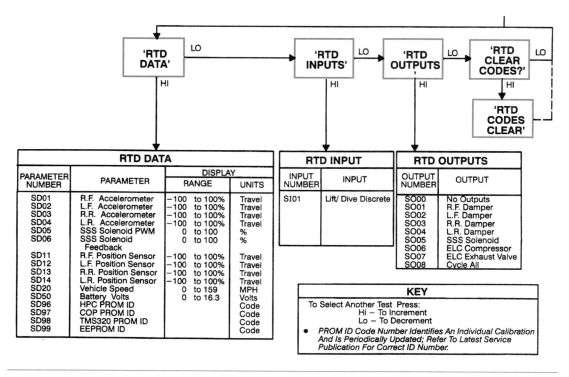

Figure 8-31 Optional test modes including RTD data, inputs, outputs, and clear codes. (Courtesy of Cadillac Motor Car Division, General Motors Corp.)

Guidelines for Servicing Computer-Controlled Suspension Systems

1. Fault codes may be obtained on a programmed ride control (PRC) system by connecting the two wires in a diagnostic connector behind the ash tray.

2. In a PRC system, the fault codes are flashed by the mode indicator light.

3. In a computer command ride (CCR) system, fault codes are obtained by connecting a jumper wire across terminals A and C in the DLC.

4. In a CCR system, both mode indicator lights flash at the same time to indicate the fault codes.

5. When an electronic air suspension system or rear load-levelling air suspension system is serviced, always turn off the air suspension switch in the trunk.

6. When hoisting, jacking, towing, hoisting, or lifting a vehicle equipped with an electronic air suspension system or a rear load-levelling air suspension system, always turn off the air suspension switch in the trunk.

7. If a defect occurs in an electronic air suspension system, the control module illuminates the suspension warning lamp with the engine running.

8. There are 10 steps in the electronic air suspension diagnostic procedure. During the first 3 tests, the suspension warning lamp flashes at 1.8 times per second if the system is normal. While test 4 through 10 are performed, the suspension warning lamp flashes the test number. The driver's door is opened and closed to move to the next test in the sequence.

9. To release air from an air spring, the air spring valve should be rotated to the first stage. Never rotate an air spring valve to the second stage until all the air is exhausted from the spring.

10. Prior to installation, an air spring membrane must be folded properly onto the lower piston.

11. After an air spring is installed there must be no creases in the spring membrane.

12. Never lower the full chassis weight onto an air spring with no air pressure in the spring.

13. In an electronic air suspension system, the front and rear height sensors may be adjusted to obtain the proper trim height.

14. A service fitting is available to splice damaged nylon air lines.

15. The rear height sensor may be adjusted to correct the trim height on a rear load-levelling air suspension system.

16. A scan tool may be connected to a data link connector (DLC) in the trunk to obtain fault codes in a rear load-levelling suspension system.

17. The test modes in the automatic air suspension diagnosis are drive cycle diagnosis, service bay diagnosis, and spring fill diagnosis.

18. A scan tool may be used to perform drive cycle diagnostics, auto test diagnostics, and functional tests on air suspension systems with speed leveling capabilities.

19. A scan tool may be used to perform module parameter identification, module diagnostics, and intermittent fault tests on automatic ride control (ARC) systems.

20. To enter the diagnostics on the road sensing suspension system, the off and warmer buttons in the climate control center are pressed simultaneously with the ignition switch on.

21. When diagnosing the road sensing suspension system, the HI and LO buttons in the climate control center are used as yes and no inputs to select fault code displays from various onboard computers.

22. When diagnosing the road sensing suspension system, the HI and LO buttons are used to select various test options, and move ahead or back up within the parameters in a specific test option.

CASE STUDY

A customer complained about excessive body lean while cornering on the left front suspension of a Thunderbird turbo coupe equipped with a programmed ride control (PRC) system. The customer also said that the programmed ride control lamp in the tachometer started flashing when the problem occurred, but this light would quit flashing after awhile. Further questioning of the customer indicated that the problem was intermittent. Sometimes the problem did not occur during hard cornering. The service writer asked the customer if there were any evidence of noise from the suspension, and the customer indicated that the flashing mode indicator lamp and the body lean on the left front suspension were the only symptoms.

Since the PRC mode indicator lamp was not flashing when the customer brought the car into the shop, the technician took the car for a road test. The technician took with him a digital volt-ohmmeter and a fabricated tool to connect the terminals in the PRC self-test connector. After several attempts at hard cornering, the car displayed the symptoms explained by the customer. While the PRC mode indicator light was still flashing the technician drove the car into a parking lot and performed a self-test on the PRC system. A code 4 was present during the self-test indicating a problem in the left front strut actuator circuit. The technician disconnected the wires from the left front actuator and connected an ohmmeter across the strut motor terminals connected to the firm and soft relays. A normal reading was obtained at these terminals. Careful examination of the wiring harness near the strut tower indicated the harness had been punctured by something. The car had a new paint job and closer examination indicated extensive repairs on the front body structure. It was logical to assume that the wiring harness had been punctured during collision damage. The technician removed some of the tape from the wiring harness and found that the insulation was damaged on the wires from the firm and soft relays to the strut actuator, and these wires were touching together at times.

The technician drove the car back to the shop and repaired the wires. Another road test indicated that the PRC system worked perfectly.

Terms to Know

Air conditioning programmer (ACP)	History code
Automatic ride control (ARC) system	Instrument panel cluster (IPC)
Auto/manual test	Intermittent fault
Auto test diagnostic mode	Light-emitting diodes (LEDs)
Climate control center (CCC)	On-demand DTCs
Computer command ride (CCR) system	Parameter identification (PID) screen
Continual strut cycling	Pinpoint test
Continuous DTCs	Programmed ride control (PRC)
Current code	Powertrain control module (PCM)
Data link connector (DLC)	Real time damping (RTD)
Diagnostic trouble codes (DTCs)	Scan tool
Drive cycle diagnosis	Service bay diagnostics
Electronically erasable programmable read only memory (EEPROM)	Spring fill diagnostics
	Supplemental inflatable restraint (SIR)
Functional tests	Traction control system (TCS)
Hard fault	Trim height

ASE Style Review Questions

1. While discussing programmed ride control (PRC) system service:
   Technician A says the self-test connector is located behind the ash tray.
   Technician B says a flashing mode indicator light indicates a system defect.
   Who is correct?
   A. A only
   B. B only
   C. Both A and B
   D. Neither A nor B

2. While discussing computer command ride (CCR) system service and diagnosis:
   A. terminals A and B should be connected in the data link connector (DLC) to obtain fault codes.
   B. if a defect occurs in the system, the control module illuminates both driver select switch light-emitting diodes (LEDs).
   C. when this system is in the diagnostic mode, the LED beside the driver select switch position flashes the DTCs.
   D. when this system is in the diagnostic mode, each DTC is flashed 4 times.

3. On a vehicle with an electronic air suspension system, a bumper jack is used to lift one corner of the vehicle to change a tire.

   Technician A says the electronic air suspension switch should be in the On position.

   Technician B says the air spring should be deflated on the corner of the vehicle being lifted.

   Who is correct?

   A. A only   **C.** Both A and B

   B. B only   **D.** Neither A nor B

4. While discussing electronic air suspension system service:

   Technician A says the suspension warning lamp flashes a 1.8 times per second during the first 3 tests in the test sequence.

   Technician B says the suspension warning lamp flashes the test number during tests 4 through 10 in the test sequence.

   Who is correct?

   A. A only   **C.** Both A and B

   B. B only   **D.** Neither A nor B

5. While discussing rear load-levelling air suspension system service:

   Technician A says a scan tool may be connected to a diagnostic connector in the trunk to obtain fault codes.

   Technician B says if there are no faults in the control module memory, a code 14 is provided.

   Who is correct?

   A. A only   **C.** Both A and B

   B. B only   **D.** Neither A nor B

6. All of these statements about road sensing suspension diagnosis are true EXCEPT:

   A. The diagnostics are entered by pressing the AUTO and DEFOG buttons simultaneously with the ignition switch on.

   B. The HI button in the climate control center may be pressed to move to the next parameter in a test option.

   C. In the output test option, the road sensing suspension module cycles various outputs.

   D. A history code represents an intermittent fault in the suspension computer memory.

7. When diagnosing an automatic ride control (ARC) system:

   A. the ARC data link connector (DLC) is located on the right front fender shield.

   B. the ignition switch should be turned on when connecting or disconnecting the scan tool.

   C. the parameter identification (PID) screen displays the operation of ARC system inputs and outputs.

   D. during the intermittent fault test, DTCs must be left in the ARC module memory.

8. When diagnosing an air suspension system with speed levelling capabilities:

   A. during the functional tests, the scan tool may be used to command the operation of air suspension system components.

   B. if DTC 10 is displayed in the auto test mode, there are no DTCs or electrical defects in the system.

   C. if DTC 12 is displayed in the auto test mode, the test sequence is completed.

   D. after driving the car during a road test, DTCs are erased from the air suspension module when the ignition switch is turned off.

9. While performing module diagnostics on an automatic ride control (ARC) system, the compressor stops operating. When this action occurs, the technician should:

   A. replace the compressor fusible link.

   B. replace the compressor.

   C. turn off the ignition switch and wait 10 minutes.

   D. check the compressor ground circuit.

10. While performing an intermittent fault test on an automatic ride control (ARC) system, CHANGED appears beside the RFSHK_E display. Under this condition, the technician should:

    A. repair the intermittent open circuit at the right front shock absorber.

    B. replace the right front shock absorber.

    C. check for fluid leaks in the right front shock absorber.

    D. check for loose mounting bushings on the right front shock absorber.

ASE Challenge Questions

1. The owner of a car with an automatic air suspension says the Check Air Suspension light is illuminated. A preliminary inspection does not reveal any air leaks and the measured trim eight is to specifications.
   Technician A says the next step is to connect a scan tool.
   Technician B says the next step is to complete the drive cycle.
   Who is correct?
   A. A only
   B. B only
   C. Both A and B
   D. Neither A nor B

2. All of these statements about a Programmed Ride Control system are true EXCEPT:
   A. an air spring is mounted at each corner of the vehicle.
   B. an electric actuator is located in each strut.
   C. a mode indicator light is positioned in the tachometer.
   D. the PRC module provides a firm ride during severe braking.

3. *Technician A* says the suspension switch must be turned off before raising any corner of a car with an electronic air suspension.
   Technician B says the ignition switch must not be turned on while any corner of a car with electronic air suspension is raised.
   Who is correct?
   A. A only
   B. B only
   C. Both A and B
   D. Neither A nor B

4. *Technician A* says if a defect occurs in a CCR system, both driver select switch LEDs will illuminate.
   Technician B says if a defect corrects itself in a GM CCR system, a DTC will be set in control module memory.
   Who is correct?
   A. A only
   B. B only
   C. Both A and B
   D. Neither A nor B

5. *Technician A* says when diagnosing the ARC system, the ignition switch must be on and the driver's door open.
   Technician B says when diagnosing the ARC system, the transfer case must be in 4WD.
   Who is correct?
   A. A only
   B. B only
   C. Both A and B
   D. Neither A nor B

Table 8-1 ASE TASK

Diagnose, inspect, adjust, repair, or replace components of electronically controlled suspension systems, including primary and supplemental air suspension systems.

Problem Area	Symptoms	Possible Causes	Classroom Manual	Shop Manual
RIDE QUALITY	Body sway, drive, or lift	Defective electronic suspension system	182	226
RIDE HEIGHT	Low or high ride height	1. Defective electronic air suspension system	182	226
		2. Trim height adjustment	187	229

Job Sheet 23

23

Name _____ Date _____

Electronic Air Suspension System Diagnosis

NATEF and ASE Correlation

This job sheet is related to NATEF and ASE Automotive Suspension and Steering Task List content area: B. Suspension System Diagnosis and Repair, 3. Miscellaneous Service, Task 4: Diagnose, inspect, adjust, repair, or replace components of electronically controlled suspension systems including primary and supplemental air suspension systems.

Tools and Materials

None

Wheel make and model year _____

Vehicle VIN _____

Procedure

1. Be sure the air suspension system switch is turned on.

2. Turn the ignition switch on for 5 seconds and then turn it off.

   Leave the driver's door open and the other doors closed.

3. Ground the diagnostic lead located near the control module and close the driver's door with the window down.

4. Turn the ignition switch on. The warning lamp should blink continuously at 1.8 times per second to indicate that the system is in the diagnostic mode.

5. Perform test 1. Open and close the driver's door. The rear suspension should be raised for 30 seconds, lowered for 30 seconds, and raised for 30 seconds.

   Did the rear suspension complete this sequence? yes _____ no _____

   If the answer to this question is no, state the suspension action that did occur.

   _____

   _____

6. Perform test 2. Open and close the driver's door. The right front suspension should be raised for 30 seconds, lowered for 30 seconds, and raised for 30 seconds.

   Perform test 3. Did the right front suspension complete this sequence?
   yes _____ no _____

   If the answer to this question is no, state the suspension action that did occur.

   _____

   _____

7. Perform test 3. Open and close the driver's door. The left front suspension should be raised for 30 seconds, lowered for 30 seconds, and raised for 30 seconds.

Did the left front suspension complete this sequence? yes _____ no _____

If the answer to this question is no, state the suspension action that did occur.

_____

_____

Did the air suspension warning light blink at 1.8 times per second during tests 1, 2, and 3? yes _____ no _____

If the answer to this question is no, state the warning light action that did occur.

_____

_____

8. Perform test 4. Open and close the driver's door. The compressor is cycled on and off at 0.25 cycles per second. This action is limited to 50 cycles.

Did the compressor cycle on and off at 0.25 times per second for 50 cycles? yes _____ no _____

If the answer to this question is no, state the compressor action that did occur.

_____

_____

9. Perform test 5. Open and close the driver's door. The vent solenoid opens and closes at 1 cycle per second.

Did the vent solenoid cycle at 1 cycle per second? yes _____ no _____

If the answer to this question is no, state the vent solenoid action that did occur.

_____

_____

10. Perform test 6. Open and close the driver's door. The left front air valve opens and closes at 1 cycle per second, and the vent solenoid is opened. When this occurs, the left front corner of the vehicle should drop slowly.

Did the left front air spring solenoid valve and the vent valve open and the left front suspension drop slowly? yes _____ no _____

If the answer to this question is no, state the left front air valve and solenoid valve action that did occur.

_____

_____

11. Perform test 7. Open and close the driver's door. The right front air valve opens and closes at 1 cycle per second, and the vent solenoid is opened. This action causes the right front corner of the vehicle to drop slowly.

Did the right front air spring solenoid valve and the vent valve open and the right front suspension drop slowly? yes _____ no _____

If the answer to this question is no, state the right front air valve and solenoid valve action that did occur.

_____

_____

12. Perform test 8. Open and close the driver's door. During this test the right rear air valve opens and closes at 1 cycle per second and the vent valve is opened. This action should cause the right rear corner of the vehicle to drop slowly.

Did the right rear air spring solenoid valve and the vent valve open and the rear suspension drop slowly? yes _____ no _____

If the answer to this question is no, state the right front air valve and solenoid valve action that did occur.

_____

_____

13. Perform test 9. Open and close the driver's door. The left rear solenoid opens and closes at 1 cycle per second and the vent valve is opened, which should cause the left rear corner of the vehicle to drop slowly.

Did the left rear air spring solenoid valve and the vent valve open and the rear suspension drop slowly? yes _____ no _____

If the answer to this question is no, state the left rear air solenoid valve and vent solenoid action that did occur.

_____

_____

Did the air suspension warning light blink the test number during tests 4 through 9?
yes _____ no _____

If the answer to this question is no, state the warning light action that did occur.

_____

_____

14. Perform test 10. Return the module from the diagnostic mode to normal operation by disconnecting the diagnostic lead from ground. This mode change also occurs if the ignition switch is turned off or when the brake pedal is depressed.

System returned to the normal mode? yes _____ no _____

List air suspension problems located during the test sequence.

_____

_____

_____

☑ **Instructor Check** _____

Job Sheet 24

24

Name _____ Date _____

Adjust Trim Height, Rear Load-Levelling Air Suspension System

NATEF and ASE Correlation

This job sheet is related to NATEF and ASE Automotive Suspension and Steering Task List content area: B. Suspension System Diagnosis and Repair, 3. Miscellaneous Service, Task 4: Diagnose, inspect, adjust, repair, or replace components of electronically controlled suspension systems including primary and supplemental air suspension systems.

Tools and Materials

Suspension height gauge

Wheel make and model year _____

Vehicle VIN _____

Procedure

1. Position the vehicle on a level floor surface over an open service pit or equivalent. Place the transmission in Park and apply the parking brake.

2. Be sure the air suspension switch is on, and then open the driver's door and turn the ignition switch to the Run position.

3. Add weight to the rear bumper and wait for the suspension air compressor to run and restore the rear suspension to the trim height.

   Is the rear suspension restored to trim height? yes _____ no _____

   If the answer to this question is no, state the rear suspension and compressor action that did occur.

   _____

   _____

4. After the compressor stops and the rear suspension is at trim height, remove the weight from the rear bumper and close the driver's door. Allow the rear suspension drop to trim height.

   Is the rear suspension lowered to trim height? yes _____ no _____

   If the answer to this question is no, state the rear suspension action that did occur.

   _____

   _____

   State the necessary repairs to correct the rear air suspension problems indicated in steps 1 through 4.

   _____

   _____

5. Rock the rear suspension laterally and allow the suspension to stabilize.

6. Turn off the air suspension switch.

7. Remove weld flash and debris from the frame in the suspension height measurement location.

8. Install the suspension height gauge between the top of the axle housing and the lower side of the frame on the left side of the vehicle. Release the thumb screw on the gauge, and slide the gauge post upward until the post contacts the frame. Tighten the thumb screw in this position.

   Is the suspension height gauge properly positioned? yes _____ no _____

   Is the thumb screw on gauge properly tightened? yes _____ no _____

   Instructor check _____

9. Remove the suspension height gauge and read the dimension on the gauge post at the edge of the gauge body.

   Specified rear suspension trim height _____

   Actual rear suspension trim height _____

10. If a trim height adjustment is required, the upper height sensor mounting bolts may be loosened and the bracket moved downward to increase trim height or upward to decrease trim height.

    State the necessary action to correct the rear suspension trim height.

    _____

    _____

11. After the suspension height adjustment, repeat steps 2 through 9 to be sure the adjustment is accurate.

    Trim height _____

    Specified trim height _____

✔ Instructor Check _____

Job Sheet 25

25

Name _____ Date _____

Diagnose Automatic Ride Control (ARC) System

NATEF and ASE Correlation

This job sheet is related to NATEF and ASE Automotive Suspension and Steering Task List content area: B. Suspension System Diagnosis and Repair, 3. Miscellaneous Service, Task 4: Diagnose, inspect, adjust, repair or replace components of electronically controlled suspension systems including primary and supplemental air suspension systems.

Tools and Materials

Scan tool
Battery charger

Wheel make and model year _____

Vehicle VIN _____

Procedure

1. Connect a battery charger to the vehicle battery terminals with the correct polarity. Select a low charging rate on the battery charger.

2. Open any vehicle door.

3. Place the transfer case selector switch in 4X4 LOW if the vehicle has a 4.0L engine. This switch position is identified as 4X4 OFF ROAD if the vehicle is equipped with a 5.0L engine.

   Is the transfer case switch in 4X4 LOW or 4X4 OFF ROAD? yes _____ no _____

   Instructor check _____

4. Connect the scan tool to the ARC DLC.

   Is the scan tool connected to DLC? yes _____ no _____

   Instructor check _____

5. Select the on-demand self-test on the scan tool display.

6. Observe and record all the on-demand DTCs on the scan tool display. On-demand DTCs represent intermittent ARC system faults.

   On-demand DTCs displayed on scan tool: _____
   _____

7. Use the clear codes function on the scan tool to clear the DTCs.

   Are the DTCs cleared? yes _____ no _____

   Instructor check _____

WARNING: If the compressor runs for more than 2 minutes in functional test 221, the internal circuit breaker in the compressor may open the compressor circuit and cause the compressor to stop. If this action occurs, turn off the ignition switch for 10 minutes to allow this circuit breaker to reset.

8. Select these active commands on the scan tool to raise the vehicle: front fill, rear fill, gate valve, and compressor. Operate the compressor for 30 seconds.

    Did the suspension height increase? yes _____ no _____

    If the answer to this question is no, state the compressor and suspension action.

    _____

    _____

9. Select these active commands on the scan tool to lower the vehicle: front fill, rear fill, gate valve, and vent.

    Did the suspension height decrease? yes _____ no _____

    If the answer to this question is no, state the suspension action.

    _____

    _____

10. Observe and record all DTCs displayed on the scan tool.

    DTCs displayed on scan tool: _____

    _____

11. Use the clear codes function on the scan tool to clear the DTCs.

    Are the DTCs cleared? yes _____ no _____

    Instructor check _____

12. On the basis of the DTCs displayed and the suspension action during the diagnosis, state the necessary suspension repairs and explain the reasons for your diagnosis.

    _____

    _____

    _____

☑ **Instructor Check** _____

Steering Column and Linkage Diagnosis and Service

Upon completion and review of this chapter, you should be able to:

❏ Diagnose steering columns.

❏ Remove and replace steering wheels on air-bag-equipped vehicles and non-air-bag-equipped vehicles.

❏ Remove and replace air bag deployment modules and clock spring electrical connectors.

❏ Remove and replace steering columns.

❏ Inspect collapsible steering columns for damage.

❏ Disassemble steering columns.

❏ Inspect steering column components and replace necessary parts.

❏ Assemble steering columns.

❏ Diagnose and service flexible couplings and universal joints.

❏ Diagnose steering linkage mechanisms.

❏ Diagnose, remove, and replace tie-rod ends.

❏ Diagnose, remove, and replace pitman arms.

❏ Diagnose, remove, and replace center links.

❏ Diagnose, remove, and replace idler arms.

❏ Diagnose, remove, and replace steering dampers.

❏ Diagnose steering arms.

Air Bag Deployment Module, Steering Wheel, and Clock Spring Electrical Connector Removal and Replacement

⚠ **WARNING:** Always disconnect the negative battery terminal and wait one minute before diagnosing or servicing any air bag system component. Failure to observe this precaution may result in accidental air bag deployment. Air bag service precautions vary depending on the vehicle. Always follow the vehicle manufacturer's air bag service precautions in the service manual.

⚠ **WARNING:** On an air-bag-equipped vehicle, the wait time prior to servicing electrical components after the negative battery terminal is disconnected varies depending on the vehicle make and model year. Always follow the wait time and all other service precautions recommended in the vehicle manufacturer's service manual.

⚠ **WARNING:** Always use the vehicle manufacturer's recommended tools when diagnosing or servicing an air bag system or the electrical system on an air-bag-equipped vehicle. Do not use a 12V test light or a self-powered test light to diagnose air bag systems or electrical systems on air-bag-equipped vehicles. **Accidental air bag deployment** may result from improper use of tools.

☑ **SERVICE TIP:** Before disconnecting the battery negative cable, note how the customer has the radio stations programmed, and reset the radio and clock after the battery negative cable is reconnected.

Prior to working on an air bag system, always disconnect the negative battery cable and wait one minute before proceeding with the diagnostic or service work. Many air bag systems have a **backup power supply** circuit designed into the air bag computer or located in a separate module. This backup power supply provides power to deploy the air bag for a specific length of time after the battery power is disconnected in a collision. One minute after the negative battery terminal is disconnected, this backup power supply is powered down, and no power is available to deploy the air bag while the battery remains disconnected.

An air bag warning light in the instrument panel indicates the status of the air bag system. On some vehicles, this warning light should be illuminated for a few seconds when the ignition switch is turned on. The warning light should remain off while cranking the engine, and it should be on for a few seconds after the engine starts. After the engine has been running for a few seconds, the air bag warning light should remain off. On other vehicles, the air bag warning light flashes 7 to 9 times when the ignition switch is turned on and after the engine is started. The air bag warning light operation varies depending on the make and model year of the vehicle. Always check the vehicle manufacturer's service manual for the exact air bag warning light operation. When the air bag warning light does not operate as specified by the vehicle manufacturer, the air bag system is defective, and the air bag or bags will probably not deploy if the vehicle is involved in a collision.

Air bag module and steering wheel removal and replacement procedures vary depending on the vehicle. Always follow the vehicle manufacturer's recommended procedure in the service manual. The following is a typical air bag module and steering wheel removal and replacement procedure.

1. Turn the ignition switch to the Lock position and place the front wheels facing straight ahead.

2. Remove the negative battery terminal and wait one minute.

3. Loosen the three air-bag-retaining Torx screws under the steering wheel (Figure 9-1).

4. Loosen the other two air-bag-retaining Torx screws under the steering wheel (Figure 9-2). Loosen all five Torx screws until the groove along the screw circumference catches on the screw case.

An air bag deployment module may be referred to as a steering wheel pad.

CAUTION: When an air bag deployment module is temporarily stored on the workbench, always place this module face upward. If the air bag deployment module accidentally deployed when facing downward, the module would become a projectile, and personal injury might result.

5. Pull **the air bag deployment module** from the steering wheel and disconnect the air bag module electrical connector (Figure 9-3). Do not pull on the air bag wires in the steering column. Place the air bag deployment module face upward on the workbench.

6. Disconnect the air bag wiring retainer in the steering wheel (Figure 9-4).

7. Use the proper size socket and a ratchet to remove the steering wheel retaining nut.

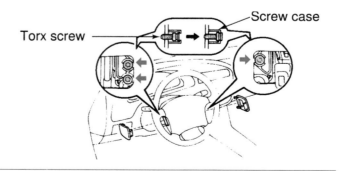

Figure 9-1 Three air bag retaining Torx screws. (Reprinted with permission)

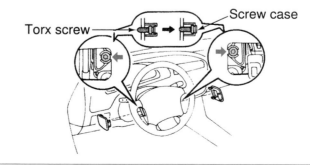

Figure 9-2 Two air bag retaining Torx screws. (Reprinted with permission)

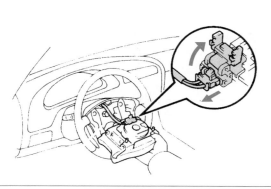

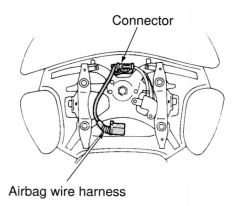

Connector

Airbag wire harness

Figure 9-3 Disconnecting the air bag module electrical connector. (Reprinted with permission)

Figure 9-4 Disconnecting the air bag wiring retainer (Reprinted with permission)

8. Observe the matching alignment marks on the steering wheel and the steering shaft. If these alignment marks are not present, place alignment marks on the steering wheel and steering shaft with a center punch and a hammer.

⚠ **WARNING:** Do not hammer on the top of the steering shaft to remove the steering wheel. This action may damage the shaft.

⚠ **WARNING:** If the steering wheel puller bolts are too long they will extend through the steering wheel and damage the clock spring.

■ **CAUTION:** Do not pull on the steering wheel in an attempt to remove it from the steering shaft. Photo Sequence 7 shows a typical procedure for removing a steering wheel. The steering wheel may suddenly come off, resulting in personal injury, or the steering wheel may be damaged by the pulling force.

9. Install a steering wheel puller with the puller bolts threaded into the bolt holes in the steering wheel. Tighten the puller nut to remove the steering wheel (Figure 9-5). Visually check the steering wheel condition. If the steering wheel is bent or cracked, replace the wheel.

Matchmarks

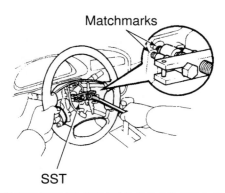

SST

Figure 9-5 Removing steering wheel with the proper steering wheel puller. (Reprinted with permission)

A clock spring electrical connector may be called a coil, **spiral cable,** or cable reel.

10. Disconnect the four retaining screws and remove the clock spring electrical connector (Figure 9-6).

11. Be sure the front wheels are facing straight ahead. **Turn the clock spring electrical connector** counterclockwise by hand until becomes harder to turn as it becomes fully wound in that direction.

 WARNING: Failure to center a clock spring prior to installation may cause a broken conductive tape in the clock spring.

12. Turn the clock spring electrical connector clockwise three turns, and align the red mark on the center part of the spring face with the notch in the cable circumference (Figure 9-7). This action centers the clock spring electrical connector.

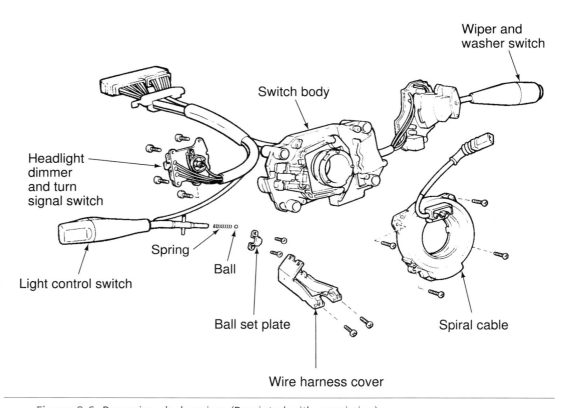

Figure 9-6 Removing clock spring. (Reprinted with permission)

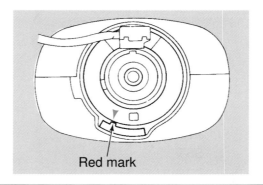

Figure 9-7 Centering clock spring. (Reprinted with permission)

Photo Sequence 7
Typical Procedure for Removing a Steering Wheel

P7-1 Check and record the radio stations programmed in the stereo system.

P7-2 Look up the car manfacturer's steering wheel removal procedure in the service manual.

P7-3 Disconnect the negative battery cable and wait for the car manufacturer's specified length of time.

P7-4 Remove the air bag deployment module retaining screws under the steering wheel.

P7-5 Lift the air bag deployment module upward from the steering wheel, and disconnect the module wiring connector.

P7-6 Set the air bag deployment module face upward on the workbench.

P7-7 Loosen and remove the steering wheel retaining nut.

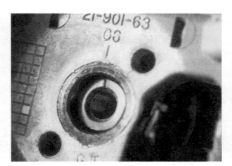

P7-8 Observe the alignment marks on the steering wheel and shaft.

P7-9 Connect the proper steering wheel puller to the steering wheel.

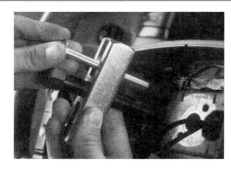

P7-10 Turn the puller screw to loosen the steering wheel on the shaft.

P7-11 Remove the puller from the steering wheel.

P7-12 Lift the steering wheel off the shaft.

13. Install the clock spring electrical connector and tighten the four retaining screws to the specified torque.

14. Align the marks on the steering wheel and the steering shaft, and install the steering wheel on the shaft.

15. Install the steering wheel retaining nut and tighten this nut to the specified torque.

16. Install the air bag wiring retainer in the steering wheel.

17. Hold the air bag deployment module near the top of the steering wheel, and connect the air bag module connector.

18. Install the air bag deployment module in the top of the steering wheel, and tighten the five retaining Torx screws.

19. Reconnect the negative battery cable.

20. Reset the clock and radio.

If the vehicle is not equipped with an air bag, the steering wheel removal and replacement procedure is basically the same, but all steps pertaining to the air bag module and clock spring are not required. On a non-air-bag-equipped vehicle, the center steering wheel cover must be removed to access the steering wheel retaining nut.

⬤ **CUSTOMER CARE:** While servicing a vehicle, always check the operation of the indicator lights or gauges in the instrument panel. These lights or gauges may indicate a problem that the customer has been ignoring. For example, if the air bag warning light is not operating properly, the air bag or bags may not deploy in a collision, resulting in serious injury to the driver and/or passenger. If the air bag warning light is not working properly, always advise the customer that he or she will not be protected by the air bag in a collision, and the vehicle should not be driven under this condition.

Steering Column Removal and Replacement

Steering column removal and replacement procedures vary depending on the vehicle make, type of steering column, and gearshift lever position. Always follow the vehicle manufacturer's recommended procedure in the service manual. The following is a typical steering column removal and replacement procedure.

1. Disconnect the battery negative cable. If the vehicle is equipped with an air bag, wait one minute.

2. Install a seat cover on the front seat.

3. Place the front wheels in the straight-ahead position, and remove the ignition key from the switch to lock the steering column.

4. Remove the cover under the steering column, and remove the lower finish panel if necessary.

5. Disconnect all wiring connectors from the steering column.

6. If the vehicle has a column-mounted gearshift lever, disconnect the gearshift linkage at the lower end of the steering column. If the vehicle has a floor-mount gearshift, disconnect the shift interlock.

7. Remove the retaining bolt or bolts in the lower universal joint or flexible coupling.

8. Remove the steering-column-to-instrument-panel mounting bolts.

9. Carefully remove the steering column from the vehicle. Be careful not to damage upholstery or paint.

10. Install the steering column under the instrument panel, and insert the steering shaft into the lower universal joint.

11. Install the steering-column-to-instrument-panel mounting bolts. Be sure the steering column is properly positioned, and tighten these bolts to the specified torque.

12. Install the retaining bolt or bolts in the lower universal joint or flexible coupling, and tighten the bolts to the specified torque.

13. Connect the gearshift linkage, if the vehicle has a column-mounted gearshift.

14. Connect all the wiring harness connectors to the steering column connectors.

15. Install the steering column cover and the lower finish panel.

16. Reconnect the negative battery cable.

17. Road test the vehicle and check for proper steering column operation.

Collapsible Steering Column Inspection

Since steering column design varies depending on the vehicle, the **collapsible steering column** inspection procedure should be followed in the vehicle manufacturer's service manual. The following is a typical collapsible steering column inspection procedure.

1. Measure the clearance between the capsules and the slots in the steering column bracket (Figure 9-8). If this measurement is not within specifications, replace the bracket.

2. Check the contact between the bolt head and the bracket (Figure 9-9). If the bolt head contacts the bracket, the shear load is too high, and the bracket must be replaced.

3. Check the steering column jacket for **sheared injected plastic** in the openings on the side of the jacket (Figure 9-10). If sheared plastic is present, the column is collapsed. Measure the distance from the end of the bearing assembly to the lower edge of the

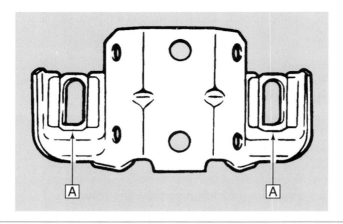

Figure 9-8 Capsules in steering column bracket. (Courtesy of Oldsmobile Motor Division, General Motors Corp.)

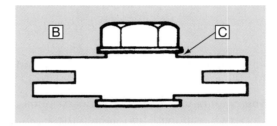

Figure 9-9 Bolt head to bracket clearance. (Courtesy of Oldsmobile Motor Division, General Motors Corp.)

upper steering column jacket (Figure 9-11). If this distance is not within the vehicle manufacturer's specification, a new jacket must be installed.

4. Visually check the gearshift tube for sheared injected plastic. If the gearshift tube indicates sheared plastic, replace the tube.

5. Remove the intermediate steering shaft between the column and the steering gear. Position a dial indicator stem against the lower end of the steering shaft and rotate the steering wheel. If the runout on the dial indicator exceeds the vehicle manufacturer's specification, the steering shaft is bent and must be replaced.

6. If the steering shaft is not bent but shows sheared injected plastic (Figure 9-12), this shaft may be repaired with a service repair package.

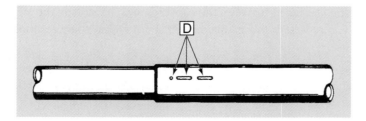

Figure 9-10 Checking for sheared injected plastic in jacket openings. (Courtesy of Oldsmobile Motor Division, General Motors Corp.)

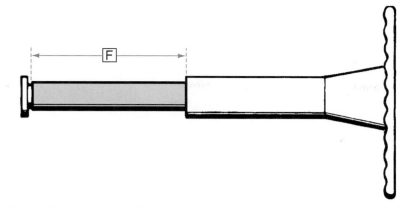

Figure 9-11 Measuring distance from the end of the bearing assembly to the upper steering column jacket. (Courtesy of Oldsmobile Motor Division, General Motors Corp.)

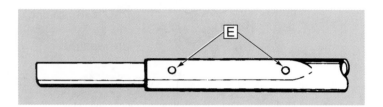

Figure 9-12 Checking for sheared injected plastic in steering shaft. (Courtesy of Oldsmobile Motor Division, General Motors Corp.)

Tilt Steering Column Disassembly

Since there are many variations in steering column design, always follow the vehicle manufacturer's recommended steering column disassembly procedure in the service manual. The following is a typical tilt steering column disassembly procedure.

1. Remove the air bag deployment module, steering wheel, and spiral cable as mentioned previously in this chapter.

2. Remove the ignition key cylinder illumination mounted on top of the ignition switch cylinder (Figure 9-13).

3. Remove the universal joint from the lower end of the steering shaft.

4. Remove the steering column protector and the wiring harness clamp mounted under the steering column.

5. Remove the steering damper mounted near the top of the column.

6. Remove the retaining screw and the combination switch with the wiring harness (Figure 9-14).

7. Mark the center of the tapered-head bolts on the steering column with a center punch. Use a 0.120 to 0.160 in (3 to 4 mm) drill bit to drill into the tapered-head bolts (Figure 9-15).

8. Remove the **tapered-head bolts** with a screw extractor and separate the upper bracket and column tube (Figure 9-16).

9. Grasp the compression spring seat with a pair of pliers and turn the seat to release the seat, compression spring, and bushing (Figure 9-17).

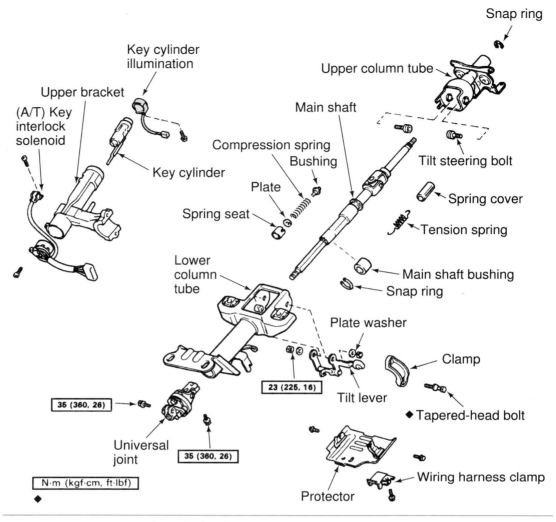

Figure 9-13 Ignition key cylinder illumination mounted above ignition switch cylinder. (Reprinted with permission)

10. Grasp the tension spring with a pair of pliers and extend this spring to remove it from the column (Figure 9-18).

11. Remove the snapring from the upper column tube with snapring pliers (Figure 9-19).

12. Use a soft hammer and a screwdriver to loosen the staked parts of the upper column tube (Figure 9-20).

13. Remove the two nuts on the upper column pivot bolts (Figure 9-21).

14. Use a plastic hammer to tap the pivot bolts out of the column (Figure 9-22).

15. Remove the upper column tube, tilt lever assembly, and plate washers (Figure 9-23).

16. Remove the steering shaft from the column (Figure 9-24).

17. Use snapring pliers to remove the snapring above the steering shaft bushing (Figure 9-25). Remove the steering shaft bushing.

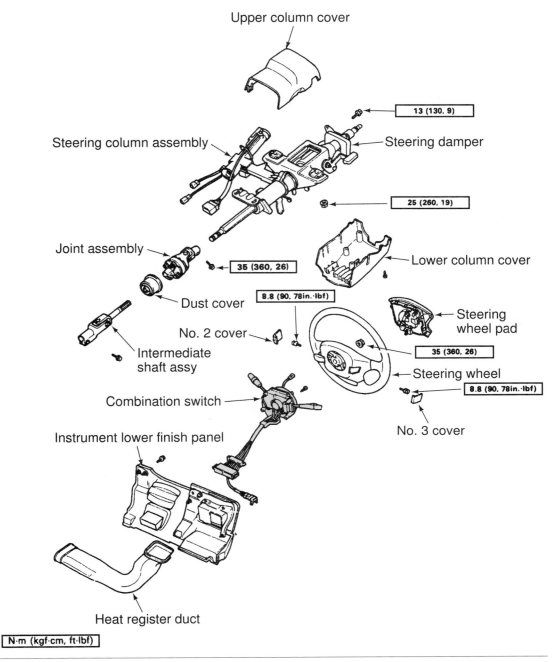

Upper column cover

13 (130, 9)

Steering column assembly

Steering damper

25 (260, 19)

Joint assembly

35 (360, 26)

Lower column cover

Dust cover

8.8 (90, 78in.·lbf)

No. 2 cover

Steering wheel pad

35 (360, 26)

Intermediate shaft assy

Steering wheel

8.8 (90, 78in.·lbf)

Combination switch

No. 3 cover

Instrument lower finish panel

Heat register duct

N·m (kgf·cm, ft·lbf)

Figure 9-14 Removing combination switch and wiring harness. (Reprinted with permission)

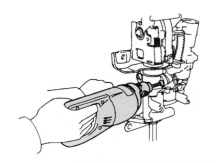

Figure 9-15 Drilling tapered-head bolts. (Reprinted with permission)

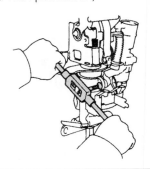

Figure 9-16 Removing tapered-head bolts with a screw extractor. (Reprinted with permission)

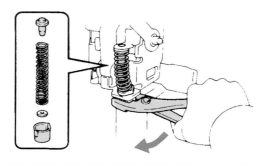

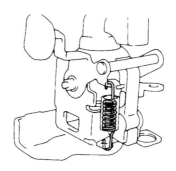

Figure 9-17 Removing compression spring, seat, and bushing. (Reprinted with permission)

Figure 9-18 Removing tension spring. (Reprinted with permission)

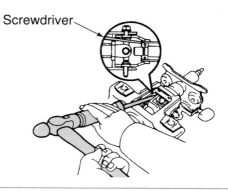

Figure 9-19 Removing snapring from upper column tube. (Reprinted with permission)

Figure 9-20 Loosening staked parts in upper column tube. (Reprinted with permission)

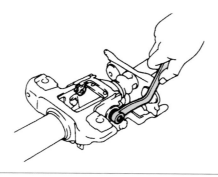

Figure 9-21 Removing nuts on upper column pivot bolts. (Reprinted with permission)

Figure 9-22 Removing pivot bolts from column. (Reprinted with permission)

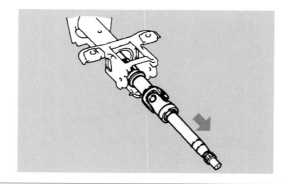

Figure 9-23 Removing upper column tube, tilt lever assembly, and plate washers. (Reprinted with permission)

Figure 9-24 Removing steering shaft from the steering column. (Reprinted with permission)

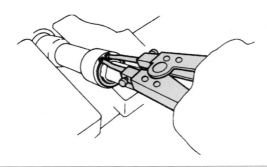

Figure 9-25 Removing snapring above steering shaft bushing. (Reprinted with permission)

Tilt Steering Column Inspection and Parts Replacement

1. Place the ignition key in the switch and move the key through all the switch positions. Be sure the operating pin moves properly in the switch mechanism (Figure 9-26). If the ignition key cylinder must be replaced, turn the key to the accessory (ACC) position. Use a small steel rod to push down on the stop pin near the bottom of the key cylinder (Figure 9-27), and pull the cylinder out of the housing. Install the new key cylinder with the key in the ACC position.

2. Rotate the upper steering shaft bearing and check for noise, looseness, and wear (Figure 9-28). If any of these conditions are present, replace the upper tube.

3. Check the lower steering shaft bearing for noise, looseness, and wear (Figure 9-29). Replace this bearing if necessary.

4. Inspect the ignition key interlock solenoid for damaged wires and loose mounting screws (Figure 9-30). Repair or replace this solenoid as required.

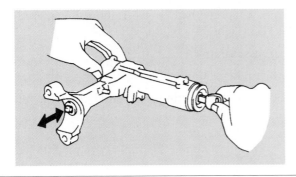

Figure 9-26 Checking ignition switch operation. (Reprinted with permission)

Figure 9-27 Depressing stop pin to release ignition key cylinder. (Reprinted with permission)

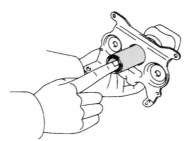

Figure 9-28 Checking upper steering shaft bearing. (Reprinted with permission)

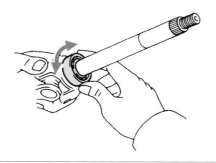

Figure 9-29 Checking lower steering shaft bearing. (Reprinted with permission)

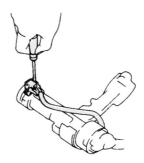

Figure 9-30 Checking ignition key interlock solenoid. (Reprinted with permission)

Tilt Steering Column Assembly

1. Coat all rubbing parts with **molybdenum disulphide lithium-based grease.** Install the steering shaft bushing and the bushing snapring (Figure 9-31).

2. Install the steering shaft in the lower tube (Figure 9-32).

3. Install the upper column tube, tilt lever mechanism, and two plate washers (Figure 9-33). Install the two pivot bolts.

4. Install the pivot bolt nuts, and tighten these nuts to the specified torque (Figure 9-34).

5. Use a brass bar and a hammer to tap the steering shaft into the upper column tube (Figure 9-35).

6. Stake the upper column tube with a pin punch and a hammer (Figure 9-36).

7. Install the snapring in the top of the upper column tube (Figure 9-37).

8. Install the tension spring (Figure 9-38). Assemble the compression spring, bushing, plate, and seat. Use a vise to install the compression spring and related components (Figure 9-39).

9. Install the upper bracket with two new tapered-head bolts. Tighten the tapered-head bolts until the tapered head breaks off (Figure 9-40).

10. Install the protector and wiring harness clamp.

11. Install the steering damper and ignition key illumination.

12. Install the universal joint on the bottom of the steering shaft, and tighten the retaining bolt to the specified torque.

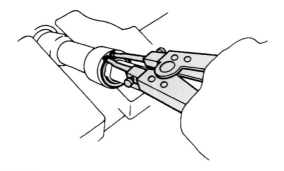

Figure 9-31 Installing steering shaft bushing and snapring. (Reprinted with permission)

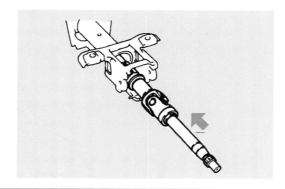

Figure 9-32 Installing the steering shaft in the lower tube. (Reprinted with permission)

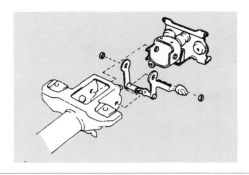

Figure 9-33 Installing upper column tube, tilt lever mechanism and two plate washers. (Reprinted with permission)

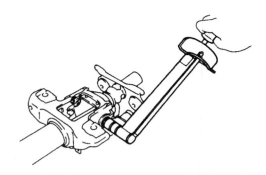

Figure 9-34 Tightening pivot bolt nuts. (Reprinted with permission)

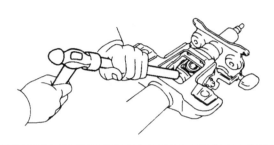

Figure 9-35 Tapping steering shaft into upper column tube. (Reprinted with permission)

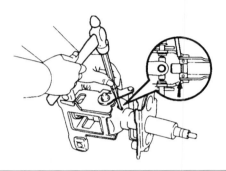

Figure 9-36 Staking upper column tube. (Reprinted with permission)

Figure 9-37 Installing snapring in upper column tube. (Reprinted with permission)

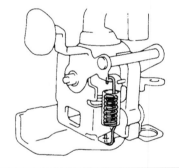

Figure 9-38 Installing tension spring. (Reprinted with permission)

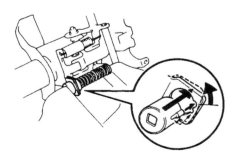

Figure 9-39 Installing compression spring. (Reprinted with permission)

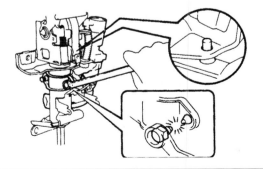

Figure 9-40 Tightening tapered-head bolts in upper bracket. (Reprinted with permission)

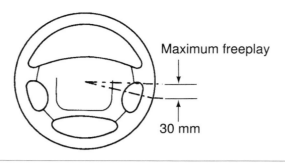

Figure 9-41 Measuring steering wheel freeplay. (Reprinted with permission)

13. Install the combination switch and wiring harness, and tighten the retaining screw to the specified torque.

14. Install the spiral cable, steering wheel, and air bag deployment module as mentioned earlier in this chapter.

Steering Column Flexible Coupling and Universal Joint Service and Diagnosis

Checking Steering Wheel Freeplay

Steering wheel freeplay is the amount of steering wheel movement before the front wheels start to turn.

With the engine stopped and the front wheels in the straight-ahead position, move the steering wheel in each direction with light finger pressure. Measure the amount of steering wheel movement before the front wheels begin to turn (Figure 9-41). This movement is referred to as **steering wheel freeplay.** On some vehicles, this measurement should not exceed 1.18 in (30 mm). Always refer to the vehicle manufacturer's specifications. Excessive steering wheel freeplay is caused by worn steering shaft universal joints or flexible coupling. Other causes of excessive steering wheel freeplay include worn steering linkage mechanisms and a worn or out of adjustment steering gear.

A worn universal joint or flexible coupling in the steering column may also cause rattling noises. The rattling noises may occur while driving the vehicle straight ahead on irregular road surfaces. With the normal vehicle weight resting on the front suspension, observe the flexible coupling or universal joint as an assistant turns the steering wheel 1/2 turn in each direction. If the vehicle has power steering, the engine should be running with the gear selector in Park. The flexible coupling or universal joint must be replaced if there is freeplay in this component.

Flexible Coupling Replacement

If the flexible coupling must be replaced, loosen the coupling-to-steering-gear-stub-shaft bolt. Disconnect the steering column from the instrument panel, and move the column rearward until the flexible coupling can be removed from the steering column shaft. Remove the coupling-to-steering-shaft bolts and disconnect the coupling from the shaft. When the new coupling and the steering column are installed on some vehicles, the clearance between the coupling clamp and the steering gear adjusting plug should be 1/16 in (1.5 mm) (Figure 9-42). This specification may vary depending on the vehicle. Always use the vehicle manufacturer's specifications in the service manual.

Steering Column Diagnosis

Classroom Manual
Chapter 9, page 228

There are variations in steering columns depending on the vehicle, type of transmission, and the transmission gear selector position. Thus different column diagnostic procedures may be required. See Table 9-1, Table 9-2, and Table 9-3 for a typical steering column diagnosis.

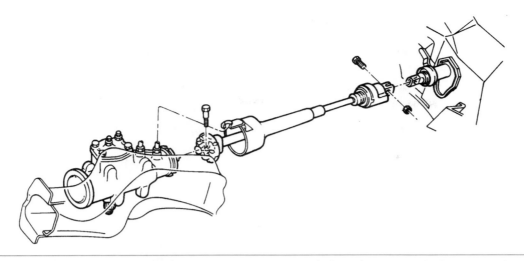

Figure 9-42 Flexible coupling installation. (Courtesy of Chevrolet Motor Division, General Motors Corp.)

Table 9-1 AUTOMATIC TRANSMISSION—STEERING COLUMN DIAGNOSIS

Condition	Possible Cause	Correction
Lock system—will not unlock	1. Lock bolt damaged	1. Replace lock bolt.
	2. Defective lock cylinder	2. Replace or repair lock cylinder.
	3. Damaged housing	3. Replace housing.
	4. Damaged or collapsed sector	4. Replace sector.
	5. Damaged rack	5. Replace rack.
	6. Shear flange on sector shaft collapsed	6. Replace.
Lock system—will not lock	1. Lock bolt spring broken or defective	1. Replace spring.
	2. Damaged sector tooth, or sector installed incorrectly	2. Replace, or install correctly.
	3. Defective lock cylinder	3. Replace lock cylinder.
	4. Burr or lock bolt or housing	4. Remove burr.
	5. Damaged housing	5. Replace housing.
	6. Transmission linkage adjustment incorrect	6. Readjust.
	7. Damaged rack	7. Replace rack.
	8. Interference between bowl and coupling (tilt)	8. Adjust or replace as necessary.
	9. Ignition switch stuck	9. Readjust or replace.
	10. Actuator rod restricted or bent	10. Readjust or replace.

Table 9-1 AUTOMATIC TRANSMISSION—STEERING COLUMN DIAGNOSIS (continued)

Condition	Possible Cause	Correction
Lock system—high effort	1. Lock cylinder defective	1. Replace lock cylinder.
	2. Ignition switch defective	2. Replace switch.
	3. Rack preload spring broken or deformed	3. Replace spring.
	4. Burrs on sector, rack, housing, support, tang of shift gate, or actuator rod coupling	4. Remove burr.
	5. Bent sector shaft	5. Replace shaft.
	6. Distorted rack	6. Replace rack.
	7. Misalignment of housing to cover (tilt)	7. Replace either or both.
	8. Distorted coupling slot in rack (tilt)	8. Replace rack.
	9. Bent or restricted actuator rod	9. Straighten, remove restriction, or replace.
	10. Ignition switch mounting bracket bent	10. Straighten or replace.
High effort lock cylinder— between Off and Off-lock positions	1. Burr on tang of shift gate	1. Remove burr.
	2. Distorted rack	2. Replace rack.
Sticks in Start position	1. Actuator rod deformed	1. Straighten or replace.
	2. Any high effort condition	2. Check items under high effort section.
Key cannot be removed in Off-lock position	1. Ignition switch not set correctly	1. Readjust ignition switch.
	2. Defective lock cylinder	2. Replace lock cylinder.
Lock cylinder can be removed without depressing retainer	1. Lock cylinder with defective retainer	1. Replace lock cylinder.
	2. Lock cylinder without retainer	2. Replace lock cylinder.
	3. Burr over retainer slot in housing cover	3. Remove burr.
Lock bolt hits shaft lock in Off and Park positions	Ignition switch not set correctly	Readjust ignition switch.
Ignition system—electrical system does not function	1. Defective fuse in "accessory" circuit	1. Replace fuse.
	2. Connector body loose or defective	2. Tighten or replace.

Table 9-1 AUTOMATIC TRANSMISSION—STEERING COLUMN DIAGNOSIS (continued)

Condition	Possible Cause	Correction
	3. Defective wiring	3. Repair or replace.
	4. Defective ignition switch	4. Replace ignition switch.
	5. Ignition switch not adjusted properly	5. Readjust ignition switch.
Switch does not actuate mechanically	Defective ignition switch	Replace ignition switch.
Switch cannot be set correctly	1. Switch actuator rod deformed	1. Repair or replace switch actuator rod.
	2. Sector to rack engaged in wrong tooth (tilt)	2. Engage sector to rack correctly.
Noise in column	1. Coupling bolts loose	1. Tighten pinch bolts to specified torgue.
	2. Column not correctly aligned	2. Realign column.
	3. Coupling pulled apart	3. Replace coupling and realign column.
	4. Sheared intermediate shaft plastic joint	4. Replace or repair steering shaft and realign column.
	5. Horn contact ring not lubricated	5. Lubricate with lubriplate.
	6. Lack of grease on bearings or bearing surfaces	6. Lubricate bearings.
	7. Lower shaft bearing tight or frozen	7. Replace bearing. Check shaft and replace if scored.
	8. Upper shaft tight or frozen	8. Replace housing assembly.
	9. Shaft lock plate cover loose	9. Tighten three screws or, if missing, replace. CAUTION: Use specified screws (15 inch-pounds).
	10. Lock plate snapring not seated	10. Replace snapring. Check for proper seating in groove.
	11. Defective buzzer dog cam on lock cylinder	11. Replace lock cylinder.
	12. One click when in Off-lock position and the steering wheel is moved	12. Normal condition: lock bolt is seating.
High steering shaft effort	1. Column assembly misaligned in vehicle	1. Realign.
	2. Improperly installed or deformed dust seal	2. Remove and replace.

Table 9-1 AUTOMATIC TRANSMISSION—STEERING COLUMN DIAGNOSIS (continued)

Condition	Possible Cause	Correction
	3. Tight or frozen upper or lower bearing	3. Replace affected bearing or bearings.
	4. Flash on ID of shift tube from plastic joint	4. Replace shift tube.
High shift effort	1. Column not aligned correctly in car	1. Realign.
	2. Improperly installed dust seal	2. Remove and replace.
	3. Lack of grease on seal or bearing areas	3. Lubricate bearings and seals.
	4. Burr on upper or lower end of shift tube	4. Remove burr.
	5. Lower bowl bearing not assembled properly (tilt)	5. Reassemble properly.
	6. Wave washer with burrs (tilt)	6. Replace wave washer.
Improper transmission shifting	1. Sheared shift tube joint	1. Replace shift tube assembly.
	2. Improper transmission linkage adjustment	2. Readjust linkage.
	3. Loose lower shift lever	3. Replace shift tube assembly.
	4. Improper gate plate	4. Replace with correct part.
	5. Sheared lower shift lever weld	5. Replace tube assembly.
Lash in mounted column assembly	1. Instrument panel mounting bolts loose	1. Tighten to specifications (20 foot-pounds).
	2. Broken weld nuts on jacket	2. Replace jacket assembly.
	3. Instrument panel bracket capsure sheared	3. Replace bracket assembly.
	4. Instrument panel to jacket mounting bolts loose	4. Tighten to specifications (15 foot-pounds).
	5. Loose shoes in housing (tilt)	5. Replace.
	6. Loose tilt head pivot pins (tilt)	6. Replace.
	7. Loose shoe lock pin in support (tilt)	7. Replace.
Miscellaneous	1. Housing loose on jacket noticed with ignition in Off-lock and a torque applied to the steering wheel	1. Tighten four mounting screws (60 inch-pounds).
	2. Shroud loose on shift bowl	2. Bend tabs on shroud over lugs on bowl.

Table 9-2 MANUAL TRANSMISSION—STEERING COLUMN DIAGNOSIS

Condition	Possible Cause	Correction
	1. Defective upper shift lever 2. Defective shift lever gate 3. Loose relay lever on shift tube 4. Wrong shift lever	1. Replace shift lever. 2. Replace shift lever gate. 3. Replace shift tube assembly. 4. Replace with current lever.
High shift effort	1. Column not aligned correctly 2. Lower bowl bearing not assembled correctly 3. Improperly installed seal 4. Wave washer in lower bowl bearing defective 5. Improper adjustment of lower shift levers 6. Lack of grease on seal, bearing areas, or levers 7. Damaged shift tube in bearing areas	1. Realign column. 2. Reassemble correctly. 3. Remove and replace. 4. Replace wave washer. 5. Readjust. 6. Lubricate seal, levers, and bearings. 7. Replace shift tube assembly.
Improper transmission shifting	Loose relay lever on shift tube	Replace shift tube assembly.

Table 9-3 MANUAL TRANSMISSION—TILT COLUMN DIAGNOSIS

Condition	Possible Cause	Correction
Housing scraping on bowl	Bowl bent or not concentric with hub	Replace bowl.
Steering wheel loose	1. Excessive clearance between holes in support or housing and pivot pin diameters 2. Defective or missing antilash spring in spheres 3. Upper bearing seat not seating in bearing 4. Upper bearing inner race seat missing 5. Loose support screws 6. Bearing preload spring missing or broken	1. Replace either or both. 2. Add spring or replace both. 3. Replace both. 4. Install seat. 5. Tighten to 60 inch-pounds. 6. Replace preload spring.

Table 9-3 MANUAL TRANSMISSION—TILT COLUMN DIAGNOSIS (continued)

Condition	Possible Cause	Correction
Steering wheel loose every other tilt position	Loose fit between shoe and shoe pivot pin	Replace both.
Noise when tilting column	1. Upper tilt bumper worn 2. Tilt spring rubbing in housing or dirt	1. Replace tilt bumper. 2. Lubricate.
Steering column not locking in any tilt position	1. Shoe seized on its pivot pin 2. Shoe grooves might have burrs 3. Shoe lock spring weak or broken	1. Replace shoe and pivot pin. 2. Replace shoe. 3. Replace lock spring.
Steering wheel fails to return to top tilt position	1. Pivot pins bound up 2. Wheel tilt spring defective 3. Turn signal switch wires too tight	1. Replace pivot pins. 2. Replace tilt spring. 3. Reposition wires.

Steering Linkage Diagnosis and Service

Diagnosis of Center Link, Pitman Arm, and Tie-Rod Ends

The vehicle should be raised and safety stands positioned under the lower control arms to support the vehicle weight. Use vertical hand force to check for looseness in all the pivots on the tie-rod ends and the center link. Check the seals on each tie-rod end and pivot on the center link or pitman arm for damage and cracks. Cracked seals allow dirt to enter the pivoted joints, which result in rapid wear. If looseness or damaged seals is found on any pivoted joint on the tie rods and center link, these components must be replaced.

The second part of this diagnosis is done with the front wheels resting on the shop floor. If the vehicle is equipped with power steering, start the engine and allow the engine to idle with the transmission in Park and the parking brake applied. While someone turns the steering wheel one-quarter turn in each direction from the straight-ahead position, observe all the pivoted joints on the tie-rod ends and center link. This test allows the technician to check the steering linkage pivots under load. If any of the pivoted joints show a slight amount of play, they must be replaced.

Front wheel shimmy may be defined as a consistent, fast, side-to-side movement of the front wheels and steering wheel. This movement is usually experienced at speeds above 40 mph (25 kph), and it may occur more frequently while driving on irregular road surfaces.

Tie-Rod End Replacement

Worn tie-rod ends result in these problems:

1. Excessive steering wheel freeplay
2. Incorrect front wheel toe setting
3. Tire squeal on turns
4. Tread wear on front tires
5. Front wheel shimmy
6. Rattling noise on road irregularities

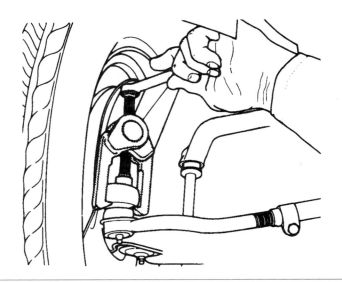

Figure 9-43 Tie-rod end removal. (Courtesy of Chevrolet Motor Division, General Motors Corp.)

The cotter pin and nut must be removed prior to tie-rod end replacement. A puller is used to remove the tie-rod end from the steering arm (Figure 9-43). Tie-rod ends with rubber encapsulated ball studs require special inspection and diagnostic procedures (Figure 9-44). On this type of tie-rod end, check for looseness of the ball stud in the rubber capsule and looseness of the stud and rubber capsule in the outer housing. If any looseness is present, replace the tie-rod end. Tie-rod end replacement is also necessary if there is any indication of the rubber capsule starting to come out of the outer housing.

After the tie-rod end is removed from the steering arm, install two nuts on the stud threads and tighten these nuts against each other. Use the proper size of socket and a torque wrench to rotate the ball stud through a 40° arc (Figure 9-45). If the ball stud turning torque is less that 20 ft-lbs (27 Nm), replace the tie-rod end.

The tie-rod clamp must be loosened before the tie-rod is removed from the sleeve. Count the number of turns required to remove the tie-rod end from the sleeve, and install the new tie rod with the same number of turns. Even when this procedure is followed, the toe must be checked after the steering linkage components are replaced. Before the new tie-rod end is installed, center

Special Tools

Tie rod end puller

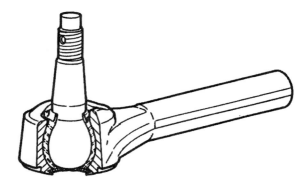

Figure 9-44 Tie-rod end with rubber encapsulated ball stud. (Courtesy of Moog Automotive, Inc.)

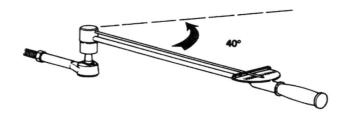

Figure 9-45 Measuring turning torque on a rubber encapsulated tie-rod end. (Courtesy of Moog Automotive, Inc.)

the stud in the tie-rod end. When the tie-rod stud is installed in the steering arm opening, only the threads should be visible above the steering arm surface. If the machined surface of the tie-rod stud is visible above the steering arm surface or if the stud fits loosely in the steering arm opening, this opening is worn or the tie-rod end is not correct for that application. The tie-rod nut must be torqued to manufacturer's specifications, and the cotter pin must be installed through the tie-rod end and nut openings (Figure 9-46).

The tie-rod nut must never be loosened from the specified torque to install the cotter pin. Another method of positioning replacement tie-rod ends is to measure the distance from the center of the tie-rod stud to the end of the sleeve prior to removal. When the new tie rod is installed, be sure this measurement is the same. The slots in the tie-rod sleeve must be positioned away from the opening in the sleeve clamps (Figure 9-47). Leave the sleeve clamps loose until the front wheel toe is checked, and then tighten the sleeve clamp bolts to the specified torque. A special tool is available to rotate the tie-rod sleeves and set the front wheel toe (Figure 9-48).

 WARNING: If rubber encapsulated tie-rod ends are tightened with the front wheels in any position but straight ahead, steering pull and wander may occur.

When rubber encapsulated tie-rod ends are installed and tightened to the specified torque, the front wheels must be straight ahead. Tightening this type of tie rod with the front wheels in any other position but straight ahead may cause steering pull or wander. After rubber encapsulated tie-rod ends are tightened to the specified torque, it is acceptable for the tie-rod housing to be tilted to one side (Figure 9-49).

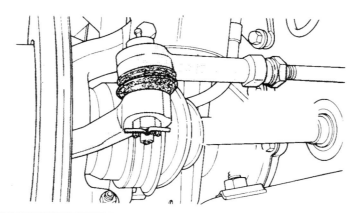

Figure 9-46 Tie-rod nut and cotter pin installation. (Courtesy of Chrysler Corp.)

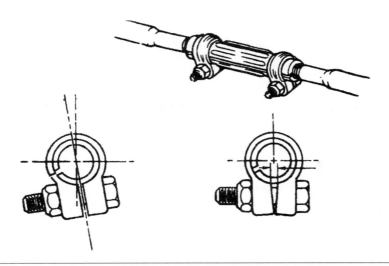

Figure 9-47 Proper slot and clamp position on tie-rod sleeves. (Courtesy of Chevrolet Motor Division, General Motors Corp.)

Figure 9-48 Tie-rod sleeve adjusting tool. (Courtesy of Mac Tools, Inc.)

Pitman Arm Diagnosis and Replacement

Some pitman arms contain a ball socket joint on the outer end. The threaded extension on this ball socket fits into the center link. On other steering linkages, the ball socket joint is in the center link, and the threaded extension fits into the pitman arm opening. If the pitman arm is bent, it must be replaced. If the pitman arm is bent, the tie rod is not parallel to the lower control arm. Under this

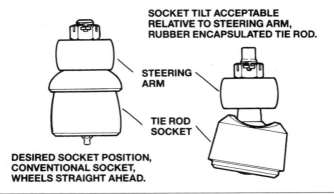

Figure 9-49 Rubber encapsulated tie-rod end housing may be tilted to one side after it is installed and tightened. (Courtesy of Moog Automotive, Inc.)

condition, excessive front wheel toe change occurs on road irregularities, and front tire wear may be excessive. The following is a typical pitman arm replacement procedure.

1. Position the front wheels straight ahead, and remove the cotter pin and nut from the ball socket joint on the outer end of the pitman arm.

2. Remove the ball socket extension from the pitman arm or center link with a tie-rod end puller.

3. Loosen the pitman-arm-to-pitman-shaft nut.

4. Use a puller to pull the pitman arm loose on the shaft.

5. Remove the nut, lock washer, and pitman arm.

6. Check the pitman shaft splines. If the splines are damaged or twisted, the shaft must be replaced.

7. Reverse steps 1 through 5 to install the pitman-arm. The pitman arm-to-shaft nut and the ball socket extension nut must be tightened to the manufacturer's specified torque. Be sure the pitman arm is installed in the correct position on the shaft splines. Install the cotter pin in the ball socket extension.

Center Link Diagnosis and Replacement

WARNING: Never attempt to straighten steering linkage components. This action may weaken the metal and cause sudden component failure, vehicle damage, and personal injury.

A bent center link must be replaced. Do not attempt to straighten this rod. If the ball socket joints are loose on either end of the rod, center link assembly replacement is necessary. If the ball stud openings in the center link are worn, replace the center link. Follow these steps for a typical center link replacement:

1. Remove the cotter pins from the tie-rod-to-center-link nuts, and the idler arm and pitman-arm-to-center-link nuts.

2. Remove the nuts on the tie-rod inner ends, idler-arm-to-center-link ball socket extension, and the pitman-arm-to-center-link ball socket extension.

3. Use a tie-rod end puller to pull the inner tie rods from the center link. Follow the same procedure to remove the center-link-to-pitman-arm ball socket extension.

4. Remove the idler arm from the center link, then remove the center link.

5. Reverse steps 1 through 4 to install the center link. Tighten all the ball socket nuts to the manufacturer's specified torque, and install cotter pins in all the nuts. If the ball sockets have grease fittings, lubricate the ball sockets with a grease gun and chassis lubricant.

Idler Arm Diagnosis

To measure idler arm vertical movement, attach the magnetic base of a dial indicator to the frame near the idler arm. Position the dial indicator stem against the upper side on the outer end of the idler arm. Preload the dial indicator stem and zero the dial. Use a pull scale to apply 25 lbs (11.34 kg) of force downward and upward on the idler arm (Figure 9-50). Observe the total vertical idler arm movement on the dial indicator. If this vertical movement exceeds the vehicle manufacturer's specifications, replace the idler arm. Typical maximum vertical idler arm movement from the downward to the upward position is 0.250 in (63.5 mm). If idler arm vertical movement is excessive, the

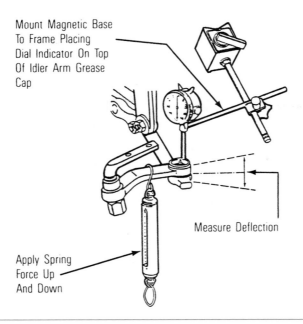

Mount Magnetic Base
To Frame Placing
Dial Indicator On Top
Of Idler Arm Grease
Cap

Measure Deflection

Apply Spring
Force Up
And Down

Figure 9-50 Measuring idler arm vertical movement. (Courtesy of Federal-Mogul Corp.)

tie rod is not parallel to the lower control arm. Excessive idler arm vertical movement causes these steering problems:

1. Excessive toe change and front tire tread wear
2. Excessive steering wheel freeplay and reduced steering control
3. Front end shimmy

 SERVICE TIP: A binding idler arm may suddenly break off and cause complete loss of steering control, vehicle damage, and personal injury.

Binding idler arm bushings result in these complaints:

1. Hard steering
2. Squawking noise when the front wheels are turned
3. Poor steering wheel returnability

Some idler arms contain steel bushings, whereas others are equipped with rubber bushings. The following is a typical idler arm removal and replacement procedure.

1. Remove the idler-arm-to-center-link cotter pin and nut.
2. Remove the center link from the idler arm.
3. Remove the idler arm bracket mounting bolts and remove the idler arm.
4. If the idler arm has a steel bushing, thread the bracket into the idler arm bushing until the specified clearance is obtained between the center of the lower bracket bolt hole and the upper idler arm surface (Figure 9-51).
5. Install the idler arm bracket to frame bolts and tighten the bolts to the specified torque. Be sure that lock washers are installed on the bolts.

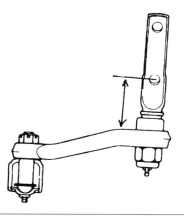

Figure 9-51 Specified clearance between center of lower bracket bolt hole and upper idler arm surface. (Courtesy of Chevrolet Motor Division, General Motors Corp.)

6. Install the center link into the idler arm, and tighten the mounting nut to the specified torque. Install the cotter pin in the nut.

7. If the idler arm contains a grease fitting, lubricate as required.

The idler arm adjustment is very important. If this adjustment is incorrect, front wheel toe is affected. After idler arm replacement, the front wheel toe should be checked.

Steering Damper Diagnosis and Replacement

Some steering systems have a damper connected between the center link and the chassis. A damper is similar to a small shock absorber. The purpose of the damper is to prevent the transmission of steering shock and vibrations to the steering wheel. A worn-out steering damper may cause excessive steering shock and vibration on the steering wheel, especially on irregular road surfaces. A rattling noise occurs if the damper mounting bolts or brackets are loose. The following is a typical steering damper checking and replacement procedure.

1. Lift the vehicle on a hoist and grasp the damper firmly. Apply vertical and horizontal pressure to the damper and check for movement in the damper mounts. If movement exists, tighten or replace the damper mounting bushings or brackets.

2. Visually inspect the damper for oil leaks. A slight film of oil on the damper body near the shaft seal is acceptable. If there is any indication of oil dripping from the damper, the unit must be replaced.

3. Disconnect one end of the damper and pull the damper back and forth horizontally. The damper should offer a slight equal resistance to movement in either direction. When this resistance is not felt in one or both directions, replace the damper.

4. To replace the damper, remove the mounting bolts from the chassis and the center link.

5. When the new damper is installed, tighten the mounting bolts to the specified torque. Turn the steering wheel fully in each direction and be sure the damper does not restrict linkage movement.

Steering Arm Diagnosis

Classroom Manual
Chapter 9, page 244

If the front rims have been damaged, the steering arms should be checked for a bent condition. Measure the distance from the center of the tie-rod stud to the edge of the rim on each side. Unequal readings may indicate a bent steering arm. Bent steering arms must be replaced. Steering linkage diagnosis is summarized in Table 9-4.

Table 9-4   STEERING LINKAGE DIAGNOSIS

Condition	Possible Cause	Correction
Excessive play or looseness in steering system	1. Front wheel bearings loosely adjusted	1. Adjust bearings to obtain proper endplay.
	2. Worn steering shaft couplings	2. Replace part.
	3. Worn upper ball joints	3. Check and replace, if necessary.
	4. Steering wheel loose on shaft, or loose pitman arm, tie rods, steering arms, or steering linkage ball studs	4. Tighten to specified torque, or replace if necessary.
	5. Steering gear thrust bearings loosely adjusted	5. Adjust preload to specifications.
	6. Excessive over-center lash in steering gear	6. Adjust preload to specifications per shop manual.
	7. Worn intermediate rod or tie-rod sockets	7. Replace worn part.
Excessive looseness in tie rod or intermediate rod pivots, or excessive vertical lash in idler support	Seal damage and leakage resulting in loss of lubricant, corrosion, and excessive wear	Replace damaged parts as necessary. Properly position upon reassembly.
Hard steering—excessive effort required at steering wheel	1. Low or uneven tire pressure	1. Inflate to specified pressures.
	2. Steering linkage or bolt joints need lubrication	2. Lube with specified lubricant.
	3. Tight or frozen intermediate rod, tie rod, or idler socket	3. Lube, replace, or reposition as necessary.
	4. Steering gear-to-column misalignment	4. Align column.
	5. Steering gear adjusted too tightly	5. Adjust over-center and thrust bearing preload to specification.
	6. Front wheel alignment incorrect (manual gear)	6. Check alignment and correct as necessary.
Poor returnability	1. Steering linkage or ball joints need lubrication	1. Lube with specified lubricant.
	2. Steering gear adjusted too tightly	2. Adjust over-center and thrust bearing preload to specifications.
	3. Steering gear-to-column misalignment	3. Align columns.
	4. Front wheel aligment incorrect (caster)	4. Check alignment and correct as necessary.

Guidelines for Servicing Steering Columns and Steering Linkage Mechanisms

1. Prior to servicing an air bag system component, the technician must disconnect the negative battery cable and wait one minute.

2. Only the vehicle manufacturer's recommended tools should be used when servicing an air bag system or the electrical system on an air-bag-equipped vehicle. Do not use such tools as a 12V test lamp or a self-powered test lamp.

3. Always store an air bag deployment module facing upward on the workbench.

4. Do not hammer on the top of the steering shaft to remove a steering wheel.

5. Do not pull on the steering wheel to remove it from the steering shaft.

6. Prior to installing a clock spring electrical connector or spiral cable, the front wheels must be in the straight-ahead position and the clock spring must be centered.

7. When a steering wheel is installed, the marks on the steering wheel and the steering shaft must be aligned.

8. When a collapsible steering column is checked for collision damage, the column jacket, gearshift tube, and steering shaft should be checked for sheared injected plastic.

9. To check for collision damage in a collapsible steering column, the distance should be measured from the end of the bearing assembly to the lower edge of the upper column jacket.

10. Some ignition key cylinders must be removed and replaced with the key in the ACC position.

11. Some vehicle manufacturers recommend coating all rubbing steering column components with molybdenum disulphide lithium-based grease.

12. When a flexible coupling is installed, there should be a specified clearance between the coupling clamp and the steering gear adjusting plug.

13. When a tie-rod end stud is installed in the steering arm opening, only the threaded part of the stud should appear above the knuckle surface.

14. The slots in the tie-rod sleeves must be positioned away from the openings in the sleeve clamps.

15. The front wheel toe should be checked after steering linkage components are replaced.

16. After tie-rod end nuts are tightened to the specified torque, they must never be loosened to install the cotter pin.

17. Never straighten steering linkage components.

18. An idler arm with excessive vertical movement must be replaced.

19. Some idler arms with steel bushings require an adjustment to set the clearance between the idler arm bracket flange and the idler arm surface.

20. If there is any indication of oil dripping from a steering damper, this component must be replaced.

21. The distance may be measured from the center of each outer tie-rod end stud to the inside edge of each rim to check for a bent steering arm.

A customer complained about excessive steering effort and poor steering wheel returnability after a turn on a 1993 Lincoln Town Car with power steering. The technician road tested the car and found the customer's description of the problems to be accurate except for one point. The steering continually required excessive steering effort and the steering wheel did not return properly after a turn. However, the customer did not mention, or possibly did not notice, that a squawking and creaking noise was sometimes heard during a turn.

The technician checked the power steering fluid level and condition, and found this fluid to be in good condition and at the proper level. Next, the technician checked the power steering belt condition and the power steering pump pressure. The belt tension and condition were satisfactory, and the power steering pump pressure was normal. A check of the power steering pump mounting bolts and brackets indicated they were in good condition. The technician disconnected the center link from the pitman arm and rotated the steering wheel with the engine running. Under this condition, the steering wheel turned very easily. Therefore, the technician concluded the excessive steering effort and poor returnability problems were not in the steering column or steering gear. Next, the technician disconnected the idler arm from the center link. When the technician attempted to move the idler arm back and forth, the idler arm had a severe binding problem. After an idler arm replacement, a road test revealed the excessive steering effort and poor returnability problems had disappeared.

Terms to Know

Accidental air bag deployment

Air bag deployment module

Backup power supply

Clock spring electrical connector

Collapsible steering column

Molybdenum disulphide lithium-based grease

Sheared injected plastic

Spiral cable

Steering wheel freeplay

Tapered-head bolts

ASE Style Review Questions

1. While discussing steering column service on an air-bag-equipped vehicle:
    Technician A says service personnel should disconnect the negative battery cable and wait one minute prior to servicing an air bag system component.
    Technician B says a 12V test lamp may be used to diagnose an air bag system.
    Who is correct?
    A. A only
    B. B only
    C. Both A and B
    D. Neither A nor B

2. While discussing steering column service on an air-bag-equipped vehicle:
    Technician A says an air bag deployment module should be placed face downward on the workbench.
    Technician B says the backup power supply provides power to deploy the air bag if the battery is disconnected during a frontal collision.
    Who is correct?
    A. A only
    B. B only
    C. Both A and B
    D. Neither A nor B

3. While discussing steering wheel and clock spring electrical connector removal and replacement on an air-bag-equipped vehicle:

   Technician A says the steering may bind if the clock spring is improperly installed.

   Technician B says when the clock spring electrical connector is installed, the front wheels should be straight ahead and the clock spring should be turned fully clockwise.

   Who is correct?

   A. A only   **C.** Both A and B

   B. B only   **D.** Neither A nor B

4. While discussing steering wheel removal:

   Technician A says a steering wheel should be marked in relation to the steering shaft.

   Technician B says the steering wheel should be grasped firmly with both hands and pulled from the shaft.

   Who is correct?

   A. A only   **C.** Both A and B

   B. B only   **D.** Neither A nor B

5. While discussing collapsible steering column damage:

   Technician A says if the vehicle has been in a frontal collision, the injected plastic in the column jacket, gearshift tube, and steering shaft may be sheared.

   Technician B says after a collision some steering columns may be shifted on the bracket.

   Who is correct?

   A. A only   **C.** Both A and B

   B. B only   **D.** Neither A nor B

6. All these statements about steering column service are true EXCEPT:

   A. Many ignition switches must be in the ACC position before removing the switch cylinder.

   B. An improperly installed dust seal may cause excessive steering wheel freeplay.

   C. High steering effort may be caused by an improperly aligned steering column.

   D. A worn steering shaft universal joint may cause excessive steering wheel freeplay.

7. A rattling noise in the steering column and linkage may be caused by:

   A. a bent center link.

   B. a binding idler arm.

   C. a worn steering shaft U-joint.

   D. loose tie-rod sleeve clamps.

8. When diagnosing and servicing tie-rod ends:

   A. rubber encapsulated tie-rod ends should be tightened with the wheels turned fully to the right or left.

   B. the machined part of the tie-rod stud should be visible above the surface of the steering arm.

   C. the nut on the tie-rod stud may be loosened to install the cotter pin.

   D. if the turning torque on a rubber encapsulated tie-rod end is less that specified, the tie-rod end must be replaced.

9. When diagnosing and servicing steering linkages:

   A. a worn tie-rod end causes excessive front wheel toe change.

   B. bent steering linkages may be straightened in a hydraulic press.

   C. the slot in a tie-rod sleeve clamp must be positioned above the slot in the tie-rod sleeve.

   D. front wheel shimmy may be caused by a binding idler arm.

10. An idler arm has 5/8 in (15.8 mm) vertical movement. To correct this problem it is necessary to:

    A. install thicker shims between the idler arm and the bracket.

    B. install an oversize bushing in the idler arm.

    C. replace the idler arm assembly.

    D. replace the belleville washer and spacer under the bushing.

ASE Challenge Questions

1. A preliminary steering inspection reveals that the steering wheel has nearly 1/2 inch of movement in either direction before the pitman arm begins to move. All of the following could cause this problem EXCEPT:
   A. coupling joints.
   B. U joints.
   C. tight or frozen shaft bearing.
   D. worn or out of adjustment steering gear.

2. A customer complains about an ignition switch that sometimes won't unlock and the key won't turn. A preliminary inspection reveals that turning the steering wheel does not help, but repositioning the key does, after several attempts, unlock the ignition.
   Technician A says the problem could be the column assembly is misaligned.
   Technician B says the problem could be the ignition key is worn.
   Who is correct?
   A. A only
   B. B only
   C. Both A and B
   D. Neither A nor B

3. If the clock spring electrical connector is not properly centered before it is installed:
   A. the steering may pull to one side.
   B. the clock spring electrical connector may be broken.
   C. the air bag may deploy without the vehicle being in an accident.
   D. the steering may have a binding condition.

4. A customer says the steering on his vehicle is very hard to turn in both directions. A preliminary inspection reveals no problems with the manual rack and pinion steering gear.
   Technician A says the problem could be a misaligned steering column.
   Technician B says the problem could be tight or frozen tie-rod ends.
   Who is correct?
   A. A only
   B. B only
   C. Both A and B
   D. Neither A nor B

5. A customer says his steering has a shimmy problem above 50 mph (80 kmph). Preliminary inspection indicates excessive wear on the right front tire.
   Technician A says the problem could be a worn tie-rod end.
   Technician B says the problem could be worn idler arm ball joint.
   Who is correct?
   A. A only
   B. B only
   C. Both A and B
   D. Neither A nor B

Table 9-1 NATEF and ASE TASK

Diagnose steering column noises, looseness, and binding problems (including tilt mechanisms); determine needed repairs.

Problem Area	Symptoms	Possible Causes	Classroom Manual	Shop Manual
NOISES	Rattling on road irregularities	Loose, worn coupling or universal joints	229	280
	Clunk while tilting	1. Worn universal joint or spherical bearing	229	280
		2. Worn tilt bumper	232	273
BINDING	Excessive steering effort	1. Column misaligned	233	283
		2. Improperly installed or defective dust shield	238	283
		3. Worn or defective column bearings	233	284
LOOSENESS	Excessive steering wheel freeplay	1. Worn universal joints or flexible coupling	229	280
		2. Worn tilt spherical bearing or universal joint	233	284
		3. Defective column bearings	233	284

Table 9-2 NATEF and ASE TASK

Inspect and replace steering shaft U-joint(s), flexible coupling(s), collapsible columns, steering wheels (includes steering wheels with air bags and/or other steering wheel mounted controls and components).

Problem Area	Symptoms	Possible Causes	Classroom Manual	Shop Manual
NOISE	Rattling on road irregularities	Worn or loose flexible coupling or universal joint	229	280
STEERING CONTROL	Excessive steering wheel freeplay	Worn or loose flexible coupling or universal joint	229	280
	Improper steering wheel position	Collapsed column	230	272
	Damaged steering wheel	Frontal collision	227	270
AIR BAG DEPLOYMENT	Failure to deploy	Defective clock spring	230	268

Table 9-3 ASE TASK

Inspect and adjust (where applicable) front and rear steering linkage geometry, including parallelism and vehicle ride height.

Problem Area	Symptoms	Possible Causes	Classroom Manual	Shop Manual
TIRE LIFE	Excessive front tire wear	1. Bent pitman arm	238	289
	Excessive tire tread wear	2. Worn or misadjusted idler arm	239	290

Table 9-4 NATEF and ASE TASK

Inspect and replace pitman arm.

Problem Area	Symptoms	Possible Causes	Classroom Manual	Shop Manual
TIRE LIFE	Excessive front tire wear	Bent pitman arm	238	289
LOOSENESS	Excessive steering wheel freeplay	Worn ball socket in pitman arm	238	289

Table 9-5 NATEF and ASE TASK

Inspect and replace relay rod (center link/drag link/intermediate rod).

Problem Area	Symptoms	Possible Causes	Classroom Manual	Shop Manual
LOOSENESS	Excessive steering wheel freeplay	1. Worn ball sockets in center link	240	290
		2. Worn stud openings in center link	241	290

Table 9-6   NATEF and ASE TASK

Inspect, adjust (where applicable), and replace idler arm and mountings.

Problem Area	Symptoms	Possible Causes	Classroom Manual	Shop Manual
NOISE	Rattling on road irregularities	Worn idler arm	239	290
STEERING CONTROL	Excessive steering wheel freeplay	Worn idler arm or loose mounting	239	290
	Binding, excessive steering effort	Binding idler arm	240	291
TIRE LIFE	Excessive front tire wear	Worn or misadjusted idler arm	240	291

Table 9-7   NATEF and ASE TASK

Inspect, replace, and adjust tie rods, tie-rod sleeves, clamps, and tie-rod ends (sockets).

Problem Area	Symptoms	Possible Causes	Classroom Manual	Shop Manual
NOISE	Rattling noise on road irregularities	Loose tie-rod ends	236	288
TIRE LIFE	Excessive front tire wear	Loose or improperly adjusted tie-rod ends	237	289

Table 9-8   NATEF and ASE TASK

Inspect and replace steering linkage damper.

Problem Area	Symptoms	Possible Causes	Classroom Manual	Shop Manual
STEERING CONTROL	Shock, vibration on steering wheel	Worn or loose steering damper	244	292

Job Sheet 26

26

Name _____ Date _____

Remove and Replace Air Bag Inflator Module and Steering Wheel

NATEF and ASE Correlation

This job sheet is related to NATEF and ASE Automotive Suspension and Steering Task List content area: A. Steering Systems Diagnosis and Repair. 1. Steering Columns and Manual Steering Gears. Task 4: Inspect and replace steering shaft U-joint(s), flexible coupling(s), collapsible columns, steering wheels (includes steering wheels with air bags and/or other steering wheel mounted controls and components).

Tools and Materials

Steering wheel puller
Torque wrench

Vehicle make and model year _____

Vehicle VIN _____

Procedure

1. Turn the ignition switch to the lock position and place the front wheels facing straight ahead.

2. Remove the negative battery terninal and wait the time period specified by the vehicle manufacturer before servicing air bag components.

   Specified waiting period _____

   After the battery negative cable is disconnected, explain why a waiting period is necessary before working on the air bag system.

   _____

   _____

   _____

   _____

3. Loosen the air bag retaining screws under the steering wheel.

4. Loosen the other two air bag retaining screws under the steering wheel.

   Loosen all the screws until the groove along the screw circumference catches on the screw case.

   CAUTION: When an air bag deployment module is temporarily stored on the workbench, always place this module face upward. If the air bag deployment module accidentally deployed when facing downward, the module would become a projectile and personal injury may result.

5. Pull the air bag deployment module from the steering wheel and disconnect the air bag module electrical connector. Do not pullon the air bag wires in the steering column. Place the air bag deployment module face upward on the workbench.

6. Disconnect the air bag wiring retainer in the steering wheel.

7. Use the proper size socket and a ratchet to remove the steering wheel retaining nut.

8. Observe the matching alignment marks on the steering wheel and the steering shaft. If these alignment markd are not present, place alignment marks on the steering wheel and steering shaft with a center punch and a hammer.

   Alignment marks on steering wheel and steering shaft: satisfactory _____
   unsatisfactory _____

   ⚠️ **WARNING:** Do not hammer on thetop of the steering shaft to remove the steering wheel. This action may damage the shaft.

   ⬛ **CAUTION:** Do not pull on the steering wheel in an attempt to remove it from the steering shaft. The steering wheel may suddenly come off resulting in personal injury, or the steering wheel may be damaged by the pulling force.

9. Install a steering wheel puller with the puller bolts threaded into the bolt holes in the steering wheel. Tighten the puller nut to remove the steering wheel. Visually check the steering wheel condition. If the steering wheel is bent or cracked, replace the wheel.

   Steering wheel condition: satisfactory _____ unsatisfactory _____

10. Disconnect the four retaining screws and remove the clock spring electrical connector.

11. Be sure the front wheels are facing straight ahead. Turn the clock spring electrical connector counterclockwise by hand until it becomes harder to turn as it becomes fully wound in that direction.

    ⚠️ **WARNING:** Failure to center a clock spring prior to installation may cause a broken conductive tape in the clock spring.

12. Turn the clock spring electrical connector clockwise three turns and align the red mark on the center part of the spring face with the notch in the cable circumference. This action centers the clock spring electrical connector.

    Is the clock spring centered? yes _____ no _____

    Instructor check _____

    Explain why the clock spring electrical connector must be centered before it is installed.

    _____
    _____
    _____

13. Install the clock spring electrical connector and tighten the four retaining screws to the specified torque.

14. Align the marks on the steering wheel and the steering shaft, and install the steering wheel on the shaft.

    Are the marks on the steering wheel and steering shaft aligned? yes _____ no _____

    Instructor check _____

15. Install the steering wheel retaining nut and tighten this nut to the specified torque.

    Specified steering wheel retaining nut torque _____

    Actual steering wheel retaining nut torque _____

16. Install the air bag wiring retainer in the steering wheel.

17. Hold the air bag deployment module near the top of the steering wheel, and connect the air bag module connector.

18. Install the air bag deployment module inthe top of steering wheel and tighten the retaining screws.

19. Reconnect the negative battery cable.

20. Reset the clock and radio.

21. Turn on the ignition switch and start the vehicle. Check the operation of the air bag system warning light.

    Explain the air bag system warning light operation that indicates normal air bag system operation.

    _____

    _____

    _____

☑ **Instructor Check** _____

Job Sheet 27

27

Name _____ Date _____

Remove and Replace Steering Column

NATEF and ASE Correlation

This job sheet is related to NATEF and ASE Automotive Suspension and Steering Task List content area: A. Steering Systems Diagnosis and Repair. 1. Steering Columns and Manual Steering Gears. Task 4: Inspect and replace steering shaft U-joint(s), flexible coupling(s), collapsible columns, steering wheels (includes steering wheels with air bags and/or other steering wheel mounted controls and components).

Tools and Materials

Seat cover
Torque wrench

Vehicle make and model year _____
Vehicle VIN _____

Procedure

1. Disconnect the battery negative cable. If the vehicle is equipped with an air bag, wait for the time period specified by the vehicle manufacturer.

   Specified waiting period prior to servicing air bag components _____

   Explain why it is necessary to disconnect the battery before removing the steering column.

   _____

   _____

   _____

2. Install a seat cover on the front seat.

3. Place the front wheels in the straight-ahead position, and remove the ignition key from the switch to lock the steering column.

   Are the front wheels straight? yes _____ no _____

   Is the ignition key removed?

   Instructor check _____

4. Remove the cover under the steering column and remove the lower finish panel if necessary.

5. Disconnect all wiring connectors from the steering column.

6. If the vehicle has a column-mounted gearshift lever, disconnect the gearshift linkage at the lower end of the steering column.

7. Remove the retaining bolt or bolts in the lower universal joint or flexible coupling.

8. Remove the steering-column-to-instrument-panel mounting bolts.

9. Carefully remove the steering column from the vehicle. Be careful not to damage upholstery or paint.

10. Install the steering column under the instrument panel, and insert the steering shaft into the lower universal joint.

11. Install the steering column to instrument panel mounting bolts. Be sure the steering column is properly positioned, and tighten these bolts to the specified torque.

    Specified torque, steering column mounting bolts _____

    Actual torque, steering column mounting bolts _____

12. Install the retaining bolt or bolts in the lower universal joint or flexible coupling, and tighten the bolt(s) to the specified torque.

    Specified torque, lower U-joint retainig bolt(s)_____

    Actual torque, lower U-joint retaining bolt(s) _____

13. Connect the gearshift linkage, if the vehicle has column mounted gearshift.

14. Connect all the wiring harness connectors to the steering column connectors.

15. Install the steering column cover and the lower finish panel.

16. Reconnect the negative battery cable.

17. Road test the vehicle and check for proper steering column operation.

    Steering column operation: satisfactory _____ unsatisfactory _____

☑ **Instructor Check** _____

Job Sheet 28

28

Name _____ Date _____

Diagnose, Remove, and Replace Idler Arm

NATEF and ASE Correlation

This job sheet is related to NATEF and ASE Automotive Suspension and Steering Task List content area: A. Steering Systems Diagnosis and Repair. Steering Linkage Task 4: Inspect, adjust (where applicable), and replace idler arm and mountings.

Tools and Materials

Pull scale
Dial indicator
Torque wrench
Grease gun

Vehicle make and model year _____

Vehicle VIN _____

Procedure

1. Attach the magnetic base of a dial indicator to the frame near the idler arm.

2. Position the dial indicator stem against the upper side on the outer end of the idler arm. Preload the dial indicator stem and zero the dial.

3. Use a pull scale to apply 25 lbs (11.34 kg) of force downward and then upward on the idler arm. Observe the total vertical idler arm movement on the dial indicator. If this vertical movement exceeds the vehicle manufacturer's specifications, replace the idler arm.

   Specified vertical idler arm movement _____

   Actual vertical idler arm movement _____

   Recommended idler arm service _____

   Explain the effect of idler arm wear on steering quality.

   _____

   _____

   _____

4. Remove the idler arm to center link cotter pin and nut.

5. Remove the center link from the idler arm.

6. Remove the idler arm bracket mounting bolts, and remove the idler arm.

7. If the idler arm has a steel bushing, thread the bracket into the idler arm bushing until the specified clearance is obtained between the center of the lower bracket bolt hole and the upper idler arm surface.

Specified distance from center of lower bracket hole to upper idler arm surface

_____

Actual distance from center of lower bracket hole to upper idler arm surface _____

8. Install the idler arm bracket to frame bolts and tighten the bolts to the specified torque. Be sure that lock washers are installed on the bolts.

Specified idler arm retaining bolt torque _____

Actual idler arm retaining bolt torque _____

9. Install the center link into the idler arm and tighten the mounting nut to the specified torque. Install the cotter pin in the nut.

Specified center link to idler arm retaining nut torque _____

Actual center link to idler arm retaining nut torque _____

10. If the idler arm has a grease fitting, lubricate as required.

☑ **Instructor Check** _____

Job Sheet 29

29

Name _____ Date _____

Remove and Replace Outer Tie-Rod End, Parallelogram Steering Linkage

NATEF and ASE Correlation

This job sheet is related to NATEF and ASE Automotive Suspension and Steering Task List content area: A. Steering Systems Diagnosis and Repair. 3. Steering Linkage Task 5: Inspect, replace, and adjust tie rods, tie-rod sleeves, clamps, and tie-rod ends (sockets).

Tools and Materials

Tie rod puller

Vehicle make and model year _____
Vehicle VIN _____

Procedure

1. Raise the vehicle on a lft with the tires supported on the lift. If the vehicle has rubber encapsulated tie-rod ends, be sure the front wheels are straight ahead.

2. Loosen the nut on tie-rod sleeve bolt that retains the sleeve to the tie rod ends.

3. Measure fron the top center of the tie-rod end ball stud to the outer end of the tie-rod sleeve and record the measurement.

   Distance from the center of the tie-rod end ball stud to the outer end of the tie-rod sleeve

   _____

4. Remove the cotter pin from the tie-rod end retaining nut.

5. Loosen the tie-rod end retaining nut.

6. Use a tie-rod end puller to loosen the tie-rod end taper in the steering arm.

7. Remove the tie-rod end retaining nut, and remove the tie-rod end from the steering arm.

8. Count the number of turns required to thread the tie-rod end out of the tie-rod sleeve.

   Number of turns required to remove the tie-rod end from the tie-rod sleeve

   _____

   Describe the effect of worn, loose, outer tie-rod ends on steering quality.

   _____

   _____

9. Install the new tie-rod end in the tie-rod sleeve using the same number of turns recorded in step 7.

10. Push the tie-rod end ball stud fully into the steering arm opening. Only the threads on the tie-rod end ball stud should be visible above the steering arm surface.

    Is the taper on the tie-rod end ball stud visible above the steering arm surface?
    yes _____ no _____

If the answer to this question is yes, state the necessary repairs required to correct this problem.

_____

_____

Instructor check _____

11. With the tie-rod end ball stud pushed fully upward into the steering arm, measure the distance from the center of the ball stud to the outer end of the tie-rod sleeve.

Distance from the center of the tie-rod end ball stud to the outer end of the tie-rod sleeve

_____

12. If this distance is not the same as recorded in step 3, remove the tie-rod end ball stud from the steering arm and rotate the tie-rod end until this distance is the same as recorded in step 3.

13. Install the tie-rod end ball stud retaining nut, and tighten this nut to the specified torque.

Specified ball stud nut torque _____

Actual ball stud nut torque _____

⚠️ **WARNING:** Never loosen a tie-rod end ball stud nut to align the openings in the ball stud and nut to allow cotter pin installation. This action may cause the nut to loosen during operation of the vehicle, and this will result in excessive steering freeplay. Complete loss of steering control will occur if the nut comes off the tie-rod end ball stud, and this may cause a collision.

14. Install a new cotter pin through the tie-rod end ball stud and nut openings, and bend the cotter pin legs separately around this nut.

15. Tighten the nut outer tie-rod sleeve clamp bolt to the specified torque. Be sure the slot in the tie-rod sleeve is positioned away from the opening in the clamp.

16. Measure the front wheel tow with a toe bar.

Specified front wheel toe _____

Actual front wheel toe _____

Explain the type of front tire wear that is caused by improper front wheel toe.

_____

_____

_____

✔️ **Instructor Check** _____

Power Steering Pump Diagnosis and Service

Upon completion and review of this chapter, you should be able to:

❏ Check power steering belt condition and adjust belt tension.

❏ Diagnose power steering belt problems.

❏ Check power steering fluid level and add fluid as required.

❏ Drain and flush power steering system.

❏ Bleed air from power steering system.

❏ Perform power steering pump pressure test.

❏ Check power steering pump fluid leaks.

❏ Remove and replace power steering pumps, and inspect pump mounts.

❏ Remove and replace power steering pump pulleys.

❏ Remove and replace power steering pump integral reservoirs.

❏ Remove, replace, and check flow control valve and pressure relief valve.

❏ Remove, replace, and check power steering pump rotating components.

❏ Remove and replace power steering pump seals and O-rings.

❏ Check, remove, and replace power steering lines.

Power Steering Pump Belt Service

Checking Belt Condition and Tension

Power steering belt condition and tension are extremely important for satisfactory power steering pump operation. A loose belt causes low pump pressure and hard steering. The steering wheel may jerk and surge during a turn if the power steering pump belt is loose. A loose, dry, or worn belt may cause squealing and chirping noises, especially during engine acceleration and cornering.

The power steering pump belt should be checked for tension, cracks, oil-soaking, worn or glazed edges, tears, and splits (Figure 10-1). If any of these conditions are present, belt replacement is necessary.

Since the friction surfaces are on the sides of a **V-type belt,** wear occurs in this area. If the belt edges are worn, the belt may be rubbing on the bottom of the pulley. This condition requires belt

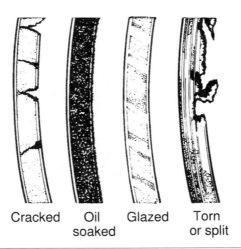

Cracked   Oil soaked   Glazed   Torn or split

Figure 10-1 Defective belt conditions. (Courtesy of Chrysler Corp.)

Basic Tools

Basic technician's tool set
Service manual
Pry bar
Floor jack
Safety stands
Oil drain pan
Crocus cloth

Special Tools

Belt tension gauge

replacement. Belt tension may be checked by measuring the belt deflection. Press on the belt with the engine stopped to measure the belt deflection, which should be 1/2 inch per foot of free span. The belt tension may be checked with a **belt tension gauge** placed over the belt (Figure 10-2). The tension on the gauge should equal the vehicle manufacturer's specifications.

If the belt requires tightening, follow this procedure:

 WARNING: Do not pry on the pump reservoir with a pry bar. This action may damage the reservoir.

1. Loosen the power steering pump bracket or tension adjusting bolt.

2. Loosen the power steering pump mounting bolts.

3. Check the bracket and pump mounting bolts for wear. If these bolts or bolt openings in the bracket or pump housing are worn, replacement is necessary.

4. Pry against the pump ear and hub with a pry bar to tighten the belt. Some pump brackets have a 1/2-inch square opening in which a breaker bar may be installed to move the pump and tighten the belt.

5. Hold the pump in the position described in step 4, and tighten the bracket or tension adjusting bolt.

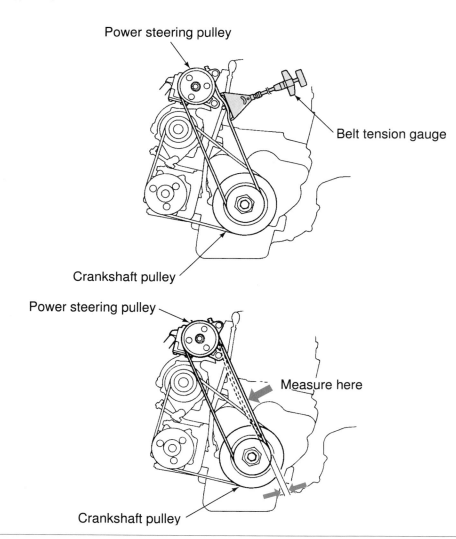

Figure 10-2 Methods of checking belt tension. (Courtesy of American Honda Company, Inc.)

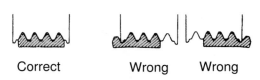

Correct   Wrong   Wrong

Figure 10-3 Proper and improper installation of ribbed V-belt on a pulley. (Reprinted with permission)

6. Recheck the belt tension with the tension gauge. If the belt does not have the specified tension, repeat steps 1 through 5.

7. Tighten the tension adjusting bolt and the mounting bolts to the manufacturer's specified torque.

Some power steering pumps have a **ribbed V-belt.** Many ribbed V-belts have an automatic tensioning pulley; therefore, a tension adjustment is not required. The ribbed V-belt should be checked to make sure it is installed properly on each pulley in the belt drive system (Figure 10-3). The tension on a ribbed V-belt may be checked with a belt tension gauge in the same way as the tension on a V-belt.

Many ribbed V-belts have a spring-loaded tensioner pulley that automatically maintains belt tension. As the belt wears or stretches, the spring moves the tensioner pulley to maintain the belt tension. Some of these tensioners have a belt length scale that indicates new belt range and used belt range (Figure 10-4). If the indicator on the tensioner is out of the used belt length range, belt replacement is required. Many belt tensioners have a 1/2-inch drive opening in which a ratchet or flex handle may be installed to move the tensioner pulley off the belt during belt replacement (Figure 10-5).

Belt pulleys must be properly aligned to minimize belt wear. The edges of the pulleys must be in line when a straightedge is placed on the pulleys (Figure 10-6). Repeated belt failure may be caused by a misaligned power steering pump pulley or extremely high pump pressure caused by a sticking pressure relief valve or a continual or intermittent restriction in the high pressure hose.

A ribbed V-belt may be referred to as a **serpentine belt.**

Classroom Manual
Chapter 10, page 249

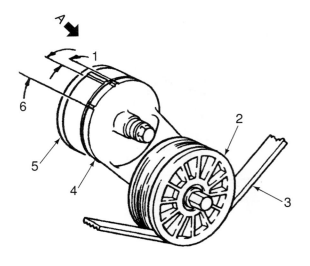
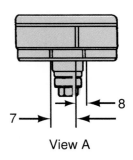

1.	New belt range	5.	Spindle
2.	Pulley	6.	Used belt acceptable wear range
3.	Belt	7.	Used belt length range
4.	Arm	8.	New belt length range

View A

Figure 10-4 Belt tensioner scale. (Courtesy of Chevrolet Motor Division, General Motors Corp.)

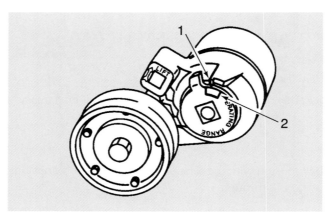

1. Cast arrow
2. Range indicator

Figure 10-5 One-half inch drive opening in the tensioner pulley. (Courtesy of Cadillac Motor Car Division, General Motors Corp.)

● **CUSTOMER CARE:** When servicing vehicles, always promote preventive maintenance to the customer. For example, if you are performing some underhood service such as replacing a battery and starting motor, always make a quick check of the belt or belts, and cooling system hoses. If you see a ribbed V-belt with chunks out of it, recommend a belt replacement to the customer. The customer will usually appreciate this service, and authorize the belt replacement. If you ignore the belt with chunks out of it, and the belt breaks the next day while the customer is driving on the freeway, the customer is inconvenienced and has to pay a tow bill. Although you did a satisfactory battery and starting motor replacement, the customer will not be happy if this problem occurs the day after you

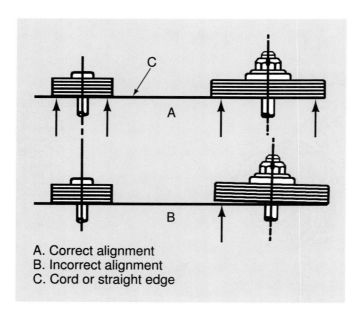

A. Correct alignment
B. Incorrect alignment
C. Cord or straight edge

Figure 10-6 Checking pulley alignment. (Courtesy of Chevrolet Motor Division, General Motors Corp.)

worked on the vehicle. The next time this unhappy customer requires automotive service, he or she may take the vehicle to another repair shop.

Power Steering Pump Fluid Service

Fluid Level Checking

Most car manufacturers recommend power steering fluid or automatic transmission fluid in power steering systems. Always use the type of power steering fluid recommended in the vehicle manufacturer's service manual. If the power steering fluid level is low, **steering effort** is increased and may be erratic. A low fluid level may cause a growling noise in the power steering pump. Some car manufacturers now recommend checking the power steering pump fluid level with the fluid at an ambient temperature of 176°F (80°C). Follow these steps to check the power steering fluid level:

1. With the engine idling at 1,000 rpm or less, turn the steering wheel slowly and completely in each direction several times to boost the fluid temperature (Figure 10-7).

2. If the vehicle has a remote power steering fluid reservoir, check for foaming in the reservoir, which indicates low fluid level or air in the system.

3. Observe the fluid level in the remote reservoir. This level should be at the hot full mark. Shut off the engine and remove dirt from the neck of the reservoir with a shop towel. If the power steering pump has an integral reservoir, the level should be at the hot level on the dipstick. When an external reservoir is used, the dipstick is located in the external reservoir (Figure 10-8).

4. Pour the required amount of the car manufacturer's recommended power steering fluid into the reservoir to bring the fluid level to the hot full mark on the reservoir or dipstick with the engine idling.

Power Steering System Draining and Flushing

If the power steering fluid is contaminated with moisture, dirt, or metal particles, the system must be drained and new fluid installed. Follow these steps to drain and flush the power steering system:

1. Lift the front of the vehicle with a floor jack and install safety stands under the suspension. Lower the vehicle onto the safety stands and remove the floor jack.

2. Remove the return hose from the **remote reservoir** that is connected to the steering gear. Place a plug on the reservoir outlet, and position the return hose in an empty drain pan (Figure 10-9).

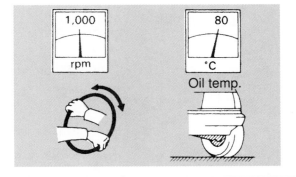

Figure 10-7 Boosting fluid temperature. (Reprinted with permission)

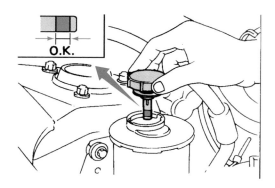

Figure 10-8 Power steering pump dipstick. (Courtesy of Nissan Motor Co., Ltd.)

3. With the engine idling, turn the steering wheel fully in each direction, and stop the engine (Figure 10-10).

4. Fill the reservoir to the hot full mark with the manufacturer's recommended fluid.

5. Start the engine and run the engine at 1,000 rpm while observing the return hose in the drain pan. When fluid begins to discharge from the return hose, shut the engine off.

6. Repeat steps 4 and 5 until there is no air in the fluid discharging from the return hose.

7. Remove the plug from the reservoir and reconnect the return hose. Bleed the power steering system.

Bleeding Air from the Power Steering System

When air is present in the power steering fluid, a growling noise may be heard in the pump, and steering effort may be increased or erratic. When a power steering system has been drained and refilled, follow this procedure to remove air from the system:

1. Fill the power steering pump reservoir as outlined previously.

2. With the engine running at 1,000 rpm, turn the steering wheel fully in each direction three or four times (Figure 10-11). Each time the steering wheel is turned fully to the right or left, hold it there for 2 to 3 seconds before turning it in the other direction.

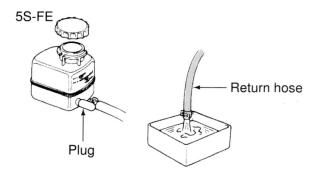

Figure 10-9 Return hose installed in drain pan for power steering draining and flushing. (Reprinted with permission)

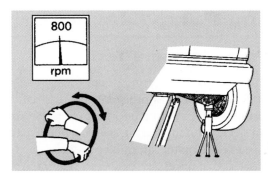

Figure 10-10 Draining fluid from the power steering system. (Reprinted with permission)

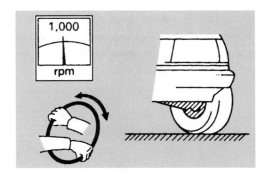

Figure 10-11 Turning steering wheel slowly to bleed air from the power steering system. (Reprinted with permission)

3. Check for foaming of the fluid in the reservoir. When foaming is present, repeat steps 1 and 2.

4. Check the fluid level with the engine running, and be sure it is at the hot full mark. Shut off the engine and make sure the fluid level does not increase more than 0.020 in (5 mm) (Figure 10-12).

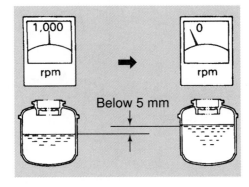

Figure 10-12 Power steering fluid level after bleeding. (Reprinted with permission)

Power Steering Pump Diagnosis

Power Steering Pump Pressure Test

Since there are some variations in power steering pump pressure test procedures and pressure specifications, the vehicle manufacturer's test procedures and specifications must be used. If the power steering pump pressure is low, steering effort is increased. Erratic power steering pump pressure causes variations in steering effort, and the steering wheel may jerk as it is turned. Since a power steering pump will never develop the specified pressure if the belt is slipping, the belt tension must be checked and adjusted, if necessary, prior to a pump pressure test. The following is a typical power steering pressure test procedure.

Special Tools

Power steering pressure test gauge

1. With the engine stopped, disconnect the pressure line from the power steering pump, and connect the gauge side of the **pressure gauge** to the pump outlet fitting. Connect the valve side of the gauge to the pressure line (Figure 10-13).

2. Start the engine and turn the steering wheel fully in each direction two or three times to bleed air from the system (Figure 10-14). Be sure the fluid level is correct and the fluid temperature is at least 176°F (80°C). A thermometer may be inserted in the pump reservoir fluid to measure the fluid temperature.

CAUTION: During the power steering pump pressure test with the pressure gauge valve closed, if this valve is closed for more than 10 seconds, excessive pump pressure may cause power steering hoses to rupture, resulting in personal injury.

CAUTION: Do not allow the fluid to become too hot during the power steering pump pressure test. Excessively high fluid temperature reduces pump pressure. Wear protective gloves and always shut the engine off before disconnecting gauge fittings, because the hot fluid may cause burns.

3. With the engine idling, close the pressure gauge valve for no more than 10 seconds, and observe the pressure gauge reading (Figure 10-15). Turn the pressure gauge valve

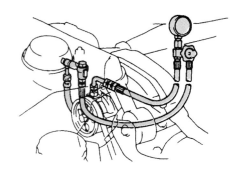

Figure 10-13 Pressure gauge connections to power steering pump. (Reprinted with permission)

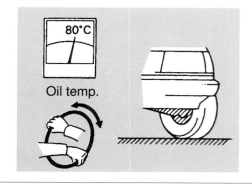

Figure 10-14 Bleeding air from the power steering pump system and checking fluid temperature. (Reprinted with permission)

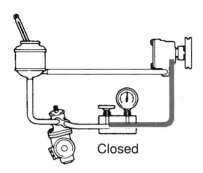

Closed

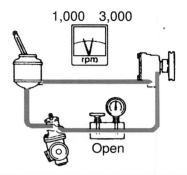

1,000   3,000

Open

Figure 10-15 Power steering pump pressure test with pressure gauge valve closed. (Reprinted with permission)

Figure 10-16 Power steering pressure test at 1,000 rpm and 3,000 rpm. (Reprinted with permission)

to the fully open position. If the pressure gauge reading does not equal the vehicle manufacturer's specifications, repair or replace the power steering pump.

4. Check the power steering pump pressure with the engine running at 1,000 rpm and 3,000 rpm, and record the pressure difference between the two readings (Figure 10-16). If the pressure difference between the pressure readings at 1,000 rpm and 3,000 rpm does not equal the vehicle manufacturer's specifications, repair or replace the **flow control valve** in the power steering pump.

5. With the engine running, turn the steering wheel fully in one direction, and observe the steering pump pressure while holding the steering wheel in this position (Figure 10-17). If the pump pressure is less than the vehicle manufacturer's specifications, the steering gear housing has an internal leak and should be repaired or replaced.

6. Be sure the front tire pressure is correct, and center the steering wheel with the engine idling. Connect a spring scale to the steering wheel and measure the steering effort in both directions (Figure 10-18). If the power steering pump pressure is satisfactory and the steering effort is more than the vehicle manufacturer's specifications, the power steering gear should be repaired. Photo Sequence 8 shows a typical procedure for pressure testing a power steering pump.

Special Tools

Spring scale

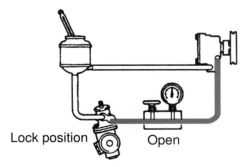

Lock position   Open

Figure 10-17 Power steering pump pressure test with the front wheels turned fully in one direction. (Reprinted with permission)

Figure 10-18 Steering effort measurement. (Reprinted with permission)

Photo Sequence 8
Typical Procedure for Pressure Testing a Power Steering Pump

P8-1 Connect the pressure gauge to the power steering pump.

P8-2 Connect a tachometer to the ignition system.

P8-3 Look up the vehicle manufacturer's specified power steering pump pressure in the service manual.

P8-4 Start the engine and turn the steering wheel from lock to lock three times to bleed air from the system and heat the power steering fluid.

P8-5 Check the power steering fluid level, and add fluid as required.

P8-6 Place a thermometer in the power steering fluid reservoir. Be sure the fluid temperature is at least 176°F (80°C).

P8-7 With the engine idling, close the pressure gauge valve for no more than 10 seconds. Observe and record the pressure gauge reading.

P8-8 Check and record the power steering pump pressure at 1,000 rpm.

P8-9 Check and record the power steering pump pressure at 3,000 rpm.

Photo Sequence 8
Typical Procedure for Pressure Testing a Power Steering Pump (continued)

P8-10 Check and record the power steering pump pressure with the steering wheel turned fully in one direction and the engine idling.

Power Steering Pump Oil Leak Diagnosis

The possible sources of power steering pump oil leaks are the drive shaft seal, reservoir O-ring seal, high-pressure outlet fitting, and the dipstick cap. If leaks occur at any of the seal locations, seal replacement is necessary. When a leak is present at the high-pressure outlet fitting, tighten this fitting to the specified torque (Figure 10-19). If the leak still occurs, replace the O-ring seal on the fitting and retighten the fitting.

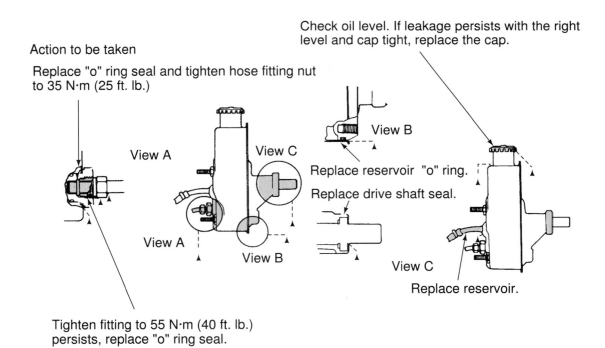

Check oil level. If leakage persists with the right level and cap tight, replace the cap.

Action to be taken

Replace "o" ring seal and tighten hose fitting nut to 35 N·m (25 ft. lb.)

View A

View C

View B

Replace reservoir "o" ring.

Replace drive shaft seal.

View A

View B

View C

Replace reservoir.

Tighten fitting to 55 N·m (40 ft. lb.) persists, replace "o" ring seal.

Figure 10-19 Power steering pump oil leak diagnosis. (Courtesy of Chrysler Corp.)

Power Steering Pump Service

Power Steering Pump Replacement

 CAUTION: If the vehicle has been driven recently, the pump, hoses, and fluid could be extremely hot. Use caution when handling components to avoid burns.

If a growling noise is present in the power steering pump after the fluid level is checked and air has been bled from the system, the pump bearings or other components are defective and pump replacement or repair is required. When the power steering pump pressure is lower than specified, pump replacement or repair is necessary. To replace the power steering pump, proceed as follows:

1. Disconnect the power steering return hose from the remote reservoir or pump. Allow the fluid to drain from this hose into a drain pan. Discard the used fluid.

2. Loosen the bracket or belt tension adjusting bolt, and the pump mounting bolt.

3. Loosen the belt tension until the belt can be removed. On some cars, it is necessary to lift the vehicle on a hoist and gain access to the power steering pump from underneath the vehicle.

4. Remove the hoses from the pump, and cap the pump fittings and hoses.

5. Remove the belt tension adjusting bolt and the mounting bolt, and remove the pump.

6. Check the pump mounting bolts and bolt holes for wear. Worn bolts must be replaced. If the bolt mounting holes in the pump are worn, pump replacement is necessary.

7. Reverse steps 1 through 5 to install the power steering pump. Tighten the belt as described previously, and tighten the pump mounting and bracket bolts to the manufacturer's specifications. If O-rings are used on the pressure hose, replace the O-rings. Be sure the hoses are not contacting the exhaust manifold, catalytic converter, or exhaust pipe during or after pump replacement.

8. Fill the pump reservoir with the manufacturer's recommended power steering fluid, and bleed air from the power steering system as described earlier.

Power Steering Pump Pulley Replacement

If the pulley wobbles while it is rotating, the pulley is probably bent, and pulley replacement is necessary. Worn pulley grooves also require pulley replacement. Always check the pulley for cracks. If this condition is present, pulley replacement is essential. A pulley that is loose on the pump shaft must be replaced. Never hammer on the pump drive shaft during pulley removal or replacement. This action will damage internal pump components. If the pulley is pressed onto the pump shaft, a special puller is required to remove the pulley (Figure 10-20), and a pulley installation tool is used to install the pulley (Figure 10-21).

If the power steering pump pulley is retained with a nut, mount the pump in a vise. Always tighten the vise on one of the pump mounting bolt surfaces. Do not tighten the vise with excessive force. Use a special holding tool to keep the pulley from turning, and loosen the pulley nut with a box end wrench (Figure 10-22). Remove the nut, pulley, and woodruff key. Inspect the pulley, shaft, and **woodruff key** for wear. Be sure the key slots in the shaft and pulley are not worn. Replace all worn components.

Remove and Replace the Power Steering Pump Reservoir

Power steering pump service procedures vary depending on the type of pump. Always follow the vehicle manufacturer's recommended service procedures in the service manual. The following service

---

Special Tools

Puller for power steering pump pulley, press-on pulley

Special Tools

Power steering pump pulley installing tool, press-on pulley

A woodruff key is a half-moon shaped key that fits snugly in a shaft opening, and the top edge of the key extends out of the shaft. The top edge of the key fits snugly in a groove cut inside the pulley hub. When the pulley is installed on the shaft, the pulley groove slides over the key to prevent pulley rotation on the shaft.

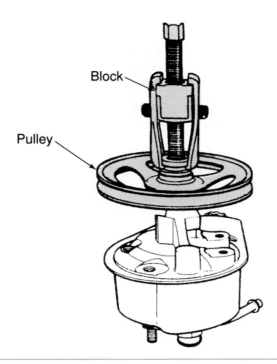

Figure 10-20 Press-on power steering pump pulley removal. (Courtesy of Chrysler Corp.)

procedures are for a Chrysler power steering pump with an **integral reservoir.** When the power steering pump has an integral reservoir, follow these steps for reservoir removal and replacement:

1. Remove the filler cap and drain the oil from the reservoir.
2. Remove the two studs and the fitting from the rear of the reservoir.
3. Rock the reservoir by hand or tap it gently with a soft hammer to remove it from the housing.

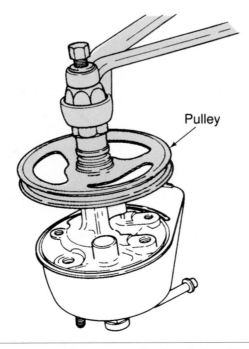

Figure 10-21 Press-on power steering pump pulley installation. (Courtesy of Chrysler Corp.)

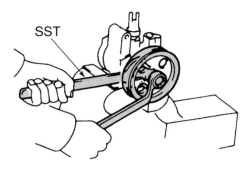

Figure 10-22 Removing power steering pump pulley and nut. (Reprinted with permission)

4. Clean all the parts and discard O-rings and seals. Inspect all O-ring and seal surfaces for damage.

5. Lubricate the new seals and O-rings with the manufacturer's recommended power steering fluid and install them (Figure 10-23). Be sure the pressure union fitting O-ring is installed in the groove nearest the fitting hex head.

Remove and Replace Flow Control Valve and End Cover

When the flow control valve and end plate are serviced, follow these steps:

1. Remove the retaining ring with a slotted screwdriver and a punch (Figure 10-24).

2. Remove the flow control valve, end cover, spring, and magnet. Check the flow control valve for burrs. Remove minor burrs with **crocus cloth,** and clean the flow control

Crocus cloth is a very fine polishing-type emery paper.

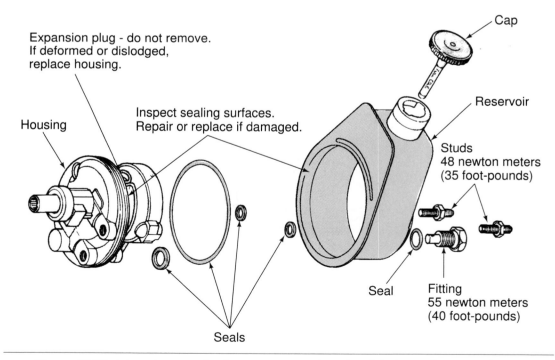

Figure 10-23 Integral power steering pump reservoir seal and O-ring replacement. (Courtesy of Chrysler Corp.)

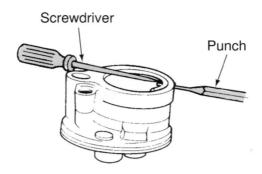

Figure 10-24 Retaining ring removal. (Courtesy of Chrysler Corp.)

valve in solvent. Damaged or worn flow control valves must be replaced. Inspect the end cover sealing surface for damage. Check the pump drive shaft for corrosion and damage. Remove corrosion with crocus cloth (Figure 10-25). Clean the magnet with a shop towel.

3. Clean all parts and lubricate the end cover with power steering fluid.

4. Install the end cover, retaining ring, and related components.

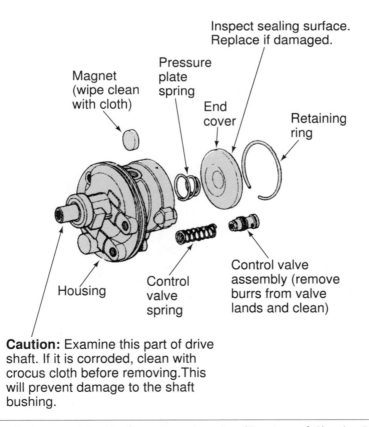

Figure 10-25 End cover and related component service. (Courtesy of Chrysler Corp.)

Remove and Replace Pressure Relief Valve

Follow this procedure to service the **pressure relief valve:**

1. Wrap a shop towel around the land end of the flow control valve and clamp this end in a soft-jawed vise. Be very careful not to mark valve lands.

2. Remove the hex-head ball seat (Figure 10-26). Clean the components in solvent. A worn or damaged pressure relief ball, spring, guide, or seat must be replaced.

Servicing Power Steering Pump Rotating Components

1. After the end cover and spring have been removed, pull the pressure plate, pump ring, rotor, and shaft out of the pump housing (Figure 10-27).

2. Remove and discard the rotor lock ring (Figure 10-28).

3. Remove the rotor and thrust plate from the shaft (Figure 10-29).

4. Examine all the wear surfaces, and replace parts that are worn, scuffed, or scored. Check vanes and vane slots in the rotor for wear.

5. Clean all the components and install the inner thrust plate, rotor, and a new lock ring on the shaft (Figure 10-30).

Special Tools

Seal drivers

6. Remove the drive shaft seal from the pump housing with a slotted screwdriver (Figure 10-31), and install a new seal with the proper driver (Figure 10-32). Lubricate the new seal with the manufacturer's recommended fluid.

7. Lubricate the new housing O-rings with the recommended fluid and position the O-rings in the housing.

8. Install the thrust plate, shaft, and pump rotor assembly in the housing, and install the pump ring (Figure 10-33).

9. Install the vanes in the rotor and lubricate all parts with power steering fluid.

10. Install and seat the pressure plate (Figure 10-34).

11. Install the pressure plate spring, end cover, and retaining ring. Reassemble the magnet and flow control valve and install the pump reservoir with a new O-ring and seals.

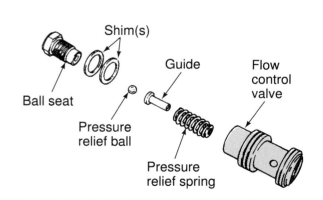

Figure 10-26 Pressure relief valve removal. (Courtesy of Chrysler Corp.)

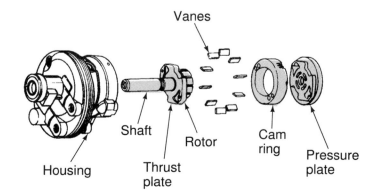

Figure 10-27 Removing pressure plate, pump ring, rotor, and shaft assembly. (Courtesy of Chrysler Corp.)

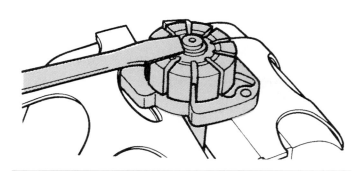

Figure 10-28 Rotor lock ring removal. (Courtesy of Chrysler Corp.)

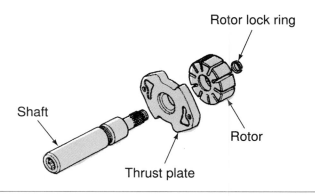

Figure 10-29 Pump rotor and thrust plate removal. (Courtesy of Chrysler Corp.)

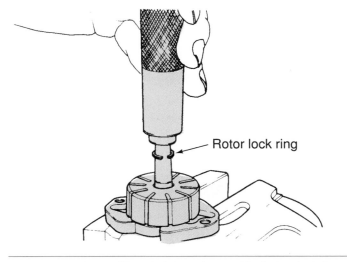

Figure 10-30 Thrust plate, rotor, and lock ring installation. (Courtesy of Chrysler Corp.)

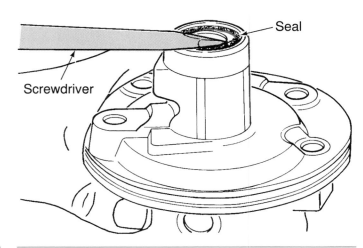

Figure 10-31 Drive shaft seal removal. (Courtesy of Chrysler Corp.)

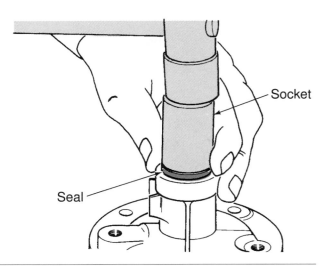

Figure 10-32 Installation of drive shaft seal. (Courtesy of Chrysler Corp.)

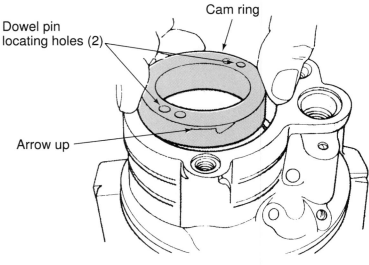

Figure 10-33 Cam ring installation. (Courtesy of Chrysler Corp.)

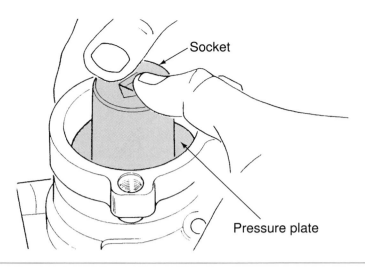

Figure 10-34 Pressure plate installation and seating. (Courtesy of Chrysler Corp.)

Checking and Servicing Power Steering Lines and Hoses

Power steering lines should be checked for leaks, dents, sharp bends, cracks, and contact with other components. Lines and hoses must not rub against other components. This action could wear a hole in the line or hose. Many high-pressure power steering lines are made from high-pressure steel-braided hose with molded steel fittings on each end (Figure 10-35).

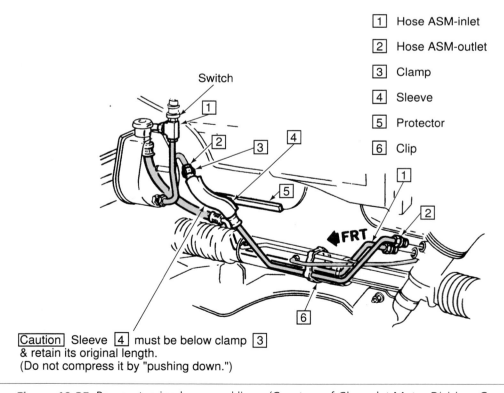

1 Hose ASM-inlet
2 Hose ASM-outlet
3 Clamp
4 Sleeve
5 Protector
6 Clip

Caution Sleeve 4 must be below clamp 3 & retain its original length. (Do not compress it by "pushing down.")

Figure 10-35 Power steering hoses and lines. (Courtesy of Chevrolet Motor Division, General Motors Corp.)

Power Steering Hose Replacement

When power steering hose replacement is required, follow these steps:

1. With the engine stopped, remove the return hose at the power steering gear, and allow the fluid to drain from this hose into a drain pan. On some cars, the vehicle must be lifted on a hoist to replace the power steering hoses.

2. Loosen and remove all hose fittings from the pump and steering gear.

3. Remove all hose-to-chassis clips.

4. Remove the hoses from the chassis, and cap the pump and steering gear fittings.

5. If O-rings are used on the hose ends, install new O-rings. TRW steering gear cylinder lines have gaskets that are serviced as a unit. The old gasket must be pried out of the fitting in the housing before the new lines and gaskets are installed. Lubricate O-rings with power steering fluid.

6. Reverse steps 1 through 4 to install the power steering hoses. Tighten all fittings to the manufacturer's specified torque. Be sure all hose-to-chassis clips are in place. Do not position hoses where they rub on other components.

7. Fill the pump reservoir to the full mark with the manufacturer's recommended fluid. Bleed air from the power steering system as mentioned previously in this chapter. Check the fluid level in the reservoir and add fluid as required.

Classroom Manual
Chapter 10, page 255

Guidelines for Servicing Power Steering Pumps

1. A loose power steering belt causes low pump pressure, hard steering, and belt chirping, or squealing, especially on acceleration.

2. Power steering belt tension may be measured with belt tension gauge or by checking the amount of belt deflection.

3. When prying on a power steering pump to tighten the belt, always pry on the pump ear and hub. Do not pry on the reservoir.

4. Most car manufacturers recommend power steering fluid or automatic transmission fluid in the power steering system.

5. Many car manufacturers recommend checking the power steering pump fluid level when the fluid is warmed up to 176°F (80°C).

6. If the power steering fluid is contaminated with moisture, dirt, or metal particles, the system should be drained and flushed, and new fluid should be installed.

7. Air may be bled from the power steering system by turning the steering wheel fully in each direction several times with the engine running. Each time the wheel is turned fully to the right or left, it should be held in that position for 2 to 3 seconds.

8. A pressure gauge and manual valve are connected in series in the power steering pump pressure hose to check pump pressure. Maximum pump pressure is checked by closing the manual valve for less than 10 seconds.

9. The power steering pump pressure may be checked at 1,000 rpm and 3,000 rpm to check the operation of the pump flow control valve.

10. After the power steering pump pressure is tested and proven to be satisfactory, steering gear leakage may be tested by turning the steering wheel fully in one direction and observing the pressure reading. If the power steering pump pressure is less than specified, the steering gear has an internal leak.

11. With the engine idling, a spring scale may be attached at the outer end of the steering wheel cross bar to measure steering effort.

12. The possible sources of power steering pump leaks are the drive shaft seal, reservoir O-ring seals, high-pressure outlet fitting, dipstick cap, and low-pressure inlet pipe.

13. When a power steering pump is replaced, the mounting bolts and bolt holes must be checked for wear. If excessive wear is present on these components, pump and belt misalignment may occur.

14. Power steering pump pulleys may be pressed onto the pump shaft or retained on the shaft with a woodruff key and nut.

15. Never tap on the pump shaft while removing or replacing the pump pulley.

16. A sticking flow control valve causes erratic power steering pump pressure. This valve may be cleaned and polished with crocus cloth.

17. A defective power steering pump may be overhauled or replaced.

18. Power steering pump lines should be checked for leaks, dents, sharp bends, cracks, and contact with other components.

19. When power steering pump components are serviced, always use caution because these components may be very hot if the engine has been running.

20. If it is necessary to drain the power steering fluid, disconnect the return hose at the remote reservoir or pump and drain the fluid from this hose into a drain pan.

CASE STUDY

The owner of a Dodge Spirit with a 2.5 L turbocharged engine requested a price on a power steering gear replacement. The service writer questioned the owner regarding the reason for the steering gear replacement, and discovered that the problem was intermittent hard steering. Further discussion with the owner also revealed that the car had been taken to another automotive service center, and the owner was informed that her car required a steering gear replacement. The owner was now looking for a second opinion. The technician road tested the vehicle and found no indication of hard steering. The owner was asked about the exact driving conditions when the hard steering was experienced, and the answer to this question provided the solution to the problem. The owner indicated that the hard steering was usually experienced while cornering and always on a rainy day. Once this information was revealed, the technician suspected a problem with the power steering belt, because a wet belt slips more easily. A check of the power steering belt indicated the belt was oil soaked from engine oil leaks and was also slightly loose. The power steering belt was replaced and adjusted, and the owner was informed that the engine oil leaks should be corrected or the belt would become oil soaked again. A short while later, after a rain, the owner brought the car to the shop to have the engine oil leaks repaired. The owner reported that the belt corrected the hard steering problem. In this situation, a brief discussion about the customer's complaint led to accurate diagnosis of the problem, and because of this accurate diagnosis the shop gained a steady customer.

Terms to Know

Belt tension guage	Pressure gauge	Steering effort
Crocus cloth	Pressure relief valve	Serpentine belt
Flow control valve	Remote reservoir	V-type belt
Integral resevoir	Ribbed V-belt	Woodruff key

ASE Style Review Questions

1. While discussing power steering belt adjustment:
   Technician A says the tension on a power steering pump V-belt should be checked with a tension gauge.
   Technician B says that some ribbed V-belts have an automatic tensioning pulley that eliminates the need for tension adjustments.
   Who is correct?
   A. A only   **C.** Both A and B
   B. B only   **D.** Neither A nor B

2. While discussing power steering belt adjustment:
   Technician A says that a power steering pump V-belt should be tightened with a pry bar installed against the pump reservoir.
   Technician B says a loose power steering pump V-belt may cause the steering wheel to jerk while turning.
   Who is correct?
   A. A only   **C.** Both A and B
   B. B only   **D.** Neither A nor B

3. While discussing power steering fluid level checking:
   Technician A says foaming in the remote reservoir may indicate air in the power steering system.
   Technician B says many car manufacturers recommend checking the power steering fluid level with the fluid warmed up.
   Who is correct?
   A. A only   **C.** Both A and B
   B. B only   **D.** Neither A nor B

4. While discussing the steering pump pressure test:
   Technician A says the pressure gauge and valve should be connected in the power steering pump return hose to check pump pressure.
   Technician B says the pressure gauge valve should be closed for 30 seconds during the power steering pump pressure test.

   Who is correct?
   A. A only   **C.** Both A and B
   B. B only   **D.** Neither A nor B

5. While discussing power steering system draining, flushing, refilling, and bleeding:
   Technician A says the return hose from the remote reservoir should be disconnected at the remote reservoir to drain the fluid.
   Technician B says the return hose should be loosened at the remote reservoir with the engine running to bleed air from the power steering system.
   Who is correct?
   A. A only   **C.** Both A and B
   B. B only   **D.** Neither A nor B

6. A vehicle experiences excessive, erratic steering effort. The most likely cause of this problem is:
   A. underinflated and worn front tires.
   B. power steering pump V-belt pump bottomed in the pulley.
   C. worn power steering pump mountings.
   D. continual low power steering pump pressure.

7. A power steering system has the specified pressure with the pressure gauge valve closed and with the engine running at 1,000 and 3,000 rpm and the pressure gauge valve open. The power steering pump pressure is lower than specified with the front wheels turned fully in either direction. The cause of this problem could be:
   A. a defective steering gear.
   B. a stuck flow control valve.
   C. a stuck pressure relief valve.
   D. worn pump vanes and cam ring.

8. All of these statements about power steering pump pulley replacement are true EXCEPT:
   A. If the pulley is retained with a nut, a woodruff key prevents pulley rotation on the shaft.
   B. If the pulley is pressed onto the shaft, a special puller must be used to remove the pulley.
   C. If the pulley is pressed onto the shaft, when reinstalling the pulley a soft hammer may by used to drive the pulley onto the shaft.
   D. If the pulley is misaligned with the engine running, the power steering pump mountings may be worn.

9. A power steering pump has the specified pressure with the pressure gauge valve closed. However, with this valve open, the pump does not have specified pressure difference between the pressure readings at 1,000 and 3,000 rpm. The cause of this problem could be:
   A. a sticking pressure relief valve.
   B. a sticking flow control valve.
   C. a slipping pump belt.
   D. a misaligned pump pulley.

10. A power steering pump has repeated belt failures. The power steering pump V-belt tension and pulley alignment are satisfactory, and the V-belt is not bottomed in the pulley. The most likely cause of this problem is:
   A. a pressure relief valve sticking open.
   B. worn pulley friction surfaces on the pulley.
   C. worn thrust plate in the power steering pump.
   D. an intermittent restriction in the high-pressure hose.

ASE Challenge Questions

1. A vehicle has intermittent excessive steering steering effort.
   Technician A says check the fluid because it probably has air in it.
   Technician B says check the belt because it is probably slipping.
   Who is correct?
   A. A only
   B. B only
   C. Both A and B
   D. Neither A nor B

2. While discussing power steering pump leaks,
   Technician A says leaks at the driveshaft seal will leave a wet and oily pump pulley backside. Easier yet, says *Technician B* says leaks at the driveshaft seal may leave oil on the hood pad above the pump.
   Who is correct?
   A. A only
   B. B only
   C. Both A and B
   D. Neither A nor B

3. A vehicle has continual excessive steering effort, but there is no noise and the fluid level in the reservoir is correct.
   Technician A says the cause of the problem could be a stuck flow control valve.
   Technician B says the cause of the problem could be air in the power steering fluid.
   Who is correct?
   A. A only
   B. B only
   C. Both A and B
   D. Neither A nor B

4. There is air in the power steering fluid. To correct this problem, you should:
   A. with the engine running, crack the pressure line at the pump to release trapped air.
   B. with the engine running, turn the steering wheel full left and right.
   C. with the engine running, drain all fluid from the system and refill.
   D. with the engine stopped, raise wheels and remove the return hose from the gear. Turn the steering wheel full left and right. Refill the reservoir and replace the return hose.

5. All of the following are true EXCEPT:
   A. A loose belt may cause the steering wheel to jerk.
   B. If V-belt edges are worn, the belt may be stretched or the wrong width.
   C. An oil-soaked V-belt may only slip in wet weather.
   D. When air is in the fluid, the fluid will never reach operating temperature.

Job Sheet 29

Name _____ Date _____

Draining and Flushing Power Steering System

NATEF and ASE Correlation

This job sheet is related to NATEF and ASE Automotive Suspension and Steering Task List content area: A. Steering Systems Diagnosis and Repair. 2. Power-Assisted Steering Units. Task 17: Flush, fill, and bleed power steering systems.

Tools and Materials

Drain pan
Specified type of power steering fluid
Floor jack
Safety stands

Vehicle make and model year _____

Vehicle VIN _____

Procedure

1. Describe the steering problems and noises that occur when there is air in the power steering system.

   _____

   _____

   _____

   _____

2. Lift the front of the vehicle with a floor jack and install safety stands under the suspension. Lower the vehicle onto the safety stands and remove the floor jack.

3. Remove the return hose from the remote reservoir that is connected to the steering gear. Place a plug on the reservoir outlet, and position the return hose in an empty drain pan.

4. With the engine idling, turn the steering wheel fully in each direction and stop the engine.

5. Fill the reservoir to the hot full mark with the manufacturer's recommended fluid.

   Specified type of power steering fluid _____

   Power steering reservoir filled to the specified level with the proper power steering fluid? yes _____ no _____

   Instructor check _____

6. Start the engine and run the engine at 1,000 rpm while observing the return hose in the drain pan. When fluid begins to discharge from the return hose, shut the engine off.

7. Repeat steps 4, 5, and 6 until there is no air in the fluid discharging from the return hose.

   Have steps 4, 5, and 6 been repeated? yes _____ no _____

   Instructor check _____

8. Remove the plug from the reservoir and reconnect the return hose.

Is the return hose connected and tightened? yes _____ no _____

Instructor check _____

9. Fill the power steering pump reservoir to the specified level.

10. With the engine running at 1,000 rpm, turn the steering wheel fully in each direction three or four times. Each time the steering wheel is turned fully to the right or left, hold it there for 2 to 3 seconds before turning it in the other direction.

Number of times the steering wheel was turned fully in each direction _____

11. Check for foaming of the fluid in the reservoir. When foaming is present, repeat steps 8 and 9.

Is foaming present in the power steering pump reservoir? yes _____ no _____

Have steps 8 and 9 been repeated? yes _____ no _____

Instructor check _____

12. Check the fluid level and be sure it is at the hot full mark.

13. Explain how you know that all the air has been bled from the power steering system.

_____
_____
_____
_____

Instructor Check _____

Job Sheet 30

30

Name _____ Date _____

Testing Power Steering Pump Pressure

NATEF and ASE Correlation

This job sheet is related to NATEF and ASE Automotive Suspension and Steering Task List content area: A. Steering Systems Diagnosis and Repair. 2. Power-Assisted Steering Units. Task 8: Perform power steering pump pressure and flow tests; determine needed repairs.

Tools and Materials

Specified type of power steering fluid

Power steering pressure test gauge

Thermometer

Vehicle make and model year _____

Vehicle VIN _____

Procedure

1. With the engine stopped, disconnect the pressure line from the power steering pump.

2. Connect the gauge side of the pressure gauge to the pump outlet fitting. Connect the valve side of the gauge to the pressure line.

   Is the power steering pressure gauge properly connected? yes _____ no _____

   Are all power steering gauge fittings tightened to the specified torque?
   yes _____ no _____

3. Start the engine and turn the steering wheel fully in each direction two or three times to bleed air from the system.

4. Install a thermometer in the pump reservoir fluid to measure the fluid temperature. Be sure the fluid level is correct and the fluid temperature is at least 176°F (80°C).

   Is the power steering fluid at the specified level? yes _____ no _____

   Is the power steering fluid temperature? yes _____ no _____

   Instructor check _____

   CAUTION: During the power steering pump pressure test with the pressure gauge valve closed, if this valve is closed for more than 10 seconds excessive pump pressure may cause power steering hoses to rupture, resulting in personal injury.

   CAUTION: Do not allow the fluid to become too hot during the power steering pump pressure test. Excessively high fluid temperature reduces pump pressure. Wear protective gloves and always shut the engine off before disconnecting gauge fittings because the hot fluid may cause burns.

5. With the engine idling, close the pressure gauge valve for no more than 10 seconds and observe the pressure gauge reading. Turn the pressure gauge valve to the fully open position. If the pressure gauge reading does not equal the vehicle manufacturer's specifications, repair or replace the power steering pump.

Specified power steering pump pressure with pressure gauge valve closed _____

Actual power steering pump pressure with the pressure gauge valve closed _____

Recommended power steering pump service. _____

_____

6. Check the power steering pump pressure with the engine running at 1,000 rpm and 3,000 rpm. Record the pressure difference between the two readings.

Specified power steering pump pressure at 1,000 rpm _____

Specified power steering pump pressure at 3,000 rpm _____

Actual power steering pump pressure at 1,000 rpm _____

Actual power steering pump pressure at 3,000 rpm _____

Recommended power steering pump service. _____

_____

7. With the engine running, turn the steering wheel fully in one direction. Observe the steering pump pressure while holding the steering wheel in this position.

Specified power steering pump pressure with the steering wheel turned fully in one direction _____

Actual power steering pump pressure with the steering wheel turned fully in one direction _____

Recommended power steering service. _____

_____

8. Be sure the front tire pressure is correct, and center the steering wheel with the engine idling. Connect a spring scale to the steering wheel and measure the steering effort in both directions.

Specified force required to turn the steering wheel to the right _____

Specified force required to turn the steering wheel to the left _____

Actual force required to turn the steering wheel to the right _____

Actual force required to turn the steering wheel to the left _____

List all the required power steering service and explain the reasons why this service is necessary. _____

_____

_____

_____

_____

Instructor Check _____

Job Sheet 31

31

Name _____ Date _____

Measure and Adjust Power Steering Belt Tension and Alignment

NATEF and ASE Correlation

This job sheet is related to NATEF and ASE Automotive Suspension and Steering Task List content area: A. Steering Systems Diagnosis and Repair. 2. Power-Assisted Steering Units. Task 4: Inspect, adjust tension and alignment, and replace power steering pump belt(s).

Tools and Materials

Belt tension gauge

Pry bar

Vehicle make and model year _____

Vehicle VIN _____

Procedure

1. Describe the power steering problems and noises that occur when the power steering pump belt is loose.

   _____

   _____

   _____

   _____

2. Inspect the power steering belt for fraying, oil soaking, wear on friction surfaces, cracks, glazing, and splits.

   Belt condition: satisfactory _____ unsatisfactory _____

3. Press on the belt at the longest belt span with the engine stopped to measure the belt deflection which should be 1/2 inch per foot of free span.

   Length of belt span where belt deflection is measured _____

   Amount of belt deflection _____

   Belt tension: satisfactory _____ unsatisfactory _____

4. Install a belt tension gauge over the belt in the center of the longest span to measure the belt tension.

   Specified belt tension _____

   Actual belt tension _____

5. Loosen the power steering pump bracket or tension adjusting bolt.

6. Loosen the power steering pump mounting bolts.

7. Remove the power steering belt.

8. Check the bracket and pump mounting bolts for wear.

Power steering pump mounting bolt holes and bolt condition:

satisfactory _____ unsatisfactory _____

Power steering pump bracket and bracket bolt condition:

satisfactory _____ unsatisfactory _____

List the required power steering pump and bracket service and explain the reasons for your diagnosis.

_____

_____

_____

_____

9. Install the new power steering pump belt over all the pulleys.

10. Pry against the pump ear and hub with a pry bar to tighten the belt. Some pump brackets have a 1/2 inch square opening in which a breaker bar may be installed to move the pump and tighten the belt.

11. Hold the pump in the position described in step 9, and tighten the bracket or tension adjusting bolt.

12. Recheck the belt tension with the tension gauge.

Specified power steering belt tension _____

Actual power steering pump belt tension _____

13. Tighten the tension adjusting bolt and the mounting bolts to the manufacturer's specified torque.

Specified power steering pump mounting bolt torque _____

Actual power steering pump mounting bolt torque _____

Specified power steering pump bracket or tension adjusting bolt torque _____

Actual power steering pump bracket or tension adjusting bolt torque _____

14. Check alignment of the power steering pump pulley in relation to the other pulleys surrounded by the power steering belt.

Power steering pump pulley alignment: satisfactory _____ unsatisfactory _____

Explain the service required to align power steering pump pulley with other related pulleys and give the reasons for your diagnosis.

_____

_____

_____

_____

Instructor Check _____

Table 10-1 NATEF and ASE TASK

Inspect power steering fluid level and condition; adjust level in accordance with vehicle manufacturer's recommendations.

Problem Area	Symptoms	Possible Causes	Classroom Manual	Shop Manual
STEERING EFFORT	Hard, erratic steering	1. Low fluid level	258	315
		2. Contaminated fluid	258	315

Table 10-2 NATEF and ASE TASK

Inspect, adjust tension and alignment, and replace power steering pump belt(s).

Problem Area	Symptoms	Possible Causes	Classroom Manual	Shop Manual
NOISE	Squealing while accelerating or cornering	Loose belt	249	311
STEERING EFFORT	Hard, erratic steering	Loose belt	249	311

Table 10-3 NATEF and ASE TASK

Remove and replace power steering pump; inspect pump mounts.

Problem Area	Symptoms	Possible Causes	Classroom Manual	Shop Manual
NOISE	Growling noise from pump with engine running	1. Low fluid level	255	315
		2. Defective pump bearing or components	255	318
STEERING EFFORT	Hard, erratic steering	Defective pump	256	322

Table 10-4 NATEF and ASE TASK

Inspect and replace power steering pump seals and gaskets.

Problem Area	Symptoms	Possible Causes	Classroom Manual	Shop Manual
FLUID LOSS	Fluid leaks	Worn pump shaft seal, O-rings	256	326

Table 10-5 NATEF and ASE TASK

Inspect and replace power steering pump pulley.

Problem Area	Symptoms	Possible Causes	Classroom Manual	Shop Manual
NOISE	Squealing while accelerating or cornering	Worn pump pulley	255	323
BELT REPLACEMENT	Rapid belt wear	Bent pulley	255	323

Table 10-6 NATEF and ASE TASK

Perform power system steering pressure and flow tests; determine needed repairs.

Problem Area	Symptoms	Possible Causes	Classroom Manual	Shop Manual
STEERING EFFORT	Hard, erratic steering	1. Low pump pressure	258	318
		2. Sticking, worn flow control valve	259	318
		3. Defective steering gear	259	318

Table 10-7 NATEF and ASE TASK

Inspect and replace power steering hoses, fittings, and O-rings.

Problem Area	Symptoms	Possible Causes	Classroom Manual	Shop Manual
STEERING EFFORT	Hard steering, loss of fluid	1. Leaking hoses	252	328
		2. Restricted hoses	253	328

Table 10-8 NATEF and ASE TASK

Flush, fill, and bleed power steering system.

Problem Area	Symptoms	Possible Causes	Classroom Manual	Shop Manual
NOISE	Growling noise from pump	Air in power steering system	258	315
STEERING EFFORT	Hard, erratic steering	Contaminated fluid	258	315

Recirculating Ball Steering Gear Diagnosis and Service

Upon completion and review of this chapter, you should be able to:

❏ Diagnose manual recirculating ball steering gear problems.

❏ Remove and replace manual recirculating ball steering gears.

❏ Adjust worm shaft bearing preload on manual recirculating ball steering gears.

❏ Adjust sector lash on manual recirculating ball steering gears.

❏ Repair oil leaks in manual recirculating ball steering gears.

❏ Disassemble, repair, and reassemble manual recirculating ball steering gears.

❏ Diagnose power recirculating ball steering gear problems.

❏ Remove and replace power recirculating ball steering gears.

❏ Adjust worm shaft thrust bearing preload in power recirculating ball steering gears.

❏ Adjust sector lash in power recirculating ball steering gears.

❏ Diagnose and repair oil leaks in power recirculating ball steering gears.

❏ Disassemble, repair, and reassemble power recirculating ball steering gears.

Manual Recirculating Ball Steering Gear Diagnosis

Basic Tools

Basic technician's tool set

Service manual

Feeler gauge set

Masking tape

CAUTION: Recirculating ball steering gears are often mounted near the exhaust manifold, which may be extremely hot. Wear protective gloves and use caution when inspecting, adjusting, and servicing the steering gear.

Loose steering gear adjustments result in excessive steering wheel freeplay and steering wander. Tight steering gear adjustments cause excessive steering effort.

When the oil-level plug is removed from the steering gear, the lubricant should be level with the bottom of the plug opening. If the oil level is low, fill the steering gear with the manufacturer's specified steering gear lubricant. A steering gear provides binding and hard, uneven steering effort if the oil level is very low. If the oil level is low, visually check the sector shaft seal and the worm shaft seal area for leaks. Leaking seals must be replaced. Hard steering effort may also be caused by wheel alignment problems. Defective worm shaft bearings cause uneven turning effort and steering gear noise. Excessive steering effort may be caused by worn steering gears. A rattling noise from the steering gear may be caused by loose mounting bolts or worn steering shaft U-joints. Excessive **steering wheel freeplay** may be caused by these items:

1. Loose worm shaft **bearing preload** adjustment

2. Loose sector shaft lash adjustment

3. Worn steering gears

4. Loose steering gear mounting bolts

5. Worn steering shaft U-joints or flexible coupling

Manual Recirculating Ball Steering Gear Replacement

WARNING: Hard steel bolts may be used for steering gear mounting. Bolt hardness is indicated by ribs on the bolt head. Harder bolts have five, six, or seven ribs on the bolt heads. Never substitute softer steel bolts in place of the original harder bolts because these softer bolts may break, allowing the steering box to detach from the frame, and resulting in loss of steering control.

⚠️ **WARNING:** When the steering linkage is disconnected from the gear, do not turn the steering wheel hard against the stops. This action may damage the ball guides in the steering gear.

Follow these preliminary steps prior to steering gear removal or adjustment:

1. Disconnect the battery ground cable.

2. Raise the vehicle with the front wheels in the straight-ahead position. If the vehicle is lifted with a floor jack, place safety stands under the chassis or suspension, and lower the vehicle onto the safety stands.

3. Remove the pitman arm nut and washer. Mark the pitman arm position in relation to the pitman shaft with a center punch and use a puller to remove the pitman arm (Figure 11-1).

4. Loosen the sector shaft backlash adjuster lock nut, and back the adjuster screw off one-quarter turn.

5. Turn the steering wheel gently in one direction until it is stopped by the gear, and then turn the steering wheel back one-half turn toward the center position.

6. Remove the center steering wheel cover, and place a socket and inch-pound (in-lb) torque wrench on the steering wheel nut (Figure 11-2). Do not use a torque wrench with a maximum scale reading above 50 in-lbs.

7. Rotate the steering wheel through a 90° arc and record the turning torque, which indicates the worm shaft bearing preload.

Follow these steps for manual steering gear removal and replacement:

1. Disconnect the flexible coupling from the worm shaft.

2. Remove the steering-gear-to-frame mounting bolts.

3. Remove the steering gear from the chassis. The steering gear may be cleaned externally with solvent or in a parts washer.

4. Reverse steps 1 through 3 to reinstall the steering gear. All bolts must be tightened to the specified torque. Be sure the steering gear is filled with the manufacturer's specified steering gear lubricant.

Special Tools

Pitman arm puller

Special Tools

Inch-pound torque wrench

Special Tools

Foot-pound torque wrench

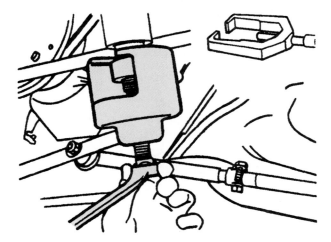

Figure 11-1 Disconnecting pitman arm from the steering gear. (Courtesy of Chevrolet Motor Division, General Motors Corp.)

Figure 11-2 Measuring worm shaft bearing preload on vehicle. (Courtesy of Buick Motor Division, General Motors Corp.)

Manual Recirculating Ball Steering Gear Adjustments

Worm Shaft Bearing Preload Adjustment

When a bearing has preload, all endplay is removed and there is a slight tension on the bearing.

If the worm shaft bearing preload adjustment is loose, steering wheel freeplay is excessive. This results in steering wander when the vehicle is driven straight ahead. Steering effort is increased if the worm shaft bearing preload adjustment is too tight. Use the following procedure to adjust the worm shaft bearing preload.

1. Follow the preliminary steps listed previously.

2. Loosen the worm shaft adjuster plug lock nut with a brass punch and a hammer, and tighten the adjuster plug until all the worm shaft endplay is removed. Then loosen the plug one-quarter turn.

 WARNING: Applying force to the worm shaft at either of the stops may damage the steering gear.

3. Turn the worm shaft fully to the right with a socket and an inch-pound torque wrench, and then turn the worm shaft one-half turn toward the center position.

4. Tighten the adjuster plug until the specified bearing preload is indicated on the torque wrench as the worm shaft is rotated (Figure 11-3). The specification on some steering gears is 5 to 8 in-lbs (0.56 to 0.896 Nm). Always use the vehicle manufacturer's specified preload.

5. Tighten the adjuster plug lock nut to 85 ft-lbs (114 Nm).

Sector Lash Adjustment

Steering wheel freeplay refers to the amount of steering wheel rotation before the front wheels begin to turn right or left.

When the sector shaft lash adjustment is too loose, steering wheel freeplay is excessive and vehicle wander occurs when the vehicle is driven straight ahead. A loose sector shaft lash adjustment decreases driver road feel. If the sector lash adjustment is too tight, steering effort is increased, especially with the front wheels in the straight-ahead position.

The following procedure may be used when the pitman shaft sector lash is adjusted.

Vehicle wander is the tendency of a vehicle to steer right or left as it is driven straight ahead.

1. Turn the pitman backlash adjuster screw outward until it stops, and then turn it inward one turn.

2. Rotate the worm shaft fully from one stop to the other stop, and carefully count the number of shaft rotations.

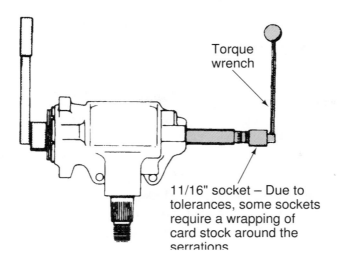

Torque
wrench

11/16" socket – Due to
tolerances, some sockets
require a wrapping of
card stock around the
serrations

Figure 11-3 Adjusting worm shaft bearing preload, manual recirculating ball steering gear.
(Courtesy of Chevrolet Motor Division, General Motors Corp.)

3. Turn the worm shaft back exactly one-half the total number of turns from one of the stops.

4. With the steering gear positioned as it was in step 3, connect an inch-pound torque wrench and socket to the worm shaft. Note the steering gear turning torque while rotating the worm shaft 45° in each direction.

5. Turn the pitman backlash adjuster screw until the torque wrench reading is 6 to 10 in-lbs (0.44 to 1.12 Nm) more than the worm shaft bearing preload torque in step 4 (Figure 11-4).

6. Tighten the pitman backlash adjuster screw lock nut to the specified torque.

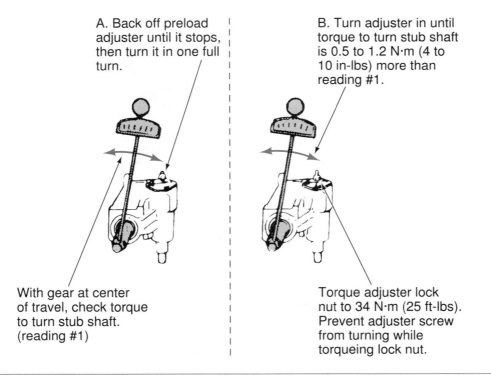

A. Back off preload
adjuster until it stops,
then turn it in one full
turn.

B. Turn adjuster in until
torque to turn stub shaft
is 0.5 to 1.2 N·m (4 to
10 in-lbs) more than
reading #1.

With gear at center
of travel, check torque
to turn stub shaft.
(reading #1)

Torque adjuster lock
nut to 34 N·m (25 ft-lbs).
Prevent adjuster screw
from turning while
torqueing lock nut.

Figure 11-4 Adjusting pitman backlash adjuster screw, manual recirculating ball steering gear.
(Courtesy of Chevrolet Motor Division, General Motors Corp.)

Road feel is
experienced by a driver
when the steering
wheel is turned, and
the driver has a positive
feeling that the front
wheels are turning in
the intended direction.

Gear tooth backlash
refers to movement
between gear teeth
that are meshed with
each other.

Manual Recirculating Ball Steering Gear Disassembly

After the steering gear is removed from the chassis, clean the housing externally with solvent and place the gear in a vise. The worm shaft seal may be changed without disassembling the steering gear. Do not damage the shaft or housing during the seal replacement procedure. Follow these steps for steering gear disassembly:

1. Rotate the worm shaft until it is in the centered position, and use a center punch to mark the worm shaft in relation to the housing.

2. Remove the sector lash adjusting screw lock nut and the cover attaching screws. Turn the lash adjusting screw clockwise down through the cover to remove the cover and gasket.

3. Remove the lash adjusting screw and **selective shim** from the end of the pitman shaft, and remove this shaft from the housing. Place all steering gear components on clean shop towels on the workbench.

4. Use a brass punch and a hammer to loosen the worm shaft bearing adjuster lock nut, and remove the adjuster and bearing. Remove the bearing retainer with a slotted screwdriver.

5. Remove the worm shaft and ball nut assembly. Remove the screws to release the ball nut return guide, and remove this guide. Turn the ball nut over and rotate the worm shaft from side to side to remove all the steel balls. Remove the ball nut from the worm shaft.

6. Clean all internal parts in solvent, and inspect all components for wear, scoring, or pitting.

7. Pry the pitman shaft seal from the housing with a slotted screwdriver.

8. Check for movement between the pitman shaft and the bearing in the housing. If movement is present, use the proper bearing driver to remove the bearing (Figure 11-5). Install

Special Tools

Pitman arm bearing puller

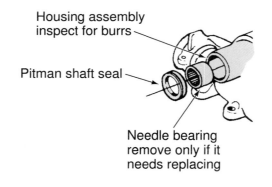

Housing assembly inspect for burrs

Pitman shaft seal

Needle bearing remove only if it needs replacing

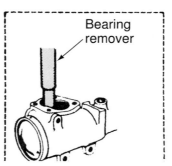

Bearing remover

Figure 11-5 Removing pitman shaft bearing with proper driving tool. (Courtesy of Chevrolet Motor Division, General Motors Corp.)

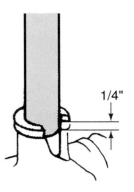

1/4"

When tool bottoms on
housing bearing is
fully installed.

Figure 11-6 Installing pitman shaft bearing with the proper driving tool. (Courtesy of Chevrolet Motor Division, General Motors Corp.)

the new bearing with the proper bearing driver (Figure 11-6). If the pitman shaft is worn or scored in the bearing contact area, shaft replacement is necessary.

9. Remove the steering shaft seal with a slotted screwdriver, and use the proper seal driver to tap the new seal into the housing.

10. If the upper and lower worm shaft bearing cups in the housing or adjuster nut are scored, pitted, or worn, use a slide-hammer-type puller to remove these cups. Install the new bearing cups with the proper cup driver.

Special Tools

Seal drivers

Manual Recirculating Ball Steering Gear Reassembly

1. Lubricate all seals, bushings, and bearings with the manufacturer's recommended steering gear lubricant.

2. Position the ball nut on the worm shaft and install the balls in the return guides and the ball nut. Place an equal number of balls in each ball nut circuit. Install the return guide clamp and tighten the screws.

3. Install the upper bearing over the worm shaft and center the ball nut on the worm gear. Slide the worm shaft assembly into the gear housing. Be careful not to damage the upper worm shaft seal.

4. Place the lower bearing on the worm shaft adjuster plug with the bearing retainer, and install the adjuster and lock nut in the housing. Tighten the adjuster enough to hold the bearing in place.

5. Install the pitman shaft adjusting screw and shim in the pitman shaft. Measure the end-play between the screw and the slot. If this measurement exceeds 0.002 inch, replace the selective shim. These shims are available in four thicknesses: 0.063, 0.065, 0.067, and 0.069 inch.

6. Install the pitman shaft and adjusting screw in the housing. The sector teeth must mesh properly with the centered ball nut teeth.

7. Install the sector cover and gasket on the adjusting screw, and turn the adjusting screw counterclockwise to thread the screw up through the cover. Continue turning the

adjusting screw until the cover and gasket contact the housing and the screw extends 5/8 to 3/4 inch above the cover. Install, but do not tighten, the adjusting screw lock nut. Install and tighten the cover screws.

8. Tighten the sector lash adjusting screw until the sector teeth and the ball nut teeth are meshed, with some movement between the teeth. Final adjustment is made later.

9. Wrap masking tape around the pitman shaft splines to protect the seal, and install the pitman shaft seal with the proper driver.

10. Fill the steering gear housing with the manufacturer's recommended steering lubricant, and turn the worm shaft fully in each direction several times to check for a binding condition. Do not apply force to the worm shaft when it reaches the end of its travel in either direction.

Classroom Manual
Chapter 11, page 264

11. Perform the worm shaft bearing preload and sector lash adjustments as mentioned previously.

Power Recirculating Ball Steering Gear Diagnosis

If the steering gear is noisy, check these items:

1. Loose pitman shaft lash adjustment—may cause a rattling noise when the steering wheel is turned.

2. Cut or worn dampener O-ring on the valve spool—when this defect is present, a squawking noise is heard during a turn.

3. Loose steering gear mounting bolts.

4. Loose or worn flexible coupling or steering shaft U-joints.

A hissing noise from the power steering gear is normal if the steering wheel is at the end of its travel, or when the steering wheel is rotated with the vehicle standing still. If the steering wheel jerks or surges when the steering wheel is turned with the engine running, check the power steering pump belt condition and tension. When excessive kick-back is felt on the steering wheel, check the poppet valve in the steering gear.

When the steering is loose, check these defects:

1. Air in the power steering system. To remove the air, fill the power steering pump reservoir and rotate the steering wheel fully in each direction several times.

2. Loose pitman lash adjustment.

3. Loose worm shaft thrust bearing preload adjustment.

4. Worn flexible coupling or universal joint.

5. Loose steering gear mounting bolts.

6. Worn steering gears.

A complaint of hard steering while parking could be caused by one of these defects:

1. Loose or worn power steering pump belt.

2. Low oil level in the power steering pump.

3. Excessively tight steering gear adjustments.

4. Defective power steering pump with low-pressure output.

5. Restricted power steering hoses.

6. Defects in the steering gear such as:
   (a) pressure loss in the cylinder because of scored cylinder, worn piston ring, or damaged backup O-ring.
   (b) excessively loose spool in the valve body.
   (c) defective or improperly installed gear check poppet valve.

Power Recirculating Ball Steering Gear Replacement

When the power steering gear is replaced, proceed as follows:

1. Disconnect the hoses from the steering gear and cap the lines and fittings to prevent dirt from entering the system.

2. Remove the pitman arm nut and washer and mark the pitman arm in relation to the shaft with a center punch. Use a puller to remove the pitman arm.

3. Disconnect the steering shaft from the worn shaft.

4. Remove the steering gear mounting bolts, and remove the steering gear from the chassis (Figure 11-7).

5. Reverse steps 1 through 4 to install the steering gear. All bolts must be tightened to the specified torque. Be sure the pitman arm is installed in the original position.

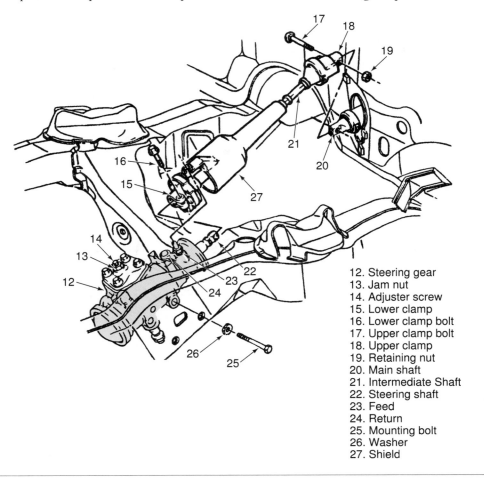

12. Steering gear
13. Jam nut
14. Adjuster screw
15. Lower clamp
16. Lower clamp bolt
17. Upper clamp bolt
18. Upper clamp
19. Retaining nut
20. Main shaft
21. Intermediate Shaft
22. Steering shaft
23. Feed
24. Return
25. Mounting bolt
26. Washer
27. Shield

Figure 11-7 Removing steering gear from the chassis. (Courtesy of Chevrolet Motor Division, General Motors Corp.)

Power Recirculating Ball Steering Gear Adjustments

Worm Shaft Thrust Bearing Preload Adjustment

A loose worm shaft thrust bearing preload adjustment or sector lash adjustment causes excessive steering wheel freeplay and steering wander. The power recirculating ball steering gear adjustment procedures may vary depending on the vehicle make and model year. Always follow the vehicle manufacturer's recommended procedure in the service manual. When the worm shaft thrust bearing preload adjustment is performed, use this procedure:

1. Remove the worm shaft thrust bearing adjuster plug lock nut with a hammer and brass punch (Figure 11-8).

2. Turn this adjuster plug inward, or clockwise, until it bottoms, and tighten the plug to 20 ft-lbs (27 Nm).

3. Place an index mark on the steering gear housing next to one of the holes in the adjuster plug (Figure 11-9).

4. Measure 0.50 in (13 mm) counterclockwise from the index mark, and place a second index mark at this position (Figure 11-10).

5. Rotate the adjuster plug counterclockwise until the hole in the adjuster plug is aligned with the second index mark placed on the housing (Figure 11-11).

6. Install and tighten the adjuster plug lock nut to the specified torque.

Photo Sequence 9 shows a typical procedure for performing a worm shaft bearing preload adjustment.

Special Tools

Worm shaft adjuster plug rotating tool

Figure 11-8 Removing worm shaft thrust bearing adjuster plug lock nut. (Courtesy of Chevrolet Motor Division, General Motors Corp.)

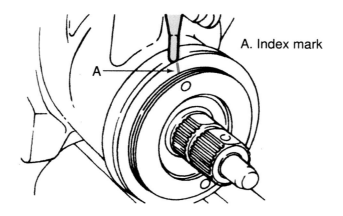

A. Index mark

Figure 11-9 Placing index mark on steering gear housing opposite one of the adjuster plug holes. (Courtesy of Chevrolet Motor Division, General Motors Corp.)

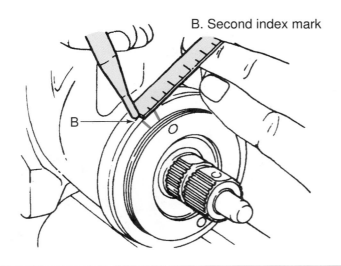

B. Second index mark

Figure 11-10 Measuring 0.50 in (13 mm) counterclockwise from the index mark on the steering gear housing. (Courtesy of Chevrolet Motor Division, General Motors Corp.)

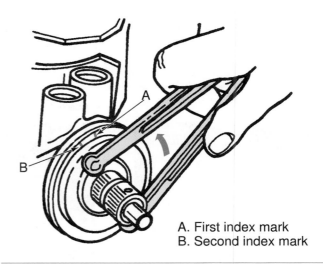

A. First index mark
B. Second index mark

Figure 11-11 Aligning adjuster plug hole with second index mark placed on steering gear housing. (Courtesy of Chevrolet Motor Division, General Motors Corp.)

Pitman Sector Shaft Lash Adjustment

When the pitman sector shaft lash adjustment is performed, proceed as follows:

1. Rotate the stub shaft from stop to stop and count the number of turns.
2. Starting at either stop, turn the stub shaft back two-thirds of the total number of turns. In this position, the flat on the stub shaft should be facing upward (Figure 11-12), and the master spline on the pitman shaft should be aligned with the pitman shaft backlash adjuster screw (Figure 11-13).

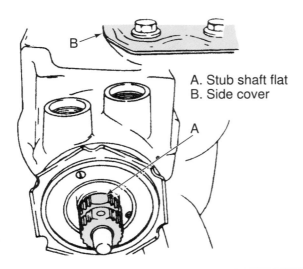

A. Stub shaft flat
B. Side cover

Figure 11-12 Stub shaft flat facing upward and parallel with the side cover. (Courtesy of Chevrolet Motor Division, General Motors Corp.)

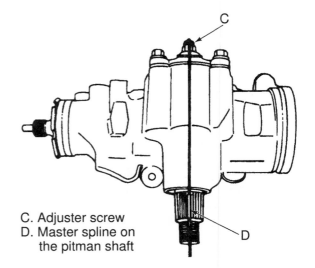

C. Adjuster screw
D. Master spline on the pitman shaft

Figure 11-13 Pitman shaft master spline aligned with the pitman backlash adjuster screw. (Courtesy of Chevrolet Motor Division, General Motors Corp.)

Photo Sequence 9
Typical Procedure for Performing a Worm Shaft Bearing Preload Adjustment

P9-1 Retain the steering gear in a vise.

P9-2 Remove the worm shaft bearing lock nut with a hammer and brass punch.

P9-3 Use the proper tool to turn the worm shaft adjuster plug clockwise until it bottoms, and tighten this plug to 20 ft-lbs (27 Nm).

P9-4 Place an index mark on the steering gear housing next to one of the adjuster plug holes.

P9-5 Measure 0.50 in (13 mm) counterclockwise from the mark placed on the housing in step P9-4, and place a second mark on the housing.

P9-6 Rotate the worm shaft adjuster plug counterclockwise until the adjuster plug hole is aligned with the second mark placed on the housing.

P9-7 Install and tighten the worm shaft adjuster plug lock nut.

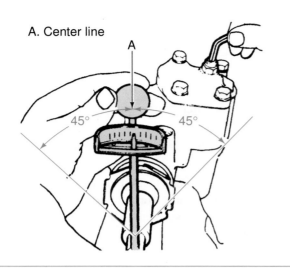

A. Center line

Figure 11-14 Measuring worm shaft turning torque to adjust pitman backlash adjuster screw. (Courtesy of Chevrolet Motor Division, General Motors Corp.)

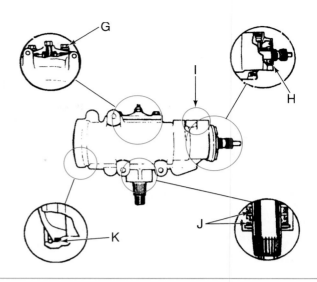

Figure 11-15 Power recirculating ball steering gear oil leak locations. (Courtesy of Chevrolet Motor Division, General Motors Corp.)

3. Turn the pitman shaft backlash adjuster screw fully counterclockwise, and then turn it clockwise one turn.

4. Use an inch-pound torque wrench to turn the stub shaft through a 45° arc on each side of the position in step 2. Read the over-center torque as the stub shaft turns through the center position (Figure 11-14).

5. Continue to adjust the pitman shaft adjuster screw until the torque is 6 to 10 in-lbs (0.6 to 1.2 Nm) more than the torque in step 4.

6. Hold the pitman shaft adjuster screw in this position and tighten the lock nut to the specified torque.

Power Recirculating Ball Steering Gear Oil Leak Diagnosis

Five locations where oil leaks may occur in a power steering gear are the following:

1. Side cover O-ring seal (Figure 11-15, view G)
2. Adjuster plug seal (H)
3. Pressure line fitting (I)
4. Pitman shaft oil seals (J)
5. End cover seal (K)

If an oil leak is present at any of these areas, complete or partial steering gear disassembly and seal or O-ring replacement is necessary.

Power Recirculating Ball Steering Gear Seal Replacement

Side-Cover O-Ring Replacement

Prior to any disassembly procedure, clean the steering gear with solvent or in a parts washer. The steering gear service procedures vary depending on the make of gear. Always follow the vehicle manufacturer's recommended procedure in the service manual. Following is a typical side-cover O-ring replacement procedure:

1. Loosen the pitman backlash adjuster screw lock nut and remove the side-cover bolts. Rotate the pitman backlash adjuster screw clockwise to remove the cover from the screw (Figure 11-16).

2. Discard the O-ring and inspect the side-cover matching surfaces for metal burrs and scratches.

3. Lubricate a new O-ring with the vehicle manufacturer's recommended power steering fluid and install the O-ring.

4. Rotate the pitman backlash adjuster screw counterclockwise into the side cover until the side cover is properly positioned on the gear housing. Turn this adjuster screw fully counterclockwise, and then one turn clockwise. Install and tighten the side-cover bolts to the specified torque. Adjust the pitman sector shaft lash as explained earlier.

End Plug Seal Replacement

Follow these steps for end plug seal replacement:

1. Insert a punch into the access hole in the steering gear housing to unseat the retaining ring, and remove the ring (Figure 11-17).

2. Remove the end plug and seal.

3. Clean the end plug and seal contact area in the housing with a shop towel.

4. Lubricate a new seal with the vehicle manufacturer's recommended power steering fluid, and install the seal.

5. Install the end plug and retaining ring.

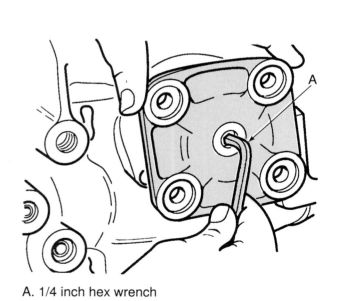

A. 1/4 inch hex wrench

Figure 11-16 Removing steering gear side cover. (Courtesy of Chevrolet Motor Division, General Motors Corp.)

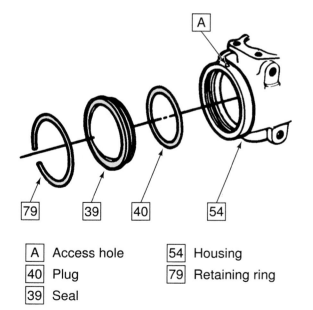

79 39 40 54

A	Access hole	54	Housing
40	Plug	79	Retaining ring
39	Seal		

Figure 11-17 Removing steering gear end plug, retaining ring, and seal. (Courtesy of Chevrolet Motor Division, General Motors Corp.)

Worm Shaft Bearing Adjuster Plug Seal and Bearing Replacement

Follow these steps for worm shaft bearing adjuster plug seal and bearing service:

1. Remove the adjuster plug lock nut, and use a special tool to remove the adjuster plug.

2. Use a screwdriver to pry at the raised area of the bearing retainer to remove this retainer from the adjuster plug (Figure 11-18).

3. Place the adjuster plug face down on a suitable support, and use the proper driver to remove the needle bearing, dust seal, and lip seal.

> **WARNING:** The bearing identification number must face the driving tool to prevent bearing damage during installation.

4. Place the adjuster plug outside face up on a suitable support, and use the proper driver to install the needle bearing dust seal and lip seal.

5. Install the bearing retainer in the adjuster plug, and lubricate the bearing and seal with the vehicle manufacturer's recommended power steering fluid.

6. Install the adjuster plug and lock nut, and adjust the worm shaft bearing preload as discussed previously.

Classroom Manual
Chapter 11, page 267

> **CUSTOMER CARE:** As an automotive technician, you should be familiar with the maintenance schedules recommended by various vehicle manufacturers. Of course, it is impossible to memorize all the maintenance schedules on different makes of vehicles, but maintenance schedule books are available. This maintenance schedule information is available in the owner's manual, but the vehicle owner may not take time to read this manual. If you advise the customer that his or her vehicle requires some service, such as

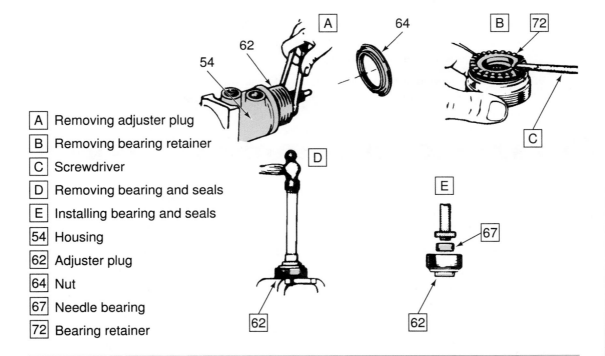

A	Removing adjuster plug
B	Removing bearing retainer
C	Screwdriver
D	Removing bearing and seals
E	Installing bearing and seals
54	Housing
62	Adjuster plug
64	Nut
67	Needle bearing
72	Bearing retainer

Figure 11-18 Removing and replacing worm shaft adjuster plug, bearing, and seal. (Courtesy of Chevrolet Motor Division, General Motors Corp.)

a cooling system flush, according to the vehicle manufacturer's maintenance schedule, the customer will often have the service performed. The customer will usually appreciate your interest in his or her vehicle, and the shop will benefit from the increased service work.

Guidelines for Servicing Manual and Power Recirculating Ball Steering Gears

1. Oil leaks may occur at the pitman shaft seal or worm shaft seal in a manual recirculating ball steering gear.

2. A rattling noise and excessive steering freeplay may be caused by loose gear mounting bolts, worn steering shaft U-joints, or a worn flexible coupling in a recirculating ball steering gear system.

3. A loose worm shaft bearing preload adjustment or sector lash adjustment on a manual or power recirculating ball steering gear causes excessive steering wheel freeplay, steering wander, and reduced road feel.

4. If the worm shaft bearing preload adjustment or sector lash adjustment is tighter than specified on a manual or power recirculating ball steering gear, steering effort is higher than normal.

5. On a manual recirculating ball steering gear, the worm shaft bearing preload adjuster plug is rotated until the correct worm shaft turning torque is obtained with the worm shaft positioned one-half turn from the full-right or full-left position.

6. On a manual recirculating ball steering gear, the sector lash adjusting screw is tightened until the correct worm shaft turning torque is obtained as the worm shaft is rotated through the center position.

7. Oil leaks in a power recirculating ball steering gear may occur at the end plug O-ring seal, side-cover O-ring seal, pitman shaft seal, or worm shaft thrust bearing seal.

8. On some power recirculating ball steering gears, the worm shaft thrust bearing adjuster plug is tightened until the correct worm shaft turning torque is obtained with the worm shaft positioned one-half turn from the full-right or full-left position.

9. On a power recirculating ball steering gear, the sector lash screw is tightened until the correct worm shaft turning torque is obtained as the worm shaft is rotated through the center position.

CASE STUDY

The owner of a rear-wheel-drive Oldsmobile 98 complained about increased and somewhat erratic steering effort. The service writer asked the customer about the conditions when this problem occurred, and the customer said that the condition was always present. During a road test, the technician discovered that the customer's description of the problem was accurate.

The technician checked the power steering fluid level, and made a careful check of the belt tension and condition without finding any problems. Next, the technician checked the power steering pump pressure, and found it to be normal. A check of the power steering hoses did not reveal any hose restrictions.

The technician removed and disassembled the steering gear, and found a severely scored cylinder bore in the gear housing. The ball nut piston ring was also worn and scored. A replacement steering gear was installed, and the system filled with the manufacturer's recommended power steering fluid. A road test indicated a new gear worked satisfactorily.

Terms to Know

Bearing preload

Gear tooth backlash *or* lash

Road feel

Selective shim

Steering wheel freeplay

Vehicle wander

ASE Style Review Questions

1. While discussing manual recirculating ball steering gear diagnosis:

   Technician A says excessive steering wheel freeplay may be caused by a worm shaft bearing preload adjustment that is tighter than normal.

   Technician B says excessive steering wheel freeplay may be caused by loose steering gear mounting bolts.

   Who is correct?

   A. A only   **C.** Both A and B

   B. B only   **D.** Neither A nor B

2. While discussing manual recirculating ball steering gear diagnosis:

   Technician A says the worm shaft bearing preload adjustment is performed with the sector lash adjuster screw backed off one-quarter turn.

   Technician B says the worm shaft bearing preload adjustment is performed with the worm shaft turned fully to the right or left.

   Who is correct?

   A. A only   **C.** Both A and B

   B. B only   **D.** Neither A nor B

3. While discussing manual recirculating ball steering gear service:

   Technician A says steering wander may be caused by a loose sector lash adjustment.

   Technician B says a loose sector lash adjustment may cause reduced feel of the road.

   Who is correct?

   A. A only   **C.** Both A and B

   B. B only   **D.** Neither A nor B

4. While discussing manual recirculating steering gear diagnosis:

   Technician A says the sector lash adjustment is performed with the worm shaft half-way between the centered position and the full-right position.

5. While discussing manual recirculating ball steering gear service:

   Technician A says if the endplay between the sector lash screw and the slot is excessive, selective shims are available to correct this endplay.

   Technician B says excessive steering effort may be caused by a wheel alignment problem.

   Who is correct?

   A. A only   **C.** Both A and B

   B. B only   **D.** Neither A nor B

   Technician B says the worm shaft bearing preload adjustment is performed before the sector lash adjustment.

   Who is correct?

   A. A only   **C.** Both A and B

   B. B only   **D.** Neither A nor B

6. A power recirculating ball steering gear has excessive kick-back on the steering wheel. The most likely cause of this problem is:

   A. a loose sector lash adjustment.

   B. a loose worm shaft bearing preload adjustment.

   C. a defective poppet valve in the steering gear.

   D. a worn pitman shaft bearing.

7. All these statements about power recirculating ball steering gear defects are true EXCEPT:

   A. A worn steering shaft U-joint may cause a growling noise when turning the front wheels.

   B. Excessive steering effort may be caused by a defective power steering pump.

   C. Excessive steering effort may be caused by a scored steering gear cylinder.

   D. Steering wheel jerking when turning may be caused by a slipping power steering belt.

8. When adjusting on a power recirculating ball steering gear:
   A. the worm shaft bearing adjuster plug should be bottomed and then backed off until the specified worm shaft turning torque is obtained.
   B. the sector lash adjustment screw should be tightened one turn before the worn shaft bearing preload is adjusted.
   C. the pitman shaft backlash adjuster screw is turned fully counterclockwise and then one turn clockwise prior to the backlash adjustment.
   D. the pitman shaft is positioned one-half turn from the fully right or left position prior to the pitman shaft backlash adjustment.

9. A power recirculating ball steering gear experiences excessive steering effort while parking. The most likely cause of this problem is:
   A. a restricted high-pressure steering hose.
   B. a misaligned power steering belt.
   C. a leaking pitman shaft seal.
   D. excessive torque on the worm shaft adjuster plug lock nut.

10. A recirculating ball steering gear experiences hard steering for a short time after the vehicle sits overnight. The most likely cause of this problem is:
    A. excessively tight worm bearing preload adjustment.
    B. a scored cylinder and worn piston ring in the steering gear.
    C. a restricted power steering return hose.
    D. a binding U-joint in the steering shaft.

ASE Challenge Questions

1. A car with a power recirculating ball steering has excessive steering kickback and a preliminary inspection shows no abnormal wear in the linkage. Which of the following could be the cause of the problem?
   A. Worn gear piston or bore
   B. Slipping pump belt
   C. Worn pump poppet valve
   D. Sticking valve spool

2. The complaint is loss of power assist, but there is no mention of any associated noise. All of the following could cause this problem EXCEPT:
   A. low fluid.
   B. improperly inflated tires.
   C. broken pump belt.
   D. steering column misalignment.

3. While discussing steering problems, *Technician A* says a "jerky" steering wheel and a "clunking" noise could indicate worn steering column U joints. *Technician B* says lack of assist and a "growling" noise in a fluid filled steering pump could indicate a hose or pump internal restriction.
   Who is correct?
   A. A only   C. Both A and B
   B. B only   D. Neither A nor B

4. A vehicle with a manual recirculating ball steering system requires much higher than normal steering effort, especially in the parking lot.
   Technician A says the cause of the problem is a worn pitman shaft seal.
   Technician B says the cause of the problem is a worn worm shaft thrust bearing.
   Who is correct?
   A. A only   C. Both A and B
   B. B only   D. Neither A nor B

5. The complaint is excessive steering wheel freeplay. All of the following could cause this problem EXCEPT:
   A. loose wormshaft bearing preload.
   B. worn steering gears.
   C. steering gear column misalignment.
   D. worn flex coupling or U joint.

Table 11-1 NATEF and ASE TASK

Diagnose manual steering gear (non-rack and pinion-type) noises, binding, uneven turning effort, looseness, hard steering, and lubricant leakage problems; determine needed repairs.

Problem Area	Symptoms	Possible Causes	Classroom Manual	Shop Manual
	Fluid leaks	1. Worn pitman shaft seal	265	342
		2. Worn worm shaft seal	264	342
	Rattling on road irregularities	1. Loose steering shaft U-joints, flexible coupling	264	342
		2. Loose steering gear mounting bolts	266	344
LOOSENESS	Excessive steering wheel freeplay	1. Loose worm bearing preload adjustment	265	344
		2. Loose sector lash adjustment	265	344
		3. Worn steering gears	265	344
		4. Loose steering gear mounting bolts	265	342
STEERING EFFORT	Excessive steering effort	1. Tight worm bearing preload adjustment	264	344
		2. Tight sector lash adjustment	264	344
		3. Low steering gear lubricant	264	342
		4. Worn steering gears or bearings	264	342
	Uneven steering effort	Worn steering gears or bearings	264	342

Table 11-2 NATEF and ASE TASK

Remove and replace manual steering gear (non-rack and pinion-type) (includes vehicles equipped with air bags and/or other steering wheel mounted controls and components).

Problem Area	Symptoms	Possible Causes	Classroom Manual	Shop Manual
STEERING EFFORT	Excessive steering effort, binding	1. Worn steering gears or bearings	265	342
		2. Tight steering gear adjustments	264	342
		3. Low steering gear lubricant	264	342
	Fluid leaks	1. Worn pitman shaft seal	264	342
		2. Worn worm shaft seal	264	342

Table 11-3 NATEF and ASE TASK

Adjust manual steering gear (non-rack and pinion-type) worm bearing preload and sector lash.

Problem Area	Symptoms	Possible Causes	Classroom Manual	Shop Manual
LOOSENESS	Excessive steering wheel freeplay	1. Loose worm bearing preload adjustment	265	344
		2. Loose sector lash adjustment	265	344
STEERING EFFORT	Excessive steering effort	1. Tight worm bearing preload adjustment	265	344
		2. Tight sector lash adjustment	265	344

Table 11-4 NATEF and ASE TASK

Diagnose power steering gear (non-rack and pinion-type)
noises, binding, uneven turning effort, looseness, hard steering,
and fluid leakage problems; determine needed repairs.

Problem Area	Symptoms	Possible Causes	Classroom Manual	Shop Manual
LOSS OF LUBRICANT	Fluid leaks	1. Worn pitman shaft seal	267	353
		2. Worn end-plug O-ring seal	267	353
		3. Worn side-cover O-ring seal	267	353
		4. Worn thrust-bearing O-ring seal	268	353
NOISE	Rattling on road irregularities	1. Loose steering shaft U-joints, flexible coupling	269	348
		2. Loose steering gear mounting bolts	269	348
LOOSENESS	Excessive steering wheel freeplay	1. Loose thrust bearing preload adjustment	267	348
		2. Loose sector lash adjustment	267	348
		3. Loose steering gear mounting bolts	266	348
		4. Worn steering gears or components	267	348
STEERING EFFORT	Excessive steering effort	1. Tight sector lash adjustment	268	348
		2. Tight thrust bearing adjustment	268	348
		3. Low fluid level	269	348
		4. Worn steering gears, bearings, or cylinder	268	348
	Uneven steering	1. Low fluid level	268	348
		2. Worn steering gears, bearings, or components	269	348

Table 11-5 NATEF and ASE TASK

Remove and replace power steering gear (non-rack and pinion-type) (includes vehicles equipped with air bags and/or other steering wheel mounted controls and components).

Problem Area	Symptoms	Possible Causes	Classroom Manual	Shop Manual
STEERING EFFORT	Excessive steering effort	1. Worn steering gears bearings, or components	267	348
		2. Tight steering gear adjustments	267	348
		3. Low fluid level	268	348
LOSS OF FLUID	Fluid leaks	1. Worn pitman shaft seal	269	353
		2. Worn thrust-bearing O-ring seal	267	353
		3. Worn side-cover O-ring	268	353
		4. Worn end-plug O-ring	269	353

Table 11-6 NATEF and ASE TASK

Adjust power steering gear (non-rack and pinion type) worm bearing preload and sector lash.

Problem Area	Symptoms	Possible Causes	Classroom Manual	Shop Manual
LOOSENESS	Excessive steering wheel freeplay	1. Loose thrust bearing preload adjustment	267	350
		2. Loose sector lash adjustment	267	351
STEERING EFFORT	Excessive steering effort	1. Tight thrust bearing preload adjustment	267	350
		2. Tight sector lash adjustment	267	351

Table 11-7   NATEF and ASE TASK

Inspect and replace power steering gear (non-rack and pinion type) seals and gaskets.

Problem Area	Symptoms	Possible Causes	Classroom Manual	Shop Manual
LOSS OF FLUID	Fluid leaks	1. Worn pitman shaft seal	266	353
		2. Worn end-plug O-ring seal	267	353
		3. Worn thrust-bearing O-ring	267	353
		4. Worn side-cover O-ring seal	267	353

Job Sheet 32

32

Name _____ Date _____

Adjust Manual Recirculating Ball Steering Gear, Worm Bearing Preload, and Sector Lash, Steering Gear Removed

NATEF and ASE Correlation

This job sheet is related to NATEF and ASE Automotive Suspension and Steering Task List content area: A. Steering Systems Diagnosis and Repair. 1. Steering Columns and Manual Steering Gears. Task 6: Adjust manual steering gear (non-rack and pinion-type) worm bearing preload and sector lash.

Tools and Materials

Torque wrench, in-lbs

Vehicle make and model year _____

Vehicle VIN _____

Make of steering gear _____

Procedure

1. Describe the results of loose steering gear adjustments on steering quality.

_____

_____

_____

2. Describe the results of tight steering gear adjustments.

_____

_____

3. Loosen the worm shaft adjuster plug lock nut with a brass punch and a hammer, and tighten the adjuster plug until all the worm shaft endplay is removed, then loosen the plug one-quarter turn.

> ⚠ **WARNING:** Applying force to the worm shaft at either of the stops may damage the steering gear.

4. Turn the worm shaft fully to the right with a socket and an inch-pound torque wrench, then turn the worm shaft one-half turn toward the center position.

Worm shaft properly positioned one-half turn from the fully right position?

yes _____ no _____

Instructor check _____

5. Tighten the adjuster plug until the specified bearing preload is indicated on the torque wrench as the worm shaft is rotated. The specification on some steering gears is 5 to 8 in-lbs (0.56 to 0.896 Nm). Always use the vehicle manufacturer's specified preload.

   Actual worm shaft bearing preload _____

6. Tighten the adjuster plug lock nut to 85 ft-lbs (114 Nm).

   Actual adjuster plug lock nut torque _____

7. Turn the pitman backlash adjuster screw outward until it stops, then turn it inward one turn.

   Is the pitman backlash adjuster screw properly positioned? yes _____ no _____

   Instructor check _____

8. Rotate the worm shaft fully from one stop to the other stop, and carefully count the number of shaft rotations.

   Total number of worm shaft turns _____

9. Turn the worm shaft back exactly one-half the total number of turns from one of the stops.

   Number of turns that worm shaft is rotated from the fully right or left position to the center position _____

10. With the steering gear positioned as it was in step 7, connect an inch-pound torque wrench and socket to the worm shaft, and note the steering gear turning torque while rotating the worm shaft 45° in each direction.

11. Turn the pitman backlash adjuster screw until the torque wrench reading is 6 to 10 in-lbs (0.44 to 1.12 Nm) more than the worm shaft bearing preload torque in step 3.

    Actual worm shaft turning torque _____

12. Tighten the pitman backlash adjuster screw lock nut to the specified torque.

    Actual pitman shaft adjuster screw lock nut torque _____

✔ Instructor Check _____

Job Sheet 33

Name _____ Date _____

Adjust Power Recirculating Ball Steering Gear Worm Bearing Preload, Steering Gear Removed

NATEF and ASE Correlation

This job sheet is related to NATEF and ASE Automotive Suspension and Steering Task List content area: A. Steering Systems Diagnosis and Repair. 2. Power Assisted Units. Task 12: Adjust power steering gear (non-rack and pinion-type) worm bearing preload and sector lash.

Tools and Materials

Torque wrench, in-lbs

Vehicle make and model year _____

Vehicle VIN _____

Make of steering gear _____

Procedure

1. Remove the worm shaft thrust bearing adjuster plug lock nut with a hammer and brass punch.

2. Turn this adjuster plug inward, or clockwise, until it bottoms and tighten the plug to 20 ft-lbs (27 Nm).

   Actual worm shaft adjuster plug nut torque _____

3. Place an index mark on the steering gear housing next to one of the holes in the adjuster plug.

   Is an index mark placed on steering gear housing beside one of the adjuster plug holes?
   yes _____ no _____

   Instructor check _____

4. Measure 0.50 in (13 mm) counterclockwise from the index mark, and place a second index mark at this position.

   Is a second index mark placed 0.50 in. (13 mm) counterclockwise from the first index mark?
   yes _____ no _____

   Instructor check _____

5. Rotate the adjuster plug counterclockwise until the hole in the adjuster plug is aligned with the second index mark placed on the housing.

Is the hole in adjuster plug properly aligned with second index mark?

yes _____ no _____

Instructor check _____

6. Install and tighten the adjuster plug lock nut to the specified torque.

Specified adjuster plug lock nut torque _____

Actual adjuster plug lock nut torque _____

☑ **Instructor Check** _____

Job Sheet 34

Name _____ Date _____

Adjust Power Recirculating Ball Steering Gear Sector Lash, Steering Gear Removed

NATEF and ASE Correlation

This job sheet is related to NATEF and ASE Automotive Suspension and Steering Task List content area: A. Steering Systems Diagnosis and Repair. 2. Power Assisted Units. Task 12: Adjust power steering gear (non-rack and pinion-type) worm bearing preload and sector lash.

Tools and Materials

Torque wrench, in-lbs

Vehicle make and model year _____

Vehicle VIN _____

Make of steering gear _____

Procedure

1. Rotate the worm shaft from stop to stop and count the number of turns.

   Total number of worm shaft turns from stop to stop _____

2. Starting at either stop turn the worm shaft back two thirds of the total number of turns.

   Number of worm shaft turns from fully right or fully left to position the worm shaft properly prior to sector lash adjustment _____

   Explain the reason for placing the worm shaft in this position prior to the sector lash adjustment. _____

   _____

   _____

   _____

3. In this position, the flat on the worm shaft should be facing upward, and the master spline on the pitman shaft should be aligned with the pitman shaft backlash adjuster screw.

   Is the worm shaft flat properly positioned? yes _____ no _____

   Is the master spline on the pitman shaft properly positioned? yes _____ no _____

   If the answer is no to either of the above questions, state the necessary corrective action.

   _____

   _____

4. Turn the pitman shaft backlash adjuster screw fully counterclockwise, and then turn it clockwise one turn.

   Is the pitman shaft backlash adjuster screw properly positioned? yes _____ no _____

   Instructor check _____

5. Use an inch-pound torque wrench to turn the worm shaft through a 45° arc on each side of the position in step 2. Read the over-center torque as the worm shaft turns through the center position.

Stub shaft turning torque _____

6. Continue to adjust the pitman shaft adjuster screw until the torque is 6 to 10 in-lbs (0.6 to 1.2 Nm) more than the torque in step 5.

Final worm shaft turning torque after adjustment _____

7. Hold the pitman shaft adjuster screw in this position and tighten the lock nut to the specified torque.

Specified pitman shaft adjuster screw lock nut torque _____

Actual pitman shaft adjuster screw lock nut torque _____

☑ **Instructor Check** _____

Rack and Pinion Steering Gear Diagnosis and Service

Upon completion and review of this chapter, you should be able to:

❏ Perform a manual or power rack and pinion steering gear inspection.

❏ Remove and replace manual or power rack and pinion steering gears.

❏ Disassemble, inspect, repair, and reassemble manual rack and pinion steering gears.

❏ Adjust manual rack and pinion steering gears.

❏ Diagnose manual rack and pinion steering systems.

❏ Diagnose oil leaks in power rack and pinion steering gears.

❏ Disassemble, inspect, and repair power rack and pinion steering gears.

❏ Adjust power rack and pinion steering gears.

❏ Diagnose electronic power steering systems.

Manual or Power Rack and Pinion Steering Gear On-Car Inspection

The wear points are reduced to four in a rack and pinion steering gear. These wear points are the inner and outer tie-rod ends on both sides of the rack and pinion assembly (Figure 12-1).

The first step in manual or power rack and pinion steering gear diagnosis is a very thorough inspection of the complete steering system. During this inspection, all steering system components such as the inner and outer tie-rod ends, **bellows boots,** mounting bushings, couplings or univer-

Basic Tools

Basic technician's tool set
Service manual
Floor jack
Safety stands
Machinist's rule

Ⓣ 59 - 78 (6.0 - 8.0, 43 - 58)

Steering gear mounting clamp

★ Outer snap ring (inside)

Gear housing assembly

★ Outer snap ring (outside)

Spacer

★ Oil seal

Pinion bearing

★ Inner snap ring

Pinion

★ Lock plate

Retainer

Tie-rod inner socket   Ⓣ 78 - 98 (8 - 10, 58 - 72)

Retainer spring

Boot clamp

Boot

Cotter pin

Adjusting screw

Front

Adjusting screw lock nut

Boot band

Ⓣ 37 - 46 (3.8 - 4.7, 27 - 34)

N

Tie-rod outer sock

★ : Always replace when disassembled.
Ⓣ : N.m (kg-m, ft-lb)

Figure 12-1 Wear points at the inner and outer tie-rod ends in a rack and pinion steering gear. (Courtesy of Nissan Motor Co., Ltd.)

sal joints, ball joints, tires, and **steering wheel freeplay** must be checked. Follow these steps for manual or power rack and pinion steering gear inspection:

1. With the front wheels straight ahead and the engine stopped, rock the steering wheel gently back and forth with light finger pressure (Figure 12-2). Measure the maximum steering wheel freeplay. The maximum specified steering wheel freeplay on some vehicles is 1.18 in (30 mm). Always refer to the vehicle manufacturer's specifications in the service manual. Excessive steering wheel freeplay indicates worn steering components.

2. With the vehicle sitting on the shop floor and the front wheels straight ahead, have an assistant turn the steering wheel about 1/4 turn in both directions. Watch for looseness in the flexible coupling or universal joints in the steering shaft. If looseness is observed, replace the coupling or universal joint.

3. While an assistant turns the steering wheel about 1/2 turn in both directions, watch for movement of the steering gear housing in the mounting bushings. If there is any movement of the housing in these bushings, replace the bushings. The steering gear mounting bushings may be deteriorated by oil soaking, heat, or age.

4. Grasp the pinion shaft extending from the steering gear and attempt to move it vertically. If there is steering shaft vertical movement, a pinion bearing preload adjustment may be required. When the steering gear does not have a pinion bearing preload adjustment, replace the necessary steering gear components.

5. Road test the vehicle and check for excessive steering effort. A bent steering rack, tight rack bearing adjustment, or damaged front drive axle joints in a front-wheel-drive car may cause excessive steering effort.

6. Visually inspect the bellows boots for cracks, splits, leaks, and proper clamp installation. Replace any boot that indicates any of these conditions. If the boot clamps are loose or improperly installed, tighten or replace the clamps as necessary. Since the bellows boots protect the inner tie-rod ends and the rack from contamination, boot condition is extremely important. Boots should be inspected each time undercar service such as oil and filter change or chassis lubrication is performed.

7. Loosen the inner bellows boot clamps and move each boot toward the outer tie rod end until the inner tie-rod end is visible. Push outward and inward on each front tire and watch for movement in the inner tie-rod end. If any movement or looseness is present, replace the inner tie-rod end. An alternate method of checking the inner tie-rod ends is to squeeze the bellows boots and grasp the inner tie-rod end socket. Movement in the inner tie-rod end is then felt as the front wheel is moved inward and outward. Hard plastic bellows boots may be found on some applications. With this type of bellows boot, remove the ignition key from the switch to lock the steering column and

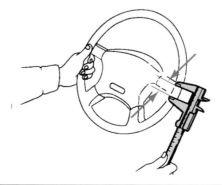

Figure 12-2 Checking steering wheel freeplay. (Courtesy of Nissan Motor Co., Ltd.)

push inward and outward on the front tire while observing any lateral movement in the tie rod. When lateral movement is observed, replace the inner tie-rod end.

> **⚠ WARNING:** Bent steering components must be replaced. Never straighten steering components because this action may weaken the metal and result in sudden component failure, serious personal injury, and vehicle damage.

8. Grasp each outer tie-rod end and check for vertical movement. While an assistant turns the steering wheel 1/4 turn in each direction, watch for looseness in the outer tie-rod ends. If any looseness or vertical movement is present, replace the tie-rod end. Check the outer tie-rod end seals for cracks and proper installation of the nuts and cotter pins. Cracked seals must be replaced. Inspect the tie rods for a bent condition. Bent tie rods or other steering components must be replaced. Do not attempt to straighten these components.

Manual or Power Rack and Pinion Steering Gear Removal and Replacement

The replacement procedure is similar for manual or power rack and pinion steering gears. This removal and replacement procedure varies depending on the vehicle. On some vehicles, the front crossmember or engine support cradle must be lowered to remove the rack and pinion steering gear. Always follow the vehicle manufacturer's recommended procedure in the service manual. The following is a typical rack and pinion steering gear removal and replacement procedure.

1. Place the front wheels in the straight-ahead position and remove the ignition key from the ignition switch to lock the steering wheel. Place the driver's seat belt through the steering wheel to prevent wheel rotation if the ignition switch is turned on (Figure 12-3). This action maintains the clock spring electrical connector or spiral cable in the centered position on air-bag-equipped vehicles.

2. Lift the front end with a floor jack, and place safety stands under the vehicle chassis. Lower the vehicle onto the safety stands. Remove the left and right fender apron seals (Figure 12-4).

3. Place punch marks on the lower universal joint and the steering gear pinion shaft so they may be reassembled in the same position (Figure 12-5). Loosen the upper universal joint bolt and remove the lower universal joint bolt, and disconnect this joint.

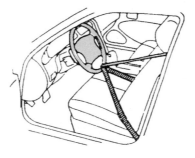

Figure 12-3 Driver's seat belt wrapped around the steering wheel to prevent wheel rotation. (Reprinted with permission)

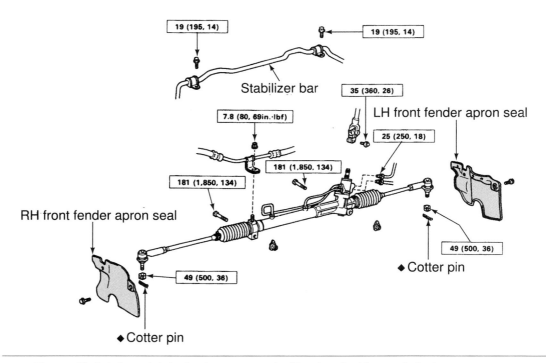

Figure 12-4 Left and right fender apron seals. (Reprinted with permission)

Tie rod end puller

4. Remove the cotter pins from the outer tie-rod ends. Loosen, but do not remove, the tie-rod end nuts. Use a tie-rod end puller to loosen the outer tie-rod ends in the steering arms (Figure 12-6). Remove the tie-rod end nuts and remove the tie-rod ends from the arms.

5. Use the proper wrenches to disconnect the high-pressure hose and return hose from the steering gear (Figure 12-7). This step is not required on a manual steering gear.

6. Remove the four stabilizer bar mounting bolts (Figure 12-8).

7. Remove steering gear mounting bolts (Figure 12-9).

8. Remove the steering gear assembly from the right side of the car (Figure 12-10).

9. Position the right and left tie rods the specified distance from the steering gear housing (Figure 12-11). Install the steering gear through the right fender apron.

10. Install the pinion shaft in the universal joint with the punch marks aligned. Tighten the upper and lower universal joint bolts to the specified torque.

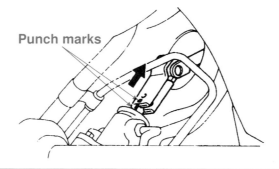

Figure 12-5 Punch marks on universal joint and pinion shaft. (Reprinted with permission)

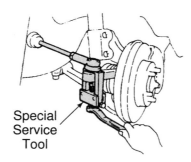

Figure 12-6 Removing outer tie-rod ends. (Reprinted with permission)

374

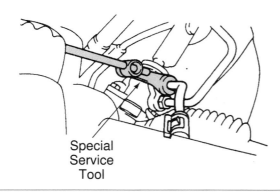

Special
Service
Tool

Figure 12-7 Removing high-pressure and return hoses from the steering gear. (Reprinted with permission)

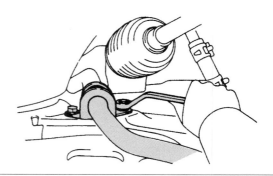

Figure 12-8 Removing stabilizer bar mounting bolts. (Reprinted with permission)

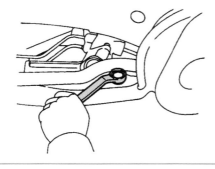

Figure 12-9 Removing steering gear mounting bolts. (Reprinted with permission)

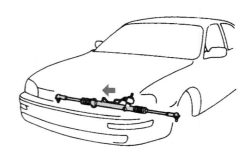

Figure 12-10 Removing steering gear from the car. (Reprinted with permission)

11. Install the steering gear mounting bolts and tighten these bolts to the specified torque.

12. Install the stabilizer bar mounting bolts and torque these bolts to specifications.

13. Install and tighten the high-pressure and return hoses to the specified torque. (This step is not required on a manual rack and pinion steering gear.)

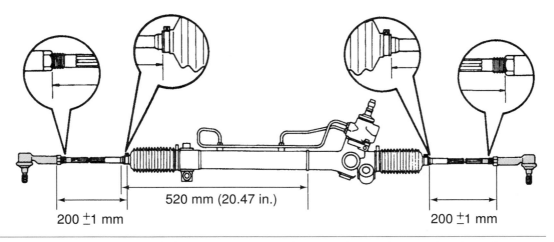

520 mm (20.47 in.)

200 ±1 mm   200 ±1 mm

Figure 12-11 Right and left tie rod position in relation to the steering gear housing prior to installation. (Reprinted with permission)

WARNING: Do not loosen the tie-rod end nuts to align cotter pin holes. This action causes improper torquing of these nuts.

14. Install the outer tie-rod ends in the steering knuckles, and tighten the nuts to the specified torque. Install the cotter pins in the nuts.

15. Check the front wheel toe and adjust as necessary. Tighten the outer tie-rod end jam nuts to the specified torque, and tighten the outer bellows boot clamps.

16. Install the left and right fender apron seals, and lower the vehicle with a floor jack.

17. Fill the power steering pump reservoir with the vehicle manufacturer's recommended power steering fluid and bleed the air from the power steering system. (Refer to Chapter 10 for these tasks. This step is not required on a manual rack and pinion steering gear.)

18. Road test the vehicle and check for proper steering gear operation and steering control.

Manual Rack and Pinion Steering Gear Diagnosis and Service

Manual Rack and Pinion Steering Gear Disassembly

Follow these steps for manual rack and pinion steering gear disassembly:

1. Clamp the center of the steering gear housing in a **soft-jaw vise.** Do not apply excessive force to the vise.

2. Insert a short piece of wire through each outer tie-rod end cotter pin hole and connect a pull scale to this wire. Pull upward on the scale to check the tie rod articulation effort (Figure 12-12). If this effort is not within specifications, replace the inner tie-rod end.

3. Mark the outer tie rod and jam nut position with masking tape wrapped around the tie-rod threads next to the jam nut, or place a dab of paint on the jam nut and tie-rod threads (Figure 12-13). Loosen the jam nuts and remove the outer tie-rod ends.

4. Remove the inner and outer bellows boot clamps and pull the bellows boots from the tie rods.

5. Hold the rack in a soft-jaw vise and straighten the lock plate where it is bent over the inner tie-rod end (Figure 12-14). Hold the rack with a wrench, and use the proper size wrench to remove the inner tie-rod from the rack (Figure 12-15). A jam nut is used in place of the lock ring on some inner tie-rod ends, and a roll pin retains some inner tie-rod ends to the rack. Various inner tie-rod socket removal procedures are required depending on the socket design.

6. Rotate the pinion shaft until the specified distance is obtained between the end of the rack and the steering gear housing. Mark the pinion shaft position in relation to the housing with a center punch. If the specified distance from the end of the rack to the housing is not available, rotate the pinion shaft fully clockwise. Rotate the pinion shaft fully counterclockwise and count the number of turns. Turn the pinion shaft clockwise one-half this number of turns. Measure the distance from each end of the rack to the housing and record this measurement. Use a center punch to mark the pinion shaft position in relation to the housing.

7. Loosen the adjuster plug lock nut and remove the adjuster plug and spring.

8. Remove the rack bearing through the adjuster plug opening.

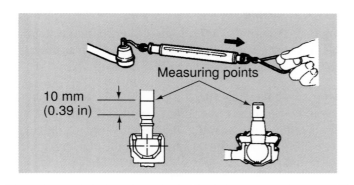

Figure 12-12 Measuring inner tie-rod end articulation effort. (Courtesy of Nissan Motor Co., Ltd.)

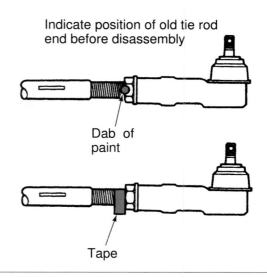

Figure 12-13 Marking outer tie-rod end and jam nut position. (Courtesy of Chrysler Corp.)

9. Clean the surface around the pinion shaft seal and pierce the pinion shaft seal at one of the two round areas on the seal surface. Pry this seal from the housing.

10. Remove the pinion shaft retaining ring with snapring pliers.

11. Place the end of the pinion shaft in a soft-jaw vise, and tap the gear housing with a soft hammer to remove the pinion shaft and bearing from the housing.

12. Remove the rack from the housing.

13. Use an approved solvent to clean all steering gear components except the inner tie-rod assemblies, and blow dry these parts with compressed air.

Inspection of Manual Rack and Steering Gear Components

Inner Tie-Rod Sockets. After the steering gear components are removed and cleaned, they must be visually inspected. If the inner tie-rod ends were loose during the on-car inspection or the inner tie-rod end articulation effort did not meet specifications, replace these ends.

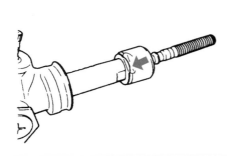

Figure 12-14 Straightening lock plate where it is bent over the inner tie-rod end. (Courtesy of Nissan Motor Co., Ltd.)

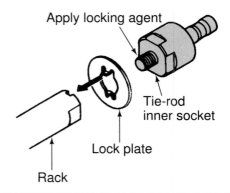

Figure 12-15 Removing inner tie-rod end from the rack. (Courtesy of Nissan Motor Co., Ltd.)

Pinion and Bearing Assembly. If the pinion bearing is loose on the pinion shaft, replace the **pinion and bearing assembly.** A pinion shaft with worn or chipped teeth must be replaced. Inspect the pilot bearing contact area on the pinion shaft. Wear, pitting, or scoring in this area indicates that a new pinion shaft is required. If the **pinion bearing** is rough, loose, or noisy, replace the pinion shaft and bearing assembly.

Pilot Bearing. Check the **pilot bearing** in the steering gear housing. If this bearing is worn or scored, replace the pilot bearing and the pinion shaft and bearing assembly. Use an appropriate bearing driver to remove and replace the pilot bearing in the steering gear housing.

Rack Bushing. If the **rack bushing** is worn, bushing replacement is necessary. Remove the bushing retaining ring prior to bushing removal. Position the puller fingers behind the bushing, and operate the slide hammer on the puller to remove the bushing. If a puller is not available, an appropriate bushing driver or socket may be used to drive the bushing out of the housing.

Mounting Bushings. If the mounting bushings are loose during the on-car inspection, replace the bushings. Always replace the bushings in pairs. If the bushings are in satisfactory condition, do not disturb them.

Manual Rack and Pinion Steering Gear Assembly

Proceed as follows to reassemble the manual rack and pinion steering gear:

1. Install the rack in the housing with the teeth parallel to the pinion shaft. Move the rack until the specified distance is obtained between the end of the rack and the housing. If this specification is not available, position the rack ends at the distance from the housing recorded during the disassembly procedure.
2. Use an **anhydrous calcium grease** to lubricate the pilot bearing and the pilot bearing contact area on the pinion shaft.
3. Install the pinion shaft with the punch marks aligned on the shaft and housing.
4. Install the pinion shaft retaining ring with the beveled edge facing upward.
5. Place a generous coating of anhydrous calcium grease on top of the pinion shaft bearing.
6. Use the proper seal driver to install the pinion shaft seal.
7. Apply a coating of **lithium-based grease** to the rack bearing and install this bearing.
8. Apply lithium-based grease to the ends of the preload spring and the adjuster plug threads, and install the preload spring and adjuster plug.
9. Turn the adjuster plug until it bottoms, then rotate the plug 45° to 60° counterclockwise.
10. Rotate the pinion shaft with a socket and a torque wrench and observe the turning torque on the torque wrench. If necessary, rotate the adjuster plug to obtain the specified turning torque. Tighten the adjuster plug lock nut to the specified torque.
11. Coat both ends of the rack with lithium-based grease and fill the rack teeth with the same lubricant. Rotate the pinion shaft fully in each direction several times. Apply more grease to the rack teeth after each complete pinion shaft rotation.
12. Rotate the inner tie rod-ends onto the rack until they bottom. Install the inner tie-rod pins or stake these ends as required. Always use a wooden block to support the opposite side of the rack and inner tie-rod end while staking. If jam nuts are located on the inner tie-rod ends, be sure they are tightened to the specified torque.
13. Place a large bellows boot clamp over each end of the gear housing. Install the bellows boots, and be sure the boots are seated in the housing and the tie-rod undercuts. Install and tighten the large inner boot clamps.
14. Install the outer bellows boot clamps on the tie rods, but do not install these clamps on the boots until the steering gear is installed and the toe is adjusted.

15. Install the jam nuts and the outer tie-rod ends. Align the marks placed on these components during the disassembly procedure. Leave the jam nuts loose until the steering gear is installed and the toe is adjusted.

16. Install the steering gear in the vehicle and check the front wheel toe. Tighten the outer bellows boot clamps and the outer tie-rod end jam nuts.

Diagnosis of Manual Steering and Suspension Systems

It is sometimes difficult to separate the diagnosis of manual steering gears, steering columns, and suspension systems because some noises and symptoms may be caused by problems in any of these components or systems. Therefore, we have attempted to provide a complete diagnosis that includes manual recirculating ball steering gears, manual rack and pinion steering gears, and some of the common steering column and suspension system complaints (Table 12-1).

Classroom Manual
Chapter 12, page 273

Table 12-1 DIAGNOSIS OF MANUAL STEERING GEARS, SUSPENSION SYSTEMS, AND STEERING COLUMNS

Condition	Possible Cause	Correction
Loose steering	1. Steering linkage connections loose	1. Tighten as required.
	2. Steering linkage ball stud rubber deteriorated	2. Replace as required.
	3. Pitman arm-to-steering-gear attaching nut loose	3. Tighten to specifications.
	4. Flex coupling-to-steering-gear attaching bolt loose	4. Tighten the bolt to specifications.
	5. Intermediate shaft-to-steering-column-shaft attaching bolt loose	5. Tighten the bolt to specifications.
	6. Flex coupling damaged or worn	6. Replace as required.
	7. Front and/or rear suspension components loose, damaged, or worn	7. Tighten or replace as required.
	8. Steering gear adjustment set too low	8. Check the steering gear adjustment and readjust as required.
Noise: Knock, clunk, rapping, squeaking noise when turning	1. Steering wheel-to-steering-column shroud interference	1. Adjust or replace as required.
	2. Lack of lubrication where speed control brush contacts steering wheel pad	2. Lubricate as required.
	3. Steering column mounting bolts loose	3. Tighten the bolt to specifications.
	4. Intermediate shaft-to-steering-column-shaft attaching bolt loose	4. Tighten the bolt to specifications.
	5. Flex coupling-to-steering-gear attaching bolt loose	5. Tighten the bolt to specifications.

Condition	Possible Cause	Correction
	6. Steering gear mounting bolts loose	6. Tighten the bolt to specifications.
	7. Tires rubbing or grounding out against body or chassis	7. Adjust/replace as required.
	8. Front suspension components loose, worn, or damaged	8. Tighten or replace as required.
	9. Steering linkage ball stud rubber deteriorated	9. Replace as required.
	10. Normal lash between sector gear teeth and ball nut gear teeth when the steering gear is in the off-center position permits gear tooth chuckle noise while turning on rough road surfaces	10. This is a normal condition and cannot be eliminated.
Vehicle pulls/drifts to one side *Note:* This condition cannot be caused by the manual steering gear.	1. Vehicle overloaded or unevenly loaded	1. Correct as required.
	2. Improper tire pressure	2. Adjust air pressure as required.
	3. Mismatched tires and wheels	3. Install correct tire and wheel combination.
	4. Unevenly worn tires	4. Replace as required (check for cause).
	5. Loose, worn, or damaged steering linkage	5. Tighten or replace as required.
	6. Steering linkage stud not centered within the socket	6. Repair or replace as required.
	7. Bent spindle or spindle arm	7. Repair or replace as required.
	8. Broken and/or sagging front and/or rear suspension springs	8. Replace as required.
	9. Loose, bent, or damaged suspension components	9. Tighten or replace as required.
	10. Bent rear axle housing	10. Replace as required.
	11. Excessive camber/caster split (excessive side-to-side variance in the caster/camber settings)	11. Adjust camber/caster split as required.
	12. Improper toe setting	12. Adjust as required.
	13. Front wheel bearings out of adjustment	13. Adjust as required.
	14. Investigate tire variance (conicity for radial tires and unequal circumference for bias-ply belted bias-ply tires)	14. Repair or replace as required.

Table 12-1 DIAGNOSIS OF MANUAL STEERING GEARS, SUSPENSION SYSTEMS, AND STEERING COLUMNS (continued)

Condition	Possible Cause	Correction
High efforts/poor returnability	1. Improper tire/wheel size	1. Install correct tire and wheel combinations.
	2. Underinflated tires	2. Adjust tire pressure as required.
	3. Misaligned steering column and/or distorted flex coupling	3. Align/replace as required.
	4. Steering wheel-to-column interference	4. Remove interference.
	5. Steering gear with insufficient grease	5. Check steering gear seals and replace, if required. Add lubricant as required.
	6. Improper steering column indexing and/or improper toe setting	6. Set front wheels in straight-ahead position and ensure the following: — Steering wheel hub and steering column shaft index marks must be aligned. — The index marks must be in 12 o'clock position. — Steering gear input shaft flat must be in 12 o'clock position. If the indexing is improper, reindex the steering column. If the steering gear input shaft flat is not in 12 o'clock position, reset the toe.
	7. Steering gear preload/mesh-load set too high	7. Check the steering gear pre-load/meshload and adjust as required.
	8. Steering linkage and rubber captured ball stud binding	8. Repair/replace as required.
	9. Spindle ball joint torque too high	9. Repair/replace as required.
Vehicle wanders (side to side) constant corrective steering is required at highway speeds.	1. Vehicle overloaded or unevenly loaded	1. Correct as required.
	2. Improper tire/wheel size	2. Install correct tire and wheel combination.
	3. Steering gear mounting bolts loose	3. Tighten the bolts to specifications.
	4. Flex coupling-to-steering-gear attaching bolt loose	4. Tighten the bolt to specifications.
	5. Flex coupling damaged or worn	5. Replace as required.

Condition	Possible Cause	Correction
	6. Improper steering column and/or improper toe setting	6. Set front wheels in straight-ahead position and ensure the following: — Steering wheel hub and steering column shaft index marks must be aligned. — The index marks must be in 12 o'clock position. — Steering gear input shaft flat must be in 12 o'clock position. If the indexing is improper, reindex the steering column. If the steering gear input shaft flat is not in 12 o'clock position, reset the toe.
	7. Steering linkage loose, worn, or damaged	7. Correct as required.
	8. Broken or sagging suspension springs	8. Replace as required.
	9. Steering gear adjustment set too low	9. Check the steering gear adjustment and adjust as required.
	10. Improper caster/camber	10. Adjust caster/camber.
	11. Front wheel bearing misadjusted	11. Adjust as required.
Uneven efforts	Steering gear input shaft not centered due to improper toe setting	Center gear and adjust toe as required.
Sticky steering: pointing, darting, sticky, or down-the-road steering feel is not smooth; seems to be sticky and cannot make minor corrections with ease	1. Steering column misaligned	1. Align as required.
	2. Flex coupling distorted	2. Replace as required.
	3. Steering gear grease leaking from input and/or output shaft seal	3. Replace the steering gear seals.
	4. Steering gear input shaft not centered due to incorrect toe settings	4. Center gear and adjust toe as required.
	5. Steering gear adjustment set too high	5. Check the steering gear adjustment and adjust as required to specifications.
	6. Excessive caster split (excessive side-to-side variance in the caster setting)	6. Adjust caster split as required.
	7. Spindle ball joint torque too high	7. Repair/replace as required.

Condition	Possible Cause	Correction
Rough turning: rough, lumpy, binding steering; rugged or catchy steering feel while turning in either direction	1. Steering column misaligned 2. Flex coupling distorted 3. Steering linkage components damaged due to impact 4. Excessive steering gear roughness	1. Align as required. 2. Replace as required. 3. Replace as required. 4. Check steering gear adjustment and readjust as required. If torque wrench indicates excessive roughness, repair the gear.
Poor center feel /poor groove feel: increase in steering efforts when the steering wheel is rotated off-center during straight-ahead highway driving. Vehicle with poor groove feel might require repeated small corrections to maintain a desired straight-ahead path.	1. Improper steering column indexing and/or improper toe setting 2. Steering gear adjustment set too high 3. Excessive caster split (side-to-side variance is too large)	1. Set front wheels in straight-ahead position and ensure the following: — Steering wheel hub and steering column shaft index marks must be aligned. — The index marks must be in 12 o'clock position. — Steering gear input shaft flat must be in 12 o'clock position. If the indexing is improper, reindex the steering column. If the steering gear input shaft flat is not in 12 o'clock position, reset the toe. 2. Check the steering gear adjustment and adjust as required. 3. Adjust as required.
Shimmy: a consistent side-to-side wobble in the steering wheel or road wheels usually felt at over 40 mph and amplified when driven over pot holes or bumps in the road surface. *Note:* This condition cannot be caused by the manual steering gear.	1. Wheels out of balance 2. Improper tire pressure 3. Excessive tire sidewall deflection 4. Irregular tire wear 5. Loose wheel lug nuts 6. Radius arm bushings deteriorated 7. Loose or leaky shock absorber 8. Broken or sagging suspension spring 9. Loose, worn, or damaged steering linkage 10. Steering gear mounting bolts loose	1. Check/correct as required. 2. Adjust as required. 3. Inspect, adjust air pressure, and replace as required. 4. Replace as required. 5. Tighten to specifications. 6. Replace as required. 7. Tighten or replace as required. 8. Replace as required. 9. Tighten/replace as required. 10. Tighten the bolts.

Condition	Possible Cause	Correction
	11. Front wheel bearings out of adjustment	11. Adjust as required.
	12. Out-of-round wheels	12. Replace as required.
	13. Out-of-round tires	13. Replace as required.
	14. Wheel and tire lateral runout not within specification	14. Check/replace as required.
	15. Worn spindle ball joints	15. Replace as required.
	16. Steering linkage ball stud rubber deteriorated	16. Replace as required.
Noise: rattle or chuck in steering gear	1. Insufficient or improper lubricant in steering gear	1. Add specified lubrication.
	2. Pitman arm loose on shaft or steering gear mounting bolt loose	2. Tighten to specified torque.
	3. Loose or worn steering shaft bearing	3. Replace steering shaft bearing.
	4. Excessive over-center lash or worm thrust bearing adjustment too loose	4. Adjust steering gear to specified preloads.
	Note: On turns, a slight rattle might occur due to the increased lash between ball nut and pitman shaft as gear moves off the center of high point position. This is normal—lash must not be reduced to eliminate the slight rattle.	
Poor returnability	1. Steering column misaligned	1. Align column.
	2. Insufficient or improper lubricant in steering gear or front suspension	2. Lubricate as specified.
	3. Steering gear adjusted too tightly	3. Adjust over-center and thrust bearing preload to specifications.
	4. Front wheel alignment incorrect (caster)	4. Adjust to specifications.
Excessive play or looseness in steering system	1. Front wheel bearings loosely adjusted	1. Adjust to obtain proper end-play.
	2. Worn upper ball joints	2. Check and replace ball joints if necessary.

Table 12-1 DIAGNOSIS OFMANUAL STEERING GEARS, SUSPENSION SYSTEMS, AND STEERING COLUMNS (continued)

Condition	Possible Cause	Correction
	3. Steering wheel loose on shaft, loose pitman arm, tie rods, steering arms, or steering linkage ball nuts	3. Tighten to specifications; replace if worn or damaged.
	4. Excessive over-center lash	4. Adjust over-center preload to specifications.
	5. Worm thrust bearings loosely adjusted	5. Adjust worm thrust bearing preload to specifications.
Hard steering: excessive effort required at steering wheel	1. Low or uneven tire pressure	1. Inflate to specified pressures.
	2. Insufficient or improper lubricant in steering gear or front suspension	2. Lubricate as specified. Relubricate at specified intervals.
	3. Steering shaft flexible coupling misaligned	3. Align column and coupling.
	4. Steering gear adjusted too tightly	4. Adjust over-center and thrust bearing preload to specifications.
	5. Front wheel alignment incorrect (manual gear)	5. Adjust to specifications.

Power Rack and Pinion Steering Gear Diagnosis and Service

Oil Leak Diagnosis

CAUTION: If the engine has been running for a length of time, power steering gears, pumps, lines, and fluid may be very hot. Wear eye protection and protective gloves when servicing these components.

If power steering fluid leaks from the cylinder end of the power steering gear, the outer rack seal is leaking (Figure 12-16).

The inner rack seal is defective if oil leaks from the pinion end of the housing when the rack reaches the left internal stop (Figure 12-17). An oil leak at one rack seal may result in oil leaks from both boots because the oil may travel through the breather tube between the boots.

If an oil leak occurs at the pinion end of the housing and this leak is not influenced when the steering wheel is turned, the pinion seal is defective (Figure 12-18).

If an oil leak occurs in the pinion coupling area, the input shaft seal is leaking (Figure 12-19). This seal and the pinion seal will require replacement because the pinion seal must be replaced if the pinion is removed.

When the rack is removed, the inner and outer rack seals and the pinion seal must be replaced (Figure 12-20). If oil leaks occur at fittings, these fittings must be torqued to manufacturer's

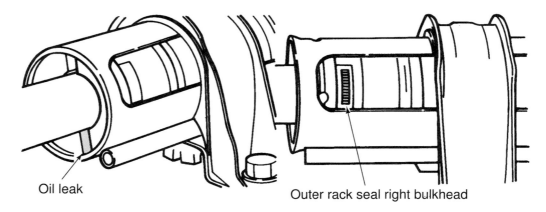

Oil leak

Outer rack seal right bulkhead

If a leak is detected at the housing cylinder end,
the origin of the leak is the outer rack seal.

Figure 12-16 Oil leak diagnosis at outer rack seal. (Courtesy of Chrysler Corp.)

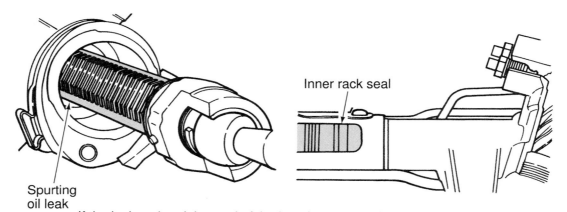

Inner rack seal

Spurting
oil leak

If the leak at the pinion end of the housing spurts when the rack reaches the
left internal stop, the inner rack seal is at fault.

Figure 12-17 Inner rack seal leak diagnosis. (Courtesy of Chrysler Corp.)

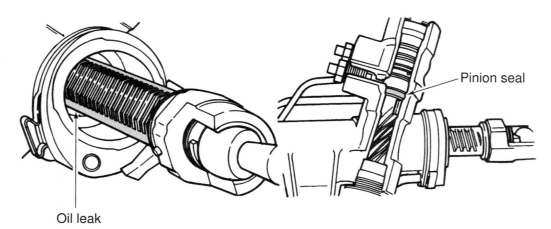

Pinion seal

Oil leak

If you detect a leak at the pinion end of the housing and it is not influenced
by the direction of the turn, the origin of the leak is the pinion seal.

Figure 12-18 Pinion seal leak diagnosis. (Courtesy of Chrysler Corp.)

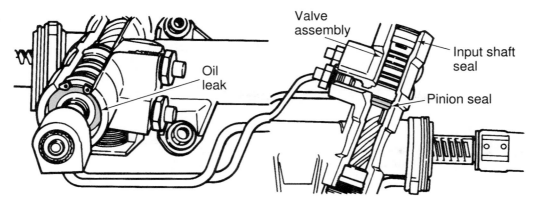

If you discover a leak at the pinion coupling area, you'll have to replace both the input shaft seal and the lower pinion seal.

Figure 12-19 Oil leak diagnosis in pinion coupling area. (Courtesy of Chrysler Corp.)

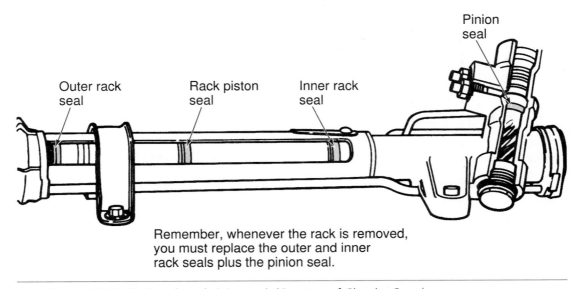

Remember, whenever the rack is removed, you must replace the outer and inner rack seals plus the pinion seal.

Figure 12-20 Rack seals and pinion seal. (Courtesy of Chrysler Corp.)

specifications. If the leak is still present, the line and fitting should be replaced. Leaks in the lines or hoses require line or hose replacement.

Turning Imbalance Diagnosis

 SERVICE TIP: To provide equal turning effort in both directions, the rack must be centered with the front wheels straight ahead.

The same amount of effort should be required to turn the steering wheel in either direction. A pressure gauge connected to the high-pressure hose should indicate the same pressure when the steering wheel is turned in each direction. **Steering effort imbalance** or lower power assist in each direction may be caused by defective rack seals (Figure 12-21). Steering effort imbalance or low power assist in both directions may also be caused by defective rotary valve rings and seals or restricted hoses and lines (Figure 12-22).

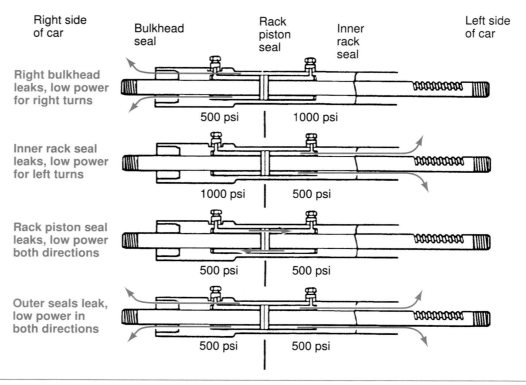

Right side of car — Bulkhead seal — Rack piston seal — Inner rack seal — Left side of car

Right bulkhead leaks, low power for right turns
500 psi   1000 psi

Inner rack seal leaks, low power for left turns
1000 psi   500 psi

Rack piston seal leaks, low power both directions
500 psi   500 psi

Outer seals leak, low power in both directions
500 psi   500 psi

Figure 12-21 Effect of defective rack seals on steering effort imbalance and low power assist. (Courtesy of Chrysler Corp.)

Power Rack and Pinion Steering Gear Disassembly

The power rack and pinion steering gear disassembly procedure varies depending on the vehicle and steering gear manufacturer. Always follow the vehicle manufacturer's recommended procedure in the service manual. The following is a typical power rack and pinion steering gear disassembly procedure:

1. Install a holding tool on the steering gear housing, and clamp this tool in a vise (Figure 12-23).

2. Use the proper wrench to disconnect the left and right turn tubes (Figure 12-24). Remove the O-rings from the turn tubes.

3. Mark the outer tie-rod ends, jam nuts, and tie rods (Figure 12-25). Loosen the jam nuts and remove the tie-rod ends and jam nuts.

4. Loosen the inner and outer clamps on both bellows boots and remove these boots (Figure 12-26).

5. Clamp the inner tie-rod socket lightly in a soft-jaw vise, and use a hammer and chisel to unstake the **claw washers** on the inner tie-rod sockets (Figure 12-27).

6. Hold the rack with an adjustable wrench and use the proper tool to loosen the inner tie rod ends (Figure 12-28). Remove the inner tie-rod ends and the claw washers.

✓ **SERVICE TIP:** In some power rack and pinion steering gears, the inner tie-rod ends are staked after they are installed on the rack (Refer to Photo Sequence 10 later in this chapter). In other power rack and pinion steering gears, a pin retains the inner tie-rod ends to the rack, and this pin must be removed before the inner tie-rod ends are removed. When the inner tie-rod ends are replaced, a new retaining pin must be installed.

7. Rotate the pinion shaft until the specified distance is obtained between the end of the rack and the steering gear housing. Mark the pinion shaft position in relation to the housing with a center punch. If the specified distance from the end of the rack to the housing is not available, rotate the pinion shaft fully clockwise. Rotate the pinion shaft

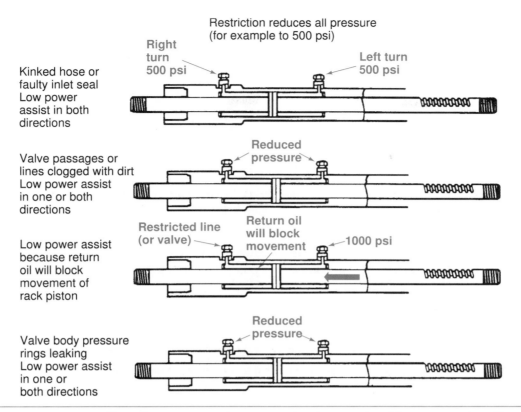

Restriction reduces all pressure
(for example to 500 psi)

Right turn 500 psi   Left turn 500 psi

Kinked hose or faulty inlet seal
Low power assist in both directions

Reduced pressure

Valve passages or lines clogged with dirt
Low power assist in one or both directions

Restricted line (or valve)   Return oil will block movement   1000 psi

Low power assist because return oil will block movement of rack piston

Reduced pressure

Valve body pressure rings leaking
Low power assist in one or both directions

Figure 12-22 Effect of worn rotary valve rings and seals or restricted lines on steering effort. (Courtesy of Chrysler Corp.)

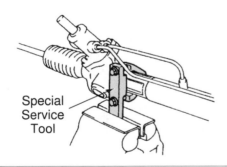

Special Service Tool

Figure 12-23 Steering gear holding tool. (Reprinted with permission)

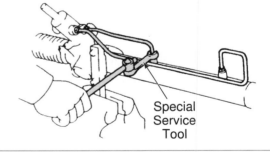

Special Service Tool

Figure 12-24 Removing left and right turn tubes. (Reprinted with permission)

fully counterclockwise and count the number of turns. Turn the pinion shaft clockwise one-half this

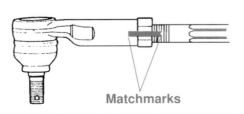

Matchmarks

Figure 12-25 Marking outer tie-rod ends, jam nuts, and tie rods prior to outer tie-rod end removal. (Reprinted with permission)

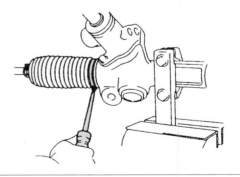

Figure 12-26 Removing bellows boot clamps. (Reprinted with permission)

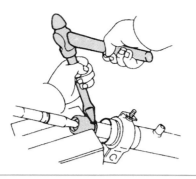

Figure 12-27 Removing claw washers from inner tie-rod sockets. (Reprinted with permission)

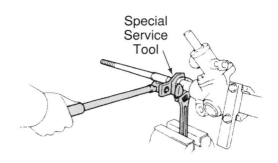

Figure 12-28 Loosening inner tie-rod ends. (Reprinted with permission)

number of turns. Measure the distance from each end of the rack to the housing and record this measurement. Use a center punch to mark the pinion shaft position in relation to the housing.

8. Use the proper tool to loosen the adjuster plug lock nut (Figure 12-29).

9. Remove the adjuster plug with the proper tool (Figure 12-30).

10. Remove the spring, rack guide, and the seat from the adjuster plug opening (Figure 12-31).

11. Use the appropriate socket and breaker bar to remove the pinion housing cap (Figure 12-32).

12. Hold the top of the pinion shaft with a socket and breaker bar, and loosen the **self-locking nut** on the end of the pinion shaft (Figure 12-33). Remove this self-locking nut.

13. Remove the dust cover from the control valve housing and mark this housing in relation to the steering gear housing (Figure 12-34).

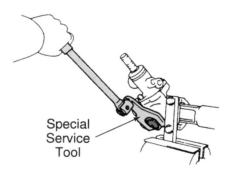

Figure 12-29 Loosening adjuster plug lock nut. (Reprinted with permission)

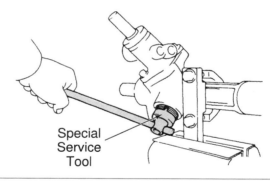

Figure 12-30 Removing adjuster plug. (Reprinted with permission)

Figure 12-31 Removing spring, rack guide, and seat. (Reprinted with permission)

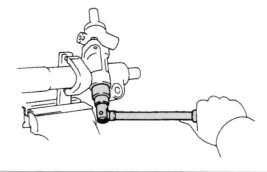

Figure 12-32 Removing pinion housing cap. (Reprinted with permission)

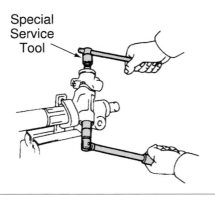

Figure 12-33 Loosening pinion shaft self-locking nut. (Reprinted with permission)

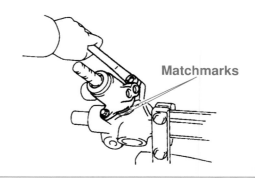

Figure 12-34 Marking control valve housing in relation to steering gear housing. (Reprinted with permission)

14. Remove the control valve housing retaining bolts, and remove this housing from the steering gear housing. Use a soft hammer to tap the pinion shaft from the control valve housing (Figure 12-35).

15. Place matching marks on the number 2 bracket and steering gear housing (Figure 12-36). Use a slotted screwdriver to pry the clasp open on the housing bracket (Figure 12-37). Remove the bushing and bracket from the housing and disconnect the bushing from the bracket.

16. Use an end stopper removal tool to rotate the cylinder end stopper clockwise until the wire end comes through the hole in the gear housing (Figure 12-38). Rotate the end stopper counterclockwise with the same tool and remove the wire through the gear housing opening.

Special Tools

End stopper removal tool

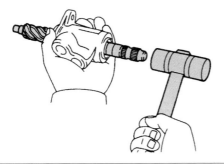

Figure 12-35 Removing pinion shaft from control valve housing. (Reprinted with permission)

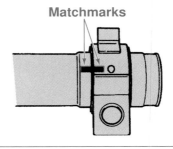

Figure 12-36 Matching marks placed on the bracket and steering gear housing. (Reprinted with permission)

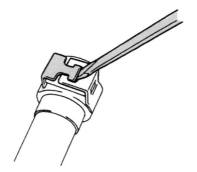

Figure 12-37 Prying the clasp open on the steering gear mounting bracket. (Reprinted with permission)

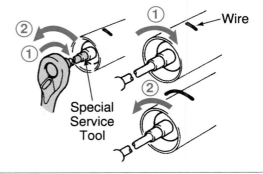

Figure 12-38 Removing wire from the end stopper. (Reprinted with permission)

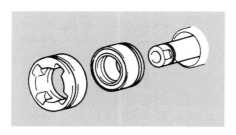

Figure 12-39 Removing rack and rack bushing. (Reprinted with permission)

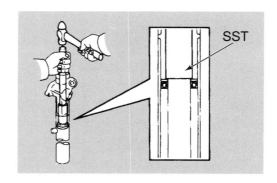

Figure 12-40 Removing cylinder side oil seal and spacer. (Reprinted with permission)

17. Use a brass bar and a hammer to tap the rack and rack bushing from the housing (Figure 12-39). Remove the O-ring from the housing.

18. Drive the cylinder side oil seal and spacer out of the housing with the appropriate driving tool and a brass bar (Figure 12-40).

Power Rack and Pinion Steering Gear Parts Inspection

Clean all steering gear components except the inner tie-rod ends and any sealed bearings with an approved solvent. Blow dry components with compressed air. The power rack and pinion steering gear inspection and parts replacement procedure may vary depending on the vehicle and steering gear manufacturer. Always follow the procedure in the vehicle manufacturer's service manual. The following is a typical inspection and parts replacement procedure.

1. Inspect the rack for tooth wear or damage. If these conditions are present, replace the rack or complete steering gear. Place the rack on V-blocks and slowly rotate the rack with a dial indicator stem positioned near the center of the rack (Figure 12-41). If the rack runout exceeds specifications, replace the rack or complete steering gear.

2. If the outer circumference of the rack piston is scored or worn, replace the rack. Visually check the rack piston contact area in the housing for scoring, pitting, and wear. When any of these conditions are present, replace the housing or complete steering gear.

3. Use the proper driving tool to drive the oil seal and upper bearing out of the control valve housing (Figure 12-42). If the control valve housing is scored, pitted, or worn in the ring contact area, replace the housing.

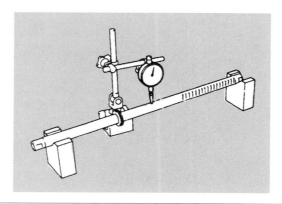

Figure 12-41 Measuring rack runout. (Reprinted with permission)

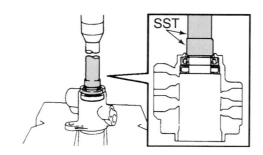

Figure 12-42 Removing upper bearing and seal in the control valve housing. (Reprinted with permission)

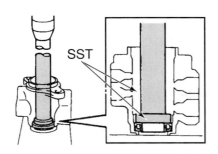

Figure 12-43 Installing control valve housing seal. (Reprinted with permission)

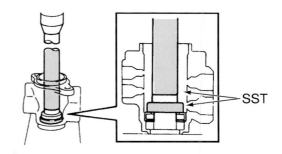

Figure 12-44 Installing upper bearing in the control valve housing. (Reprinted with permission)

4. Drive the new seal into the control valve housing with the proper driving tool (Figure 12-43). With the new seal in place, use the proper driving tool to install the new upper bearing (Figure 12-44).

5. Check the lower bearing and center bearing. If the bearings are scored or worn, replacement is necessary. Visually check the pinion shaft in the bearing contact areas. If the pinion shaft is worn, scored, or pitted in these areas, replace the shaft. When the pinion shaft teeth are worn, chipped, or pitted, replace the pinion shaft. If lower and center bearing replacement is required, drive the lower bearing out of the housing with a hammer and brass bar (Figure 12-45). Use the appropriate bearing puller to remove the center bearing (Figure 12-46).

6. Use the proper driving tool to press the new center bearing into the housing (Figure 12-47).

7. Press the new lower bearing into the steering gear housing with the proper driving tool (Figure 12-48).

8. Pull the rack bushing oil seal from the housing with a seal puller (Figure 12-49). Coat the new seal with power steering fluid and use the correct driving tool to install the seal (Figure 12-50).

> ⚠️ **WARNING:** Be careful not to score or mark the rack piston while removing and replacing the piston ring or O-ring. A scored rack piston must be replaced.

9. Pry the Teflon ring and O-ring from the rack piston with a small slotted screwdriver (Figure 12-51). Be careful not to mark the rack piston.

10. Coat the new Teflon ring with power steering fluid and install this ring on the **expanding tool** (Figure 12-52). Some steering gear manufacturers recommend placing the Teflon rings in boiling water for 10 minutes to expand the rings prior to ring installation.

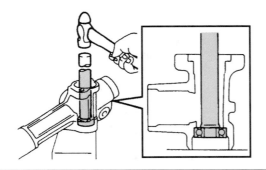

Figure 12-45 Removing lower control valve bearing. (Reprinted with permission)

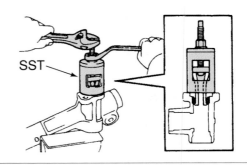

Figure 12-46 Removing center control valve bearing. (Reprinted with permission)

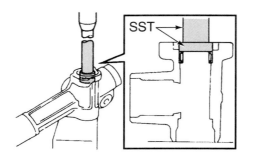

Figure 12-47 Installing center bearing. (Reprinted with permission)

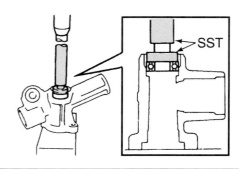

Figure 12-48 Installing lower control valve bearing. (Reprinted with permission)

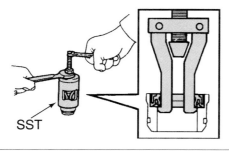

Figure 12-49 Removing rack bushing seal. (Reprinted with permission)

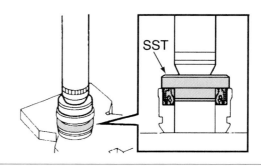

Figure 12-50 Installing rack bushing seal. (Reprinted with permission)

11. Remove the Teflon ring from the expanding tool and install it on the rack piston. Wrap your hand around the Teflon ring and squeeze it down into the rack piston groove (Figure 12-53).

⚠ **WARNING:** Do not score or mark the control valve while removing the Teflon rings. Scored control valves must be replaced.

12. Pry the Teflon rings from the control valve with a small slotted screwdriver (Figure 12-54). Be careful not to mark the control valve.

Special Tools

Teflon ring compressing tool

13. Coat one of the control valve Teflon rings with power steering fluid and install it on the expanding tool. Remove the Teflon ring from the expanding tool and install it in a control valve groove. Wrap your hand around the Teflon ring and squeeze it into the groove (Figure 12-55). Follow the same procedure for each Teflon ring. When all the

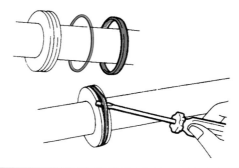

Figure 12-51 Removing Teflon ring and O-ring from the rack piston. (Reprinted with permission)

Figure 12-52 Teflon ring expanding tool. (Reprinted with permission)

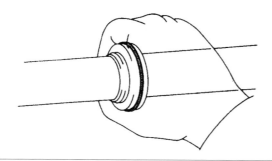

Figure 12-53 Installing the Teflon ring in rack piston groove. (Reprinted with permission)

Figure 12-54 Removing Teflon rings from the control valve. (Reprinted with permission)

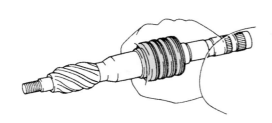

Figure 12-55 Installing rings on control valve. (Reprinted with permission)

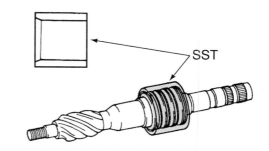

Figure 12-56 Compressing tool installed on Teflon rings. (Reprinted with permission)

Teflon rings are installed on the control valve, coat them with power steering fluid and install the **compressing tool** over the rings (Figure 12-56).

Power Rack and Pinion Steering Gear Assembly

1. Coat the new cylinder side oil seal with power steering fluid. Place one layer of vinyl tape on the seal driving tool extension, and install the seal and spacer on the tool. Press the cylinder side oil seal and spacer into the housing with the driving tool (Figure 12-57).

2. Check the rack surface and rack teeth for small metal burrs. Remove any burrs with crocus cloth and wipe the rack with a clean shop towel. Place the rack installation tool over the rack and coat the tool with power steering fluid. Slide the rack and installation tool into the housing (Figure 12-58). The installation tool prevents seal damage. Remove the installation tool.

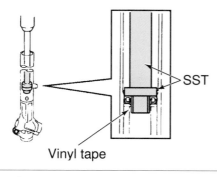

Figure 12-57 Installing cylinder side oil seal and spacer. (Reprinted with permission)

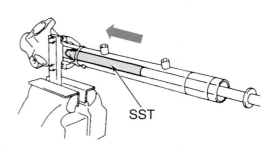

Figure 12-58 Rack installation. (Reprinted with permission)

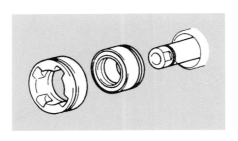

Figure 12-59 Installing rack bushing and cylinder end stopper. (Reprinted with permission)

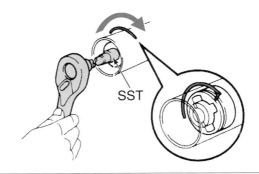

Figure 12-60 Installing retaining wire. (Reprinted with permission)

3. Coat a new rack bushing O-ring with power steering fluid and install the O-ring in the bushing. Push the rack bushing and **cylinder end stopper** into the housing until the wire installation hole appears (Figure 12-59).

4. Insert the end of a new wire into the wire installation hole. Use the proper tool and a ratchet to rotate the cylinder end stopper clockwise until the outer wire end disappears in the wire installation hole (Figure 12-60).

5. Connect a leak test hose between the left and right turn ports on the steering gear housing. Attach a vacuum pump hose to the leak test hose and operate the hand pump until 15.75 in Hg (53.3 kPa) is obtained on the vacuum gauge (Figure 12-61). After 30 seconds, observe the vacuum gauge. If the gauge reading decreased, the rack housing seal or rack bushing seal must be leaking and the rack should be removed to recheck these seals.

6. If the bushing and bracket were removed from the rack housing, install a new bushing in the bracket. Be sure the projection on the bushing aligns with the bracket hole (Figure 12-62). Install the bracket and bushing on the housing and align the marks placed on the housing and bracket during disassembly (Figure 12-63).

7. Place the housing bracket in a vise and tighten the vise to fasten the bracket clasp (Figure 12-64).

8. Wind a layer of vinyl tape over the serrations on the upper end of the pinion shaft. Coat the Teflon rings with power steering fluid, and push the control valve and pinion shaft assembly into the housing (Figure 12-65). Coat a new oil seal with power steer-

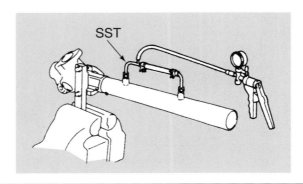

Figure 12-61 Leak testing rack cylinders. (Reprinted with permission)

Figure 12-62 Installing bushing in housing bracket. (Reprinted with permission)

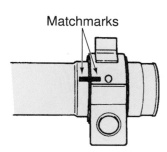

Figure 12-63 Installing bracket and bushing on steering gear housing. (Reprinted with permission)

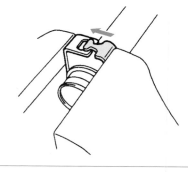

Figure 12-64 Fastening the bracket clasp. (Reprinted with permission)

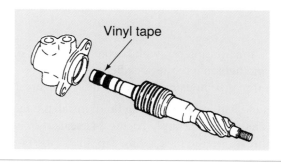

Figure 12-65 Installing control valve and pinion shaft assembly in control valve housing. (Reprinted with permission)

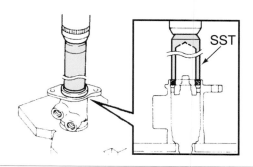

Figure 12-66 Installing oil seal in lower side of control valve housing. (Reprinted with permission)

ing fluid and press the seal into the lower side of the control valve housing with the appropriate seal driving tool (Figure 12-66).

9. Position the rack ends at the specified distance from the housing or at the distance from the housing measured during the disassembly procedure.

10. Install a new control valve housing gasket on the steering gear housing. Align the marks on the control valve and steering gear housings, and align the marks on the pinion shaft and housing. Install the control valve housing and tighten the two housing bolts to the specified torque (Figure 12-67). Install the housing dust cover.

11. Install a new self-locking nut on the lower end of the pinion shaft. Use the proper tool to hold the upper end of the pinion shaft and tighten the self-locking nut to the specified torque with a socket and foot-pound torque wrench (Figure 12-68).

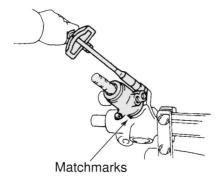

Figure 12-67 Installing control valve housing. (Reprinted with permission)

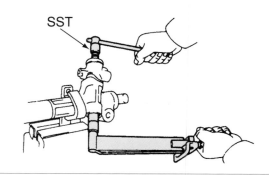

Figure 12-68 Tightening self-locking nut on pinion shaft. (Reprinted with permission)

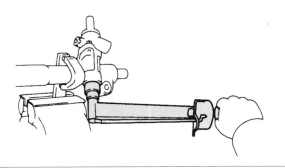

Figure 12-69 Installing steering gear housing cap. (Reprinted with permission)

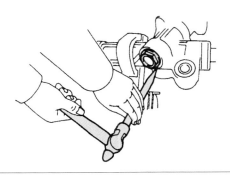

Figure 12-70 Staking steering gear housing. (Reprinted with permission)

12. Coat the first two or three threads on the housing cap with Loctite 242 sealant or its equivalent. Install the housing cap and tighten it to the specified torque (Figure 12-69). Use a center punch to stake the steering gear housing in two locations (Figure 12-70).

13. Coat the rack guide seat and rack guide with power steering fluid. Install the rack guide seat, rack guide, and spring (Figure 12-71). Apply Loctite 242 sealant or its equivalent to the first two or three threads on the rack spring cap. Install the rack spring cap and tighten it to the specified torque (Figure 12-72).

14. Use the proper tool to rotate the rack spring cap 12° counterclockwise. Turn the pinion shaft fully in each direction, and repeat this action. Loosen the rack spring cap until there is no tension on the rack guide spring. Place the proper turning tool and a foot-pound torque wrench on top of the pinion shaft. Tighten the rack spring cap while rotating the pinion shaft back and forth (Figure 12-73). Continue tightening the rack spring cap until the specified turning torque is indicated on the torque wrench.

15. Apply Loctite 242 sealant or its equivalent to the first two or three threads on the lock nut. Install the lock nut on the rack spring cap. Use the proper tool to hold the rack spring cap and tighten the lock nut to the specified torque (Figure 12-74).

16. Install a new claw washer on the inner tie-rod end and install the inner tie-rod end on the rack. Hold the rack with an adjustable wrench and use a torque wrench with the correct tool to tighten the inner tie-rod end to the specified torque (Figure 12-75). Clamp the inner tie-rod end in a soft-jaw vise and stake the claw washer with a brass bar and a hammer (Figure 12-76). Repeat this procedure on both inner tie-rod ends.

Figure 12-71 Installing rack guide seat, rack guide, and spring. (Reprinted with permission)

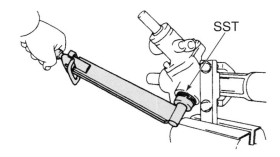

Figure 12-72 Installing rack spring cap. (Reprinted with permission)

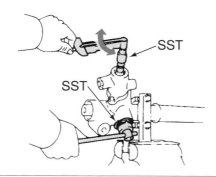

Figure 12-73 Adjusting pinion shaft turning torque. (Reprinted with permission)

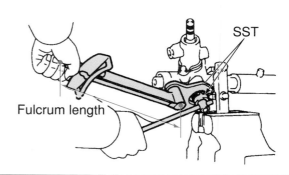

Figure 12-74 Tightening lock nut on rack spring cap. (Reprinted with permission)

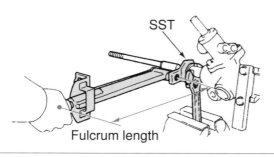

Figure 12-75 Tightening inner tie-rod end on the rack. (Reprinted with permission)

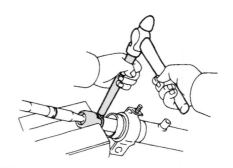

Figure 12-76 Staking claw washer on the inner tie-rod end. (Reprinted with permission)

17. Check the vent hole in each end of the rack and be sure they are not clogged with grease or other material (Figure 12-77). These vent holes allow air flow from one boot to the other during a turn.

18. Install the bellows boots, clamps, and clips. When the inner boot clamps are tightened, there must be a minimum of 0.20 in (0.08 mm) between the ends of the clamps (Figure 12-78).

19. Install the outer tie-rod ends and jam nuts. Align the marks placed on the tie-rod ends, jam nuts, and tie rods during disassembly (Figure 12-79). Leave the jam nuts loose until the steering gear is installed and the front wheel toe is adjusted.

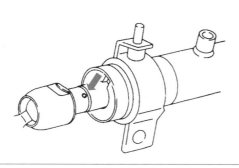

Figure 12-77 Rack vent holes must be open. (Reprinted with permission)

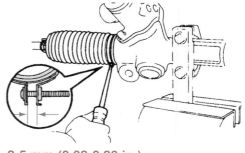

2-5 mm (0.08-0.20 in.)

Figure 12-78 Minimum clearance between boot clamp ends. (Reprinted with permission)

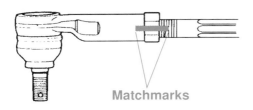

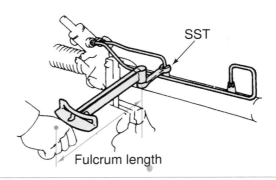

SST

Fulcrum length

Figure 12-79 Aligning marks on outer tie-rod ends, jam nuts, and tie rods. (Reprinted with permission)

Figure 12-80 Tightening turn tube fittings. (Reprinted with permission)

20. Coat the new turn tube O-rings with power steering fluid and install the O-rings on the right and left turn tubes. Tighten the left and right turn tube fittings to the specified torque with the proper wrench attached to a torque wrench (Figure 12-80).

21. Follow the steering gear installation procedure described earlier in this chapter to install the steering gear in the vehicle. Tighten all fasteners to the specified torque. (See Chapter 10 for power steering pump filling and bleeding procedure.) Check the front wheel toe and tighten the outer tie-rod end jam nuts to the specified torque. Tighten the outer bellows boot clamps. Road test the vehicle and check for proper steering operation and control.

Photo Sequence 10 shows a typical procedure for removing and replacing an inner tie-rod end on a power rack and pinion steering gear.

Photo Sequence 10
Typical Procedure for Removing and Replacing Inner Tie-Rod End, Power Rack and Pinion Steering Gear

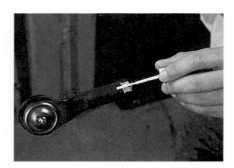

P10-1 Place an index mark on the outer tie-rod end, jam nut, and tie rod.

P10-2 Loosen the jam nut, and remove the outer tie-rod end and jam nut.

P10-3 Remove the inner and outer boot clamps.

Photo Sequence 10
Typical Procedure for Removing and Replacing Inner Tie-Rod End, Power Rack and Pinion Steering Gear (continued)

P10-4 Remove the bellows boot from the tie rod.

P10-5 Hold the rack with the proper size wrench and loosen the inner tie-rod end with the proper wrench.

P10-6 Remove the inner tie rod from the rack.

P10-7 Be sure the shock damper ring is in place on the rack.

P10-8 Install the inner tie rod on the rack and tighten the tie rod to the specified torque while holding the rack with the proper size wrench.

P10-9 Support the inner tie rod on a vise and stake both sides of the inner tie-rod joint with a hammer and punch.

P10-10 Use a feeler gauge to measure the clearance between the rack and the inner tie-rod joint housing stake.

P10-11 Install the bellows boot and new clamps.

P10-12 Install the jam nut and outer tie-rod end with the index marks aligned.

Diagnosis of Power Steering and Suspension Systems

Some power steering problems may be caused by defects in the steering gear, pump, or column. For example, heavy steering effort may be caused by a defective power steering pump, damaged rack piston and ring, or misalignment of the steering column. Therefore, we have combined the diagnoses of these components (Table 12-2).

 SERVICE TIP: A scored cylinder surface in the steering gear housing may cause temporary excessive steering effort after the car sits overnight.

Table 12-2 POWER STEERING SYSTEM AND STEERING COLUMN DIAGNOSIS

Condition	Possible Cause	Correction
Wander: vehicle wander is a condition where the vehicle wanders side to side on the roadway when it is driven straight ahead, while the steering wheel is held in a firm position. Evaluation should be conducted on a level road (little road crown).	1. Loose/worn tie-rod ends or ball socket	1. Replace tie-rod end or tie-rod assembly.
	2. Inner ball housing loose or worn	2. Replace tie-rod assemblies.
	3. Gear assembly loose crossmember	3. Tighten the two mounting nuts to specification.
	4. Excessive yoke clearance	4. Adjust yoke clearance.
	5. Loose suspension struts or ball joints	5. Tighten or replace as required.
	6. Column intermediate shaft connecting bolts loose	6. Tighten bolts to specification at gear and column.
	7. Column intermediate shaft joints loose or worn	7. Replace intermediate shaft assembly.
	8. Improper wheel alignment	8. Set alignment to specification.
	9. Tire size and pressure	9. Check tire sizes and adjust tire pressure.
	10. Vehicle unevenly loaded or overloaded	10. Adjust load.
	11. Steering gear mounting insulators and/or attachment bolts loose or damaged	11. Replace insulators and/or attachment nuts and bolts. Tighten to specification.
	12. Steering gear adjustments	12. Refer to steering gear section of shop manual.
	13. Front end misaligned	13. Check and align to specifications.
	14. Worn front end parts or wheel bearings	14. Inspect and replace affected parts.
	15. Unbalanced or badly worn steering gear control valve	15. Inspect and replace affected parts.

Condition	Possible Cause	Correction
Pulls to one side: condition in which the vehicle tends to pull to one side when driven on a level surface.	1. Improper tire pressure	1. Adjust tire pressure.
	2. Improper tire size or different type	2. Replace as required.
	3. Vehicle unevenly or excessively loaded	3. Adjust load.
	4. Improper wheel alignment	4. Adjust as required.
	5. Damaged front suspension components	5. Refer to front suspension section of shop manual.
	6. Damaged rear suspension components	6. Refer to rear suspension section of shop manual.
	7. Steering gear valve effort out of balance	7. Place transmission in Neutral while driving and turn engine off (coasting). — If vehicle does not pull, replace the steering gear valve assembly. Refer to steering gear section of shop manual. — If vehicle does drift : • Cross switch front tire wheel assemblies. • If vehicle pulls to opposite side, cross switch tire/wheel assemblies that were on the rear to the same on the front. • If vehicle pull direction is not changed, check front suspension components and wheel alignment.
	8. Front and rear brakes operation	8. Adjust if necessary.
	9. Bent rear axle housing, or damaged or sagging springs in the front and/or rear suspension	9. Replace if necessary.
	10. Loose/worn shock absorber or suspension attaching fasteners in rear suspension.	10. Tighten all attaching fasteners to specification.
Feedback: rattle, chuckle, knocking noises in the steering gear. Feedback is a condition where roughness is felt in the steering wheel by the driver when the vehicle is driven over rough pavement.	1. Column U-joints loose	1. Replace if bad.
	2. Loose tie-rod ends	2. Replace tie-rod end assemblies.
	3. Loose or worn tie-rod ball	3. Replace tie-rod assemblies.
	4. Gear assembly loose on cross member	4. Tighten the two mounting nuts to specification.

Condition	Possible Cause	Correction
	5. Loose pinion bearing cap	5. Adjust cap to specification.
	6. Loose yoke plug or lock nut	6. Adjust yoke preload to specification, and tighten lock nut to specification.
	7. Loose pinion bearing lock nut	7. Tighten lock nut to specification.
	8. Piston disengaged or loose on rack	8. Replace rack assembly.
	9. Oversize pinion shaft bushing	9. Replace gear housing.
	10. Steering gear yoke worn	10. Replace yoke assembly.
	11. Column intermediate shaft connecting bolts loose	11. Tighten bolts to specification.
	12. Loose suspension bushings/ fasteners or ball joints	12. Adjust or replace as necessary.
	13. Column intermediate shaft joints loose/worn	13. Replace intermediate shaft assembly.
	14. Steering gear mounting insulators and/or attaching bolts loose or damaged	14. Replace insulators and/or tighten attachment nuts to specification.
Poor steering wheel returnability: sticky feel. Poor returnability is when the steering fails to return to center following a turn without manual effort from the driver. In addition, when the driver returns the steering to center, it might have a sticky or catchy feel	1. Misaligned steering column or column flange rubbing steering wheel and/or flange	1. Align column. Refer to steering column section of shop manual.
	2. Binding intermediate shaft joints.	2. If binding, replace intermediate shaft.
	3. Yoke plug too tight	3. Adjust yoke preload to specification.
	4. Tight inner tie-rod ball joints	4. Replace tie-rod as required.
	5. Tight tie-rod end ball studs	5. Replace tie-rod end assemblies.
	6. Undersized pinion shaft bushing in the housing	6. Replace gear housing assemblies.
	7. Binding in valve assembly	7. Replace input shaft valve assembly.
	8. Bent or damaged rack	8. Replace rack assembly.
	9. Bent or damaged crossmember	9. Replace as necessary.
	10. Column bearing binding	10. Replace bearing.
	11. Tight suspension struts or lower control arm ball joints	11. Adjust or replace as required.
	12. Improper wheel alignment	12. Set to specification. Adjust toe as required.
	13. Contamination in system	13. Flush power steering system.
	14. Deformed engine mounts	14. Replace as required.

Condition	Possible Cause	Correction
	15. Improper tire pressure	15. Adjust tire pressures.
	16. Improper tire size or different type	16. Replace as required.
	17. Column intermediate shaft universal joints binding	17. Replace intermediate shaft assembly.
	18. Boot tears and/or binding or damage to tie-rod ends or ball joints	18. Replace as necessary.
	19. Damaged/worn front suspension components	19. Inspect struts and lower control arm ball joints.
	20. Steering gear adjustments	20. Refer to steering gear section of shop manual.
	21. Front end misaligned	21. Check and align to specifications.
	22. Lack of lubrication in steering linkage or ball joints	22. Inspect, lubricate, or replace affected parts.
	23. Sticky or plugged steering gear spool valve	23. Clean spool valve or replace.
	24. Internal leakage in steering gear	24. Inspect seals, replace as required, or overhaul steering gear.
Steering wheel: excessive steering wheel return or loose steering	1. Air in the system	1. Add oil to reservoir and bleed the system. Check all connections.
	2. Excessive over-center lash	2. Adjust to specifications.
	3. Loose steering gear mounting	3. Retorque mounting bolts to specifications.
	4. Steering linkage worn or damaged	4. Inspect and replace affected parts.
	5. Steering coupler loose or damaged	5. Inspect, tighten, or replace as required.
	6. Loose thrust bearing preload adjustment	6. Adjust to specifications.
	7. Loose or damaged front wheel bearings	7. Inspect, adjust, or replace as required.
Steering wheel: surges or jerks with engine running, especially at slow speeds	1. Pump belt loose	1. Tighten or replace belt.
	2. Low oil level in reservoir	2. Check level; fill as needed.
	3. Engine idle too low; air in the system	3. Raise idle as required, check all hose connections and pump for leaks.
	4. Insufficient pump pressure	4. Check pump pressure. Replace affected parts or replace pump.
	5. Flow control valve sticks	5. Check valve for dirt or burrs. Replace as needed.
	6. Steering linkage hitting obstruction	6. Locate obstruction and repair.
	7. Sticky flow control valve	7. Inspect for varnish or damage, replace if necessary.

Table 12-2 POWER STEERING SYSTEM AND STEERING COLUMN DIAGNOSIS (continued)

Condition	Possible Cause	Correction
Steering wheel: momentary increase in steering wheel effort when turned rapidly	1. Fluid level low in reservoir 2. Pump belt slipping 3. High internal leakage	1. Check level; fill as needed. 2. Tighten or replace belt. 3. Check pump pressure. Replace affected parts or replace pump.
Steering wheel: heavy or hard steering effort, poor or loss of assist. A heavy effort and poor assist condition is recognized by the driver while turning corners and especially while parking. A road test will verify this condition.	1. Leakage or loss of fluid 2. Low pump fluid 3. Pump flow/pressure not to specifications 4. Valve plastic ring cut or twisted 5. Damaged or worn plastic piston ring 6. Loose or missing rubber backup piston O-ring 7. Loose rack piston 8. Gear assembly oil passages restricted 9. Bent or damaged rack assembly 10. Valve assembly internal leakage 11. Improper drive belt tension 12. Hose or cooler external leakage 13. Improper engine idle speed 14. Pulley loose or warped 15. Hose or cooler line restrictions 16. System contamination 17. Plugged spool valve screen 18. Steering gear adjustments 19. Low tire pressure 20. Improper front suspension alignment 21. Low oil level in reservoir	1. Refer to external diagnosis for service. 2. Fill as necessary. 3. Refer to pump section of shop manual. 4. Replace ring. 5. Replace ring. 6. Replace or install O-ring. 7. Replace rack assembly. 8. Clear or service as required. 9. Replace rack assembly. 10. Replace valve assembly. 11. Readjust belt tension. 12. Replace as necessary. 13. Readjust idle. 14. Replace pulley. 15. Clear or replace as required. 16. Inspect system for foreign objects, kinked hose, etc. — Flush system. — Refer to power steering pump section of shop manual. 17. Prior to rebuilding a pump, examine the spool valve screen for contamination. Replace all valves that have plugged or contaminated valve screens. 18. Refer to steering gear section of shop manual. 19. Inflate to correct pressure. 20. Realign to specifications. 21. Add P/S oil to proper level.

Condition	Possible Cause	Correction
	22. Pump belt loose or glazed	22. Tighten or replace belt.
	23. Lack of lubricant in suspension or ball joints; over-center adjustment too tight	23. Lubricate or replace parts as needed. Readjust to specifications.
	24. Steering gear coupler to column misaligned	24. Align steering column.
Fluid leakage	1. Overfilled system	1. Correct fluid level as required.
	2. Component leakage	2. Locate suspected component, and refer to appropriate section for service.
Power steering pump leaks	1. Excessive fluid fill	1. Adjust fluid to proper level.
	2. Dipstick missing, loose, or damaged or missing O-ring	2. Service or replace if required.
	3. Broken/cracked reservoir	3. Replace reservoir.
	4. Loose or damaged hose fittings	4. Service or replace.
	5. Leakage at shaft seal — Shaft seal not pressed in flush with housing surface — Seal damage — Rotor shaft damage, such as helical grooving, or the OD has an axial scratch — Worn shaft bushing — Plugged drainback hole	5. Service shaft seal. — Use an appropriate seal installation tool to correct installation; if not possible or still leaking, replace seal. — Replace seal. — Replace shaft and seal. — Replace plate and bushing assembly. — Disassemble pump and clean drainback hole. If hole is not drilled through, replace plate and bushing assembly.
	6. Damaged or missing reservoir O-ring	6. Replace O-ring.
	7. Damaged or missing outlet fitting O-rings	7. Replace O-rings.
	8. Loose outlet fitting	8. Tighten outlet fitting.
	9. Excessive pump assembly bracket vibration	9. Correct bracket alignment and tighten bracket bolts to specification.
	10. Plate and bushing reservoir seal groove porosity/damage, metal chips, or foreign material on seal or in seal groove	10. Clean groove; replace plate and bushing assembly if damaged.
	11. Outlet fitting damage	11. Replace outlet fitting.

Condition	Possible Cause	Correction
Noise: in system: chirp or squeal when steering wheel is cycled lock-to-lock; in pump: whine or moan	1. Loose or worn pump belt	1. Adjust to specification, or replace as required.
	2. Fluid aeration	2. Purge system of air. Refer to purging procedures in shop manual.
	3. Low fluid	3. Check fluid level. Correct as required.
	4. Pump brackets loose or misaligned	4. Check bracket(s), bolt torques, and bracket alignment. Correct as required.
	5. Power steering hose(s) grounded	5. Check for component grounding, and correct as required.
	6. Fluid level low in reservoir	6. Check level and fill as needed.
	7. Air in the system	7. Check hose connections and bleed the system.
	8. Pump shaft bushing or bearing is scored or worn	8. Inspect pump shaft, bushing, or bearing. Replace affected parts.
Noise: rattle in steering column	1. Loose bolts/attaching brackets	1. Tighten to specification.
	2. Looseness of ball bearings or insufficient lube	2. Lube or replace bearings.
	3. Steering shaft insulators cracked or dry	3. Replace or lube insulators as required.
	4. Flex coupling compressed or extended	4. Reposition shaft assembly to flatten flex coupling.
	5. Pressure hose coming in contact with other parts of the car	5. Reposition power steering pressure hose.
	6. Loose pump or steering gear mounting	6. Retorque to specifications.
	7. Loose steering linkage	7. Inspect steering linkage for excessive play; replace affected parts as necessary.
	8. High point on steering gear improperly adjusted	8. Readjust to proper setting.
	9. Improperly installed pump vanes	9. Remove burrs and varnish from rotor and vanes. Replace as necessary.
	10. Pump vanes sticking in rotor slots	10. Remove burrs and varnish from rotor and vanes. Replace as necessary.

Condition	Possible Cause	Correction
Noise: squeak or cracks in steering column	1. Dry bushings	1. Lube shaft seal and shift tube seal.
	2. Loose or mispositioned shrouds	2. Tighten or reposition shrouds as required.
	3. Steering wheel rubbing against shrouds	3. Replace shroud(s) or wheel or reposition shrouds as required.
	4. Dry shift lever grommets	4. Lube grommets.
	5. Insufficient lube on speed control slip ring	5. Lube slip ring.
	6. Upper or lower bearing sleeve out of position	6. Reposition bearing sleeve.
Gear squawk	Cut damper O-ring on spool valve	Replace the O-ring.
Growling	1. Restriction in the power steering system	1. Locate restriction and correct. Replace parts as necessary.
	2. Extreme wear on cam ring	2. Replace affected parts.
	3. Scored pump pressure plate, thrust plate, or rotor	3. Replace affected parts.
Click	A noise caused by the pump slippers being too long, broken slipper springs, excessive wear, nicked rotors, and damaged cam contour	Check for excessive wear or slipper, rotor, or cam damage. Replace cam rotor assembly if necessary.
Power steering pump: noise, moan, or whine	1. Fluid aeration	1. Purge the power steering system to reduce aeration noise. Refer to general steering service section purging procedure of shop manual.
	2. Low fluid	2. Check fluid level.
	3. Hose grounded	3. Check for hose being grounded.
	4. Column grounded	4. Check column alignment.
	5. Valve cover O-ring missing/damaged	5. Check valve cover O-ring damage. Replace if necessary.
	6. Valve cover baffle missing/damaged	6. Check valve cover for baffle missing/damage. Replace if necessary.
	7. Interference of components in pumping elements	7. Check rotor assembly and pressure plates for indication of interference. Replace if necessary.
	8. Loose or poor bracket alignment	8. Check bracket alignment and bolt torque. Correct if required.

Condition	Possible Cause	Correction
Swish	1. Noise created by the flow of excessive fluid into the bypass port of the pump valve housing (with temperature below 130°). The shearing effect of the cooler (heavier) oil is not detrimental to pump operation.	1. A normal condition. Noise will diminish with fluid temperature increase.
	2. Defective pump control valve	2. Inspect valve for burrs or nicks. Replace as necessary.
Gear hiss: some hissing noise is normal in all power-steering systems. Under certain conditions, such as tight parking maneuvers or turning the wheel from stop-to-stop, a hissing sound is most noticeable. There is no direct relationship between the hissing noise and performance of the steering system.	1. Column intermediate shaft alignment	1. Adjust column and/or intermediate shaft. Refer to steering column section of shop manual.
	2. Grounded or loose boot at dash panel	2. Align boot or tighten fasteners as required. Refer to steering column section of shop manual.
	3. Input shaft and control valve assembly	3. Replace the pump control valve only if the hissing noise is extremely objectionable. Check the coupler on the steering column for any metal-to-metal contact that may transmit the sound to the passenger compartment.

Diagnosis of Electronic Power Steering

When the ignition switch is turned on, the electronic power steering (EPS) light in the instrument panel should be illuminated (Figure 12-81). This EPS light action proves the bulb and related circuit are functioning normally. When the engine starts, the EPS light is turned off by the EPS control unit if there are no electrical defects in the EPS system. The EPS light should remain off while the engine is running.

If an electrical defect occurs in the EPS system, the EPS control unit turns on the EPS light to alert the driver that a defect exists in the EPS system. Under this condition, the EPS control unit shuts down the EPS system, and electric assist is no longer available. Manual steering with increased steering effort is provided in this mode. When the EPS control unit senses an electrical defect in the EPS system, a diagnostic trouble code (DTC) is stored in the control unit memory. The EPS control unit memory is erased if battery voltage is disconnected from the control unit. The 7.5 ampere clock fuse may be disconnected for 10 seconds to disconnect battery voltage from the EPS control unit and erase the DTCs from the control unit memory.

If the EPS light is on with the engine running, shut the ignition switch off and disconnect the clock fuse for 10 seconds, then road test the car to determine if the EPS light comes back on. If the EPS light comes on again with the engine running, proceed with the diagnosis. Locate the two-wire EPS service check connector under the glove box. With the ignition switch off, connect a special jumper wire to the service check connector (Figure 12-82). This connects the two wires together in

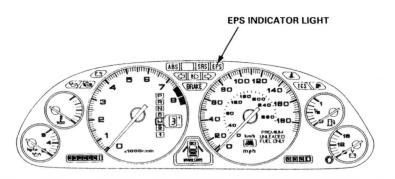

EPS INDICATOR LIGHT

Figure 12-81 Electronic power steering (EPS) warning light. (Courtesy of American Honda Motor Company, Inc.)

the service check connector. When the ignition switch is turned on, the EPS light begins to flash DTCs. Each DTC is indicated by a long light flash or flashes followed by a short flash or flashes. For example, DTC 11 is one long flash followed by a short flash, and DTC 12 is one long flash followed by two short flashes (Figure 12-83). If more than one DTC is stored in the EPS control unit memory, these DTCs are flashed in numerical order. The DTC sequence continues until the ignition switch is turned off. Be sure to remove the special jumper wire after the ignition switch is turned off.

> **SERVICE TIP:** If the special jumper wire is left in the EPS service connector and the engine is started, the malfunction indicator light (MIL) for the engine computer system is illuminated.

> **SERVICE TIP:** If DTC 33 is set in the EPS control unit memory, the EPS system remains functional, but the EPS light is illuminated.

All the possible DTCs in the EPS system are provided in Figure 12-84. A DTC indicates a defect in a certain area. For example, DTC 11 indicates high or low voltage from the torque sensor. The defect may be in the sensor itself, or the problem could be in the connecting wires from the

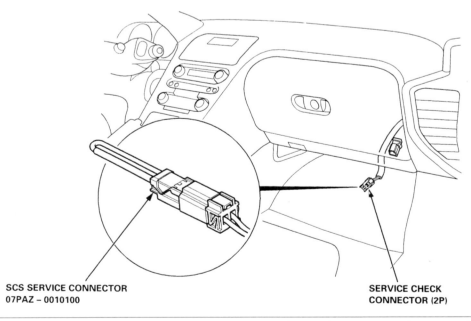

SCS SERVICE CONNECTOR
07PAZ – 0010100

SERVICE CHECK
CONNECTOR (2P)

Figure 12-82 EPS system service check connector and special jumper wire. (Courtesy of American Honda Motor Company, Inc.)

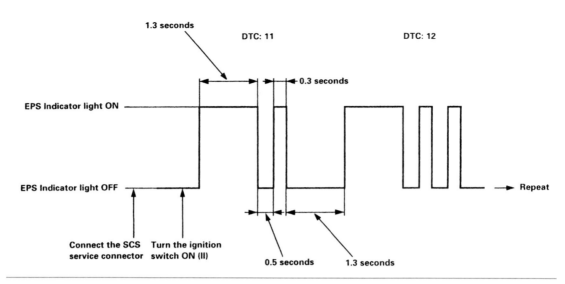

Figure 12-83 DTC 11 is one long flash followed by a short flash, and DTC 12 is one long flash followed by two short flashes. (Courtesy of American Honda Motor Company, Inc.)

DIAGNOSTIC TROUBLE CODE (LDTC)	EPS INDICATOR LIGHT	DESCRIPTION / SYMPTOM	DIAGNOSTIC PERIOD			AFTER DETECTING FOR SYSTEM	RESET	
			INDIVIDUAL DIAGNOSIS	INDIVIDUAL DIAGNOSIS	REGULAR DIAGNOSIS		INDIVIDUAL DIAGNOSIS	INDIVIDUAL DIAGNOSIS
—	O	EPS indicator light does not come on when ignition switch is turned ON (II)				—		17-27
—	O	EPS indicator light does not go off after engine is started				System OFF		17-29
3	O	A problem with the current sensor offset	O			↑	O	17-36
4	O	A problem with the current sensor offset	O	O		↑	O	
5	O	A problem with the current sensor fixed		O		↑	O	
6	O	A problem with the current sensor fixed		O		↑	O	
11	O	A problem with the high voltage or low voltage of the torque sensor (TRQ1 and TRQ2)		O		↑	O	17-31
12	O	A problem with the voltage for torque sensor (TRQ3)		O		↑	O	
13	O	A problem with the average of voltage on TRQ1 and TRQ2		O		↑	O	
14	O	A problem with the 2.5 V reference voltage (VREF)		O		↑	O	
21	O	A problem with the circuit for input motor voltage in the EPS control unit	O	O	O	↑	O	17-36
22	O	A problem with the lower current			O	↑	O	
23	O	A problem with the circuit for check function in the EPS control unit	O	O		↑	O	
24	O	The fail safe relay or the power relay is stuck ON	O			↑	O	
25	O	The lower FET is stuck ON	O			↑	O	
26	O	The upper FET is stuck ON	O			↑	O	
31	O	A problem with the voltage for IG1	O	O		↑	O	—
33	O	A problem with the average of VSS1 and VSS2			O	↑	O	17-38
34	O	A problem with the CPU in the EPS conrol unit	O	O	O	↑	O	Replace the EPS control unit

- Initial diagnosis: Performed right after the engine starts until the EPS indicator light goes off.
- Regular diagnisis: Continuously performed (under some conditions) after the EPS indicator light goes off until the engine stops.
- Individual part/system diagnosis: Diagnoses a specific part/system under its operating conditions.
- CPU: Central Processing Unit.

Figure 12-84 DTC list for EPS system. (Courtesy of American Honda Motor Company, Inc.)

sensor to the control unit. Tests usually have to be performed with a voltmeter or ohmmeter to locate the exact cause fo the defect. Detailed diagnostic charts are provided in the car manufacturer's service manual.

CUSTOMER CARE: Diagnosing is an extremely important part of a technician's job on today's high-tech vehicles. Always take time to diagnose a customer's vehicle accurately. Fast, inaccurate diagnosis of automotive problems leads to unnecessary, expensive repairs, and unhappy customers who may take their business to another shop. Accurate diagnosis may take more time, but in the long term it will improve customer relations, and bring customers back to the shop.

Classroom Manual
Chapter 12, page 291

Guidelines for Servicing Manual and Power Rack and Pinion Steering Gears

1. During a manual or power rack and pinion steering gear on-car inspection, components to be checked include the bellows boots, inner and outer tie-rod ends, tie rods, mounting bushings, flexible coupling or universal joint, and tires.

2. Since bellows boots protect the inner tie-rod ends from contamination, boot condition should be checked each time undercar service such as oil and filter change or chassis lubrication is performed.

3. Bellows boots that are cracked, split, leaking, or deteriorated must be replaced.

4. Do not straighten bent steering components such as tie rods.

5. If there is any movement or looseness in the inner tie-rod ends, these ends must be replaced.

6. Outer tie-rod ends with any looseness or cracked seals must be replaced.

7. Prior to steering gear removal, the steering wheel should be centered and the ignition key removed from the ignition switch. This action locks the steering column and maintains the clock spring or spiral cable in the centered position on an air-bag-equipped vehicle. The driver's seat belt may be wrapped through the steering wheel to maintain the steering wheel centered position if the ignition switch is turned on.

8. The flexible coupling or lower universal joint and the steering gear pinion shaft should be punch marked prior to steering gear removal.

9. Prior to outer tie-rod end removal, a dab of paint may be used to mark the jam nut position on the tie rods. An alternate method of marking this position is to wrap masking tape around the tie rod directly behind the jam nut.

10. Always hold the rack while loosening the inner tie-rod ends.

11. Prior to steering gear disassembly, the ends of the rack should be positioned at the specified distance from the steering gear housing and the pinion shaft position should be marked in relation to the housing. If this rack specification is not available, center the steering gear and measure the distance from the ends of the rack to the housing. After recording this distance, mark the pinion shaft in relation to the housing.

12. The rack adjuster plug must be adjusted until the correct turning torque is obtained on the pinion shaft.

13. The outer bellows boot clamps and the outer tie-rod end jam nuts should not be tightened until the steering gear is installed in the vehicle and the front wheel toe checked.

A customer complains about excessive steering effort on a 1998 Olds Aurora. The technician road tested the car and found no evidence of hard steering. Further questioning of the customer revealed that the problem only occurred when the engine was first started in the morning, after the car had been parked all night. The service writer informed the customer that the car would have to be parked overnight in the shop, then diagnosed the following morning. The customer complied with this request.

Prior to starting the engine the next morning, the technician connected a power steering pressure test gauge in series with the high-pressure hose. When the engine was started, the technician discovered the steering effort was very high, and the steering felt much like manual steering while the steering wheel was turned. However, the power steering pump pressure on the pressure gauge was higher than specified. Once the steering wheel was turned in one direction, the steering effort quickly decreased to normal. Since the power steering pump pressure was higher than specified, the technician concluded the pump and drive belt must be in satisfactory condition. The technician also reasoned that other causes of excessive steering effort, such as binding in the steering column or flexible coupling, would be constant. Therefore, the technician decided that the problem must be in the steering gear.

After receiving the customer's approval to remove and inspect the steering gear, the gear was removed and disassembled. The technician found the rack cylinder was severely ridged and scored in the center area, and the Teflon ring on the rack piston was badly worn. The customer was advised that the steering gear required replacement. A replacement steering gear was installed, and the system was flushed and refilled with power steering fluid. When the engine was started, the steering effort and the power steering pump pressure were normal.

Terms to Know

Anhydrous calcium grease	Expanding tool	Self-locking nut
Bellows boots	Lithium-based grease	Steering effort imbalance
Claw washer	Pilot bearing	Steering wheel freeplay
Compressing tool	Pinion and bearing assembly	
Cylinder end stopper	Rack bushing	

ASE Style Review Questions

1. While discussing steering wheel freeplay:
   Technician A says excessive steering wheel freeplay may be caused by a worn outer tie-rod end.
   Technician B says excessive steering wheel freeplay may be caused by loose steering gear mounting bushings.
   Who is correct?
   A. A only
   B. B only
   C. Both A and B
   D. Neither A nor B

2. While discussing steering effort:
   Technician A says a tight rack bearing adjustment may cause excessive steering effort.
   Technician B says a loose inner tie-rod end may cause excessive steering effort.
   Who is correct?
   A. A only
   B. B only
   C. Both A and B
   D. Neither A nor B

3. While discussing steering gear removal and replacement on air-bag-equipped vehicles:

   Technician A says the steering column should be locked in the centered position to prevent air bag deployment.

   Technician B says the rack should be centered in the housing prior to steering gear installation.

   Who is correct?

   A. A only **C.** Both A and B

   B. B only **D.** Neither A nor B

4. While discussing power rack and pinion steering gear adjustment:

   Technician A says the self-locking nut on the pinion shaft should be rotated to adjust the pinion shaft bearing preload.

   Technician B says the rack spring cap should be rotated to obtain the specified turning torque on the pinion shaft.

   Who is correct?

   A. A only **C.** Both A and B

   B. B only **D.** Neither A nor B

5. While discussing manual rack and pinion steering gear lubrication:

   Technician A says pilot bearing should be lubricated with anhydrous calcium grease.

   Technician B says the rack and rack teeth should be coated with lithium-based grease.

   Who is correct?

   A. A only **C.** Both A and B

   B. B only **D.** Neither A nor B

6. All these statements about diagnosing and servicing power rack and pinion steering gears are true EXCEPT:

   A. The inner rack seal may be defective if oil leaks from the pinion end of the housing when the rack contacts the left inner stop.

   B. A leaking rack piston seal causes increased steering effort in one direction.

   C. Leaking Teflon rings on the rotary valve may increase steering effort in both directions.

   D. If the pinion shaft is removed, the pinion seal and the input shaft seal must be replaced.

7. When diagnosing and servicing power rack and pinion steering gears:

   A. the rack should be not held when the outer tie-rod ends are loosened on the rack.

   B. in some steering gears claw washers lock the tie rods in position on the rack.

   C. a straightedge must be used to check the rack for a bent condition.

   D. bent tie rods may be straightened in a hydraulic press and then reinstalled.

8. When diagnosing and servicing power rack and pinion steering gears:

   A. Teflon rings on the rack piston may be expanded by heating them with a propane torch.

   B. a bent rack may cause excessive steering wheel freeplay.

   C. a compressing tool may be used to compress the Teflon rings on the rotary valve prior to installation.

   D. a scored cylinder surface in the steering housingmay cause excessive steering wheel kick-back.

9. When diagnosing electronic power steering (EPS) systems:

   A. the EPS system enters the diagnostic mode when the ignition switch is cycled three times in a 5-second interval.

   B. if the EPS warning light is on with the engine shut off and the ignition switch on, there is a defect in the EPS system.

   C. the EPS warning light flashes diagnostic trouble codes (DTCs) in numerical order.

   D. the EPS control unit memory only has the capability to store one DTC.

10. When diagnosing electronic power steering (EPS) systems:

    A. the DTCs may be erased from the EPS control nit by disconnecting the stop light fuse for 10 seconds.

    B. a DTC indicates the exact cause of an electrical defect in the EPS system.

    C. if a DTC is stored in the EPS control unit memory, the EPS system is usually inoperative and manual steering is available.

    D. in the EPS diagnostic mode if the EPS warning light provides two long flashes followed by two short flashes, code 2 is indicated.

1. While discussing power rack and pinion steering gears, *Technician A* says inner rack seal leaks can cause power rack and pinion steering to turn easier in one direction than another.
   Technician B says a hissing noise when the steering is turned to the limit on either side is normal.
   Who is correct?
   A. *Technician A*
   B. *Technician B*
   C. Both A and B
   D. Neither A nor B

2. Wandering, or poor straight ahead tracking may be caused by all of the following EXCEPT:
   A. Worn or loose rack mounting brushings
   B. Rack piston seal leaks*
   C. Steering gear off center
   D. Loose or worn tie rod ends

3. A customer says his front-wheel-drive car with power rack and pinion steering is hard to steer for several minutes the first time the car is driven during the day.
   Technician A says the cause of the problem could be a plugged spool valve spring.
   Technician B says the cause of the problem could be a scored steering gear cylinder.
   Who is correct?
   A. *Technician A*
   B. *Technician B**
   C. Both A and B
   D. Neither A nor B

4. A front-wheel-drive car requires very heavy steering effort in the parking lot, but seems not too bad on the highway. Tire pressure is fine and, after checking the fluid and finding the level is OK, you should:
   A. check engine idle speed.
   B. look for steering system leaks.
   C. check tie-rod ends.
   D. inspect the rack for damage.

5. *Technician A* says an oil leak at one rack seal may result in oil leaking from both boots.
   Technician B says the inner rack seal is defective if oil leaks from the pinion end of the housing when the rack reaches the left internal stop.
   Who is correct?
   A. A only
   B. B only
   C. Both A and B
   D. Neither A nor B

Table 12-1 NATEF and ASE TASK

Diagnose rack and pinion steering gear noises, vibration, looseness, and hard steering problems; determine needed reapairs.

Problem Area	Symptoms	Possible Causes	Classroom Manual	Shop Manual
NOISES	Knocks, clunks, rattles	1. Loose steering gear mounting bolts	273	379
		2. Flexible coupling loose or worn	273	379
		3. Universal joint loose or worn	273	379
		4. Loose tie-rod ends	273	380
		5. Lack of lubrication on horn bushing contact	273	380
STEERING CONTROL	Looseness, wander	1. Worn inner or outer tie-rod ends	275	384
		2. Worn steering gear mounting bushings	275	384
		3. Improper steering gear adjustment	276	385
		4. Worn flexible coupling or universal joint	276	385
STEERING EFFORT	Hard steering	1. Tight steering gear adjustment	276	381
		2. Lack of lubrication in steering gear	277	381

Table 12-2 NATEF and ASE TASK

Remove and replace rack and pinion steering gear (includes vehicles equipped with air bags and/or other steering wheel mounted controls and components).

Problem Area	Symptoms	Possible Causes	Classroom Manual	Shop Manual
STEERING CONTROL	Lack of steering control, wander	1. Worn inner tie-rod ends	279	373
		2. Worn rack or pinion teeth	280	373
		3. Improper steering gear adjustment	281	373
	Binding, uneven turning effort	Worn rack or pinion teeth	282	373

Table 12-3 NATEF and ASE TASK

Adjust rack and pinion steering gear.

Problem Area	Symptoms	Possible Causes	Classroom Manual	Shop Manual
STEERING CONTROL	Looseness	Loose steering gear adjustment	274	398
	Hard steering	Tight steering gear adjustment	274	398

Table 12-4 NATEF and ASE TASK

Inspect and replace rack and pinion steering gear inner tie-rod ends (sockets) and bellows boots.

Problem Area	Symptoms	Possible Causes	Classroom Manual	Shop Manual
STEERING CONTROL	Looseness	Worn inner tie-rod ends	274	398
PREMATURE INNER TIE-ROD END FAILURE	Looseness	Cracked or split bellows boots	274	398

Table 12-5 NATEF and ASE TASK

Inspect and replace rack and pinion steering gear mounting bushings and brackets.

Problem Area	Symptoms	Possible Causes	Classroom Manual	Shop Manual
STEERING CONTROL	Looseness	Worn steering gear mounting bushings	275	396

Table 12-6 NATEF and ASE TASK

Diagnose power rack and pinion steering gear noises, vibration, looseness, hard steering, and fluid leakage problems; determine needed repairs.

Problem Area	Symptoms	Possible Causes	Classroom Manual	Shop Manual
NOISES	Rattles, clunks, knocks, squeaks	1. Loose steering gear mounting bolts	274	403
		2. Worn steering gear mounting bushings	273	404
		3. Loose tie-rod ends	274	404
		4. Worn flexible coupling or universal joints	274	404
		5. Lack of lubrication on horn bushing contact	274	404
STEERING CONTROL	Looseness	1. Worn inner or outer tie-rod ends	275	402
		2. Worn steering gear mounting bushings	275	402
		3. Worn flexible coupling or universal joint	276	402
		4. Loose steering gear adjustment	276	402
	Hard steering	1. Tight steering gear adjustment	275	406
		2. Low fluid level	277	406
	Fluid leaks	1. Cylinder end rack seal	277	407
		2. Pinion end rack seal	277	407
		3. Pinion seal, upper or lower	277	407

Table 12-7 NATEF and ASE TASK

Remove and replace rack and pinion steering gear; (Includes vehicles equipped with air bags and/or other steering wheel mounted controls and components).

Problem Area	Symptoms	Possible Causes	Classroom Manual	Shop Manual
STEERING CONTROL	Looseness, wander	1. Worn inner tie-rod ends	280	373
		2. Improper steering gear adjustment	280	373
		3. Worn rack or pinion teeth	281	373
		4. Worn steering gear mounting bushings	281	373
	Uneven steering effort	Worn rack or pinion teeth	282	374
	Steering effort imbalance	1. Leaking rack seals	282	374
		2. Worn control valve Teflon rings	283	374

Table 12-8 NATEF and ASE TASK

Adjust rack and pinion steering gear.

Problem Area	Symptoms	Possible Causes	Classroom Manual	Shop Manual
STEERING CONTROL	Looseness	Loose steering gear adjustment	275	378
	Hard steering	Tight steering gear adjustment	274	378

Table 12-9 NATEF and ASE TASK

Inspect and replace rack and pinion steering gear inner tie-rod ends (sockets) and bellows boots.

Problem Area	Symptoms	Possible Causes	Classroom Manual	Shop Manual
STEERING CONTROL	Looseness	Worn inner tie-rod ends	274	398
LOW FLUID LEVEL	Fluid leaks	1. Worn rack or pinion seals	274	398
		2. Defective control valve housing gasket	274	398
PREMATURE INNER TIE-ROD END FAILURE	Looseness	Cracked or split bellows boots	274	398

Table 12-10 NATEF and ASE TASK

Diagnose, inspect, adjust, repair or replace components of electronically controlled steering systems.

Problem Area	Symptoms	Possible Causes	Classroom Manual	Shop Manual
IMPROPER STEERING OPERATION	No Steering Assist, EPS warning light on, engine running	Electrical defect in EPS steering system	277	412

Job Sheet 35

35

Name _____ Date _____

Inspect Manual or Power Rack and Pinion Steering Gear and Tie Rods

NATEF and ASE Correlation

This job sheet is related to NATEF and ASE Automotive Suspension and Steering Task List content area: A. Steering Systems Diagnosis and Repair. 2. Power-Assisted Steering Units. Task 15: Inspect and replace power rack and pinion steering gear inner tie-rod ends (sockets), seals, gaskets, O-rings, and bellows boots.

Tools and Materials

Tape measure

Vehicle make and model year _____

Vehicle VIN _____

Make of steering gear _____

Procedure

1. With the front wheels straight ahead and the engine stopped, rock the steering wheel gently back and forth with light finger pressure and measure the maximum steering wheel freeplay with a tape measure.

   Specified steering wheel freeplay _____

   Actual steering wheel freeplay _____

   State the necessary repairs to correct steering wheel freeplay _____

   _____

   _____

   _____

   _____

2. With the vehicle sitting on the shop floor and the front wheels straight ahead, have an assistant turn the steering wheel about 1/4 turn in both directions. Watch for looseness in the flexible coupling or universal joints, and steering shaft.

   Looseness in U-joints or flexible coupling: satisfactory _____ unsatisfactory _____

   Looseness in steering shaft: satisfactory _____ unsatisfactory _____

   State the necessary repairs and explain the reasons for your diagnosis.

   _____

   _____

   _____

   _____

3. While an assistant turns the steering wheel about 1/2 turn in both directions, watch for movement of the steering gear housing in the mounting bushings.

Looseness in steering gear mounting bushings: satisfactory _____ unsatisfactory _____

State the necessary repairs and explain the reasons for your diagnosis.

_____

_____

_____

_____

4. Grasp the pinion shaft extending from the steering gear and attempt to move it vertically. If there is steering shaft vertical movement, a pinion bearing preload adjustment may be required. When the steering gear does not have a pinion bearing preload adjustment, replace the necessary steering gear components.

Excessive pinion shaft vertical endplay: satisfactory _____ unsatisfactory _____

State the necessary repairs and explain the reasons for your diagnosis.

_____

_____

_____

_____

5. Road test the vehicle and check for excessive steering effort. A bent steering rack, tight rack bearing adjustment, or damaged front drive axle joints in front-wheel-drive cars may cause excessive steering effort.

Steering effort: satisfactory _____ excessive _____

State the necessary repairs and explain the reasons for your diagnosis.

_____

_____

_____

_____

6. Visually inspect the bellows boots for cracks, splits, leaks, and proper clamp installation. Replace any boot that indicates any of these conditions.

Condition of tie rod bellows boots and clamps:

Left side boot: satisfactory _____ unsatisfactory _____

Right side boot: satisfactory _____ unsatisfactory _____

State the necessary repairs and explain the reasons for your diagnosis.

_____

_____

_____

_____

7. Loosen the inner bellows boot clamps and move each boot toward the outer tie-rod end until the inner tie-rod end is visible. Push outward and inward on each front tire and watch for movement in the inner tie-rod end. An alternate method of checking the inner tie-rod ends is to squeeze the bellows boots and grasp the inner tie-rod end socket. Movement in the inner tie-rod end is then felt as the front wheel is moved inward and outward. Hard plastic bellows boots may be found on some applications. With this type of bellows boot, remove the ignition key from the switch to lock the steering column and push inward and outward on the front tire while observing any lateral movement in the tie rod.

Condition of inner tie-rod ends:

Left side inner tie-rod end: satisfactory _____ unsatisfactory _____

Right side inner tie-rod end: satisfactory _____ unsatisfactory _____

State the necessary repairs and explain the reasons for your diagnosis.

_____

_____

_____

_____

WARNING: Bent steering components must be replaced. Never straighten steering components because this action may weaken the metal and result in sudden component failure, serious personal injury, and vehicle damage.

8. Grasp each outer tie-rod end and check for vertical movement. While an assistant turns the steering wheel 1/4 turn in each direction, watch for looseness in the outer tie-rod ends. Check the outer tie-rod end seals for cracks and proper installation of the nuts and cotter pins. Cracked seals must be replaced. Inspect the tie-rods for a bent condition:

Tie rod and outer tie-rod end condition:

Left side tie rod and outer tie-rod end condition: satisfactory _____ unsatisfactory _____

Right side tie rod and outer tie-rod end condition: satisfactory _____ unsatisfactory _____

List the necessary repairs and explain the reasons for your diagnosis.

_____

_____

_____

_____

Instructor Check _____

Job Sheet 36

36

Name _____ Date _____

Remove and Replace Manual or Power Rack and Pinion Steering Gear

NATEF and ASE Correlation

This job sheet is related to NATEF and ASE Automotive Suspension and Steering Task List content area: A. Steering Systems Diagnosis and Repair. 2. Power-Assisted Steering Units. Task 11: Remove and replace power rack and pinion steering gear: inspect and replace mounting bushings and brackets (includes vehicles equipped with air bags and/or other steering wheel mounted controls and components).

Tools and Materials

Floor jack

Safety stands

Outer tie-rod end puller

Tape measure

Torque wrench

Power steering fluid

Vehicle make and model year _____

Vehicle VIN _____

Procedure

1. Place the front wheels in the straight-ahead position and remove the ignition key from the ignition switch to lock the steering wheel. Place the driver's seat belt through the steering wheel to prevent wheel rotation if the ignition switch is turned on. This action maintains the clock spring electrical connector or spiral cable in the centered position on air-bag-equipped vehicles.

Are the front wheels straight ahead and steering wheel locked? yes _____ no _____

Instructor check _____

2. Lift the front end with a floor jack and place safety stands under the vehicle chassis. Lower the vehicle onto the safety stands. Remove the left and right fender apron seals.

Is the chassis properly supported on safety stands? yes _____ no _____

Are the fender apron seals removed? yes _____ no _____

Instructor check _____

3. Place punch marks on the lower universal joint and the steering gear pinion shaft so they may be reassembled in the same position.

Are the pinion shaft and U-joint punchmarked? yes _____ no _____

Instructor check _____

4. Loosen the upper universal joint bolt and remove the lower universal joint bolt, and disconnect this joint.

Is the U-joint disconnected from pinion shaft? yes _____ no _____

Instructor check _____

5. Remove the cotter pins from the outer tie-rod ends. Loosen, but do not remove, the tie-rod end nuts.

Are the outer tie-rod end nuts loosened? yes _____ no _____

Instructor check _____

6. Use a tie-rod end puller to loosen the outer tie-rod ends in the steering arms. Remove the tie-rod end nuts and remove the tie-rod ends from the arms.

Are the outer tie-rod ends removed from the steering arms? yes _____ no _____

Instructor check _____

7. Use the proper wrenches to disconnect the high-pressure hose and return hose from the steering gear. (This step is not required on a manual steering gear.)

Are the high pressure and return hoses disconnected from steering gear?

yes _____ no _____

Instructor check _____

8. Remove the four stabilizer bar mounting bolts.

Are the stabilizer mounting bolts removed? yes _____ no _____

Instructor check _____

9. Remove the steering gear mounting bolts.

Are the steering gear mounting bolts removed? yes _____ no _____

Instructor check _____

10. Remove the steering gear assembly from the right side of the car.

11. Position the right and left tie rods the specified distance from the steering gear housing. Measure this distance with a tape measure.

Are the right and left tie-rods positioned at equal distances from the steering gear housing? yes _____ no _____

Instructor check _____

12. Install the steering gear through the right fender apron.

13. Install the pinion shaft in the universal joint with the punch marks aligned. Tighten the upper and lower universal joint bolts to the specified torque.

Are the punch marks on pinion shaft and U-joint properly aligned? yes _____ no _____

Instructor check _____

Specified U-joint bolt torque _____

Actual U-joint bolt torque _____

Instructor check _____

14. Install the steering gear mounting bolts and tighten these bolts to the specified torque.

Specified steering gear mounting bolt torque _____

Actual steering gear mounting bolt torque _____

15. Install the stabilizer bar mounting bolts and torque these bolts to specifications.

Specified stabilizer mounting bolt torque _____

Actual stabilizer mounting bolt torque _____

16. Install and tighten the high-pressure and return hoses to the specified torque. (This step is not required on a manual rack and pinion steering gear.)

Specified high-pressure and return hose fitting torque _____

Actual high-pressure and return hose fitting torque _____

⚠️ **WARNING:** Do not loosen the tie-rod end nuts to align cotter pin holes. This action causes improper torquing of these nuts.

17. Install the outer tie-rod ends in the steering knuckles, and tighten the nuts to the specified torque. Install the cotter pins in the nuts.

Specified outer tie-rod end nut torque _____

Actual outer tie-rod end nut torque _____

Are the cotter pins properly installed? yes _____ no _____

18. Tighten the outer tie-rod end jam nuts to the specified torque, then tighten the outer bellows boot clamps.

Specified outer tie-rod end jam nut torque _____

Actual outer tie-rod end jam nut torque _____

Are the bellows boot clamps properly installed and tightened? yes _____ no _____

Instructor check _____

19. Install the left and right fender apron seals, then lower the vehicle with a floor jack.

20. Fill the power steering pump reservoir with the vehicle manufacturer's recommended power steering fluid and bleed air from the power steering system. (This step is not required on a manual rack and pinion steering gear.)

Is the power steering pump reservoir filled to the specified level with the proper power steering fluid? yes _____ no _____

Is the air bled from the power steering system? yes _____ no _____

Instructor check _____

21. Road test the vehicle and check for proper steering gear operation and steering control.

Steering operation: satisfactory _____ unsatisfactory _____

If the steering operation in unsatisfactory, state the necessary repairs and explain the reason for your diagnosis. _____

_____

_____

_____

✓ **Instructor Check** _____

Job Sheet 37

37

Name _____ Date _____

Diagnose Power Rack and Pinion Steering Gear Oil Leakage Problems

NATEF and ASE Correlation

This job sheet is related to NATEF and ASE Automotive Suspension and Steering Task List content area: A. Steering Systems Diagnosis and Repair. 2. Power-Assisted Steering Units. Task 2: Diagnose power rack and pinion steering gear noises, vibration, looseness, hard steering, and fluid leakage problems; determine needed repairs.

Tools and Materials

Power steering fluid

Vehicle make and model year _____

Vehicle VIN _____

Procedure

CAUTION: If the engine has been running for a length of time, power steering gears, pumps, lines, and fluid may be very hot. Wear eye protection and protective gloves when servicing these components.

1. Be sure the power steering reservoir is filled to the specified level with the proper power steering fluid.

2. Be sure the power steering fluid is at normal operating temperature. If necessary, rotate the steering wheel several times from lock-to-lock to bring the fluid to normal operating temperature.

   Is the power steering reservoir filled to the specified level? yes _____ no _____

   Is the power steering fluid at normal temperature? yes _____ no _____

   Instructor check _____

3. Inspect the cylinder end of the steering gear for oil leaks.

   Oil leaks at cylinder end of the power steering gear: satisfactory _____ unsatisfactory _____

   If there are oil leaks in this area, state the necessary repairs and explain the reasons for your diagnosis. _____

   _____

   _____

   _____

4. Inspect the steering gear for oil leaks at the pinion end of the housing with the engine running and the steering wheel turned so the rack is against the left internal stop.

   Oil leaks at the pinion end of the steering gear with the rack against the left inner stop:

   satisfactory _____ unsatisfactory _____

If there are oil leaks in this area, state the necessary repairs and explain the reasons for your diagnosis. _____

_____

_____

_____

> ☑ **SERVICE TIP:** An oil leak at one rack seal may result in oil leaks from both boots, because the oil may travel through the breather tube between the boots.

5. Inspect the pinion end of the power steering gear housing for oil leaks with the steering wheel turned in either direction.

Oil leak at the pinion end of the housing with the steering wheel turned in either direction: satisfactory _____ unsatisfactory _____

If there are oil leaks in the area, state the necessary repairs and explain the reasons for your diagnosis. _____

_____

_____

_____

6. Inspect the pinion coupling area for oil leaks.

Oil leaks in the pinion coupling area: satisfactory _____ unsatisfactory _____

If there are oil leaks in the area, state the necessary repairs and explain the reasons for your diagnosis. _____

_____

_____

_____

7. Inspect the all the lines and fittings on the steering gear for oil leaks as an assistant turns the wheels with the engine running.

Oil leaks at steering gear fittings: satisfactory _____ unsatisfactory _____

Oil leaks in steering gear hoses and lines: satisfactory _____ unsatisfactory _____

If there are oil leaks in the power steering fittings, lines, or hoses, state the necessary repairs and explain the reasons for your diagnosis. _____

_____

_____

_____

8. With the engine running, turn the steering wheel fully in each direction and check the steering effort.

Steering effort in both directions: normal _____ excessive _____

If the steering effort is not equal in both directions, state the direction in which steering effort is highest: right _____ left _____

If the steering effort is excessive or unequal, state the necessary repairs and explain the reason for your diagnosis. _____

_____

_____

_____

☑ **Instructor Check** _____

Electronic Four Wheel Steering Service and Diagnosis

Upon completion and review of this chapter, you should be able to:

- ❏ Perform a preliminary inspection on a four wheel steering (4WS) system.
- ❏ Perform a trouble code diagnosis on a 4WS system with the ignition switch on.
- ❏ Perform a trouble code diagnosis on a 4WS system with the engine running.
- ❏ Remove and replace the rear steering actuator.
- ❏ Remove and replace the tie-rod ends on the rear steering actuator.
- ❏ Remove and replace tie-rods and boots on the rear steering actuator.

- ❏ Remove and replace the rear main steering angle sensor.
- ❏ Remove and replace the rear sub steering angle sensor.
- ❏ Perform an electronic neutral check on a 4WS system.
- ❏ Adjust the front main steering angle sensor.
- ❏ Adjust the front sub steering angle sensor.
- ❏ Adjust the rear sub steering angle sensor.

Preliminary Inspection

Prior to any four wheel steering diagnosis, the following concerns should be considered:

1. Have any suspension modifications been made that would affect steering?
2. Are the tire sizes the same as specified by the vehicle manufacturer?
3. Are the tires inflated to the pressure specified by the vehicle manufacturer?
4. Is the power steering belt adjusted to the vehicle manufacturer's specified tension?
5. Is the power steering pump reservoir filled to the proper level with the type of fluid specified by the vehicle manufacturer?
6. Is the engine idling at the speed specified by the vehicle manufacturer? Is the idle speed steady?
7. Is the steering wheel original equipment?

Fail-Safe Function

If the four wheel steering (4WS) control unit senses a failure in the system, the control unit switches to a **fail-safe mode.** In this mode, the control unit stores a trouble code or codes, and illuminates the 4WS indicator light to inform the driver that a problem exists in the system. When this mode is entered, the 4WS control unit shuts off voltage to the rear steering unit, and the rear wheels remain in the straight-ahead position.

Damper Control

When the 4WS control unit enters the fail-safe mode, a quick return of the rear wheels to the straight-ahead position would adversely affect steering under certain steering wheel and rear wheel

Basic Tools

Basic technician's tool set
Service manual
Jumper wire
Floor jack
Safety stands
Wax marker
Silicone grease
Chassis lubricant
Cotter pins
Length of stiff wire

The four wheel steering control unit enters the fail-safe mode if a defect occurs in the system. In this mode, the rear wheels move to the centered position.

The Society of Automotive Engineers (SAE) J1930 terminology is an attempt to standardize electronics terminology in the automotive industry.

In the SAE J1930 terminology, the term malfunction indicator light (MIL) replaces other terms for computer system indicator lights.

positions. To prevent this action, the 4WS control unit energizes the damper relay when it enters the fail-safe mode. The rear steering actuator motor is spun by the steering shaft movement as this shaft is moved to the centered position by centering spring force. This action causes the motor armature to act as a voltage generator. The voltage generated by the armature is fed back through the damper relay to the motor armature. Under this condition, the motor rotation is slowed and the return spring slowly moves the rear steering shaft to the straight-ahead position. Without the action of the damper relay, the return spring would move the rear steering shaft quickly to the straight-ahead position.

Trouble Code Diagnosis

Road Test

● **CUSTOMER CARE:** While discussing customers' automotive problems, always remain polite and never make statements that make customers feel uninformed about their vehicles.

The 4WS control unit stores a fault code and illuminates the 4WS indicator light if a defect occurs in the system, even if the defect is temporary. Always ask the customer about the conditions that caused the 4WS indicator light to come on, and duplicate this condition during a road test. If the 4WS light is not illuminated during the road test, the system is satisfactory electronically and does not require further electronic diagnosis. The troubleshooting procedures in the vehicle manufacturer's service manual assume that the problem is present at the time of diagnosis.

Trouble Code Display with Ignition Switch On

▲ **WARNING:** When diagnosing a computer system, never connect or ground any terminals unless instructed to do so in the vehicle manufacturer's service manual.

▲ **WARNING:** When diagnosing a computer system, never disconnect or connect any computer system component with the ignition switch on unless instructed to do so in the vehicle manufacturer's service manual. This action may damage the computer or system components.

▲ **WARNING:** When performing electronic diagnosis on a vehicle equipped with an air bag, most vehicle manufacturers recommend turning the ignition switch off, disconnecting the negative battery cable, and waiting one minute before proceeding with electronic component diagnosis or service.

■ **CAUTION:** When performing electronic diagnosis on a vehicle equipped with an air bag, follow all the service precautions recommended in the vehicle manufacturer's service manual.

Always follow the exact 4WS service and diagnostic procedures in the vehicle manufacturer's service manual. These procedures vary depending on the make and year of the vehicle. The following are typical procedures for a Honda Prelude. These procedures should be avoided until after the diagnosis is complete because any of these procedures will erase trouble codes.

1. Disconnect the battery terminals.
2. Disconnect the 4WS control unit connector.
3. Remove the number 43 clock-radio 10-A fuse from the underhood fuse/relay box.

Follow these steps to obtain the trouble codes:

1. Remove the dual-terminal **service check connector** located behind the center console, and connect the two terminals in this connector with a jumper wire (Figure 13-1).

2. Turn on the ignition switch, but do not start the engine.

3. Observe the 4WS indicator light to read the trouble codes. Three longer flashes followed by a brief pause and one quicker flash indicates code 31. The codes are given in numerical order.

4. Record the fault codes.

Trouble Code Display with Engine Running

The 4WS control unit actually contains two processing units that are referred to as the main and sub processing units. Each processing unit can store a maximum of ten trouble codes. If the trouble code diagnosis is performed with the engine running, the code display indicates whether the codes are stored in the main or sub processor. When the service connector terminals are connected with a jumper wire and the engine is started, the 4WS indicator light follows this sequence if there are trouble codes in the main and sub processors:

1. Blinks quickly once when the ignition switch is turned on

2. Pauses for 3 seconds

3. Displays codes stored in the main processor

4. Pauses for 1.6 seconds

5. Blinks quickly for 3 seconds to indicate a separation between the main and sub processor codes

6. Pauses for 1.6 seconds

7. Displays codes stored in the sub processor

8. Pauses for 3 seconds and then repeats the cycle (Figure 13-2)

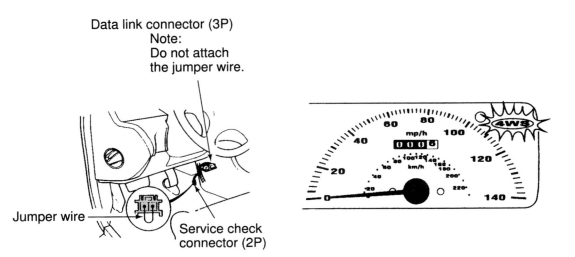

Figure 13-1 Dual-terminal service check connector positioned behind the center console. (Courtesy of American Honda Motor Company, Inc.)

DTC indication pattern

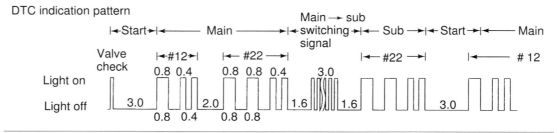

Figure 13-2 Trouble codes in main and sub processors obtained with the engine running. (Courtesy of American Honda Motor Company, Inc.)

DTC	Operation	4WS indicator light
70	The ignition is turned from OFF to ON while driving.	ON
71	The car is driven aggressively with the driver and three passengers on board, or the steering wheel is turned with a rear wheel blocked by the curb, etc.	————
73	The engine is started while quick-charging the battery.	————
74	Driving the car with the parking brake ON	ON 5 minutes after detection

Figure 13-3 Codes that represent problems caused by abnormal or harsh driving conditions. (Courtesy of American Honda Motor Company, Inc.)

Main Steering Angle Sensor Trouble Code

If a defect occurs in the main steering angle sensor system, the clock-radio 10-A fuse must be disconnected to cancel the 4WS indicator light. When defects occur in other parts of the electronic 4WS system, the 4WS indicator light is cancelled when the ignition switch is turned off. However, the 4WS indicator light is illuminated again when the ignition switch is turned on, and the 4WS control unit detects the problem again.

Classroom Manual
Chapter 13, page 304

Trouble Codes Representing Temporary Driving Conditions

Codes 70, 71, 73, and 74 represent problems resulting from abnormal or harsh driving conditions (Figure 13-3). When the 4WS control unit detects one of these problems, it does not illuminate the 4WS indicator light, but these codes are flashed during the diagnostic procedure.

Rear Steering Actuator Service

Rear Steering Actuator Removal

 WARNING: Many steering service and diagnostic procedures require the installation of the rear steering center lock pin in the rear steering actuator to lock this unit in the centered position.

WARNING: Do not start the engine with the rear steering center lock pin in place. This action may damage the lock pin and rear steering actuator.

WARNING: Do not attempt to disassemble the rear steering actuator other than tie rods, tie-rod ends, and sensors. This actuator is serviced as an assembly.

WARNING: Do not damage the tie-rod boot when using the tie-rod end removal tool.

Follow these steps to remove the rear steering actuator:

1. Raise the vehicle on a hoist or lift the rear of the vehicle with a floor jack and support the chassis with safety stands placed in the vehicle manufacturer's recommended locations.

2. Remove the cotter pin and nut from each tie-rod end.

3. Install a 12 millimeter (mm) nut on each tie-rod end until the nuts are flush with the tie-rod stud.

4. Install the special tool on the tie-rod end, and with the tool arms parallel, tighten the screw on the tool to loosen the tie-rod end (Figure 13-4). Repeat the procedure on both tie-rod ends.

5. Remove the nuts from the tie-rods and remove the tie rods from the steering arms.

6. Remove the rear steering actuator cover (Figure 13-5).

Special Tools

Tie-rod end removal tool

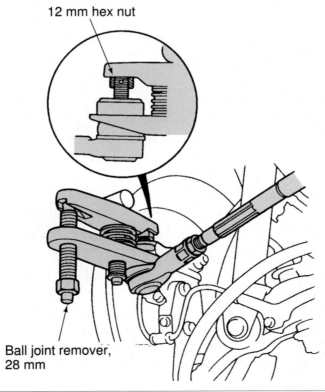

Caution: avoid damaging the ball joint boot.

12 mm hex nut

Ball joint remover, 28 mm

Figure 13-4 Removing a tie-rod end with a special tool. (Courtesy of American Honda Motor Company, Inc.)

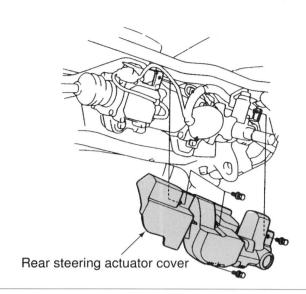

Rear steering actuator cover

Figure 13-5 Removing the rear steering actuator cover. (Courtesy of American Honda Motor Company, Inc.)

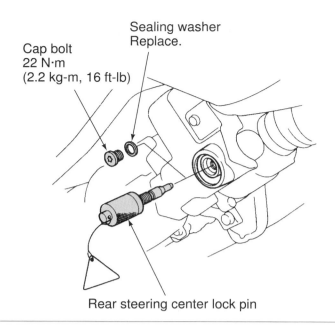

Cap bolt
22 N·m
(2.2 kg-m, 16 ft-lb)

Sealing washer
Replace.

Rear steering center lock pin

Figure 13-6 Removing the cap bolt and washer and installing the rear steering center lock pin. (Courtesy of American Honda Motor Company, Inc.)

7. Remove the cap bolt and washer and install the **rear steering center lock pin** (Figure 13-6).

8. Remove the ground cable connector and all wiring harness connectors on the rear steering actuator (Figure 13-7).

9. Remove the four mounting bolts and bracket, and remove the rear steering actuator (Figure 13-8).

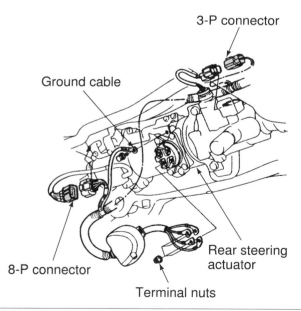

3-P connector

Ground cable

8-P connector

Rear steering actuator

Terminal nuts

Figure 13-7 Removing the ground cable connector and all wiring harness connectors on the rear steering actuator. (Courtesy of American Honda Motor Company, Inc.)

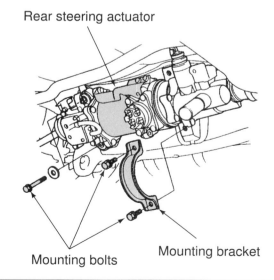

Rear steering actuator

Mounting bolts

Mounting bracket

Figure 13-8 Removing the four mounting bolts, bracket, and rear steering actuator. (Courtesy of American Honda Motor Company, Inc.)

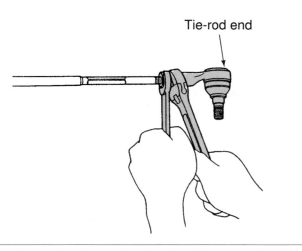

Figure 13-9 Loosening the tie-rod lock nut. (Courtesy of American Honda Motor Company, Inc.)

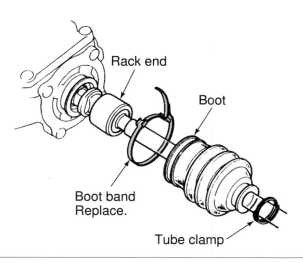

Figure 13-10 Removing boot bands and clamps from the inner tie-rod end. (Courtesy of American Honda Motor Company, Inc.)

Tie Rod and Tie-Rod End Removal

Follow these steps for tie-rod end and tie rod removal:

1. Mark the relative position of the tie-rod end, lock nut, and tie rod with a wax marker.
2. Hold the tie-rod end with a wrench and loosen the lock nut (Figure 13-9).
3. Remove the tie-rod end.
4. Remove the boot bands and clamps from the inner tie-rod ends (Figure 13-10).
5. Place the flat side of the rack holding tool toward the actuator housing and drive the special rack holding tool between the actuator housing and the stop washer with a soft hammer (Figure 13-11).
6. Straighten the tabs on the tie-rod lock washer.

Special Tools

Rack holding tool

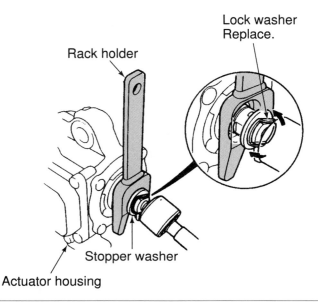

Figure 13-11 Installing the special rack holding tool between the actuator housing and the stop washer. (Courtesy of American Honda Motor Company, Inc.)

SERVICE TIP: Hold the special holding tool firmly while loosening the tie rod to avoid applying rotational force to the shaft screw in the actuator.

7. Hold the shaft screw with the holding tool and loosen the tie rod with a wrench (Figure 13-12).

8. Thread the tie rod off the shaft screw and repeat this procedure on each tie-rod end.

Tie-Rod End Boot Removal and Replacement

Tie-rod end boots must be replaced if they are cracked, split, deteriorated, or loose. Follow these steps to remove and replace the tie-rod end boots:

 WARNING: Do not put grease on the boot installation shoulder and tapered section of the ball pin in the tie-rod end.

 WARNING: Do not allow dust, dirt, or foreign material to enter the tie-rod end ball joint or boot.

1. Use a large screwdriver to pry the old boot from the tie-rod end.

2. Pack the interior of the new boot with the vehicle manufacturer's recommended grease, and place a light coating of grease on the boot lip.

3. Wipe the grease off the sliding surfaces of the ball pin with a shop towel, then pack the lower area around the ball pin and body with grease (Figure 13-13).

Special Tools

Tie-rod boot driving tool

4. Use the special driving tool to install the new boot on the tie-rod end (Figure 13-14).

5. Wipe any grease from the tapered section of the ball pin with a shop towel. Apply sealant around the lower edge of the boot and tie-rod body.

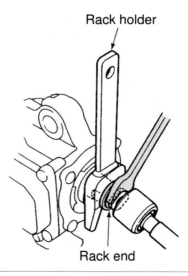

Figure 13-12 Removing the tie rod from the shaft screw in the rear steering actuator. (Courtesy of American Honda Motor Company, Inc.)

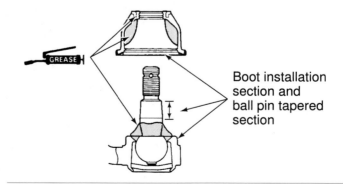

Figure 13-13 Packing boot and tie-rod end with grease. (Courtesy of Honda American Motor Company, Inc.)

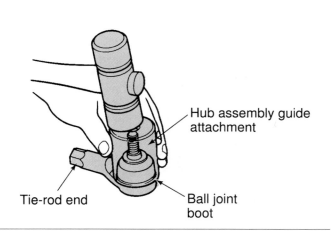

Figure 13-14 Driving the boot onto the tie-rod end with a special driving tool. (Courtesy of American Honda Motor Company, Inc.)

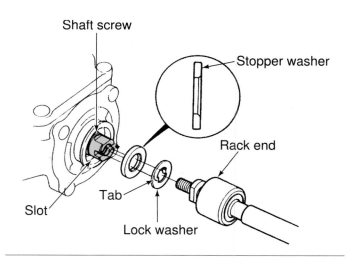

Figure 13-15 Installing the inner tie-rod end on the shaft screw. (Courtesy of American Honda Motor Company, Inc.)

Installation of Tie Rods and Tie-Rod Ends

 WARNING: Never apply axial impact or rotational force to the shaft screw in the rear steering actuator. Either of these actions may cause internal actuator damage.

Inner tie-rod boots must be replaced if they are cracked, split, deteriorated, or damaged. Follow these steps for tie-rod end and tie rod installation:

1. Install the tie-rod ends so the marks on the tie-rod ends, nuts, and tie-rods are aligned, and tighten the tie-rod nuts to the specified torque.

2. Screw each inner tie-rod onto the shaft screw while holding the lock washer so its tabs are in the inner tie-rod end. The stop washer must be installed on the shaft screw with the chamfered side facing outward (Figure 13-15).

3. Drive the special holding tool between the actuator housing and the stop washer with a soft hammer (Figure 13-16).

4. Hold the shaft screw with the holding tool and tighten the inner tie-rod end to the specified torque.

5. Bend the lock washer tabs against the flat on the inner tie-rod end.

6. Remove the special holding tool and apply silicone grease to the sliding surface of the tie rod (Figure 13-17). Place a light coating of silicone grease inside the tie-rod boot.

7. Apply the vehicle manufacturer's recommended grease to the circumference of the inner tie-rod joint housing.

8. Install the boots on the actuator housing, then install the boot bands with the locking tabs properly positioned in relation to the actuator housing (Figure 13-18).

 WARNING: While staking the boot clamps, be careful not to damage the inner tie-rod boots.

9. Tighten the boot bands and bend both sets of locking tabs over the band (Figure 13-19). Tap lightly on the doubled over portion of the band to reduce its height, and stake the locking tabs firmly.

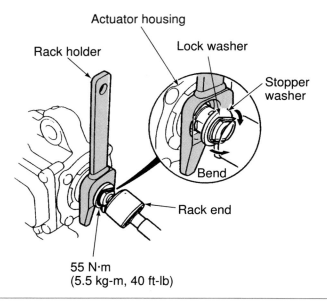

Figure 13-16 Installing the special tool to hold the shaft screw while tightening the inner tie-rod end. (Courtesy of American Honda Motor Company, Inc.)

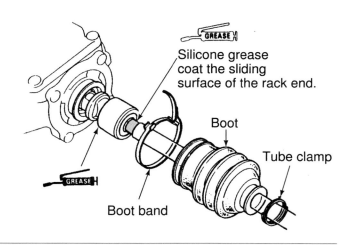

Figure 13-17 Lubrication of the inner tie-rod joint housing. (Courtesy of American Honda Motor Company, Inc.)

Remove and Replace Rear Steering Actuator Sensors

Follow these steps to remove and replace the **rear sub steering angle sensor** and the **rear main steering angle sensor:**

1. Loosen the rear sub steering angle sensor lock nut, and rotate the sensor to thread it out of the housing (Figure 13-20). Discard the sensor O-ring.

2. Remove the two mounting bolts in the rear main steering angle sensor, and remove the sensor from the actuator housing (Figure 13-21). Note the position of the dowel pins, and discard the O-ring.

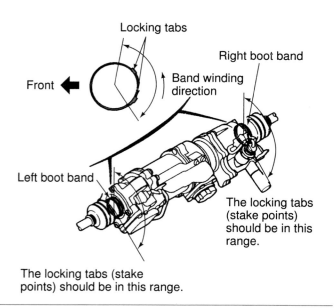

Figure 13-18 Proper boot band position in relation to the actuator housing. (Courtesy of American Honda Motor Company, Inc.)

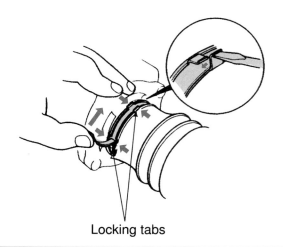

Figure 13-19 Tightening and staking the inner tie-rod boot clamps. (Courtesy of American Honda Motor Company, Inc.)

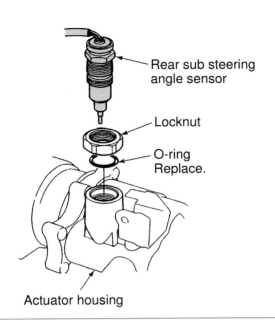

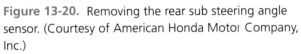

Figure 13-20. Removing the rear sub steering angle sensor. (Courtesy of American Honda Motor Company, Inc.)

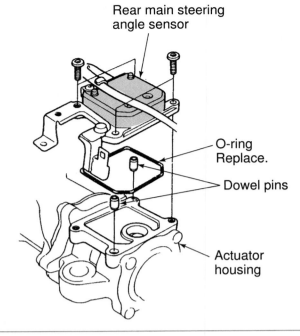

Figure 13-21 Removing the rear main steering angle sensor. (Courtesy of American Honda Motor Company, Inc.)

⚠ **WARNING:** Cover the rear main steering sensor and rear sub steering sensor openings in the actuator housing with masking tape or its equivalent to keep dirt and foreign material out of the actuator housing.

3. Install the lock nut and a new O-ring on the rear sub steering angle sensor.

4. Place a light coating of grease on the O-ring, and install the sensor in the actuator housing.

5. Rotate the sensor until it touches the tapered shaft, and back it out one-half turn. Tighten the lock nut finger tight. Final adjustment of the rear sub steering angle sensor is completed with the actuator installed in the vehicle.

6. Place a light coating of grease on the rear main steering angle sensor O-ring and install this O-ring on the sensor.

7. Install the rear main steering angle sensor and O-ring in the actuator housing with the dowel pins properly positioned, and tighten the mounting bolts to the specified torque.

Installing Rear Steering Actuator

Follow these steps for rear steering actuator installation:

1. Install the rear steering actuator and the four mounting bolts and bracket. The arrow on the bracket must face upward (Figure 13-22).

2. Tighten the rear steering actuator mounting bolts to the specified torque.

⚠ **WARNING:** Tighten the castle nut on the tie-rod ends to the specified torque, and then tighten these nuts enough to align the slots in the nut with the hole in the tie-rod pin. Do not loosen the nut to align the nut slots with the tie-rod pin hole.

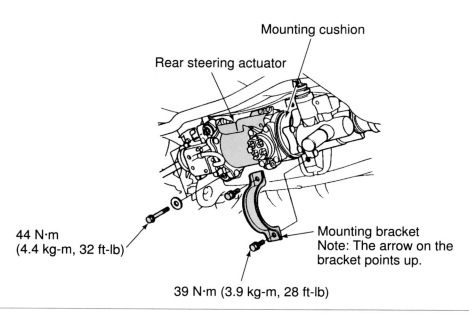

Mounting cushion

Rear steering actuator

44 N·m
(4.4 kg-m, 32 ft-lb)

Mounting bracket
Note: The arrow on the
bracket points up.

39 N·m (3.9 kg-m, 28 ft-lb)

Figure 13-22 Installing the rear steering actuator, mounting screws, and bracket. (Courtesy of American Honda Motor Company, Inc.)

3. Reconnect the tie-rod ends to the steering arms and tighten the castle nut to the specified torque. If necessary, tighten the nut slightly to align the nut slots with the tie-rod pin hole.

4. Install the cotter pin in the nut and tie-rod end pin openings, and bend one leg of the cotter pin downward over the nut. Bend the other cotter pin leg upward over the top of the tie-rod end pin (Figure 13-23).

5. Check all the wiring connectors for contamination and clean as necessary. Install all the wiring connectors on the rear steering actuator, and tighten all the terminal nuts to the specified torque (Figure 13-24).

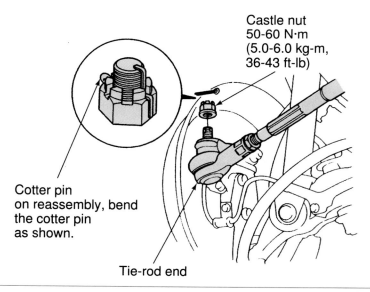

Castle nut
50-60 N·m
(5.0-6.0 kg-m,
36-43 ft-lb)

Cotter pin
on reassembly, bend
the cotter pin
as shown.

Tie-rod end

Figure 13-23 Proper installation of cotter pin in the tie-rod end. (Courtesy of American Honda Motor Company, Inc.)

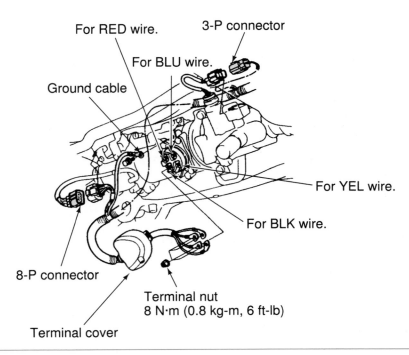

For RED wire.

3-P connector

For BLU wire.

Ground cable

For YEL wire.

For BLK wire.

8-P connector

Terminal nut
8 N·m (0.8 kg-m, 6 ft-lb)

Terminal cover

Figure 13-24 Installing wiring connectors on the rear steering actuator. (Courtesy of American Honda Motor Company, Inc.)

6. Install the terminal cover on the rear main steering sensor terminals. Remove the rear steering lock pin and install the cap bolt and washer. Leave the steering actuator cover removed until after the final rear steering actuator adjustments.

Rear Steering Actuator Adjustment

Electronic Neutral Check

Preliminary Checks. If the power to the 4WS control unit has been shut down for any of the following operations, start the engine and turn the steering wheel fully right and left.

1. Battery cables have been disconnected.

2. The 4WS control unit connector has been disconnected.

3. The number 43 clock-radio fuse has been disconnected.

 WARNING: Do not start the engine with the rear steering actuator lock pin in place. This action may damage the lock pin and rear steering actuator.

Prior to the electronic neutral check, be sure the steering wheel spoke is at the designated angle while driving straight ahead. Be sure the rear wheels are in the straight-ahead driving position before the electronic neutral check.

Steering Wheel Marking and Diagnostic Mode Entry. Follow these steps to mark the steering wheel and enter the front steering sensor diagnostic mode:

1. Drive the vehicle on an alignment rack and place all four wheels on turning radius gauge turn tables. Be sure the wheels are in the center of the turn tables with the wheels straight ahead and the turn tables in the zero-degree position (Figure 13-25).

Special Tools

Turning radius gauge turn tables

Turning radius gauge
turn table

Figure 13-25 All four wheels must be on radius gauge turn tables with the wheels straight ahead and the turn tables in the zero position. (Courtesy of American Honda Motor Company, Inc.)

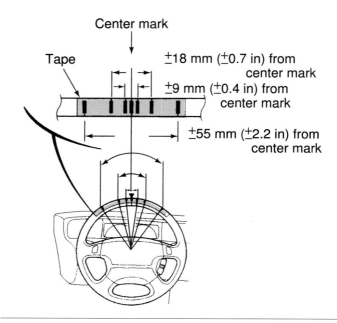

Center mark

Tape

±18 mm (±0.7 in) from center mark

±9 mm (±0.4 in) from center mark

±55 mm (±2.2 in) from center mark

Figure 13-26 Steering wheel marking for electronically neutral check. (Courtesy of American Honda Motor Company, Inc.)

2. Place a piece of masking tape 12.0 in (300 mm) long on top of the steering wheel. Mark the tape at each of these locations on the outer circumference of the wheel (Figure 13-26):
   - center, highest point on the wheel
   - 0.4 in (9 mm) right and left of the center mark
   - 0.7 in (18 mm) right and left of the center mark
   - 2.2 in (55 mm) right and left of the center mark

3. Bend a stiff piece of wire so it can be taped to the top or the dash with the outer end of the wire positioned over the steering wheel marks. Be sure the front wheels are straight ahead with the tip of the wire over the center mark on the steering wheel (Figure 13-27). Be sure the wire is securely taped to the top of the instrument panel.

4. Connect a jumper wire to the 4WS service check connector terminals. This is the same connection for obtaining 4WS control unit trouble codes. Check and verify the trouble codes prior to the electronic neutral check. The 4WS indicator light will not indicate the electronic neutral check and trouble codes at the same time.

5. Pull the parking brake fully on until the parking brake warning light is on, and turn on the ignition switch to set the front steering sensor test mode.

Front Sensor Inspection, Electronic Neutral Check. Follow these steps to check the front main steering sensor:

1. With the ignition switch on, turn the steering wheel slowly to the left and slowly to the right until the 4WS indicator light comes on. Repeat this step several times to find the exact steering wheel position where the 4WS indicator light is illuminated for more than two seconds.

2. The 4WS indicator light should be illuminated when the steering wheel is 9 mm (0.4 in) to the left and right of the center mark on the steering wheel. If the 4WS indicator light comes on at a point outside of this specified range, the 4WS system requires adjustment.

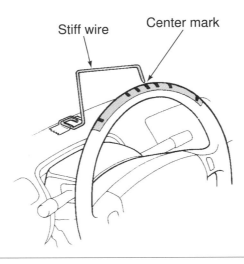

Stiff wire   Center mark

Figure 13-27 Front wheels straight ahead and wire pointer positioned over the center mark on the steering wheel. (Courtesy of American Honda Motor Company, Inc.)

Use the following procedure to check the **front sub steering angle sensor:**

1. Slowly turn the steering wheel to the left and right of the center position until the 4WS indicator light blinks at intervals of 0.2 seconds. Repeat this procedure several times to locate the exact steering wheel position where the indicator light begins blinking. The light should begin blinking within 55 mm (2.2 in) to the left or right of the center mark on the steering wheel.

2. If the 4WS indicator light does not begin flashing within this specified range, a 4WS system adjustment is necessary. After adjusting the front sub steering sensor, the 4WS indicator light should begin flashing when the steering wheel is turned 18 mm (0.7 in) to the left or right of the center mark on the steering wheel.

Rear Sensor Inspection, Electronic Neutral Check. Use the following procedure to complete the rear sensor inspection:

1. Release the parking brake and be sure the parking brake warning light is off. This causes the 4WS control unit to enter the rear steering sensor inspection mode.

2. Remove the rear cap bolt and sealing washer from the rear steering actuator and install the rear steering center lock pin until it bottoms in the actuator (Figure 13-28).

3. Position the front wheels in the straight-ahead position to prevent the rear wheels from steering if the engine is started by mistake.

4. Turn the ignition switch on and push the left rear wheel fully to the right by hand. Then push this wheel fully to the left by hand while a coworker observes the 4WS indicator light (Figure 13-29). The 4WS indicator light should begin to flash at 0.2 second intervals when the left rear wheel is pushed to the left a small amount. If the 4WS indicator light does not flash, adjust the rear sub steering angle sensor.

5. With the ignition switch on, push the left rear wheel fully to the left by hand, then slowly push it to the right. The 4WS indicator light should be illuminated for more than two seconds when the left rear wheel is pushed to the right (Figure 13-30). If the 4WS indicator light is not illuminated, remove the rear main steering angle sensor and check it for damage.

6. Turn off the ignition switch.

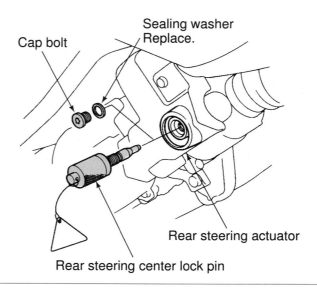

Figure 13-28 Removing the rear cap bolt and sealing washer and installing the rear steering center lock pin. (Courtesy of American Honda Motor Company, Inc.)

7. Remove the rear steering center lock pin and install the cap bolt and washer. Tighten the cap bolt to the specified torque.

8. Remove the jumper wire from the service check connector.

9. Install the rear steering actuator cover.

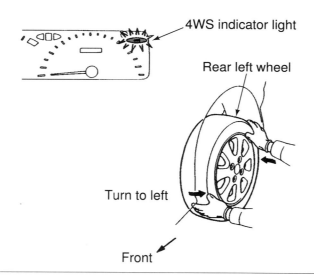

Figure 13-29 The 4WS indicator light should flash when the left rear wheel is pushed to the left, if the rear sub steering angle sensor is properly adjusted. (Courtesy of American Honda Motor Company, Inc.)

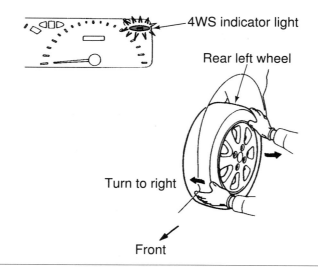

Figure 13-30 Pushing the left rear wheel to the right and observing the 4WS indicator light checks the rear main steering angle sensor. (Courtesy of American Honda Motor Company, Inc.)

Front Main Steering Angle Sensor Adjustment

Proceed as follows for the **front main steering angle sensor** adjustment:

1. Place the car on an alignment rack with each wheel on a turning radius gauge turn table. Turn the steering wheel fully to the right and then fully to the left, counting the number of turns from fully right to fully left.

2. Turn the steering wheel back from full left exactly one-half the number of turns from fully right to fully left. This action centers the front steering rack. The steering wheel spoke should be within the vehicle manufacturer's specified number of degrees from the horizontal position. If the steering wheel is not within this specified position, proceed with the front main steering angle sensor adjustment and spoke angle adjustment.

3. Set the steering wheel so the front wheels are straight ahead, and remove the steering wheel retaining nut. Use a steering wheel puller to remove the steering wheel (Figure 13-31).

4. Check to see if the yellow paint mark on the front main steering angle sensor is facing straight down (Figure 13-32). When this paint mark is facing down, the front main steering angle sensor is in the electronically neutral position.

5. If the yellow paint mark on the front main steering angle sensor is not facing downward, temporarily install the steering wheel with the spokes in the horizontal position. Turn the steering wheel until this yellow paint mark is facing downward.

6. Return the steering wheel to the horizontal position and remove the steering wheel.

7. Install the steering wheel, aligning it with the serration that makes the spoke angle closest to horizontal. Be sure the steering wheel openings fit over the pins on the **cable reel** for the air bag system (Figure 13-33). Do not push down hard on the steering wheel until the serrations and cable reel pins are properly aligned. When the serrations and cable reel pins are properly aligned, push the steering wheel down into place and install the retaining nut.

8. Hold the steering wheel and tighten the retaining nut to the specified torque.

The cable reel contains a conductive ribbon that connects the air bag module on top of the steering wheel to the air bag electrical system, while allowing steering wheel rotation.

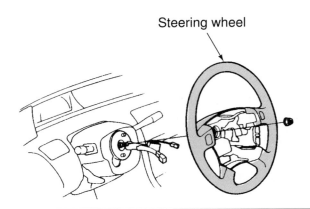

Figure 13-31 Removing the steering wheel. (Courtesy of American Honda Motor Company, Inc.)

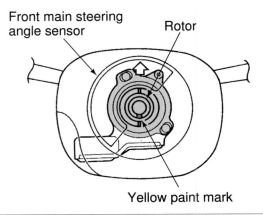

Figure 13-32 Yellow paint mark on the front main steering angle sensor indicating the electronically neutral sensor position. (Courtesy of American Honda Motor Company, Inc.)

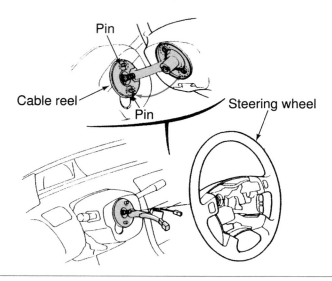

Figure 13-33 Proper alignment of steering wheel openings and cable reel pins. (Courtesy of American Honda Motor Company, Inc.)

Front Sub Steering Angle Sensor Adjustment

Use this procedure for the front sub steering angle sensor adjustment:

1. Raise the front and rear suspension with a floor jack, and place safety stands under the proper chassis locations specified by the car manufacturer. All four wheels must be off the floor.

2. Set the steering wheel in the straight-ahead driving position.

3. Connect a jumper wire across the 4WS system service check connector terminals.

4. Pull the parking brake on fully and turn on the ignition switch. Be sure the parking brake warning light is illuminated.

5. Turn the ignition switch off.

6. Cut the tie strap off the front sub steering angle sensor cover and remove this cover (Figure 13-34).

7. Remove the wiring harness from the clamp and disconnect the wiring harness connector.

8. Loosen the front sub steering angle sensor lock nut, then tighten the lock nut fully by hand. Back this lock nut off three quarters of a turn and connect the connector.

9. Be sure the front wheels are in the straight-ahead driving position and turn the steering wheel until the 4WS indicator light is illuminated. Keep the steering wheel in this position.

10. Slowly turn the front sub steering angle sensor clockwise until the 4WS indicator light goes off, and mark the sensor position in relation to the housing.

11. Slowly rotate the front sub steering angle sensor counterclockwise until the 4WS indicator light begins to blink, and mark the sensor in relation to the housing (Figure 13-35). Set the front sub steering angle sensor in the center of the range from where the light went off to where the light began to blink. Hold the sensor in this position and tighten the lock nut to the specified torque.

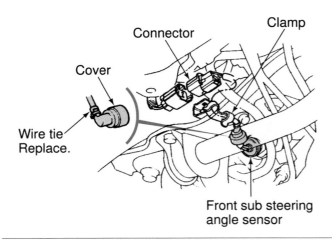

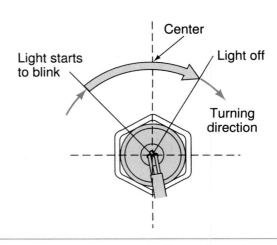

Figure 13-34 Cutting the tie strap and removing the front sub steering angle sensor cover. (Courtesy of American Honda Motor Company, Inc.)

Figure 13-35 Adjusting the front sub steering angle sensor. (Courtesy of American Honda Motor Company, Inc.)

12. Turn off the ignition switch. If the front sub steering angle sensor harness is twisted, disconnect the connector and straighten the harness. Install the harness in the clamp and install the sensor cover. Secure the cover with a new tie strap.

13. Perform the electronic neutral check described earlier in this chapter.

Rear Sub Steering Angle Sensor Adjustment

The rear main steering angle sensor is not adjustable.

Proceed with these steps to adjust the rear sub steering angle sensor:

1. Raise the front and rear suspension with a floor jack, and place safety stands under the proper chassis locations specified by the car manufacturer. All four wheels must be off the floor.

2. Connect a jumper wire across the terminals in the 4WS system service check connector, and be sure any trouble codes have been displayed.

3. Release the parking brake and turn the ignition switch on. Be sure the parking brake warning light goes off.

4. Turn off the ignition switch.

5. Remove the cap bolt and washer and install the rear steering center lock pin.

6. Remove the rear sub steering angle sensor wire from the clamp and disconnect the wiring harness connector.

7. Loosen the rear sub steering angle sensor lock nut. Tighten this lock nut fully by hand, then back it off approximately one-half turn.

8. Connect the rear sub steering angle sensor connector and set the front wheels in the straight-ahead driving position.

9. Turn on the ignition switch.

10. Push the left rear wheel fully to the left by hand, and then push this wheel slowly to the right until the 4WS indicator light comes on. This action places the main rear steering angle sensor in the electronically neutral position.

11. Slowly turn the rear sub steering angle sensor counterclockwise until the 4WS indicator light goes off, and mark the sensor in relation to the housing.

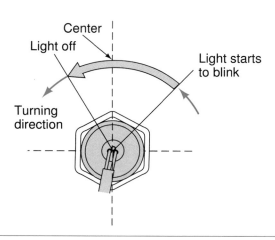

Center
Light off
Light starts
to blink
Turning
direction

Figure 13-36 Adjusting the rear sub steering angle sensor. (Courtesy of American Honda Motor Company, Inc.)

12. Slowly rotate this sensor clockwise until the 4WS indicator light starts to blink, and mark the sensor in relation to the housing. Turn the sensor to the center position between where the indicator light went off and the light started to blink (Figure 13-36). Hold the sensor in this position and tighten the lock nut to the specified torque.

13. Turn off the ignition switch.

14. If the rear sub steering angle sensor wiring is twisted, disconnect the connector, straighten the harness, and reconnect the connector.

15. Disconnect the jumper wire from the service check connector.

16. Remove the rear steering center lock pin and install the cap bolt and washer. Tighten the cap bolt to the specified torque.

17. Install the rear steering actuator cover and perform the electronic neutral check described earlier in this chapter.

Classroom Manual
Chapter 13, page 304

Guidelines for Four Wheel Steering Service and Adjustment

1. Prior to 4WS system diagnosis, a preliminary system inspection should be performed.

2. If the 4WS control unit detects a problem in the system, the control unit enters the fail-safe mode.

3. The 4WS system provides a damper control mode that gradually moves the rear wheels to the centered position when the fail-safe mode is entered.

4. A jumper wire may be connected across the service check connector terminals to obtain the 4WS system trouble codes.

5. If the trouble codes are obtained with the engine running, the code display informs the technician if the trouble codes are in the main or sub processor in the control unit.

6. When the control unit stores a trouble code representing the main steering angle sensor system, the clock-radio fuse must be removed to cancel the 4WS indicator light.

7. Some trouble codes represent problems caused by abnormal or harsh driving conditions. The 4WS control unit does not illuminate the 4WS indicator light when these codes are present.

8. The rear steering actuator is serviced as a unit except for sensor and tie rod service.

9. Axial impact or rotational force must not be applied to the shaft screw in the rear steering actuator.

10. During many rear steering actuator service operations, a rear center lock pin must be installed in the actuator.

11. An electronic neutral check may be performed to determine if the front and rear steering sensors are adjusted properly.

12. The front main steering sensor, front sub steering sensor, and rear sub steering sensor are adjustable.

13. The rear main steering sensor is not adjustable.

Photo Sequence 11 on the following page shows a typical procedure for diagnosing an electronically controlled, four wheel steering system.

CASE STUDY

A customer complained about the 4WS indicator light coming on intermittently on a Honda Prelude with an electronic 4WS system. The technician asked the customer about any other steering problems, and the customer reported the car steered normally. The technician road tested the car, but the 4WS light did not come on, which indicated there were no electronic problems in the system. The customer was concerned about a possible safety hazard while driving this vehicle with the 4WS indicator light illuminated. In reply to this concern, the technician explained to the customer about the fail-safe function in the 4WS system and the rear wheels being centered in this mode.

The technician asked the customer about any recent service work completed on the vehicle. In response to this question, the customer replied that the car had been in a rear end collision recently, and since the body work was completed the 4WS indicator light problem started occurring. The technician informed the customer that a 4WS system diagnosis and inspection should be performed.

Since the 4WS indicator light was not illuminated, the technician concluded that a trouble code diagnosis would probably not provide any diagnostic answers. However, the technician checked the system for codes in case there was a code caused by abnormal or harsh driving, which would not cause the indicator light to be illuminated.

When the technician raised the vehicle on a hoist, it was clearly visible that many of the rear suspension and body parts had been replaced recently. Even the rear steering actuator cover had been replaced. The technician removed the rear steering actuator cover to inspect the wiring on the actuator. All the wiring connectors were inspected, including the terminals on the rear main steering angle sensor. When the technician inspected the rear sub steering angle sensor wiring harness, he found this harness had been punctured by a sharp object near the sensor. The technician probed the sensor wires at the sensor, and connected a pair of ohmmeter leads from each wire at the sensor to the corresponding colored wire in the sensor connector. Each wire showed a normal, zero-ohm resistance. The technician repeated these ohmmeter connections and wiggled the wires at the damaged location. On one of the wires, the ohmmeter reading went to infinite while wiggling the wires, indicating an intermittent open circuit.

The technician replaced the rear sub steering angle sensor and performed the electronic neutral check and rear sub steering angle sensor adjustment. During a road test, the 4WS indicator light did not come on.

Photo Sequence 11
Typical Procedure for Diagnosing an Electronically Controlled, Four Wheel Steering System

P11-1 Road test the vehicle to check 4WS operation and indicator light.

P11-2 Raise the vehicle on a lift, and check all electrical connectors on the front steering gear and rear steering actuator.

P11-3 Lower the vehicle and locate the service check connector behind the center console.

P11-4 Look up the diagnostic procedure and trouble codes in the car manufacturer's service manual.

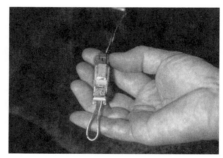

P11-5 Connect a jumper wire between the terminals in the service check connector.

P11-6 Turn on the ignition switch.

P11-7 Observe the 4WS indicator light flashes to obtain the fault codes.

P11-8 Turn off the ignition switch.

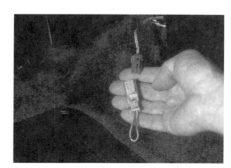

P11-9 Remove the jumper wire from the service check connector.

Terms to Know

Cable reel	Front sub steering angle sensor	Rear sub steering angle sensor
Fail-safe mode	Rear main steering angle sensor	Service check connector
Front main steering angle sensor	Rear steering center lock pin	

ASE Style Review Questions

1. While discussing the fail-safe function:
   Technician A says the 4WS indicator light is illuminated during the fail-safe function.
   Technician B says the rear wheels steer normally when the 4WS control unit enters the fail-safe mode.
   Who is correct?
   - **A.** A only
   - **B.** B only
   - **C.** Both A and B
   - **D.** Neither A nor B

2. While discussing the fail-safe function and damper control:
   Technician A says the rear wheels move instantly to the centered position when the 4WS control unit enters the fail-safe mode.
   Technician B says the return spring moves the rear wheels away from the centered position.
   Who is correct?
   - **A.** A only
   - **B.** B only
   - **C.** Both A and B
   - **D.** Neither A nor B

3. While discussing trouble code diagnosis:
   Technician A says the 4WS system service check connector is located under the driver's seat.
   Technician B says when one of the service check connector terminals is grounded, the 4WS system enters the diagnostic mode.
   Who is correct?
   - **A.** A only
   - **B.** B only
   - **C.** Both A and B
   - **D.** Neither A nor B

4. While discussing trouble code diagnosis:
   Technician A says many 4WS system trouble codes are cancelled when the ignition switch is turned off.
   Technician B says codes representing problems caused by abnormal or harsh driving conditions do not illuminate the 4WS indicator light.
   Who is correct?
   - **A.** A only
   - **B.** B only
   - **C.** Both A and B
   - **D.** Neither A nor B

5. While discussing rear steering actuator service:
   Technician A says the rear steering actuator is a replacement unit except for tie rods and sensors.
   Technician B says the arrows on the rear steering actuator brackets must face downward.
   Who is correct?
   - **A.** A only
   - **B.** B only
   - **C.** Both A and B
   - **D.** Neither A nor B

6. All of these statements about rear steering actuators and actuator service are true EXCEPT:
   - **A.** Axial impact on the shaft screw may damage the actuator.
   - **B.** Rotational force on the shaft screw may damage the actuator.
   - **C.** The engine may be started with the rear steering lock pin in place.
   - **D.** The shaft screw must be held with a special tool while loosening the tie rods.

7. While servicing and adjusting electronically controlled four wheel steering systems:
   - **A.** the steering center lock pin must be installed in the left rear tie rod for many 4WS adjustments.
   - **B.** the electronic neutral check determines if the front or rear steering sensors require adjustment.
   - **C.** the electronic neutral check is performed with the rear wheels fixed in the straight-ahead position.
   - **D.** the 4WS indicator light is illuminated with the engine running if the inner tie rods are loose in the rear steering actuator.

8. While servicing and adjusting electronically controlled four wheel steering systems:
   A. the front main steering angle sensor is not adjustable.
   B. the rear main steering angle sensor is adjustable.
   C. removing the clock-radio fuse erases the diagnostic trouble codes.
   D. if the 4WS light provides 3 long flashes followed by 3 short flashes in the diagnostic mode, code 3 is indicated.

9. While servicing and adjusting electronically controlled four wheel steering systems:
   A. in the diagnostic mode the 4WS indicator light flashes quickly for 10 seconds between the main and sub processor DTCs.
   B. a DTC may be set in the processor memory if the steering wheel is turned with a rear wheel against a curb and the engine running.

   C. if the steering wheel is turned with the engine running and a rear wheel against a curb, the 4WS indicator light is illuminated.
   D. driving the car with the parking brake on has no effect on the 4WS indicator light.

10. While servicing and adjusting electronically controlled four wheel steering systems:
    A. the 4WS indicator light will indicate DTCs and the electronic neutral check at the same time.
    B. the parking brake must be released during the front steering sensor test mode.
    C. the ignition switch must be on during the front steering test mode.
    D. the brake pedal must be depressed during the rear steering test mode.

ASE Challenge Questions

1. While discussing electronic 4WS, *Technician A* says jumping the two terminals of the service check connector with the engine off will display DTCs. *Technician B* says jumping the service check connector then starting the engine displays the processor in which the codes are stored.
   Who is correct?
   A. A only
   B. B only
   C. Both A and B
   D. Neither A nor B

2. The Honda Prelude 4WS system uses a main and a sub processing unit, each storing 10 trouble codes. If the 4WS light on the dash blinks quickly and repeatedly for three seconds, it means:
   A. a DTC is stored in the main processor.
   B. a DTC is stored in the sub processor.
   C. the system is moving from the main to the sub processor memory.
   D. a DTC sequence will be repeated.

3. Honda Prelude temporary "abnormal or harsh driving" 4WS DTCs range from _____ to _____ .
   A. 07/14
   B. 70/74
   C. 17/24
   D. 44/47

4. The Honda Prelude 4WS light has gone on and remains on.
   Technician A says that before performing anu diagnostic tests, the 10A fuse for the clock radio should be removed.
   Technician B says to retrieve the DTC, the 4WS control unit connector must be disconnected.
   A. A only
   B. B only
   C. Both A and B
   D. Neither A nor B

5. While discussing electronic 4WS, *Technician A* says that after repairing a defect of the main steering angle sensor in the Honda Prelude 4WS system, fuse #43 must be removed to cancel the code.
   Technician B says the battery terminal must be removed to cancel DTCs in parts of the Honda Prelude 4WS system other than the main steering angle sensor.
   Who is correct?
   A. A only
   B. B only
   C. Both A and B
   D. Neither A nor B

Table 13-1 NATEF and ASE Task

Diagnose, inspect, adjust, repair, or replace components of electronically controlled steering systems.

Problem Area	Symptoms	Possible Causes	Classroom Manual	Shop Manual
IMPROPER STEERING OPERATION	4WS indicator light on, no rear wheel steering	Electrical defect in 4WS system	302	433

Job Sheet 38

38

Name _____ Date _____

Retrieve Diagnostic Trouble Codes (DTCs), Four Wheel Steering (4WS) System

NATEF and ASE Correlation

This job sheet is related to NATEF and ASE Automotive Suspension and Steering Task List content area: A. Steering Systems Diagnosis and Repair. 2. Power-Assisted Steering Units. Task 16: Diagnose, inspect, adjust, repair, or replace components of electronically controlled steering systems.

Tools and Materials

Jumper wire

Vehicle make and model year _____

Vehicle VIN _____

Procedure

1. Be sure the ignition switch is off and remove the dual terminal service check connector located behind the center console. Connect the two terminals in this connector with a jumper wire.

   Is the jumper wire properly connected? yes _____ no _____

   Instructor check _____

2. Turn the ignition switch on, but do not start the engine.

3. Observe the 4WS indicator light to read the diagnostic trouble codes (DTCs). Three longer flashes followed by a brief pause and one quicker flash indicates code 31. The codes are given in numerical order.

4. List the DTCs provided with ignition switch on, engine not running. Include DTC interpretation.

   1. _____

   2. _____

   3. _____

5. Turn the ignition switch off and start the engine while observing the 4WS indicator light in the instrument panel.

6. Does the 4WS indicator light blink once quickly when the ignition switch is turned on?

   4WS light operation: satisfactory _____ unsatisfactory _____

   If the 4WS light operation is unsatisfactory, describe the light operation.

   _____

   _____

7. After the quick flash in step 6, did the 4WS indicator light pause for 3 seconds?

   4WS light operation: satisfactory _____ unsatisfactory _____

If the 4WS light operation is unsatisfactory, describe the light operation.

_____

_____

8. List the main processor DTCs displayed after the pause in step 7 and include the DTC interpretation.

   1. _____

   2. _____

   3. _____

9. Did the 4WS indicator light pauses for 1.6 seconds after the DTC's displayed in step 8?
yes _____ no _____

4WS light operation: satisfactory _____ unsatisfactory _____

If the 4WS light operation is unsatisfactory, describe the light operation.

_____

_____

10. Did the 4WS indicator light blink quickly for 3 seconds to indicate a separation between the main and sub processor codes? yes _____ no _____

4WS light operation: satisfactory _____ unsatisfactory _____

If the 4WS light operation is unsatisfactory, describe the light operation.

_____

_____

11. Did the 4WS indicator light pause for 1.6 seconds? yes _____ no _____

4WS light operation: satisfactory _____ unsatisfactory _____

If the 4WS light operation is unsatisfactory, describe the light operation.

_____

_____

12. List the sub processor DTCs displayed after the pause in step 11, and include the DTC interpretation.

   1. _____

   2. _____

   3. _____

13. Did the 4WS indicator light pause for 3 seconds and then repeat the cycle?
yes _____ no _____

4WS light operation: satisfactory _____ unsatisfactory _____

If the 4WS light operation is unsatisfactory, describe the light operation.

_____

_____

14. On the basis of all the DTCs displayed, state the required diagnostic procedure to locate the exact cause of the defect(s), and explain the reasons for your diagnosis.

_____

_____

_____

_____

✔ **Instructor Check** _____

Job Sheet 39

Name _____ Date _____

Remove and Replace Rear Steering Actuator

NATEF and ASE Correlation

This job sheet is related to NATEF and ASE Automotive Suspension and Steering Task List content area: A. Steering Systems Diagnosis and Repair. 2. Power-Assisted Steering Units. Task 16: Diagnose, inspect, adjust, repair, or replace components of electronically controlled steering systems.

Tools and Materials

Tie-rod end puller
Torque wrench

Vehicle make and model year _____
Vehicle VIN _____

Procedure

1. Raise the vehicle on a hoist, or lift the rear of the vehicle with a floor jack, and support the chassis with safety stands placed under the chassis at the vehicle manufacturer's recommended locations.

2. Remove the cotter pin and nut from each tie-rod end.

3. Install a 12-millimeter (mm) nut on each tie-rod end until the nuts are flush with the tie-rod stud.

4. Install the special tool on the tie-rod end and with the tool arms parallel, tighten the screw on the tool to loosen the tie-rod end. Repeat the procedure on both tie-rod ends.

5. Remove the nuts from the tie rods and remove the tie rods from the steering arms.

6. Remove the rear steering actuator cover.

7. Remove the the cap bolt and washer and install the rear steering center lock pin.

   Is the rear steering center lock pin installed? yes _____ no _____

   Instructor check _____

8. Remove the ground cable connector and all wiring harness connectors on the rear steering actuator.

   Are the ground cable connector and all wiring harness connectors removed?
   yes _____ no _____

   Instructor check _____

9. Remove the four mounting bolts and bracket and remove the rear steering actuator.

10. Install the rear steering actuator and the four mounting bolts and bracket. The arrow on the bracket must face upward.

    Is the arrow on the bracket facing upward? yes _____ no _____

    Instructor check _____

11. Tighten the rear steering actuator mounting bolts to the specified torque.

Specified rear steering actuator mounting bolt torque _____

Actual rear steering actuator mounting bolt torque _____

> ⚠️ **WARNING:** Tighten the castle nut on the tie-rod ends to the specified torque, then tighten these nuts enough to align the slots in the nut with the hole in the tie-rod pin. Do not loosen the nut to align the nut slots with the tie-rod pin hole.

12. Reconnect the tie-rod ends to the steering arms, and tighten the castle nut to the specified torque. If necessary, tighten the nut slightly to align the nut slots with the tie-rod pin hole.

Specified tie-rod end castle nut torque _____

Actual tie-rod end castle nut torque _____

13. Install the cotter pin in the nut and tie-rod end pin openings, and bend one leg of the cotter pin downward over the nut. Bend the other cotter pin leg upward over the top of the tie-rod end pin.

Are the cotter pins properly installed in the tie-rod end nuts? yes _____ no _____

Instructor check _____

14. Check all the wiring connectors for contamination and clean as necessary. Install all the wiring connectors on the rear steering actuator, and tighten all the terminal nuts to the specified torque.

Wiring terminal condition: satisfactory _____ unsatisfactory _____

Specified wiring terminal nut torque _____

Actual wiring terminal nut torque _____

15. Install the terminal cover on the rear main steering sensor terminals. Remove the rear steering lock pin and install the cap bolt and washer. Leave the steering actuator cover removed until after the final rear steering actuator adjustments.

Is the rear steering lock pin removed? yes _____ no _____

Are the cap bolt and washer properly installed in rear steering lock pin hole?

Instructor check _____

✔️ **Instructor Check** _____

Job Sheet 40

Name _____ Date _____

Remove and Replace Rear Steering Actuator Tie Rods and Tie-Rod Ends

NATEF and ASE Correlation

This job sheet is related to NATEF and ASE Automotive Suspension and Steering Task List content area: A. Steering Systems Diagnosis and Repair. 2. Power-Assisted Steering Units. Task 16: Diagnose, inspect, adjust, repair, or replace components of electronically controlled steering systems.

Tools and Materials

Torque wrench
Wax marker
Rack holding tool

Vehicle make and model year _____
Vehicle VIN _____

Procedure

1. Road test the vehicle and describe the steering problems that indicate the rear steering actuator tie rods and tie-rod ends that require replacement.

   _____
   _____
   _____
   _____

2. Visually inspect the rear steering actuator tie rods and tie-rod ends and list the parts that require replacement. Explain the reasons for your diagnosis.

   _____
   _____
   _____
   _____

3. Mark the relative position of the tie-rod end, lock nut, and tie rod with a wax marker.

Are the tie rod, tie-rod end, and lock nut properly marked?

yes _____ no _____

4. Hold the tie-rod end with a wrench and loosen the lock nut.

5. Remove the tie-rod end.

6. Remove the boot bands and clamps from the inner tie-rod ends.

7. Place the flat side of the rack holding tool toward the actuator housing, and drive the special rack holding tool between the actuator housing and the stop washer with a soft hammer.

Is the rack holding tool properly installed with flat side toward the actuator housing?

yes _____ no _____

8. Straighten the tabs on the tie-rod lock washer.

Are the tabs straightened on tie-rod lock washer? yes _____ no _____

SERVICE TIP: Hold the special holding tool firmly while loosening the tie rod to avoid applying rotational force to the shaft screw in the actuator.

9. Hold the shaft screw with the holding tool, and loosen the tie rod with a wrench.

10. Thread the tie rod off the shaft screw, and repeat this procedure on each tie-rod end.

WARNING: Do not allow dust, dirt, or foreign material to enter the tie-rod end ball joint or boot.

WARNING: Never apply axial impact or rotational force to the shaft screw in the rear steering actuator. Either of these actions may cause internal actuator damage.

11. Install the tie-rod ends so the marks on the tie-rod ends, lock nuts, and tie rods are aligned, and tighten the tie-rod nuts to the specified torque.

Marks on tie rods, tie-rod ends, and lock nuts properly aligned?

yes _____ no _____

Specified tie-rod end lock nut torque _____

Actual tie-rod end lock nut torque _____

Instructor check _____

12. Screw each inner tie rod onto the shaft screw while holding the lock washer so its tabs are in the inner tie-rod end. The stop washer must be installed on the shaft screw with the chamfered side facing outward.

   Is the stop washer properly installed with chamfered side facing outward?

   yes _____ no _____

   Instructor check _____

13. Drive the special holding tool between the actuator housing and the stop washer with a soft hammer.

14. Hold the shaft screw with the holding tool and tighten the inner tie-rod end to the specified torque.

   Specified inner tie rod end torque _____

   Actual inner tie rod end torque _____

15. Bend the lock washer tabs against the flat on the inner tie-rod end.

   Are the lock washer tabs bent against the flat on the inner tie-rod end?

   yes _____ no _____

   Instructor check _____

16. Remove the special holding tool and apply silicone grease to the sliding surface of the tie rod. Place a light coating of silicone grease inside the tie-rod boot.

   Is the sliding surface or the tie rod properly lubricated? yes _____ no _____

   Is the inside surface of tie-rod boot properly lubricated? yes _____ no _____

   Instructor check _____

17. Apply the vehicle manufacturer's recommended grease to the circumference of the inner tie rod joint housing.

   Is the circumference of inner tie rod joint housing properly lubricated?

   yes _____ no _____

   Instructor check _____

18. Install the boots on the actuator housing, and install the boot bands with the locking tabs properly positioned in relation to the actuator housing.

   Are the locking tabs properly positioned in relation to the actuator housing?

   yes _____ no _____

   Instructor check _____

   CAUTION: While staking the boot clamps be careful not to damage the inner tie-rod boots.

19. Tighten the boot bands and bend both sets of locking tabs over the band. Tap lightly on the doubled over portion of the band to reduce its height, and stake the locking tabs firmly.

Are the boot bands properly installed? yes _____ no _____

Are the locking tabs properly staked? yes _____ no _____

Instructor check _____

✓ **Instructor Check** _____

Frame Diagnosis and Service

Upon completion and review of this chapter, you should be able to:

❏ Take the necessary precautions to avoid frame damage.

❏ Diagnose the causes of frame damage.

❏ Follow safety precautions when measuring and welding frames.

❏ Visually inspect frames.

❏ Perform basic frame welding.

❏ Measure frames with a plumb bob and diagnose frame damage.

❏ Measure frames with a tram gauge and diagnose frame damage.

❏ Perform unitized body measurements with a tram gauge.

❏ Perform unitized body measurements with a dedicated bench system.

Indications of Frame Damage

The most common cause of frame damage on cars and light-duty trucks is collision damage. In some cases, the collision damage may be repaired to make the cosmetic appearance of the vehicle satisfactory, but the frame damage may not always be correctable. When driving the vehicle, some indications of frame damage are:

1. Excessive tire wear when the front suspension alignment angles are correct.

2. Steering pull when the front suspension alignment angles are correct.

3. Steering wheel not centered when driving straight ahead, but the steering wheel was centered in the shop.

Frame Diagnosis

Preventing Frame Damage

Since frame problems affect wheel alignment, technicians must be able to diagnose frame defects so they are not confused with wheel alignment problems. Follow these precautions to minimize frame damage:

1. Do not overload the vehicle.

2. Place the load evenly in a vehicle.

3. Do not operate the vehicle on extremely rough terrain.

4. Do not mount equipment such as a snow plow on a vehicle unless the frame is strong enough to carry the additional load and force.

Diagnosis of Frame Problems

Side Sway. The causes of frame side sway are the following:

1. Collision damage

2. Fire damage

3. Use of equipment on the vehicle for which the frame was not designed

Basic Tools

Basic technician's tool set

Service manual

Tape measure

Chalk

Floor jack

Safety stands

Sag. The causes of frame sag are:

1. Vehicle loads that exceed the load-carrying capacity of the frame
2. Uneven load distribution
3. Sudden changes in **section modulus**
4. Holes drilled in the **frame flange**
5. Too many holes drilled in the **frame web**
6. Holes drilled too close together in the web
7. Welds on the frame flange
8. Cutting holes in the frame with a cutting torch
9. Cutting notches in the frame rails
10. A fire involving the vehicle
11. Collision damage
12. The use of equipment for which the frame was not designed

Buckle. The causes of frame buckle are:

1. Collision damage
2. Using equipment such as a snow plow when the frame was not designed for this type of service
3. A fire involving the vehicle

Diamond-Shaped. Diamond-shaped frame damage may be caused by:

1. Collision damage
2. Towing another vehicle with a chain attached to one corner of the frame
3. Being towed by another vehicle with a chain attached to one corner of the vehicle frame

Twist. Frame twist may be caused by:

1. An accident or collision, especially one involving a rollover
2. Operating the vehicle on extremely rough terrain

Checking Frame Alignment

Safety Concerns

While servicing vehicle frames, always wear the proper safety clothing and safety items for the job being performed. This includes proper work clothing, safety goggles, ear protection, respirator, proper gloves, welding shield, and safety shoes (Figure 14-1).

If arc welding is necessary on a vehicle frame, follow these precautions:

1. Remove the negative battery cable before welding (Figure 14-2).
2. Remove the fuel tank before welding (Figure 14-3).
3. Protect the interior and exterior of the vehicle as necessary (Figure 14-4).

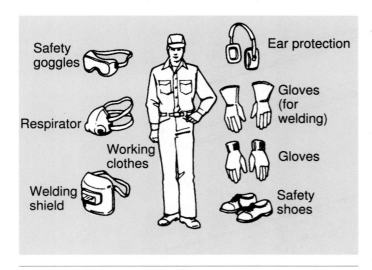

Figure 14-1 Safety items for frame service. (Courtesy of Oldsmobile Motor Division, General Motors Corp.)

Figure 14-2 Remove the negative battery cable before arc welding on a vehicle. (Courtesy of Oldsmobile Motor Division, General Motors Corp.)

Visual Inspection

Prior to frame measurement, the frame and suspension should be visually inspected. Check for wrinkles on the upper flange of the frame, which indicate a sag problem. Visually inspect the lower flange for wrinkles, which are definite evidence of buckle. Since suspension or axle problems may appear as frame problems, the suspension components should be inspected for wear and damage. For example, an offset rear axle may appear as a diamond-shaped frame. Check all suspension mounting bushings and inspect leaf-spring shackles and center bolts.

The frame should be inspected for cracks, bends, and severe corrosion. Minor frame bends are not visible, but severe bends may be visible. Straight cracks may occur at the edge of the frame flange, and **sunburst cracks** may radiate from a hole in the frame web or crossmember (Figure 14-5).

Figure 14-3 Remove the fuel tank before arc welding on a vehicle. (Courtesy of Oldsmobile Motor Division, General Motors Corp.)

Figure 14-4 Protect the interior and exterior of the vehicle while welding. (Courtesy of Oldsmobile Motor Division, General Motors Corp.)

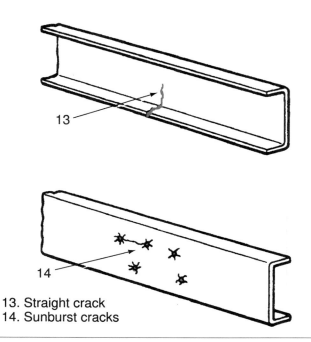

13. Straight crack
14. Sunburst cracks

Figure 14-5 Straight cracks may occur in the frame flange; sunburst cracks may radiate from a hole in the frame web or crossmember. (Courtesy of Chevrolet Motor Division, General Motors Corp.)

Frame Welding

 WARNING: Always follow the vehicle manufacturer's recommendations in the service manual regarding frame welding and reinforcing, or mounting additional equipment on the frame.

 WARNING: If the frame is cracked, frame alignment is often necessary.

Follow these steps for a typical frame welding procedure:

1. Remove any components that would interfere with the weld or be damaged by heat.
2. Find the extreme end of the crack and drill a 0.250-in (6-mm) hole at this point.
3. V-grind the entire length of the crack from the starting point to the drilled hole.
4. The bottom of the crack should be opened 0.062 in (2 mm) to allow proper weld penetration. A hacksaw blade may be used to open the crack.
5. Arc weld with the proper electrode and welding procedure.

Frame Measurement, Plumb Bob Method

Photo Sequence 12 shows a typical procedure for performing frame measurement using the plumb bob method.

Photo Sequence 12 Typical Procedure for Performing Frame Measurement, Plumb Bob Method

P12-1 Park the vehicle on a level area of the shop floor.

P12-2 Raise the front suspension with a floor jack and lower the chassis onto safety stands positioned at the manufacturer's recommended lifting points.

P12-3 Raise the rear suspension with a floor jack and lower the chassis onto safety stands positioned at the manufacturer's recommended lifting points.

P12-4 Suspend a plumb bob at the manufacturer's recommended frame measurement locations and place a chalk mark on the floor directly under the plumb bob.

P12-5 Use a floor jack to lift the vehicle, remove the safety stands, and lower the vehicle.

P12-6 Drive the vehicle away from the chalk-marked area.

P12-7 Use a tape measure to measure the vehicle's frame measurements between the chalk marks on the floor.

P12-8 Compare the frame measurements obtained to the vehicle manufacturer's specifications in the service manual.

Locating the Frame Centerline. Some vehicle manufacturers recommend measuring the frame with the **plumb bob** method. Follow these steps to complete a plumb bob measurement for frame damage:

1. Place the vehicle on a level area of the shop floor, and use a floor jack to raise the front and rear suspension off the floor. Support the chassis on safety stands at the manufacturer's recommended locations.

2. Suspend a plumb bob at locations 1, 2, 11, and 12 on the inside of the frame web, and allow the plumb bob to almost touch the floor surface (Figure 14-6). Points 1 and 2 are at the centerline of the slotted hole in each front bumper bracket, and points 11 and 12 are on the inside of the rear frame web. Mark these plumb bob locations on the floor with chalk.

 WARNING: The points on the left and right frame webs must be at the same location on each web for accurate measurements.

3. Use a plumb bob to transfer points 3 through 10 from the frame to chalk marks on the floor.

4. Raise the vehicle with the floor jack, remove the safety stands, and lower the vehicle onto the floor. Move the vehicle away from the chalk-marked area.

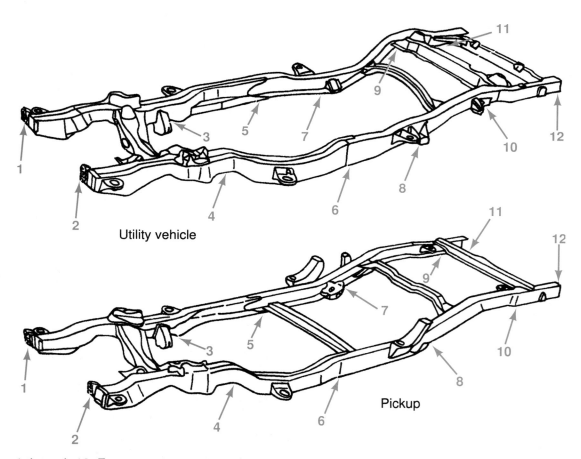

Utility vehicle

Pickup

1 through 12. Frame measurement points

Figure 14-6 Suspend a plumb bob at the locations shown on the frame and mark these locations on the floor with chalk. (Courtesy of Chevrolet Motor Division, General Motors Corp.)

SERVICE TIP: When using a tape measure, avoid twists and bends in the tape to provide accurate readings.

5. Measure the distance between points 1 and 2 with a tape measure and chalk mark the exact half-way point in this distance. This mark is the frame centerline at the front.

6. Measure the distance between points 11 and 12 and chalk mark the exact center of this measurement. This chalk mark is the frame centerline at the rear.

7. Draw a straight chalk line from the centerline at the front of the frame to the centerline at the rear of the frame. This chalk line is the complete centerline of the frame.

Horizontal Frame Measurements. The procedure used for horizontal frame measurements follows:

1. Measure the distance from the frame centerline to points 3 through 10. Each pair of these points should be equal within 0.125 in (3 mm). For example, the distance from point 5 to the centerline should be equal to the distance from point 6 to the centerline.

2. Measure diagonally from point 1 to 6 and point 2 to 5. These distances should be equal or within the manufacturer's specified tolerance. Place a straight chalk mark from points 1 to 6 and points 2 to 5. These diagonal lines should cross each other at the frame centerline.

3. Repeat step 2 at all the other diagonal frame measurements, such as points 3 to 10, 4 to 9, 5 to 12, and 6 to 11. All these diagonal measurements should be equal or within the vehicle manufacturer's specified tolerance. Each pair of diagonal lines must cross each other at the vehicle centerline (Figure 14-7).

When frame problems such as side sway, buckle, or diamond condition are present, some of the horizontal frame measurements are not within specifications, and some of the diagonal chalk lines do not cross at the frame centerline, depending on the location of the damaged area.

Frame Measurement, Tram Gauge Method

TRADE JARGON: A tram gauge may be called a tracking or track gauge.

A **tram gauge** is a long, straight bar with two adjustable pointers. The distance between the pointers is adjustable, and the height of the pointers is also adjustable (Figure 14-8).

The horizontal frame measurements may be completed with a tram gauge rather than using the plumb bob method. When measuring a frame, horizontal measurements are completed straight across the frame and diagonally forward and rearward on the frame as we have discussed previously.

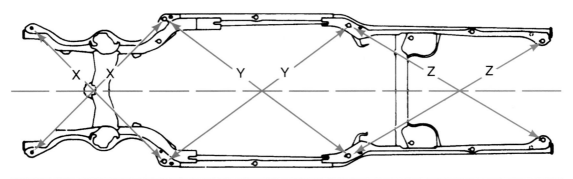

Figure 14-7 Each pair of diagonal measurement lines on a frame must cross each other at the frame centerline. (Courtesy of Chevrolet Motor Division, General Motors Corp.)

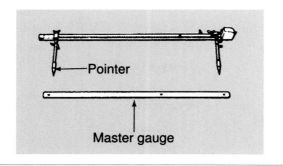

Figure 14-8 Tram gauge for frame measurements. (Reprinted with permission)

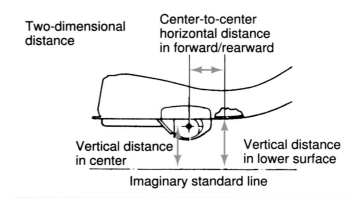

Figure 14-9 Horizontal and vertical frame measurements. (Reprinted with permission)

An imaginary horizontal line parallel to the frame is used for vertical frame measurements. This line is called the datum line.

Vertical measurements from specific locations on the frame to an imaginary standard line are also necessary to detect such conditions as frame sag and twist (Figure 14-9). This imaginary horizontal line used for vertical frame measurements is called the **datum line.**

 SERVICE TIP: When the tram gauge is installed on the frame, be sure the pointers are seated properly in the frame openings (Figure 14-10).

When vertical frame measurements are completed with the tram gauge, the gauge pointers must be set at the manufacturer's specified distance for each frame location (Figure 14-11). With the tram gauge pointers set at the specified height and the gauge properly installed across the frame in the recommended frame openings, the tram gauge should be level if the vertical frame measurements are within specifications. If the frame is twisted, the tram gauge is not level when adjusted to specifications and installed in the twisted area.

To measure frame sag, three tram gauges may be installed at various locations for vertical frame measurement. When viewed from the front or the rear, the tram gauges must be level with each other. If the tram gauge near the center of the frame is lower than the tram gauges at the front and rear of the frame, the frame is sagged.

Special Tools

Tram gauge

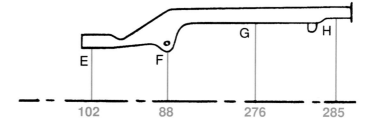

All tolerances ±3 mm
All dimensions are calculated on a horizontal plane

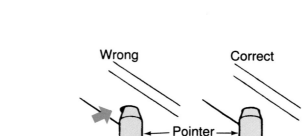

Figure 14-10 Correct and incorrect tram gauge pointer seating in the frame openings. (Reprinted with permission)

Figure 14-11 Tram gauge setting and locations for vertical frame measurements, rear subframe. (Courtesy of Cadillac Motor Car Division, General Motors Corp.)

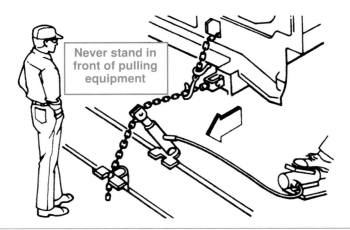

Figure 14-12 Do not stand in front of pulling equipment while it is in operation. (Courtesy of Oldsmobile Motor Division, General Motors Corp.)

Frame Straightening

Frame straightening is usually done with special hydraulically operated bending equipment. This equipment must be operated according to the equipment manufacturer's recommended procedures. Since frame straightening is usually done by experienced body technicians, we will not discuss this service in detail. All safety precautions must be observed while operating frame-straightening equipment. Never stand in front of pulling equipment when it is in operation (Figure 14-12).

Classroom Manual
Chapter 14, page 316

Measuring Unitized Body Alignment

Tram Gauge

A tram gauge may be used to perform various measurements on unitized bodies such as the upper strut towers (Figure 14-13) and front crossmember (Figure 14-14).

Since the lower control arms are attached to the front crossmember or cradle, front wheel alignment angles are affected by a bent crossmember. The engine and transaxle mounts are also connected to the front crossmember. Therefore, a bent crossmember may cause improper engine

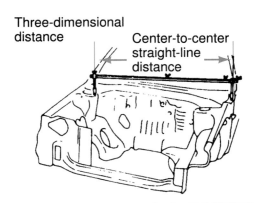

Figure 14-13 Performing an upper strut tower measurement with a tram gauge on a unitized body vehicle. (Reprinted with permission)

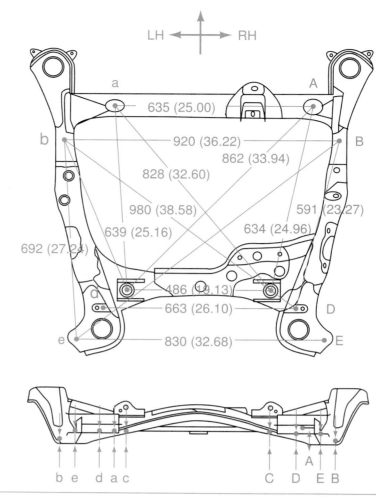

LH ← → RH

a — 635 (25.00) — A

b — 920 (36.22) — B

862 (33.94)

828 (32.60)

980 (38.58)   591 (23.27)

639 (25.16)   634 (24.96)

692 (27.24)

486 (19.13)

663 (26.10) — D

e — 830 (32.68) — E

b e   d a c   C D E B

A

Figure 14-14 Tram gauge measurements on the front crossmember of a unitized body vehicle. (Reprinted with permission)

and transaxle position, which results in front drive axle vibration. Loose or worn front crossmember mounts may cause improper front wheel alignment and engine position. The tram gauge may be used to perform vertical measurements on the front and rear subframes on a unitized body vehicle.

Bench

A **dedicated bench system** is necessary to check many unitized body measurements after medium or heavy collision damage. The dedicated bench system has three main parts: the bench, transverse beams, and the dedicated fixtures. When used together, this equipment will perform many undercar unitized body measurements, including length, width, and height at the same time. The bench contains strong steel beams mounted on heavy casters (Figure 14-15). The top of the bench acts as a datum line.

Transverse Beams

The transverse beams are mounted perpendicular to the bench. There are a variety of holes in the bench for proper transverse beam attachment (Figure 14-16). Various holes in the transverse beams provide the correct dedicated fixture positions.

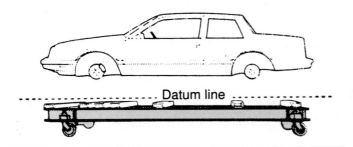

Figure 14-15 Bench from a dedicated bench system. (Courtesy of Oldsmobile Motor Division, General Motors Corp.)

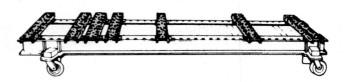

Figure 14-16 Transverse beams placed on the bench. (Courtesy of Oldsmobile Motor Division, General Motors Corp.)

Dedicated Fixtures

The dedicated fixtures are specific to the body style being measured (Figure 14-17). These fixtures are bolted to the transverse beams, and the holes in the upper end of the fixtures must be aligned with specific body openings (Figure 14-18).

The unitized body must be straightened until the holes in the dedicated fixtures and the body openings are aligned. If the holes in the unitized body are behind and above the dedicated fixture openings, the unitized body must be pulled forward and downward in that area (Figure 14-19).

Universal benches are now available with computer measuring systems. Measuring a unitized body with a bench system and straightening these bodies with special body pulling equipment is usually done in an autobody shop rather than an automotive repair shop.

Unitized Body Straightening

CAUTION: Always follow all the recommended precautions and procedures in the vehicle manufacturer's service manual and in the equipment manufacturer's operator's manual when straightening a unitized body. Failure to follow these precautions and procedures may result in severe personal injury and property damage.

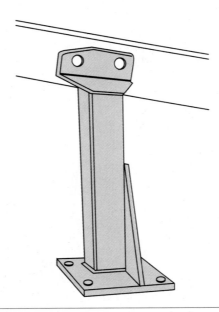

Figure 14-17 Dedicated fixtures are specific to the body style being measured. (Courtesy of Oldsmobile Motor Division, General Motors Corp.)

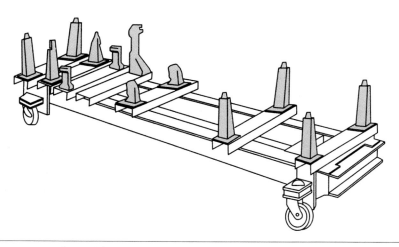

Figure 14-18 Dedicated fixtures are bolted to the transverse beams in specific locations depending on the body style. (Courtesy of Oldsmobile Motor Division, General Motors Corp.)

Classroom Manual
Chapter 14, page 317

Since unitized body straightening is usually performed by experienced body technicians, our discussion is brief on this subject. During the straightening process on a unitized body, the body is securely bolted to the dedicated bench at all the possible locations except those where misalignment is present. The dedicated bench is securely chained to special holding fixtures in the shop floor. Hydraulically operated pulling equipment is then attached securely to the unitized body to pull the body in the intended direction. Pulling equipment may be **single pull** in one location (Figure 14-20), or **multipull** in several locations at the same time (Figure 14-21). The unitized body is pulled until the holes in the body are aligned with the holes in the dedicated fixtures.

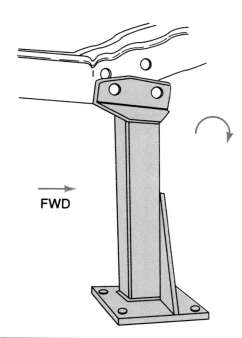

Figure 14-19 When the holes in the body are behind and above the openings in the dedicated fixture, the unitized body must be pulled forward and downward in that area. (Courtesy of Oldsmobile Motor Division, General Motors Corp.)

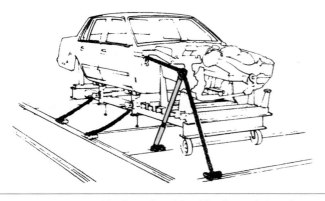

Figure 14-20 Single pull unitized body straightening system. (Courtesy of Oldsmobile Motor Division, General Motors Corp.)

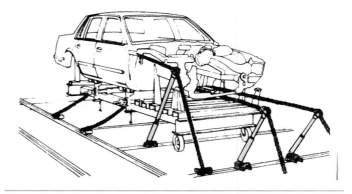

Figure 14-21 Multipull unitized body straightening system. (Courtesy of Oldsmobile Motor Division, General Motors Corp.)

Guidelines for Diagnosing and Measuring Frames and Unitized Bodies

1. Frame damage may be caused by vehicle overloading or uneven loading.

2. Frame damage may result from operating the vehicle on extremely rough terrain or mounting additional equipment, such as a snow plow, on the vehicle when the frame was not designed for this type of equipment.

3. Frame side sway may be caused by collision damage, fire damage, or the use of equipment for which the frame was not designed.

4. There are many causes of frame sag, including excessive or uneven loads, drilling too many holes in the frame, cutting holes or notches in the frame with a torch, collision or fire damage, and sudden changes in section modulus.

5. Frame buckle may be caused by collision or fire damage, and additional equipment attached to the frame for which the frame was not designed.

6. A diamond-shaped frame may be caused by collision damage or by towing and/or being towed with the chain attached to one corner of the frame.

7. Frame twist may result from collision damage or operating the vehicle on extremely rough terrain.

8. When arc welding on a vehicle, follow the basic safety precautions, including wearing proper protective clothing, removing the negative battery cable, removing the fuel tank, and protecting the interior and exterior of the vehicle.

9. Frames should be visually inspected for wrinkles on the upper and lower flanges, and for cracks or bends.

10. A plumb bob may be suspended from specific points on the vehicle frame, and these points marked with chalk. The distance between these points may be measured with a tape measure to check the frame.

11. Horizontal and vertical frame measurements may be performed with a tram gauge.

12. When the tram gauge is installed to measure vertical frame measurements, the pointers should be set at the specified distance and the tram gauge should be level.

13. A tram gauge may be used for various measurements in unitized body vehicles.

14. A dedicated bench system is necessary to check many unitized body measurements after medium or heavy collision damage.

15. The holes in the dedicated bench fixtures must be aligned with the openings in the unitized body if the body measurements are correct.

CASE STUDY

A customer complained about a vibration problem on an Oldsmobile Cutlass Supreme with front-wheel drive. Further questioning of the customer revealed that the vibration occurred when accelerating and decelerating at 40 to 60 mph (64 to 96 kmph). During a road test, the technician found the customer's description of the problem was accurate. The car did have a vibration problem at the speed indicated by the customer. While road testing the car, the technician thought about the causes of this problem and decided the most likely cause of the vibration would be inner front drive axle joints. The technician informed the customer that a drive axle joint inspection was necessary.

The technician raised the vehicle on a hoist and checked the front drive axle joints. He was surprised to discover the inner and outer joints were all in good condition with no cracked boots or looseness. An inspection of the engine cradle indicated the crosspiece at the rear of the cradle was severely damaged near the left side of the cradle. It appeared this crosspiece may have struck an object, such as a rock, during off-road driving. A check of the cradle mounts indicated the cradle openings were jammed into the rubber mounts on the left side of the cradle. The rubber mounts were bulged at the rear of the cradle openings, indicating the left side of the cradle was driven rearward.

The technician informed the customer that a cradle removal, measurement, and possible replacement was necessary. After the customer approved this service operation, the technician removed the cradle and measured it with a tram gauge. The diagonal measurements on the cradle were not within the manufacturer's specifications. Since the cradle was severely distorted, a replacement cradle was installed, and all damaged cradle mounting bushings were replaced. A road test indicated the vibration problem was corrected.

Terms to Know

Datum line	Multipull	Sunburst cracks
Dedicated bench system	Plumb bob	Tram gauge
Frame flange	Section modulus	Yield strength
Frame web	Single pull	

ASE Style Review Questions

1. While discussing frame sag:
   Technician A says frame sag may be caused by holes drilled in the frame flange.
   Technician B says frame sag may be caused by holes drilled too close together in the frame web.
   Who is correct?

   A. A only   **C.** Both A and B
   B. B only   **D.** Neither A nor B

2. While discussing frame side sway:
   Technician A says frame side sway may be caused by uneven load distribution.
   Technician B says frame side sway may result from too many holes drilled in the frame web.
   Who is correct?

   A. A only   **C.** Both A and B
   B. B only   **D.** Neither A nor B

3. While discussing a diamond-shaped frame condition:
   Technician A says this condition may be caused when the vehicle is involved in a fire.
   Technician B says this condition may be caused by towing another vehicle with the chain attached to one corner of the frame.
   Who is correct?
   A. A only
   B. B only
   C. Both A and B
   D. Neither A nor B

4. When discussing the plumb bob method of frame measurement:
   Technician A says the distance from the frame centerline to the same point on each side of the frame should be equal.
   Technician B says diagonal frame measurement lines on each side of the vehicle should cross each other at the frame centerline.
   Who is correct?
   A. A only
   B. B only
   C. Both A and B
   D. Neither A nor B

5. When discussing vertical frame measurements:
   Technician A says the datum line is an imaginary horizontal line parallel to the vehicle frame that is used for vertical frame measurements.
   Technician B says if three tram gauges are properly adjusted and installed at specified locations on the frame, these gauges should be level with each other.
   Who is correct?
   A. A only
   B. B only
   C. Both A and B
   D. Neither A nor B

6. All of these statements about frame problems are true EXCEPT:
   A. Frame sag may be caused by overloading the vehicle.
   B. Frame sag may be caused by the use of equipment for which the frame was not designed.
   C. Frame buckle may be caused by using a snow plow on the front of the vehicle.
   D. A diamond-shaped frame may be caused by operating the vehicle on extremely rough terrain.

7. While diagnosing and inspecting vehicle frames:
   A. wrinkles on the lower frame flange indicate frame buckle.
   B. fire damage may cause frame twist.
   C. unequal vehicle load distribution may cause frame side sway.
   D. frame twist may be caused by cutting notches in the frame flanges.

8. When welding a vehicle frame crack:
   A. a 0.250-in hole should be drilled in the center of the crack.
   B. the crack should be V-ground at both ends of the crack.
   C. disconnect the negative battery cable before welding the crack.
   D. use a cutting torch to open the crack for proper weld penetration.

9. While performing unitized body measurements:
   A. a dedicated bench system performs measurements on upper strut towers.
   B. fixtures on a dedicated bench system are specific for the vehicle body being straightened.
   C. the fixtures are bolted to the bench in a dedicated bench system.
   D. the transverse beams on a dedicated bench system are specific for the body being straightened.

10. While inspecting and measuring a front cradle on a front-wheel-drive car:
    A. front wheel alignment angles may be changed by a bent front cradle.
    B. engine mounting position is not affected by a bent front cradle.
    C. a bent front cradle has no effect on front drive axle vibration.
    D. a bent front cradle may cause premature front wheel bearing failure.

ASE Challenge Questions

1. When making horizontal frame measurements to determine if the frame or unibosy is true, all of the measurments should be made with reference to which one of the following?
   A. Vertical
   B. Centerline
   C. Horizontal
   D. Body bushings

2. A bent front crossmember affects all of the following alignments EXCEPT:
   A. centerline.
   B. suspension.
   C. axle.
   D. engine.

3. While discussing frame damage, *Technician A* says a possible indication of frame damage is an uncentered steering wheel when driving straight ahead. *Technician B* says suspension alignment may be impossible if the frame is bent.
   A. A only
   B. B only
   C. Both A and B
   D. Neither A nor B

4. Which of the following makes this statement true? Measurement points of a frame or unibody using a tram gauge are . . .
   A. seam notches.
   B. calculated by computer.
   C. based on a vertical datum plane.
   D. manufacturer specified frame openings.

5. Referring to Figure 14-14, *Technician A* says the measurement between "e" and "b" should be no more that ±1.0 inch different from the measurement between "E" and "B."
   Technician B says frame damage problems, such as buckling, can be determined if the measurements between points "e" and "E" are more than ±0.25 inch different than between "b" and "B."
   Who is correct?
   A. A only
   B. B only
   C. Both A and B
   D. Neither A nor B

Table 14-1 NATEF and ASE TASK

Check front cradle (subframe) alignment; determine needed repairs or adjustments.

Problem Area	Symptoms	Possible Causes	Classroom Manual	Shop Manual
STEERING CONTROL	Steering pull while driving straight ahead	1. Improper camber caused by improperly positioned cradle	318	476
		2. Improper setback caused by improperly positioned cradle	318	476

Job Sheet 41

Name _____ Date _____

Frame Measurement, Plumb Bob Method

Tools and Materials

Floor jack
Safety stands
Plumb bob
Chalk

Vehicle make and model year _____

Vehicle VIN _____

Procedure

1. Place the vehicle on a level area of the shop floor and use a floor jack to raise the front and rear suspension off the floor. Support the chassis on safety stands at the manufacturer's recommended locations.

   Is the vehicle chassis supported securely on safety stands? yes _____ no _____

   Instructor check _____

2. Locate the frame measurement locations in the vehicle manufacturer's service manual.

3. Suspend a plumb bob at each frame measurement location and place a chalk mark on the floor directly below the tip of the plumb bob.

   Are all frame measurement locations chalk marked on the shop floor? yes _____ no _____

   Instructor check _____

    WARNING: The measurement points on the left and right frame webs must be at the same location on each web to obtain accurate measurements.

4. Raise the vehicle with the floor jack, remove the safety stands, and lower the vehicle onto the floor. Move the vehicle away from the chalk-marked area.

    SERVICE TIP: When using a tape measure, avoid twists and bends in the tape to provide accurate readings.

5. Measure the distance between the two chalk marks directly across from each other at the front of frame, and chalk mark the exact half-way point in this distance. This mark is the frame centerline at the front.

   Distance between the two chalk marks directly across from each other at the front of the frame _____

Mid-point in the distance between the two chalk marks directly across from each other at the front of the frame _____

6. Measure the distance between the two chalk marks directly across from each other at the rear of the frame, and chalk mark the exact half-way point in this distance. This mark is the frame centerline at the rear.

Distance between the two chalk marks directly across from each other at the rear of the frame _____

Mid-point in the distance between the two chalk marks directly across from each other at the rear of the frame _____

7. Draw a straight chalk line from the centerline at the front of the frame to the centerline at the rear of the frame. This chalk line is the complete centerline of the frame.

Is the vehicle centerline chalk marked on shop floor? yes _____ no _____

Instructor check _____

8. Use a tape measure to measure the distance from the frame centerline to all frame measurement points directly opposite each other. Each pair of these points should be equal within 0.125 in (3 mm).

Frame measurements between specified measurement locations directly opposite each other and vehicle centerline starting at the front of the frame:

A. left side _____   right side _____   difference _____

B. left side _____   right side _____   difference _____

C. left side _____   right side _____   difference _____

D. left side _____   right side _____   difference _____

Necessary frame repairs: _____

_____

_____

_____

9. Use a tape measure to complete all diagonal frame measurements at the vehicle manufacturer's recommended locations. These distances should be equal or within the manufacturer's specified tolerances.

Distances between diagonal frame measurement points:

Diagonal frame measurements at the front of the frame:

diagonal A _____ diagonal B _____
difference between diagonal A and B _____

Diagonal frame measurements at the center of the frame:

diagonal C _____ diagonal D _____
difference between diagonal C and D _____

Diagonal frame measurements at the rear of the frame:

diagonal E _____ diagonal F _____
difference between diagonal E and F _____

Necessary frame service: _____

_____

_____

_____

10. Use a long, straight steel bar to place a straight chalk mark between all frame diagonal measurement locations. These diagonal lines should cross each other at the frame centerline.

If all diagonal lines do not cross the frame centerline at the same location, state the required frame repairs. _____

_____

_____

☑ Instructor Check _____

Job Sheet 42

42

Name _____ Date _____

Inspect and Measure Front Cradle

NATEF and ASE Correlation

This job sheet is related to NATEF and ASE Automotive Suspension and Steering Task List content area: B. Suspension Systems Diagnosis and Repair. C. Wheel Alignment Diagnosis, Adjustment, and Repair. Task 15: Check front cradle (subframe) alignment; determine needed repairs or adjustments.

Tools and Materials

Tram gauge

Vehicle make and model year _____

Vehicle VIN _____

Procedure

1. Raise the vehicle on a lift using the vehicle manufacturer's specified lifting points.

2. Inspect front cradle mounts for looseness, damage, oil soaking, wear, and deterioration.

   Cradle mount condition: right front _____ right rear _____

   left front _____ left rear _____

3. Inspect front cradle for visible bends and damage.

   Front cradle condition indicated in visible inspection: _____

   _____

   _____

   _____

4. Check front cradle alignment.

   Is the front cradle aligning hole(s) properly aligned with the matching hole in the chassis?

   yes _____ no _____

   Recommended cradle service _____

   _____

   _____

   _____

5. Use a tram gauge to complete all measurements across the width of the cradle starting at the front of the cradle.

   Width measurement A _____

   Specified width measurement A _____

   Width measurement B _____

   Specified width measurement B _____

Width measurement C _____

Specified width measurement C _____

Width measurement D _____

Specified width measurement D _____

Width measurement E _____

Specified width measurement E _____

Necessary cradle service: _____

_____

_____

_____

6. Use a tram gauge to complete all front-to-rear measurements on the cradle starting on the left side of the cradle.

Front-to-rear measurement A _____

Specified front-to-rear measurement A _____

Front-to-rear measurement B _____

Specified front-to-rear measurement B _____

Front-to-rear measurement C _____

Specified front-to-rear measurement C _____

Front-to-rear measurement D _____

Specified front-to-rear measurement D _____

Necessary cradle service: _____

_____

_____

_____

7. Use a tram gauge to complete all diagonal cradle measurements starting at the left side of the cradle.

Diagonal measurement A _____

Specified diagonal measurement A _____

Diagonal measurement B _____

Specified diagonal measurement B _____

Diagonal measurement C _____

Specified diagonal measurement C _____

Diagonal measurement D _____

Specified diagonal measurement D _____

Necessary cradle service: _____

_____

_____

_____

☑ **Instructor Check** _____

Job Sheet 43

43

Name _____ Date _____

Inspect and Weld Vehicle Frame

Tools and Materials

Hand pump
Arc welder
Welding hammer
Welding rods

Vehicle make and model year _____
Vehicle VIN _____

Procedure

> ■ **CAUTION:** Always wear eye protection while working in the shop, and use the specified eye protection while arc welding to prevent eye injury.

1. Raise the vehicle on a lift using the vehicle manufacturer's specified lifting points.

2. Read the instructions regarding disconnecting the negative battery cable in the vehicle manufacturer's service manual. These instructions may include connecting a 12V power source to the cigarette lighter socket to maintain power to computer memories, radio-stereo, and other electronic equipment.

   List the vehicle manufacturer's instructions regarding disconnecting the negative battery cable.
   _____
   _____
   _____

   Have these instructions been completed on the vehicle being serviced?
   yes _____ no _____

   Instructor check _____

3. If the vehicle is air-bag-equipped, determine the specified waiting period after the negative battery cable is disconnected before performing service work.
   Waiting period _____

4. Disconnect the negative battery cable.

   Is the negative battery cable disconnected? yes _____ no _____

   Has the specified waiting period been completed before servicing the vehicle?
   yes _____ no _____

   Instructor check _____

5. Use a hand pump to pump all the fuel from the fuel tank. Place the fuel in the proper fuel safety containers and store this fuel away from the work area and other ignition sources.

   Has all the fuel been pumped from the tank? yes _____ no _____

   Is this fuel placed in proper fuel safety containers and stored away from the work area?
   yes _____ no _____

   Instructor check _____

6. Remove the fuel tank from the vehicle and store the fuel tank away from the work area and other ignition sources.

Is the fuel tank removed? yes _____ no _____

Fuel tank stored away from the work area? yes _____ no _____

Instructor check _____

7. Protect the interior and exterior of the vehicle from heat or sparks generated by the welding process.

Are the vehicle's interior and exterior protected? yes _____ no _____

Instructor check _____

8. Inspect the entire frame for cracks, bends, severe corrosion, and wrinkles in the flanges.

Defective frame conditions: _____

_____

_____

_____

9. Remove any components that would interfere with the weld or be damaged by heat.

Have all components been removed that would interfere with the weld or be damaged by the heat? yes _____ no _____

Instructor check _____

10. Find the extreme end of the crack and drill a 0.250-in (6-mm) hole at this point.

Is a 0.250-in hole drilled at the end of the crack? yes _____ no _____

Instructor check _____

11. V-grind the entire length of the crack from the starting point to the drilled hole.

Is the entire length of the crack V-ground? yes _____ no _____

Instructor check _____

12. The bottom of the crack should be opened 0.062-in (2-mm) to allow proper weld penetration. A hacksaw blade may be used to open the crack.

Is the bottom of the crack open 0.062-in (2-mm)? yes _____ no _____

Instructor check _____

13. Arc weld the crack with the electrodes specified by the frame and/or vehicle manufacturer. Complete several passes with the arc weld to fill the crack.

Is the crack filled with arc weld material? yes _____ no _____

Instructor check _____

14. Use a pointed welding hammer to remove slag from the exterior of the weld.

Is the slag removed from the weld? yes _____ no _____

Instructor check _____

☑ **Instructor Check** _____

Four Wheel Alignment Procedure

Upon completion and review of this chapter, you should be able to:

❏ Perform a prealignment inspection.

❏ Diagnose wheel alignment and tire wear problems.

❏ Position a vehicle on the alignment rack and connect the wheel units on a computer wheel aligner to the wheel rims.

❏ Select the specifications in the computer wheel aligner for the vehicle being aligned.

❏ Perform a preliminary wheel alignment inspection with the preliminary inspection screen.

❏ Perform a ride height inspection and measurement with the ride height screen.

❏ Perform an automatic or manual wheel runout compensation procedure.

❏ Measure front wheel camber and caster.

❏ Measure front and rear wheel setback.

❏ Measure steering axis inclination (SAI).

❏ Measure front wheel toe and turning radius.

❏ Measure toe change during front wheel jounce and rebound to check steering linkage height.

❏ Diagnose bent front struts.

❏ Recognize the symptoms of improper rear wheel alignment.

❏ Measure rear wheel camber and toe.

Wheel Alignment Preliminary Diagnosis and Inspection

Customer Complaints Related to Suspension or Brakes

Drivers may experience a variety of symptoms related to incorrect wheel alignment or defective brakes (Figure 15-1). Not all the symptoms are related to wheel alignment, but the customer may request an alignment to correct these problems. The technician must diagnose and correct the cause of the specific complaint. Wheel alignment angles that affect tire wear are toe and camber. The most likely symptoms of incorrect toe or camber settings are pull to one side while driving, wander, feathered tire wear, or tire wear on one side.

Road Test

In many cases, a road test is necessary to verify customer complaints. During a road test, the vehicle should be driven under the conditions when the customer complaint occurred. While road testing the vehicle, the technician should listen for any unusual noises related to suspension and steering. The technician should check for these steering problems during a road test:

1. Excessive vertical chassis oscillations

2. Chassis waddle

3. Steering wander, pull, or drift

4. High steering effort and binding

5. Tire squeal while cornering

6. Bump steer

7. Torque steer

Basic Tools

Basic technician's tool set

Service manual

A road test may be necessary to verify customer complaints prior to a wheel alignment.

Which result from wheel alignment?

- Vibration, shimmy
- Pull while driving
- Pull while braking
- Wander
- Slow returnability
- Wheel off-center
- Tire wear
- High effort

Are any symptoms really misunderstandings of normal characteristics?

Figure 15-1 Customer complaints related to wheel alignment or brakes. (Courtesy of Chrysler Corp.)

8. Memory steer

9. Steering wheel return

Chassis waddle may be caused by excessive wheel or tire runout. Excessive vertical chassis oscillations are the result of defective shock absorbers or struts. (Refer to Chapter 4 for wheel and tire runout diagnosis, and Chapter 5 for shock absorber and strut diagnosis.)

Most steering systems are designed so the steering wheel returns to within 30° to 60° of center after a 180° turn (Figure 15-2). If the steering wheel returnability is not satisfactory and **memory steer** is evident, the steering shaft dash seal or the steering shaft universal joints may be binding. When memory steer occurs on a front-wheel-drive car, the upper strut mounts may be binding. High steering effort or binding can also be caused by a loose power steering belt, low power steering fluid level, worn front tires, or a defective steering gear.

A front-wheel-drive vehicle with unequal drive axles produces some **torque steer** on hard acceleration. Some front-wheel-drive vehicles, especially those with higher horsepower, have equal front drive axles to reduce torque steer. On a front-wheel-drive vehicle, torque steer is aggravated by different tire tread designs on the front tires or uneven wear on the front tires.

Bump steer occurs if the tie rods are not the same height, meaning one of the tie rods is not parallel to the lower control arm. This condition may be caused by a bent or worn idler arm or pitman arm on a parallelogram steering linkage. On a rack and pinion steering system, worn steering gear mounting bushings may cause this unparallel condition between one of the tie rods and the lower control arm.

Steering pull may be caused by improper caster or camber adjustments.

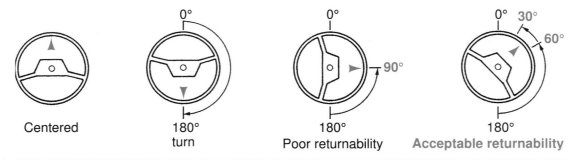

Figure 15-2 Steering wheel returnability. (Courtesy of Chrysler Corp.)

SERVICE TIP: If steering pull or drift is evident during the road test, drive the vehicle on the same road in the opposite direction to determine if a crosswind is causing the steering pull.

During a road test, any of the following symptoms may indicate improper rear wheel alingment.

1. The front wheel **toe-in** is set correctly with the steering wheel centered on the alignment rack, but steering wheel is not centered when the vehicle is driven straight ahead.

2. When there are no worn suspension parts or defective tires and all front suspension angles are within specifications, but the vehicle wanders or drifts to one side.

3. The vehicle is not overloaded, all front suspension angles are within specifications and there are no worn suspension parts, but tire wear recurs.

Prealignment Inspection

The technician should identify the exact suspension or steering complaint before a wheel alignment or other suspension work is performed. The technician must diagnose the cause or causes of the complaint and correct this problem during the wheel alignment (Figure 15-3). Since collision

Condition	Probable Cause
Tire wear:	Too much positive camber
Outer shoulder	Too much negative camber
Inner shoulder	Too little toe-in
Sawtooth pattern:	Too much toe-in
Sharp edge toward center	Under inflation, high-speed cornering,
Sharp edge away from center	overloading
	High speed in curves
Both shoulders	Wheel imbalance, radial runout
Tread roll under	
Cupping, dishing	
Shimmy:	Too much positive caster, unequal caster
With or without vehicle instability	between wheels
Without instability	Tire imbalance, tire runout, driveline vibration
Vibration	Driveline misalignment, driveline imbalance,
(Caster/camber not a probable cause.	vehicle shake (accompanied by a characteristic
The condition is listed because it is	moan), tire runout,, unequal weight distri-
sometimes misdiagnosed as shimmy.)	between wheels.
Steering wander/pull	Incorrect camber
	Unequal caster
	Overload, which elevates front end
Brake pull	Too much negative caster
	Unequal tire pressures, brake line damage
	that impedes hydraulic action on one side
Hard steering	Incorrect caster
	Damaged steering linkage, worn steering
	linkage, damaged spindles, rear end
	overload, bent steering arm causing
	incorrect turning angle.

Figure 15-3 Diagnosis of wheel alignment problems.

damage may affect wheel alignment angles, an inspection for collision damage on the vehicle should be completed prior to a wheel alignment (Figure 15-4).

Worn suspension and steering components must be replaced before a wheel alignment is performed. A wheel alignment should be performed after suspension components such as struts have been replaced. The following components and measurements should be checked in a prealignment inspection:

1. Curb weight
2. Tires
3. Suspension height
4. Steering wheel freeplay
5. Shock absorbers or struts
6. Wheel bearing adjustment
7. Ball joint condition
8. Control arms and bushings
9. Steering linkages and tie-rod ends
10. Stabilizers and bushings
11. Full fuel tank

Components to be checked in a **prealignment inspection** and wheel alignment measurements are summarized in Figure 15-5. These checks should be completed with the car on the floor:

1. Check for excessive mud adhered to the chassis. Remove heavy items from the trunk and passenger compartment that are not considered in the vehicle curb weight. If heavy items such as tool display cases or merchandise are normally carried in the vehicle, these items should be left in the car during a wheel alignment.

2. Inflate tires to the recommended pressure and note any abnormal tread wear or damage on each tire. Be sure all tires are the same size.

Look at the complete vehicle

Note: If you suspect collision damage to the inner body panels, measure the reference points of body and frame alignment as described in the service manual.

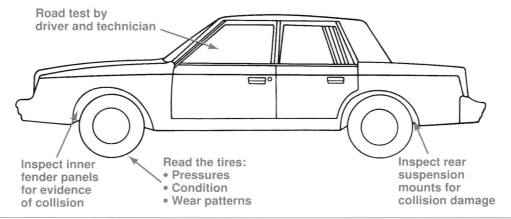

Road test by driver and technician

Inspect inner fender panels for evidence of collision

Read the tires:
• Pressures
• Condition
• Wear patterns

Inspect rear suspension mounts for collision damage

Figure 15-4 Collision damage inspection. (Courtesy of Chrysler Corp.)

PREALIGNMENT INSPECTION CHECKLIST

Owner _____ Phone _____ Date __|__|__
        LAST   FIRST

Address _____ Ser. No. _____

Make _____ Model _____ Yr. _____ Lic. No. _____ Mileage _____

1. Road Test Disclosed

Vehicle Pulls		Yes	No	Right	Left
	Above 30 MPH				
	Below 30 MPH				
	Bump Steer				
	When Braking				

	Yes	No	Right	Left
Steering Wheel Movement When Stopping From 2-3 MPH—Front Brakes				
Vehicle Steers Hard				
Steering Wheel Returns Normally				
Steering Wheel Position				

	Yes	No	Front	Rear
Vibration				

2. Tire Pressure SPECS Front ____ Rear ____

RECORD PRESSURE FOUND
RF _____ LF _____ RR _____ LR _____

3. Chassis Height SPECS Front ____ Rear ____

RECORD HEIGHT FOUND
RF _____ LF _____ RR _____ LR _____

	YES	NO
Springs Sagged		
Torsion Bars Adjusted		

4. Rubber Bushings OK

	OK
Upper Control Arm	
Lower Control Arm	
Sway Bar/Stabilizer Link	
Strut Rod	
Rear Bushing	

5. Shock Absorbers/Struts

	Front	Rear

6. Steering Linkage

	Rear OK	Front OK
Tie-Rod Ends		
Idler Arm		
Center Link		
Sector Shaft		
Pitman Arm		
Gearbox/Rack Adjustment		
Gearbox/Rack Mounting		

7. Ball Joints OK

	OK
Load Bearings	

SPECS	READINGS
Right ____ Left ____	Right ____ Left ____

Follower	
Upper Strut Bearing Mount	
Rear	

8. Power Steering OK

	OK
Belt Tension	
Fluid Level	
Leaks/Hose Fittings	
Spool Valve Centered	

9. Tires/Wheels OK

	OK
Wheel Runout	
Condition	
Equal Tread Depth	
Wheel Bearing	

10. Brakes Operating Properly

11. Alignment

	Specs		Initial Readings		Adj. Readings	
	Right	Left	Right	Left	Right	Left
Camber						
Caster						
Toe						

Bump Steer	TOE CHANGE Right Wheel		TOE CHANGE Left Wheel	
	Amount	Direction	Amount	Direction
Chassis DOWN 3"				
Chassis UP 3"				

	Specs		Initial Readings		Adj. Readings	
	Right	Left	Right	Left	Right	Left
Toe-Out On Turns						
SAI						
Rear Camber						
Rear Total Toe						
Rear Indiv. Toe						
Wheel Balance	Front _____ Rear _____					
Radial Tire Pull	Yes _____ No _____					

FLAG READINGS

COMMENTS:

Figure 15-5 Prealignment inspection and wheel alignment checklist.

3. Check the front tires and wheels for radial runout. (Refer to Chapter 4 for this measurement.)

4. Check the suspension **ride height.** If this measurement is not within specifications, check for broken or sagged springs. On a torsion bar suspension system, check the torsion bar adjustment.

5. When the wheels are in the straight-ahead position, rotate the steering wheel back and forth to check for play in the steering column, steering gear, or linkages.

6. Check the shock absorbers or struts for loose mounting bushings and bolts. Examine each shock absorber or strut for leakage.

7. Check the condition of each shock absorber or strut with a bounce test at each corner of the vehicle. (This test is explained in Chapter 5.)

The following checks should be performed with the vehicle raised and the suspension supported:

1. Check the front wheel bearings for lateral movement. (This check is described in Chapter 3.) On front-wheel-drive cars, perform this check on all four wheel bearings. Wheel bearings must be properly adjusted prior to a wheel alignment. Clean, repack, or adjust the wheel bearings as necessary.

2. Measure the ball joint radial and axial movement. If excessive movement exists in either direction, ball joint replacement is required. (Ball joint diagnosis and replacement is explained in Chapter 6.) Be sure the suspension is supported correctly during ball joint diagnosis.

3. Inspect the control arms for damage and check the control arm bushings for wear.

4. Check all the steering linkages and tie-rod ends for looseness.

5. Check for worn stabilizer mounting links and bushings.

6. Check for loose steering gear mounting bolts and worn mounting brackets and bushings.

Tire Wear Diagnosis

Improper camber settings cause wear on one side of the front tire treads.

Various types of tire tread wear patterns indicate specific alignment or balance defects (Figure 15-6).

Condition	Rapid wear at shoulders	Rapid wear at center	Cracked treads	Wear on one side	Feathered edge	Bald spots	Scalloped wear
Effect							
Cause	Under-inflation or lack of rotation	Over-inflation or lack of rotation	Under inflation or excessive speed	Excessive camber	Incorrect toe	Unbalanced wheel / Or tire defect	Lack of rotation of tires or worn or out-of-alignment suspension
Correction	Adjust pressure to specifications when tires are cool rotate tires			Adjust camber to specifications	Adjust toe-in to specifications	Dynamic or static balance wheels	Rotate tires and inspect suspension

Figure 15-6 Tire tread wear patterns and causes. (Courtesy of Chrysler Corp.)

Front and Rear Wheel Alignment with Computer Alignment Systems

Preliminary Procedure

CAUTION: When using a computer wheel aligner, always follow the equipment manufacturer's recommended procedures to provide accurate readings and to avoid equipment and vehicle damage or personal injury.

The vehicle should be on an **alignment ramp** with a **turn table** under each front tire and conventional **slip plates** under the rear tires (Figure 15-7). Slip plates under the rear tires allow unrestricted movement of the rear wheels during rear wheel camber and toe adjustments (Figure 15-8). If suspension adjustments are made with the tires contacting the alignment ramp or the shop floor, the tires cannot move when the adjustment is completed. This action causes inaccurate suspension adjustments. The center of the front wheel spindles should be in line with the zero mark on the turning plates. The locking pins must be left in these plates and the parking brake applied.

Mount the **rim clamps** on each wheel. It may be necessary to remove the chrome discs or hub caps before these clamps can be installed. The adjustment knob on the rim clamp should be pointing toward the top of the wheel, and the bubble on the wheel unit should be centered (Figure 15-9). A set knob on the wheel unit allows the service technician to lock the wheel unit in this position. Some

Special Tools

Computer wheel aligner

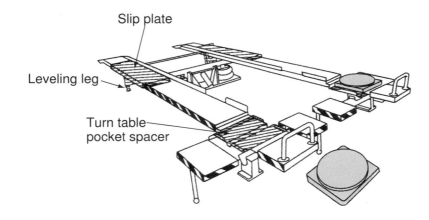

Slip plate

Leveling leg

Turn table
pocket spacer

Figure 15-7 Alignment ramp with turn tables. (Courtesy of Snap-On Tools Corp.)

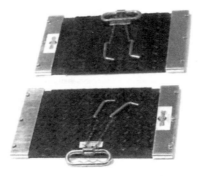

Figure 15-8 Rear wheel slip plates. (Courtesy of Snap-On Tools Corp.)

497

Figure 15-9 Rim clamp and wheel sensor mounting. (Courtesy of Hunter Engineering Company)

computer wheel aligners have a display on the monitor that indicates if any of the wheel sensors are not level (Figure 15-10). These sensors may be levelled by pressing a control knob on each wheel sensor. Some **digital signal processor (DSP)** wheel sensors contain a microprocessor and a **high-frequency transmitter** that acquire measurements and process data and then send these data to a **receiver** mounted on top of the wheel alignment monitor. This type of wheel sensor does not require any cables connected between the sensors and the computer wheel aligner. The data from this type of wheel sensor signal are virtually uninterruptible, even by solid objects. When these wheel sensors are stored on the computer wheel aligner, a "docking station" feature charges the batteries in the wheel sensors.

Some wheel sensors contain infrared emitter/detectors or light-emitting diodes (LEDs) between the front and rear wheel sensors to perform alignment measurements. Older wheel sensors have strings between the front and rear wheel sensors. Each wheel sensor contains a microprocessor with a preheat circuit that stabilizes the readings in relation to temperature. Some wheel sensors have touch keypads on the wheel units to allow the entry of commands from each unit. The front wheel units have arms that project toward the front of the vehicle to transmit signals between these units. When a blocked beam prompt appears on any screen, the beam between two wheel units is blocked. This could be caused by a person standing in the beam, an open car door, or a suspension jack. The blocked beam prompt must be eliminated before any tests are completed.

The manufacturers of various types of computer alignment systems publish detailed operator's manuals for their specific equipment. Our objective here is to discuss some of the general screens that a technician uses while measuring wheel alignment angles with a typical computer aligner.

Figure 15-10 Wheel sensor level indicator on the monitor screen. (Courtesy of Hunter Engineering Company)

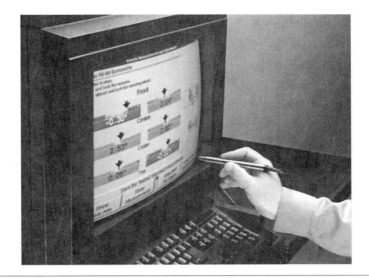

Figure 15-11 Making screen selections with the light pen. (Courtesy of Hunter Engineering Company)

Main Menu

One of the first items displayed on the computer aligner screen is the **main menu** screen. From this screen the technician makes a selection by touching the desired selection on the monitor screen with the light pen (Figure 15-11) or by pressing the number on the key pad that matches the number beside the desired procedure. A cursor may also be used to select the type of alignment.

Specifications Menu

Some computer wheel aligners have specifications contained on CDs. The appropriate CD must be placed in the aligner. Several methods may be used to enter the vehicle specifications. The technician may scroll through the **specifications menu** and press Enter when the appropriate vehicle is highlighted on the screen (Figure 15-12). The technician may also select the proper vehicle specifications by pointing with the light pen or by typing in the vehicle identification number (VIN) or the first letter in the vehicle make.

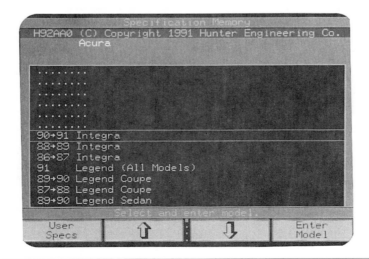

Figure 15-12 Entering vehicle specifications. (Courtesy of Hunter Engineering Company)

On other computer wheel aligners, the specifications are contained on a floppy disc and the technician selects the appropriate make, model, and year of vehicle with the cursor on the screen.

On early computer alignment systems, the vehicle specifications were contained on bar graphs. The technician moved a wand across the appropriate bar graph to enter the specifications for the vehicle being aligned.

Preliminary Inspection Screen

The **preliminary inspection screen** allows the technician to enter the condition of all front and rear suspension components that must be checked during a prealignment inspection. On some computer alignment systems the technician may enter checked, marginal, and repair or replace for each component (Figure 15-13). These entered conditions may be printed out at any time or with the complete wheel alignment results.

Ride Height Screen

Some computer wheel aligners have optical encoders in the wheel sensors that measure ride height when this selection is displayed on the screen and the ride height attachment on each wheel sensor is lifted until it touches the lower edge of the fender (Figure 15-14). The ride height measurement is displayed on the screen. If the screen display is green, the ride height is within specifications. A red ride height display indicates this measurement is not within specifications. Improper ride height may be caused by sagged or broken springs, bent components such as control arms, or worn components such as control arm bushings.

▲ **WARNING:** The ride height must be within specifications before proceeding with the wheel alignment. Improper curb riding height affects many of the other suspension angles.

Some computer wheel aligners provide a ride height screen with a graphic display indicating the exact location where the ride height should be measured (Figure 15-15). The computer aligner compares the ride height measurements entered by the technician to the specifications.

Tire Condition Screen

When the **tire inspection screen** is displayed, the technician may enter various tire wear conditions for each tire (Figure 15-16). The cursor on the screen is moved to the tire being inspected.

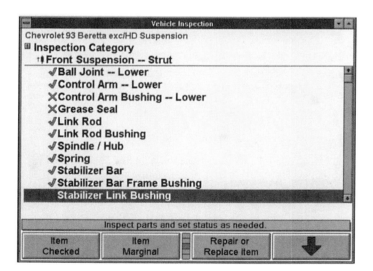

Figure 15-13 Preliminary inspection screen. (Courtesy of Hunter Engineering Company)

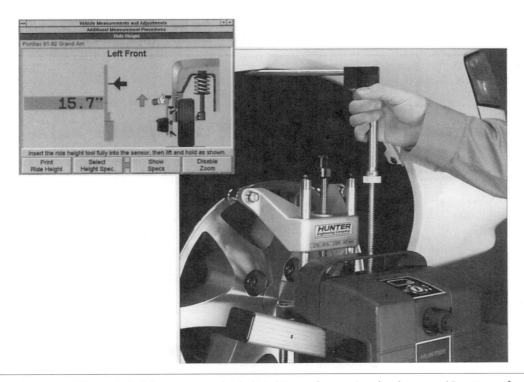

Figure 15-14 Ride height screen and ride height attachment in wheel sensor. (Courtesy of Hunter Engineering Company)

The tire conditions are printed out with the preliminary inspection results and the alignment report. On other computer wheel aligners tire condition is included in the preliminary inspection screens.

Wheel Runout Compensation

As mentioned previously, screen indicators on some computer wheel aligners inform the technician if any of the wheel sensors require levelling or compensating for wheel runout. Runout compensation is accomplished by pressing the appropriate button on each wheel sensor. Some wheel sensors provide continuous compensation. This feature provides accurate alignment angles even when a wheel is rotated after the compensation button on the wheel sensor has been pressed. When the wheel runout screen is displayed, the technician is directed to level and lock each wheel unit and then press the compensation button on each wheel sensor to provide automatic wheel runout compensation (Figure 15-17).

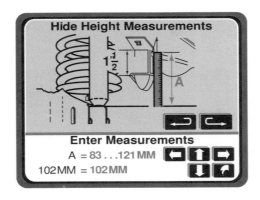

Figure 15-15 Ride height screen. (Courtesy of Hunter Engineering Company)

```
ENTER TIRE CONDITION

1  EVEN WEAR
2  REPLACEMENT REQUIRED
3  INSIDE EDGE WEAR
4  OUTSIDE EDGE WEAR
5  BOTH EDGE WEAR
6  CENTER WEAR
7  CUPPING
8  MISCELLANEOUS TIRE FAULTS


LEFT FRONT              RIGHT FRONT


LEFT REAR              RIGHT REAR
```

Figure 15-16 Tire condition screen. (Courtesy of Bear Automotive Service Equipment Company)

Figure 15-17 Wheel runout compensation screen. (Courtesy of Bear Automotive Service Equipment Company)

If the computer aligner does not have automatic wheel runout compensation, a manual wheel runout procedure must be followed. This type of computer aligner displays a wheel runout measurement screen. During this procedure the wheel being checked for runout is lifted with the hydraulic jack on the alignment rack and the wheel is rotated until the rim clamp knob faces downward. Level and lock the wheel unit in this position, then push Yes on the wheel unit as instructed on the screen. Rotate the wheel until the rim clamp knob faces upward, then level and lock the wheel unit. After this procedure press Yes on the wheel unit. This same basic procedure is followed at each wheel.

Wheel Alignment Screens

Front and Rear Wheel Alignment Angle Screen

SERVICE TIP: Many computer wheel aligners can print out any single screen. It may be helpful to print out the wheel alignment angle screen before and after the alignment angles are adjusted to the vehicle manufacturer's specifications. These printouts may be presented to the customer; many customers appreciate this service.

Figure 15-18 Brake pedal depressor. (Courtesy of Snap-On Tools Corp.)

Figure 15-19 Steering wheel holder. (Courtesy of Snap-On Tools Corp.)

Prior to a display of the **front and rear wheel alignment angle screen** on some computer wheel aligners, the screen display directs the technician to position the front wheels straight ahead, lock the steering wheel, apply the **brake pedal depressor,** and level and lock the wheel units, Photo Sequence 13. The brake pedal depressor is an adjustable rod installed between the front edge of the front seat and the brake pedal (Figure 15-18). If the vehicle has power brakes, the engine should be running when the depressor is used to apply the brakes. Some steering wheel holders are installed between the steering wheel and the top of the front seat (Figure 15-19). A ratchet and handle on the steering wheel holder allow extension of this holder.

After the wheel runout compensation procedure is completed at each wheel sensor, some computer wheel aligners automatically display **camber,** toe, **setback,** and **thrust line** measurements. After this procedure the screen directs the technician to swing the front wheels outward a specific amount (Figure 15-20). The steering wheel holder must be released and the lock pins removed from the front turn tables before the wheel swing procedure. This front wheel swing may be referred to as a caster/SAI swing. The amount of front wheel swing varies depending on the make of the computer aligner. Older computer wheel aligners required the technician to read the degrees of wheel swing on the turn table degree scales. On some newer computer wheel aligners, the required amount of wheel swing is illustrated on the screen (Figure 15-21).

After the front wheel swing procedure, all the front and rear wheel alignment angles are displayed, including **caster, steering axis inclination (SAI),** and **included angle** (Figure 15-22). Some computer wheel aligners highlight any wheel alignment angles that are not within specifications. If cross camber and cross caster are displayed on the screen, these readings indicate the

The steering axis inclination (SAI) line is an imaginary line through the centers of the upper and lower ball joints, or through the center of the lower ball joint and the upper strut mount.

The SAI angle is the angle between the SAI line the true vertical center line of the tire and wheel.

The included angle is the sum of the SAI angle and the camber angle if the camber is positive. If the camber is negative, the camber setting must be subtracted from the SAI angle to obtain the included angle.

Toe-in occurs when the distance between the front edges of the front wheels is less than the distance between the rear edges of the front wheels.

Figure 15-20 Wheel swing instructions on the screen. (Courtesy of Bear Automotive Service Equipment Company)

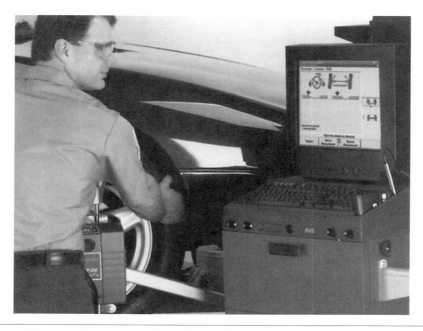

Figure 15-21 Wheel swing procedure illustrated on the screen. (Courtesy of Hunter Engineering Company)

Toe-out is present when the distance between the front edges of the front wheels is more than the distance between the rear edges of the front wheels.

Total toe is the sum of the toe settings on the front wheels.

Turning radius may be referred to as cornering angle or turning angle.

Turning radius or cornering angle is the amount of toe-out on turns.

maximum difference allowed between the right and left side readings. Alignment angles within specifications are highlighted in green; alignment angles that are not within specifications are highlighted in red.

Turning Angle Screen

Some computer wheel aligners have a turning angle screen. When this screen is displayed, the technician removes the locking pins from the turn tables. Each front wheel must be turned outward a specified amount, and the turning angle on the opposite front wheel must be entered with the

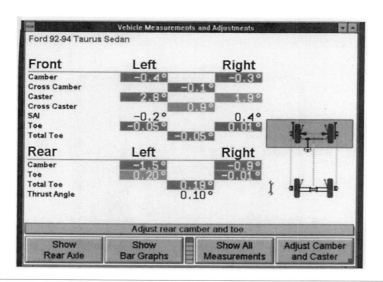

Figure 15-22 Front and rear wheel alignment angles. (Courtesy of Hunter Engineering Company)

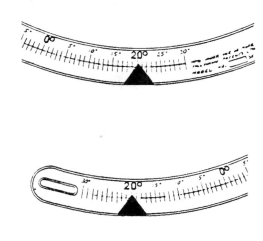

```
TURNING ANGLE

HIT CONTINUE TO SKIP THIS TEST


ENTER RIGHT TURNING ANGLE

 20.0

TURN RIGHT FRONT WHEEL OUT 20.0
                        DEGREES

 ENTER LEFT FRONT TURNING ANGLE

 20.0
```

Figure 15-23 Turning angle screen. (Courtesy of Bear Automotive Service Equipment Company)

keypad as directed on the screen (Figure 15-23). A toe-out-on-turns option is available on some DSP wheel sensors. This option has optical encoders in the wheel sensors that measure the turning angle electronically rather than reading the degree scales on the front turn tables.

Adjustment Screens

The technician may select adjustment screens that provide bar graph readings of camber, caster, and toe. An arrow on the bar graph shows the amount and direction the actual measurement is from the preferred specification. As the alignment angle is adjusted, the arrow moves. If an alignment angle moves from the out-of-specification range to within the specification range, the bar graph color changes from red to green (Figure 15-24). When the arrow on the bar graph is in the zero position, the alignment angle is at the preferred specification. A zoom feature on some computer wheel aligners provides enlarged bar graphs so they may be seen at a distance while performing the actual suspension adjustments (Figure 15-25). Some computer aligners have a jack and hold feature that allows the suspension to be lifted on the alignment rack to perform an adjustment, while maintaining accurate displays on the adjustment screen. Other computer wheel aligners have a remote display that may be connected to the aligner and taken under the car for close-up viewing while performing suspension adjustments (Figure 15-26). The remote display duplicates the bar graphs shown on the monitor screen.

Some computer wheel aligners provide **symmetry angle measurements** that help the technician determine if out-of-specification readings may have been caused by collision or frame damage. These symmetry angle measurements display **axle offset,** right or left **lateral axle sideset** and track width difference, front and rear wheel setback, and wheelbase difference (Figure 15-27). Setback is an angle formed by a line drawn at a 90° angle to the centerline and a line connecting the centers of the front or rear wheels (Figure 15-28). Wheelbase difference is an angle created by a line through the rear wheel centers and a line through the front wheel centers (Figure 15-29). Right or left lateral offset is an angle between the thrust line and a line connecting the centers of the left front and left rear wheels or right front and right rear wheels (Figure 15-30). Track width difference is an angle created by the line connecting the centers of the right rear and right front wheels and the line connecting the centers of the left rear and left front wheels (Figure 15-31). Axle offset is an angle formed by the line that bisects the track width difference angle and the thrust line (Figure 15-32).

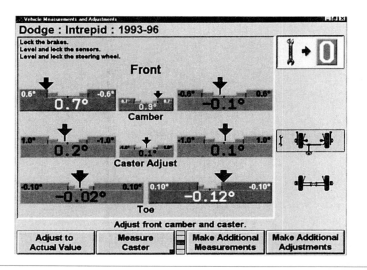

Figure 15-24 Alignment angle adjustment screens. (Courtesy of Hunter Engineering Company)

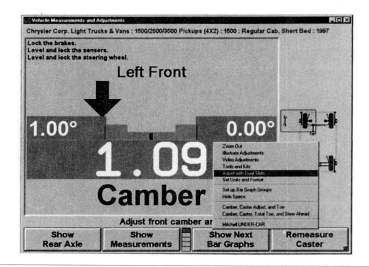

Figure 15-25 Zoom feature on adjustment screens. (Courtesy of Hunter Engineering Company)

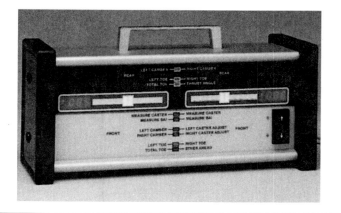

Figure 15-26 Remote display for computer wheel aligner. (Courtesy of Hunter Engineering Company)

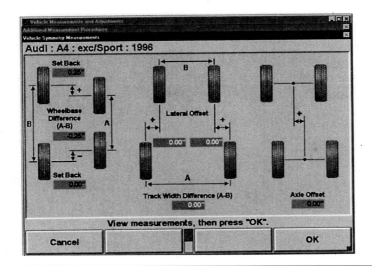

Figure 15-27 Symmetry angle measurements. (Courtesy of Hunter Engineering Company)

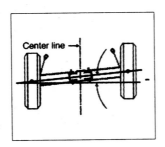

Figure 15-28 Front or rear wheel setback. (Courtesy of Hunter Engineering Company)

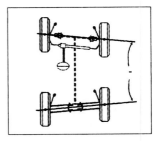

Figure 15-29 Wheelbase difference. (Courtesy of Hunter Engineering Company)

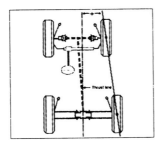

Figure 15-30 Right or left lateral offset. (Courtesy of Hunter Engineering Company)

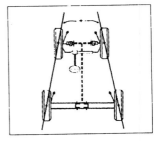

Figure 15-31 Track width difference. (Courtesy of Hunter Engineering Company)

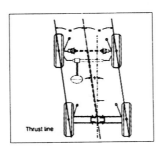

Figure 15-32 Axle offset. (Courtesy of Hunter Engineering Company)

Photo Sequence 13 Four Wheel Alignment with Computer Wheel Aligner

P17-1 Position the vehicle on the alignment ramp.

P17-2 Be sure the front tires are positioned properly on the turn tables.

P17-3 Position the rear wheels on slip plates.

P17-4 Attach the wheel units.

P17-5 Select the vehicle make and model year.

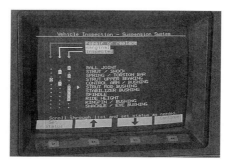

P17-6 Check items on the screen during preliminary inspection.

P17-7 Display the ride height screen.

P17-8 Check tire condition for each tire on the tire condition screen.

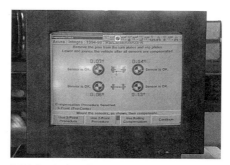

P17-9 Display the wheel runout compensation screen.

Photo Sequence 13 Four Wheel Alignment with Computer Wheel Aligner (continued)

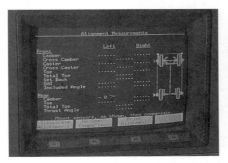

P17-10 Display the front and rear wheel alignment angle screen.

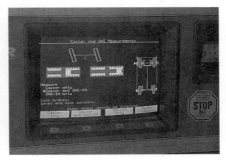

P17-11 Display the turning angle screen and perform turning angle check.

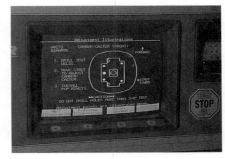

P17-12 Display adjustment screens.

Diagnostic Drawing and Text Screens

The technician may select tools and kits for the vehicle being serviced from the tools and kits database. When this feature is selected, the monitor screen displays the necessary wheel alignment adjustment tools for the vehicle selected at the beginning of the alignment (Figure 15-33). The kits displayed on the monitor screen are special components such as adjustment shims that are available for alignment adjustments on the car being serviced.

The technician may select **digital adjustment photos** that indicate how to perform wheel alignment adjustments (Figure 15-34). These digital photos include photos for cradle inspection and correction of cradle to body alignment (Figure 15-35). Live action videos can also be selected. These CD videos provide suspension component inspection procedures (Figure 15-36).

A **part finder database** is available in some computer wheel aligners. This database allows the technician to access part numbers and prices from many undercar parts manufacturers.

On many front-wheel-drive cars, the rear wheel camber and toe are adjusted with shims. Some computer aligners have **shim display screens** that indicate the thickness of shim required

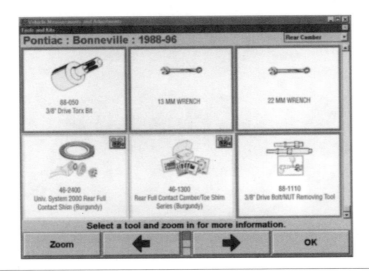

Figure 15-33 Wheel alignment tools and kits display. (Courtesy of Hunter Engineering Company)

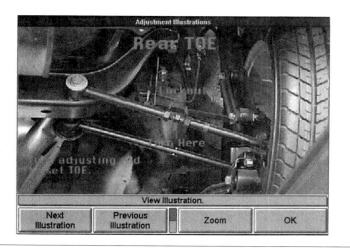

Figure 15-34 Wheel alignment adjustment photos. (Courtesy of Hunter Engineering Company)

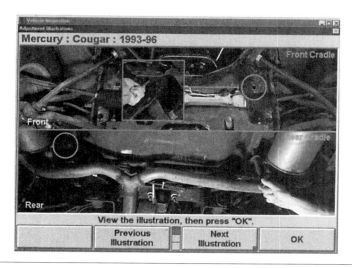

Figure 15-35 Cradle inspection and correction photos. (Courtesy of Hunter Engineering Company)

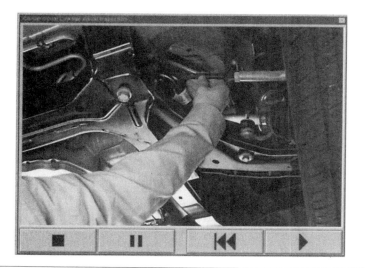

Figure 15-36 Live action videos of suspension inspection and service. (Courtesy of Hunter Engineering Company)

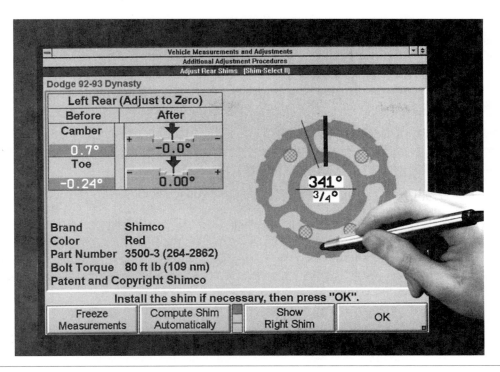

Figure 15-37 Shim select screen. (Courtesy of Hunter Engineering Company)

and the proper position of the shim. On some computer wheel aligners the technician may use the light pen to change the orientation angle of the shim on the monitor screen while observing the resulting change in camber and toe (Figure 15-37).

A **control arm movement monitor** is available on some computer wheel aligners. On short-and-long arm front suspensions, this feature indicates the required shim thickness to provide the specified camber and caster (Figure 15-38).

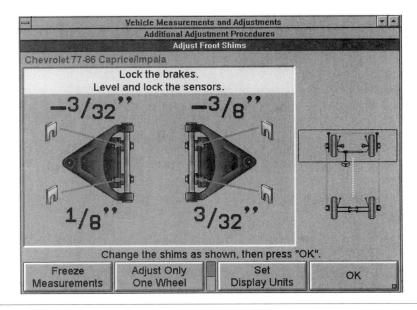

Figure 15-38 Control arm movement monitor indicates the required shim thickness to provide the specified camber and caster. (Courtesy of Hunter Engineering Company)

On some computer wheel aligners the technician may select Print at any time, and print out the displayed screen including diagnostic drawings. Another optional procedure is to print out the wheel alignment report before and after the adjustment of wheel alignment angles.

Measurement of Front Suspension Alignment Angles with Mechanical Alignment Equipment

Camber Measurement Procedure

Special Tools

Magnetic gauges

Prior to the use of computer wheel aligners, magnetic gauges were sometimes used to measure front wheel camber, caster, and SAI. The **magnetic gauge** may be mounted directly to the wheel hub (Figure 15-39) or attached to a rim clamp (Figure 15-40).

⚠️ **WARNING:** When rim clamps are used to mount magnetic gauges on aluminum or magnesium rims, be sure the clamps are designed for this type of rim. Rim clamps designed for steel rims only may damage aluminum or magnesium rims.

If rim clamps are used to mount the magnetic gauges, the rims must be checked for runout. This is done by checking the camber reading, then rotating the wheel 180° and taking another camber reading. A variation in the two readings indicates a bent rim.

Use this procedure to measure front wheel camber:

1. Perform a prealignment inspection.

2. Position the vehicle with the front wheels on the turn tables.

3. Install the magnetic gauges on the front wheels and remove the turn table locking pins.

✓ **SERVICE TIP:** Clean the outer lips on the front wheel hubs and check these surfaces for metal burrs before installing the magnetic gauges. Any dirt or irregularities on these surfaces causes inaccurate magnetic gauge readings.

Figure 15-39 Magnetic gauge mounted on wheel hub. (Courtesy of Snap-On Tools Corp.)

Figure 15-40 Magnetic gauge mounted on a rim clamp. (Courtesy of Snap-On Tools Corp.)

WARNING: If the magnetic gauges can be rocked vertically or horizontally after they are installed on the wheel hubs or rim clamps, recheck the gauge mounting surfaces for dirt or metal irregularities. Any gauge rocking action causes inaccurate camber or caster readings.

4. Place the front wheels in the straight position with the turn table pointers in the zero position.

5. Push up and down on the front bumper to jounce the vehicle several times.

6. Read the camber setting on each front wheel. The bubble on the gauge indicates the camber on each wheel.

7. Compare the camber setting on each front wheel to the manufacturer's specifications.

Caster Measurement Procedure

Follow these steps to measure the front wheel caster.

1. Install a brake pedal jack between the front seat and the brake pedal. Operate this jack to apply the brakes. If the vehicle has power brakes, the engine must be running when the brakes are applied.

2. Install the magnetic gauges on the front wheel hubs and remove the turn table locking pins. If the wheel hubs are inaccessible, for the magnetic gauges, use the rim clamps to mount the magnetic gauges.

3. Be sure the turn table pointers are in the zero position, then turn the wheel being checked 20° outward as indicated by the turn table pointer.

4. Adjust the caster vial to zero on the wheel being checked (Figure 15-41).

5. Turn the wheel being checked to the 20° inward position on the turn table dial and record the caster reading (Figure 15-42).

6. Measure the caster on the opposite front wheel using the same procedure.

Figure 15-41 Adjusting caster vial to zero with wheel turned 20° outward. (Courtesy of Snap-On Tools Corp.)

SAI Measurement

Some magnetic gauges measure SAI. The magnetic gauge calculates the angle between the SAI line and the true vertical position. Follow this procedure to measure SAI with a magnetic gauge:

1. Measure the camber angle of the front wheel as described previously.

2. Attach the magnetic gauges to the front wheel hubs and apply the brakes with a pedal jack installed between the brake pedal and the front seat.

3. Turn the front wheel 20° inward and set the SAI scale to 0°.

Figure 15-42 Measuring caster with front wheel turned 20° inward. (Courtesy of Snap-On Tools Corp.)

4. Turn the wheel to the straight-ahead position and continue turning the wheel to the 20° outward position.

5. Record the SAI reading on the gauge.

6. If the camber reading is positive, add this reading to the SAI reading to obtain the included angle. A negative camber reading must be subtracted from the SAI angle.

7. Compare the SAI or included angle readings to the manufacturer's specifications.

Front Wheel Toe Measurement

Special Tools

Toe gauge

On older wheel aligners, toe is measured with a toe gauge placed between the front wheels. The toe gauge is an adjustable rigid bar that may be referred to as a **tram gauge.** Some alignment equipment manufacturers recommended toe measurement between chalk lines placed on each tire tread near the center. Other suggested points for toe measurement are at the inside of the tire sidewalls or between the inside edges of the rims. Toe is measured at the front and rear inside edges of the tire or rim at spindle height. Many alignment equipment manufacturers specify that a steering wheel locking device be installed during the toe measurement. The toe-in or **toe-out** setting is the difference in the distance between the inside front edges of the tires or rims, and the distance between the inside rear edges of the tires and rims. If the toe setting is not within specifications, feathered tread wear occurs on the front tires.

An **optical toe gauge** is available from some alignment equipment manufacturers. This type of toe gauge attaches magnetically to the front wheel hubs and projects a light beam to a screen on the opposite gauge. The optical toe gauge changes angular deflection of the wheels to inches of toe (Figure 15-43).

Always follow the toe gauge manufacturer's recommended procedure to measure the front wheel toe. The following is a typical front wheel toe measurement procedure.

1. Perform a prealignment inspection as outlined earlier.

2. Check the camber and caster, and adjust to specifications if necessary.

3. Position the vehicle with the front wheels centered on the turn tables and remove the turn table locking pins. Be sure the front wheels are in the straight-ahead position.

4. Use a steering wheel clamp to lock the steering wheel in the straight-ahead position. If the vehicle has power steering, operate the engine when the steering wheel is locked.

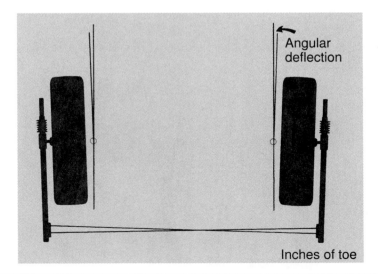

Figure 15-43 Optical toe gauge. (Courtesy of Ammco Tools, Inc.)

5. Install the brake pedal jack between the front seat and the brake pedal. If the vehicle has power brakes, operate the engine while the brakes are locked.

6. Remove the wheel covers and grease caps. Clean the surfaces on the hub faces and the optical gauge magnetic faces. These mating surfaces must be free from grease, dirt, and metal burrs.

7. Jounce the front suspension several times and allow it to stabilize. Be sure the front wheels are still in the straight-ahead position.

8. If the steering linkage is mounted at the rear edge of the front wheels, use hand pressure to push the rear edge of the front wheels outward simultaneously. When the steering linkage is positioned at the front edge of the front wheels, use hand pressure to push the front edge of the front wheels outward simultaneously.

9. Mount the optical gauges on the hub faces with the optical reflector facing toward the front of the vehicle. If the hub face is not accessible, use a rim clamp to attach the optical gauge to the hub face.

10. Use the level indicators on each optical toe gauge to locate the gauges in the level position.

11. Read the toe on the screen at the outer end of the toe gauges and add the toe on each front wheel to obtain the total toe (Figure 15-44). Compare the total toe reading to the total toe specifications and adjust the toe as required.

Changes in other alignment angles will alter the toe setting. For example, if the positive caster is increased, the top of the spindle is moved rearward. If the steering linkage is at the rear edge of the front wheels, this action rotates the rear of the steering arm downward, which moves the wheel toward a toe-out position. Therefore, other alignment angles such as camber, caster, and SAI must be adjusted within manufacturer's specifications before toe is checked. If the toe setting is not within manufacturer's specifications, a toe adjustment is required.

Classroom Manual
Chapter 15, page 335

Turning Radius Measurement

If the **turning radius** is not within specifications, excessive front tire tread wear occurs while cornering. Tire squeal may also occur while cornering if the turning radius is improper. Follow these steps to measure turning radius.

1. Perform a prealignment inspection.

2. Measure and adjust the other suspension angles in this order: camber, caster, steering axis inclination (SAI), and toe.

3. Position the front wheels on the center of the turn tables and apply the brakes with a brake pedal jack.

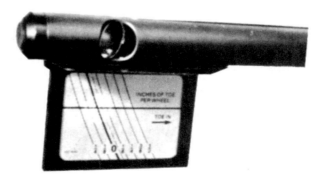

Figure 15-44 Optical toe gauge screen. (Courtesy of Ammco Tools, Inc.)

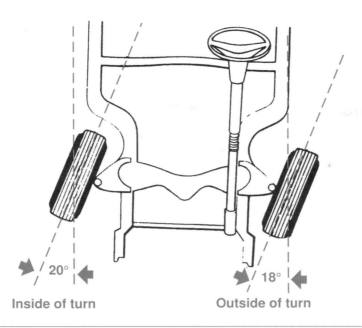

20° — Inside of turn

18° — Outside of turn

Figure 15-45 Toe-out-on-turns. (Courtesy of Hunter Engineering Company)

4. Remove the turn table locking pins and be sure the turning radius gauges are in the zero position with the front wheels straight ahead.

5. Turn the right front wheel inward toward the center of the vehicle until the turning radius gauge indicates 20° (Figure 15-45). Read the turning radius on the left front wheel and record the reading. On many steering systems the turning radius on the wheel on the inside of the turn is 2° more compared to the turning radius on the wheel on the outside of the turn when the inside wheel is in the 20° outward position.

6. Turn the left front wheel inward toward the center of the vehicle until the turning radius gauge reads 20°. Read the turning radius on the right front turn table radius gauge and record the readings.

7. Compare the turning radius readings on each front wheel to manufacturer's specifications. Turning radius or toe-out-on-turns is not adjustable. If the toe-out-on-turns is not within manufacturer's specifications, the steering arms are bent or components such as tie-rod ends are worn.

Checking Toe Change and Steering Linkage Height

If the steering linkage on each side is not the same height, a condition called bump steer may occur. When bump steer occurs, the steering suddenly swerves to one side when one of the front wheels strikes a road irregularity. When this steering linkage condition is present, abnormal toe changes occur during wheel jounce and rebound. Follow this procedure to check for abnormal toe changes during wheel jounce and rebound.

1. Be sure the toe is adjusted to specifications with the front wheels centered and the steering wheel locked.

2. Pull the chassis downward approximately 3 in (7.6 cm) and observe the toe change. It is acceptable if the toe setting on each front wheel remains at the original reading or if each front wheel toes in or toes out an equal small amount (Figure 15-46).

Bump steer is the tendency of the steering to veer suddenly in one direction when one or both front wheels strike a bump.

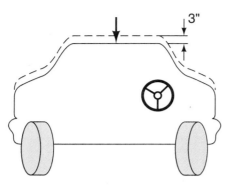

Figure 15-46 Normal toe change during wheel jounce.

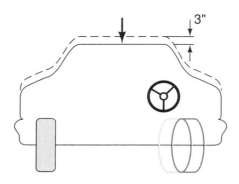

Figure 15-47 If the toe change on one front wheel remains at the original setting while the toe on the opposite wheel toes in or out during wheel jounce, unequal steering linkage height is indicated. (Courtesy of Ammco Tools, Inc.)

3. During wheel jounce if one front wheel toes inward or outward while the opposite wheel remains at the original setting, the steering linkage height is not equal and must be corrected (Figure 15-47).

4. During wheel jounce if one front wheel toes in and the opposite front wheel toes out, the steering linkage height is unequal and must be corrected (Figure 15-48).

5. Push the chassis upward 3 in (7.6 cm) and check the toe change (Figure 15-49). If the steering linkage height is equal the toe setting on each front wheel remains the same or moves the same amount to a toe-in or toe-out position.

6. If the toe on one front wheel remains at the original setting while the opposite front wheel toe changes to a toe-in or toe-out setting, the steering linkage height is unequal. When one front wheel moves to a toe-in position, and the opposite front wheel moves to a toe-out setting, unequal steering linkages are indicated.

On some imported cars the steering linkage height can be adjusted with shims between the rack and pinion steering gear and the chassis (Figure 15-50). The idler arm may be moved upward or downward on some domestic vehicles to adjust the steering linkage height. On other vehicles if the steering linkage height is unequal, steering components are worn or bent. These components include tie rods, tie-rod ends, idler arms, pitman arms, and rack and pinion steering gear mounting bushings.

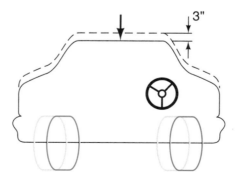

Figure 15-48 If the toe on one front wheel toes in and the toe on the opposite front wheel toes out during wheel jounce, unequal steering linkage height is indicated. (Courtesy of Ammco Tools, Inc.)

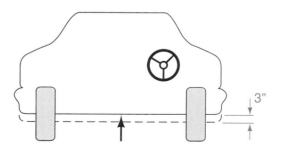

Figure 15-49 Pull the chassis upward 3 in (7.6 cm) to check for improper toe change indicating unequal steering linkage height. (Courtesy of Ammco Tools, Inc.)

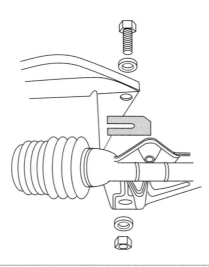

Figure 15-50 On some imported vehicles, shims between the rack and pinion steering gear and the chassis adjust the steering linkage height. (Courtesy of Ammco Tools, Inc.)

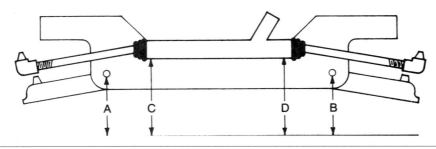

Figure 15-51 Measuring rack and pinion steering gear height. (Courtesy of Sealed Power Corp.)

Another method of checking for equal tie rod height on rack and pinion steering gears is to measure the distance from the center of the inner retaining bolt on the lower control arm to the alignment ramp on each side of the front suspension (distances A and B in Figure 15-51). If this distance is not equal on each side of the front suspension, one of the front springs may be sagged or components such as control arm bushings may be worn. When distances A and B are not equal, this problem must be corrected before performing the steering gear measurements. If distances A and B are equal, measure from the outer ends of the rack and pinion steering gear to the alignment ramp (distances C and D). If these distances are not equal, the rack and pinion steering gear mountings may be worn or distorted.

Bent Front Strut Diagnosis

Diagnostic Procedure

Front struts that are bent forward or rearward may be diagnosed as follows:

1. With the vehicle on the aligning rack and the magnetic gauges in position for the camber measurement, swing one front wheel out 10° and read the camber.

2. Turn the same wheel inward 10° and read the camber. The difference between the two readings is usually less than 4 1/2°. If the strut is bent forward or rearward, the difference in the camber readings will be excessive. A 10° difference is not uncommon.

3. Repeat steps 1 and 2 at the other front wheel.

Front struts could also be bent inward or outward. To diagnose this condition, use the following procedure:

1. With the wheel aligner in operation, sit on the front fender to load the suspension downward, and read the camber.

2. Unload the suspension and lift up on the car while the camber is recorded. The two camber readings should be within 1/2°. If the strut is bent inward or outward, the difference in the two camber readings will be excessive. A 4° to 6° camber change is not uncommon.

3. Repeat steps 1 and 2 on the other front wheel.

CUSTOMER CARE: Always concentrate on quality workmanship and customer satisfaction. Most customers do not mind paying for vehicle repairs if the work is done properly and their vehicle problem is corrected. A follow-up phone call to determine customer satisfaction a few days after repairing his or her car indicates that you are interested in the car and that you consider quality work and satisfied customers a priority.

Guidelines for Measuring and Adjusting Camber and Caster

1. A road test may be necessary to verify customer complaints prior to a wheel alignment.

2. Worn suspension and steering components must be replaced before a wheel alignment is performed.

3. After suspension components such as struts have been replaced, a wheel alignment should be performed.

4. A prealignment inspection of all steering and suspension components must be completed before a wheel alignment procedure.

5. Improper camber settings cause wear on one side of the front tire treads.

6. Front wheel camber is measured with the front wheels in the straight-ahead position.

7. The camber and caster must be within specifications before checking the front wheel toe.

8. While checking the front wheel toe, the wheels must be straight ahead with the steering wheel locked and the brakes applied

9. The main menu on a computer wheel aligner allows the technician to select the desired precedure and the type of wheel alignment.

10. The preliminary inspection menu allows the technician to select all suspension components as satisfactory or unsatisfactory from a list on the screen.

11. The ride height measurement location is displayed on a ride height screen, and the technician enters the ride height measurement. The computer compares this measurement to specifications.

12. If the wheel sensors have a ride height attachment, this attachment may be used to measure the ride height.

13. An automatic or manual wheel runout compensation procedure must be performed at each wheel because wheels with excessive runout affect the position of the rim clamps and the alignment readings.

14. The front and rear wheel alignment angle screen displays all the front and rear wheel alignment angles.

15. Adjustment screens display bar graphs of the alignment angles with the specified angle highlighted. An arrow points to the actual position of the alignment angle. When the alignment angle is adjusted, the arrow moves to indicate the new setting.

16. Shim select screens show the proper location and thickness of shims required to bring some alignment angles within specifications.

17. Digital photos illustrate adjustment locations and provide instructions regarding adjustments.

18. During a turn, the turning radius on the inside wheel is more than the turning radius on the outside wheel.

19. If the toe on both front wheels remains the same during wheel jounce and rebound, or if both front wheels move an equal amount toward a toe-in or toe-out position the tie rod height is equal.

20. If one front wheel remains at the original toe setting during wheel jounce and rebound, and if the opposite front wheel moves toward a toe-in or toe-out position, the steering arm height is unequal.

21. If one front wheel moves toward a toe-in position and the opposite front wheel moves toward a toe-out setting during wheel jounce and rebound, the tie rod height is unequal.

22. Unequal tie rod height may be caused by loose rack and pinion steering gear mounting bushings; worn or bent idler arms and pitman arms; or bent tie rods and tie-rod ends.

23. If the tie rod height is unequal, one tie rod is no longer parallel to the lower control arm. This condition causes improper toe changes during front wheel jounce and rebound, which causes the steering to suddenly veer to one side. The term bump steer is applied to this steering problem.

24. Magnetic wheel alignment gauges may be mounted directly to the wheel hubs or on rim clamps.

25. The magnetic gauge mounting surfaces on the wheel hub or rim must be clean and smooth to prevent vertical or horizontal magnetic gauge movement.

CASE STUDY

A customer complained about erratic steering on a front-wheel-drive Dodge Intrepid. A road test revealed the car steered reasonably well on a smooth road surface, but the steering would suddenly swerve to the right or left while driving on irregular road surfaces.

The technician performed a preliminary wheel alignment inspection and found the right tie-rod end was loose, but all the other suspension and steering components were in satisfactory condition. The technician replaced the loose tie-rod end, but a second road test indicated that the bump steer problem was still present. After advising the customer that a complete wheel alignment was necessary, the technician drove the vehicle on the wheel aligner and carefully checked all front and rear alignment angles. Each front and rear wheel alignment angle was within specifications. The technician realized that somehow he had not diagnosed this problem correctly.

While thinking about this problem the technician remembered a general diagnostic procedure he learned while studying Automotive Technology. This procedure stated: Listen to the customer complaints, be sure the complaint is identified, think of the possible causes, test to locate the exact problem, and be sure the complaint is eliminated. The technician realized he had not thought much about the causes of the problem and so he began to recall the wheel alignment theory he learned in college. He remembered that the tie rods must be parallel to the lower control arms, and if the tie rod height is unequal, this parallel condition no longer exists. The technician also recalled that unequal tie rod height causes improper toe changes during wheel jounce and rebound, which result in bump steer.

A check of the toe during front wheel jounce and rebound indicated the toe on the right front wheel remained the same during wheel jounce and rebound, but the toe on the left front wheel moved to a toe-out position. Since the tie rods had been inspected during the preliminary alignment inspection, the technician turned his attention to the rack and pinion steering gear mounting. He found the bushing on the right end of the steering gear was worn and loose. This bushing was replaced and all the steering gear mounting bolts tightened to the specified torque. A check of the toe change during wheel jounce and rebound revealed a normal toe change.

From this experience the technician learned two things:

1. His understanding of wheel alignment theory was very important in diagnosing steering problems.

2. Always be thorough! During a prealignment inspection check all suspension and steering components including rack and pinion steering gear mountings.

Terms to Know

Alignment ramp	Lateral axle sideset	Specifications menu
Axle offset	Magnetic gauges	Steering axis inclination (SAI)
Brake pedal depressor	Main menu	Symmetry angle measurements
Bump steer	Memory steer	Thrust line
Camber	Optical toe gauge	Tire inspection screen
Caster	Part finder database	Toe-in
Control arm movement monitor	Prealignment inspection	Toe-out
Digital adjustment photos	Preliminary inspection screen	Torque steer
Digital signal processor (DSP)	Ride height screen	Tram gauge
Front and rear wheel alignment angle screen	Rim clamps	Turning radius
High-frequency transmitter	Setback	Turn table
Included angle	Shim display screen	
	Slip plates	

ASE Style Review Questions

1. While discussing a front suspension height that is 1 in (2.54 cm) less than specified:
   Technician A says the suspension height must be correct before a wheel alignment is performed.
   Technician B says the lower front suspension height may be caused by worn lower control arm bushings.
   Who is correct?
   - **A.** A only
   - **B.** B only
   - **C.** Both A and B
   - **D.** Neither A nor B

2. While performing a prealignment inspection:
   Technician A says improper front wheel bearing adjustment may affect wheel alignment angles.
   Technician B says worn ball joints have no effect on wheel alignment angles.
   Who is correct?
   - **A.** A only
   - **B.** B only
   - **C.** Both A and B
   - **D.** Neither A nor B

3. While discussing a front suspension system in which the right front wheel has 2° positive camber and the left front wheel has 1/2° positive camber:
   Technician A says when the vehicle is driven straight ahead the steering will pull to the left.
   Technician B says there will be excessive wear on the inside edge of the left front tire tread.
   Who is correct?
   - **A.** A only
   - **B.** B only
   - **C.** Both A and B
   - **D.** Neither A nor B

4. When discussing unsatisfactory steering wheel returnability:
   Technician A says the rack and pinion steering gear mounts may be worn.
   Technician B says this problem may be caused by interference between the dash seal and the steering shaft.
   Who is correct?
   - **A.** A only
   - **B.** B only
   - **C.** Both A and B
   - **D.** Neither A nor B

5. While discussing turning radius measurement:
   Technician A says a bent steering arm will cause the turning radius to be out-of-specification.
   Technician B says if the turning radius is not within specifications tire tread wear is excessive while cornering.
   Who is correct?
   - **A.** A only
   - **B.** B only
   - **C.** Both A and B
   - **D.** Neither A nor B

6. While measuring and adjusting front wheel toe:
   - **A.** if the positive caster is increased on the right front wheel, this wheel moves toward a toe-in position.
   - **B.** improper front wheel toe setting causes steering wander and drift.
   - **C.** the front wheel toe should be checked with the front wheels straight ahead.
   - **D.** improper front wheel toe setting causes wear on the inside edge of the tire tread.

7. All these statements about front wheel toe change during wheel jounce and rebound are true EXCEPT:
   - **A.** If one front wheel toes-in and the opposite front wheel toes-out during front wheel jounce and rebound, the tie rod height is unequal.
   - **B.** If both front wheels toe-in or toe-out a small, equal amount during front wheel jounce and rebound, the tie rods are parallel to the lower control arms.
   - **C.** The improper toe change during front wheel jounce and rebound may cause bump steer.
   - **D.** Improper toe change during front wheel jounce and rebound may be caused by a worn upper strut mount.

8. While using a computer wheel aligner:
   - **A.** The technician may select defective suspension components from a list on the screen.
   - **B.** It is not necessary to check or compensate for wheel runout.
   - **C.** If the camber bar graph is red, the camber setting is within specifications.
   - **D.** A wheel sensor containing a high-frequency transmitter requires a cable connected to the aligner computer.

9. While using computer wheel aligners:
   A. On many computer wheel aligners the technician may only print out the four wheel alignment results.
   B. Some wheel sensors have the capability to measure ride height and display this reading on the screen.
   C. A front wheel swing is necessary before reading the front wheel camber.
   D. If the computer aligner contains a control arm movement monitor, the technician has to estimate the necessary shim thickness.

10. While using computer wheel aligners:
   A. Symmetry angle measurements display thrust angle, rear wheel toe, and rear wheel camber.
   B. Setback is the angle formed by a line at 90° to the vehicle centerline at the axle attachment point and a line through the left and right wheel centers.
   C. Wheelbase difference is an angle created by a line through the rear wheel centers and the thrust line.
   D. Right or left lateral offset is an angle between a line through the left front and left rear tires and a line between the right front and right rear tires.

ASE Challenge Questions

1. While discussing turning radius: *Technician A* says incorrect turning radius may often be noted by tire squeal while cornering.
   Technician B says to adjust turning radius toe-out-on-turns with the inner wheel turned to the stop.
   Who is correct?
   A. A only   C. Both A and B
   B. B only   D. Neither A nor B

2. The customer says her car sometimes suddenly swerves to one side on a bump. All of the following could cause this problem EXCEPT:
   A. loose steering gear.
   B. worn tie rods.
   C. sagging front springs.
   D. steering gear lash adjustment.

3. While discussing steering diagnosis: *Technician A* says uneven half-shaft axle lengths may cause a vehicle to pull to one side when accelerating.
   Technician B says abnormal toe changes can cause a vehicle to pull to one side on road irregularities.
   Who is correct?
   A. A only   C. Both A and B
   B. B only   D. Neither A nor B

4. A customer returns to your shop and says the car continues to drift after you have completed a front end alignment. To correct this problem, you should:
   A. check the manual steering gear for possible miscentering of the sector gear.
   B. check rear wheel alignment.
   C. ask the customer to fill the fuel tank.
   D. check the steering column flex coupling.

5. While performing a prealignment inspection:
   Technician A says a prealignment inspection should include checking the vehicle interior for heavy items.
   Technician B says tools and other items normally carried in the vehicle should be included during an alignment.
   Who is correct?
   A. A only   C. Both A and B
   B. B only   D. Neither A nor B

Table 15-1 **NATEF AND ASE TASK**

Diagnose vehicle wander, drift, pull, hard steering, bump steer, memory steer, torque steer, and steering return problems; determine needed repairs.

Problem Area	Symptoms	Possible Causes	Classroom Manual	Shop Manual
STEERING QUALITY	Wander, pull, drift	1. Incorrect camber	339	505
		2. Unequal caster	344	504
		3. Rear suspension overload	348	501
	Excessive steering effort	Excessive positive caster	346	503
STEERING QUALITY	Bump steer	1. Worn or bent idler arm or pitman arm	349	517
		2. Worn steering gear mounting bushings	349	517
STEERING QUALITY	Torque steer	1. Unequal wear on front tires	349	492
		2. Different types of front tires or tire treads	349	492
STEERING QUALITY	Steering wheel return problems	1. Improper caster setting	347	492
		2. Binding in steering column or universal joints	349	492

Table 15-2 **NATEF AND ASE TASK**

Check and adjust front and rear wheel camber on suspension systems with a camber adjustment.

Problem Area	Symptoms	Possible Causes	Classroom Manual	Shop Manual
TIRE LIFE	Excessive wear on the edges of tire tread	Excessive positive or negative camber	339	503
STEERING CONTROL	Steering pull	Excessive positive camber, one front wheel	343	503

Table 15-3 NATEF AND ASE TASK

Check and adjust front and rear wheel camber on nonadjustable suspension systems; determine needed repairs.

Problem Area	Symptoms	Possible Causes	Classroom Manual	Shop Manual
TIRE LIFE	Excessive wear on the edges of the tire tread	Excessive positive or negative camber	340	503
STEERING CONTROL	Steering pull	Excessive positive camber, one front wheel	340	503

Table 15-4 NATEF AND ASE TASK

Check and adjust caster on suspension systems with a caster adjustment.

Problem Area	Symptoms	Possible Causes	Classroom Manual	Shop Manual
STEERING CONTROL	Steering pull	Unequal caster on front wheels	344	506
STEERING EFFORT	Excessive steering effort	Excessive positive caster on both front wheels	345	506
STEERING WHEEL RETURN	Excessive steering wheel returning force	Excessive positive caster on both front wheels	347	506

Table 15-5 NATEF AND ASE TASK

Check caster on nonadjustable suspension systems; determine needed repairs.

Problem Area	Symptoms	Possible Causes	Classroom Manual	Shop Manual
STEERING CONTROL	Steering pull	Unequal caster on front wheels	344	506
STEERING EFFORT	Excessive steering effort	Excessive positive caster on both front wheels	347	506
STEERING WHEEL RETURN	Excessive steering wheel returning force	Excessive positive caster on both front wheels	347	506

Table 15-6 NATEF and ASE TASK

Check front wheel setback; determine needed repairs or adjustments.

Problem Area	Symptoms	Possible Causes	Classroom Manual	Shop Manual
STEERING CONTROL	Steering pull while driving straight ahead	Severe front wheel setback condition	331	503

Table 15-7 NATEF and ASE TASK

Check SAI/KPI (steering axis inclination/king pin inclination); determine needed repairs

Problem Area	Symptoms	Possible Causes	Classroom Manual	Shop Manual
STEERING CONTROL	Steering pull during hard acceleration	Improper SAI on one side of front suspension	339	503
	Steering pull while braking	Improper SAI on one side of front suspension	340	503
	Erratic steering on road irregularities	Improper SAI on one side of front suspension	343	503

Table 15-8 NATEF and ASE TASK

Check included angle; determine needed repairs.

Problem Area	Symptoms	Possible Causes	Classroom Manual	Shop Manual
STEERING CONTROL	Steering pull while braking or accelerating	Improper SAI on one side of front suspension	343	504
	Erratic steering on road irregularities	Improper SAI on one side of front suspension	343	504

Table 15-9 NATEF and ASE TASK

Check and adjust front wheel toe.

Problem Area	Symptoms	Possible Causes	Classroom Manual	Shop Manual
TIRE LIFE	Feathered wear on tire treads	Improper front wheel front toe	329	515

Table 15-10 NATEF and ASE TASK

Check toe-out-on-turns (turning radius/angle); determine needed repairs.

Problem Area	Symptoms	Possible Causes	Classroom Manual	Shop Manual
TIRE LIFE	Excessive front tire tread wear, scuffing	Improper turning radius	330	516
TIRE NOISE	Tire squeal while cornering	Improper turning radius	330	516
STEERING WHEEL POSITION	Steering wheel not centered while driving straight ahead	Improper tie-rod sleeve adjustment	330	516

Table 15-11 NATEF and ASE TASK

Check and adjust front and rear wheel camber on suspension systems with a camber adjustment.

Problem Area	Symptoms	Possible Causes	Classroom Manual	Shop Manual
TIRE LIFE	Wear on the outside edges of rear tire treads	Excessive negative or positive camber	339	504
STEERING CONTROL	Steering pull	Improper rear wheel camber	343	504
	Oversteer or understeer while cornering at high speed	Improper rear wheel camber	341	504

Table 15-12 NATEF and ASE TASK

Check front and rear wheel camber on nonadjustable suspension systems; determine needed repairs.

Problem Area	Symptoms	Possible Causes	Classroom Manual	Shop Manual
TIRE LIFE	Feathered wear on rear tires	Improper rear wheel toe setting	330	504
STEERING CONTROL	Steering pull	Improper toe setting on one rear wheel	330	504

Table 15-13 NATEF and ASE TASK

Check rear wheel thrust angle; determine needed repairs or adjustments.

Problem Area	Symptoms	Possible Causes	Classroom Manual	Shop Manual
STEERING	Steering pull	Thrust line positioned to the left, or right, of the geometric centerline	330	504

Job Sheet 44

44

Name _____ Date _____

Road Test Vehicle and Diagnose Steering Operation

NATEF and ASE Correlation

This job sheet is related to NATEF and ASE Automotive Suspension and Steering Task List content area: C. Wheel Alignment Diagnosis, Adjustment, and Repair. Task 1: Diagnose vehicle wander, drift, pull, hard steering, bump steer, memory steer, torque steer, and steering return problems; determine needed repairs.

Tools and Materials

None

Vehicle make and model year _____

Vehicle VIN _____

Procedure

1. Road test vehicle using a variety of driving conditionns from slow speed driving and cornering to normal cruising speed driving on a straight, level road surface. Check for the following abnormal steering conditions:

2. Vertical chassis oscillations: satisfactory _____ unsatisfactory _____

3. Chassis lateral waddle: satisfactory _____ unsatisfactory _____

4. Steering pull to right: satisfactory _____ unsatisfactory _____

5. Steering pull to left: satisfactory _____ unsatisfactory _____

6. Steering effort: satisfactory _____ unsatisfactory _____

7. Tire sqeal while cornering: satisfactory _____ unsatisfactory _____

8. Bump steer: satisfactory _____ unsatisfactory _____

9. Torque steer: satisfactory _____ unsatisfactory _____

10. Memory steer: satisfactory _____ unsatisfactory _____

11. Steering wheel return: satisfactory _____ unsatisfactory _____

12. Steering wheel freeplay: satisfactory _____ unsatisfactory _____

13. Return the vehicle to the shop and inspect suspension and steering to determine the cause of abnormal conditions. List the necessary repairs and/or adjustments to correct all abnormal conditions that occurred during the road test.

_____
_____
_____
_____

☑ **Instructor Check** _____

Job Sheet 45

45

Name _____ Date _____

Measure Front and Rear Wheel Alignment Angles with a Computer Wheel Aligner

Natef and ASE Correlation

This job sheet is related to NATEF and ASE Automotive Suspension and Steering Task List content area: C. Wheel Alignment Diagnosis, Adjustment, and Repair

Task 2:   Measure vehicle ride height; determine needed repairs.

Task 3:   Check and adjust front and rear wheel camber on suspension systems with a camber adjustment.

Task 4:   Check front and rear wheel camber on nonadjustable suspension systems; determine needed repairs.

Task 5:   Check and adjust caster on suspension systems with a caster adjustment.

Task 6:   Check caster on nonadjustable suspension ysytems; determine needed repairs.

Task 7:   Check and adjust front wheel toe.

Task 9:   Check toe-oput-on-turns (turning radius/angle); determine needed repairs.

Task 10:   Check SAI/KPI (steering axis inclination/king pin inclination); determine needed repairs.

Task 11:   Check included angle; determine needed repairs.

Task 12:   Check rear wheel toe; determine needed repairs or adjustments.

Task 13:   Check rear wheel thrust angle; determine neede repairs or adjustments.

Task 14:   Check for front wheel setback; determine needed repairs or adjustments.

Tools and Materials

Computer wheel aligner

Vehicle make and model year _____

Vehicle VIN _____

Procedure

   CAUTION: When driving a vehicle onto an alignment ramp, do not stand in front of the vehicle. This may cause personal injury.

1. Lock front turn tables and drive vehicle onto the alignment ramp. Apply the parking brake.

Are the front tires properly positioned on the front turn tables? yes _____ no_____

Are the rear tires properly positioned on the slip plates? yes _____ no_____

Instructor check _____

2. Install the rim clamps and wheel sensors. Perform wheel sensor levelling and wheel runout compensation procedures.

Are the wheel sensors level? yes _____ no_____

Is the wheel runout compensation procedure completed? yes _____ no_____

Instructor check _____

3. Select specifications for the vehicle being aligned.

Are the specifications selected? yes _____ no_____

Instructor check _____

4. Perform a prealignment inspection using the checklist in the computer wheel aligner. List any components that must be repaired or replaced, and explain the reasons for your diagnosis.

_____

_____

_____

_____

_____

5. Measure ride height.

Left front ride height _____ Specified ride height _____

Right front ride height _____ Specified ride height _____

Left rear ride height _____ Specified ride height _____

Right rear ride height _____ Specified ride height _____

State the necessary repairs to correct ride height and explain the reasons for your diagnosis.

_____

_____

_____

_____

6. Measure front and rear suspension alignment angles following the prompts on the computer wheel aligner screen.

Left front camber _____ Right front camber _____

Cross camber _____ Specified front wheel camber _____

Specified cross camber _____

Left front caster _____ Right front caster _____

Cross caster _____ Specified front wheel caster _____

Specified cross caster _____

Left front SAI _____ Right front SAI _____

Specified SAI _____ Included angle _____

Thrust angle _____

Specified thrust angle _____

Left front toe _____ Right front toe _____

Total toe _____ Specified front wheel toe _____

Left rear camber _____ Right rear camber _____

Specified camber _____

Left rear toe_____ Right rear toe _____

Total toe _____ Specified rear wheel toe _____

State the necessary adjustments and repairs to correct front and rear suspension alignment angles and explain the reasons for your diagnosis.

_____

_____

_____

_____

_____

7. Measure turning radius.

Left turn, turning radius right front wheel _____

Turning radius left front wheel _____

Specified turning radius _____

Right turn, turning radius left front wheel _____

Turning radius right front wheel _____

Specified turning radius _____

State the necessary repairs to correct turning radius, and explain the reasons for your diagnosis. _____

_____

_____

_____

✔️ **Instructor Check** _____

Job Sheet 46

Name _____ Date _____

Check Proper Steering Linkage Height by Measuring Toe Change

NATEF and ASE Correlation

This job sheet is related to NATEF and ASE Automotive Suspension and Steering Task List content area: C. Wheel Alignment Diagnosis, Adjustment, and Repair. Task 1: Diagnose vehicle wander, drift, pull, hard steering, bump steer, memory steer, torque steer, and steering return problems; determine needed repairs.

Tools and Materials

Computer wheel aligner

Vehicle make and model year _____

Vehicle VIN _____

Procedure

1. Be sure the toe is adjusted to specifications with the front wheels centered and the steering wheel locked.

   Toe, left front _____

   Toe, right front _____

   Total toe, front wheels _____

2. Pull the chassis downward approximately 3 in (7.6 cm) and record the toe change.

   Toe change left front from _____ to _____

   Toe change right front from _____ to _____

3. During the wheel jounce in step 2, did the toe remain at the original setting on the left front? Amount of toe change from the original setting on the left front wheel _____

   During the wheel jounce in step 2, did the toe remain at the original setting on the right front? Amount of toe change from the original setting on the right front wheel _____

4. During the wheel jounce in step 2, did the right and the left front wheels move an equal amount toward a toe-in position? Amount of toe change toward a toe-in position on both front wheels _____

5. During the wheel jounce in step 2, did one front wheel move toward a toe-in or toe-out position while the opposite front wheel remained at the original position? Amount that one front wheel moved toward a toe-in or toe-out position while the opposite front wheel remained at the original position _____

6. During the wheel jounce in step 2, did one front wheel move toward a toe-in and the opposite front wheel move toward a toe-out position?
Amount that one front wheel moved toward a toe-in position _____
Amount the opposite front wheel moved toward a toe-out position _____

7. Push the chassis upward 3 in (7.6 cm) and check the toe change.

Left front toe change: from _____ to _____

Right front toe change: from _____ to _____

8. During the wheel rebound in step 7, did one front wheel remain at the original setting and the opposite front wheel move toward a toe-in or toe-out position?
Amount that one front wheel moved toward a toe-in or toe-out position while the opposite front wheel remained at the original setting _____

9. During the wheel rebound in step 7, did one front wheel move toward a toe-in position while the opposite front wheel moved toward a toe-out position?
Amount that one front wheel moved toward a toe-in position, and the amount the opposite front wheel moved toward a toe-out position _____

10. List all the abnormal conditions that indicate unequal steering arm height.

_____

_____

_____

_____

11. Inspect the front steering linkages and steering gear mountings and list the necessary repairs to correct the improper toe change during wheel jounce and rebound. Explain the reasons for your diagnosis

_____

_____

_____

Instructor Check _____

Four Wheel Alignment Adjustments

Upon completion and review of this chapter, you should be able to:

❑ Adjust front wheel camber on various front suspension systems.

❑ Adjust front wheel caster on various front suspension systems.

❑ Correct setback conditions.

❑ Check and correct front engine cradle position.

❑ Correct SAI angles that are not within specifications.

❑ Adjust front wheel toe.

❑ Center steering wheel.

❑ Recognize the symptoms of improper rear wheel alignment.

❑ Diagnose the causes of improper rear wheel alignment.

❑ Perform rear wheel camber adjustments.

❑ Perform rear wheel toe adjustments.

❑ Use a track gauge to measure rear wheel tracking.

❑ Diagnose rear wheel tracking problems from the track gauge measurements.

Wheel Alignment Procedure

The proper procedure for front and rear wheel alignment is important since adjusting one wheel alignment angle may change another angle. For example, adjusting front wheel caster changes front wheel toe. The wheel alignment adjustment procedure is especially critical on four wheel independent suspension systems. A front wheel adjustment procedure is provided in Figure 16-1, and a typical rear wheel adjustment procedure is given in Figure 16-2. Always follow the wheel alignment procedure in the vehicle manufacturer's service manual.

✔ **SERVICE TIP:** Since rear wheel alignment plays a significant role in aiming or steering the vehicle, the rear wheel alignment should be corrected before adjusting front suspension angles. The rear wheel camber, toe, and thrust line must be within specifications before adjusting the front suspension angles.

Camber Adjustment

Shims

Various methods are provided by car manufacturers for camber adjustment. Some car manufacturers provide a shim-type **camber adjustment** between the upper control arm mounting and the inside of the frame (Figure 16-3). In this type of camber adjustment, increasing the shim thickness moves the camber setting to a more negative position, whereas decreasing shim thickness changes the camber toward a more positive position. Shims of equal thickness should be added or removed on both upper control arm mounting bolts to change the camber setting without affecting the caster setting.

Other vehicle manufacturers provide a shim-type camber adjustment between the upper control arm mounting and the outside of the frame (Figure 16-4). On this shim-type camber adjustment, increasing the shim thickness increases positive camber.

Basic Tools

Basic technician's tool set

Service manual

Chalk

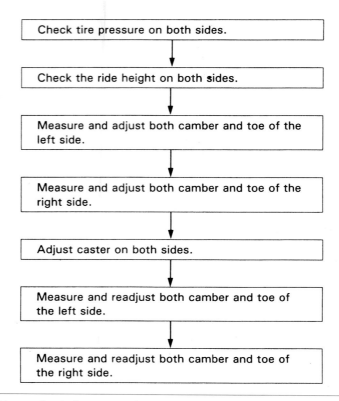

Figure 16-1 Front wheel alignment adjustment procedure. (Courtesy of American Honda Motor Company, Inc.)

Eccentric cams are out-of-round pieces of metal mounted on a retaining bolt with the shoulder of the cam positioned against a component. When the cam is rotated the component contacting the cam shoulder is moved.

Eccentric Cams

Some car manufacturers provide a cam-type camber adjustment on the inner ends of the upper control arm (Figure 16-5), whereas other suspension systems have a cam adjustment on the inner ends of the lower control arms (Figure 16-6). If the original cam adjustment on the inner ends of the upper control arms does not provide enough camber adjustment, aftermarket upper control arm shaft kits are available that provide an extra 1 1/2° of camber adjustment (Figure 16-7). The suspension adjustments shown in Figures 16-3 through 16-7 are used on rear-wheel-drive cars with short-and-long arm front suspension systems.

Some MacPherson strut front suspension systems on front-wheel-drive cars have a cam on one of the steering-knuckle-to-strut bolts to adjust camber (Figure 16-8). Aftermarket camber adjustment

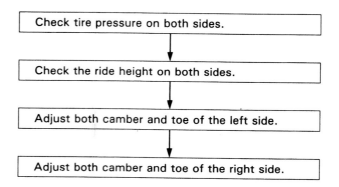

Figure 16-2 Rear wheel alignment adjustment procedure. (Courtesy of American Honda Motor Company, Inc.)

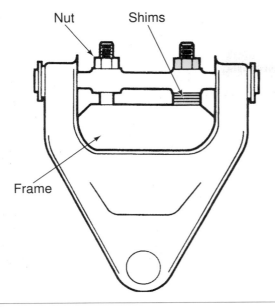

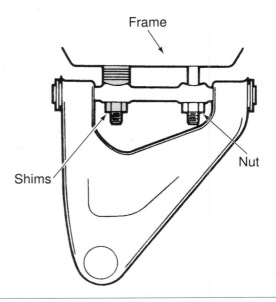

Figure 16-3 Shim-type camber adjustment between upper control arm mounting and the inside of the frame. (Courtesy of Hunter Engineering Company.)

Figure 16-4 Shim-type camber adjustment between upper control arm mounting and the outside of the frame. (Courtesy of Hunter Engineering Company.)

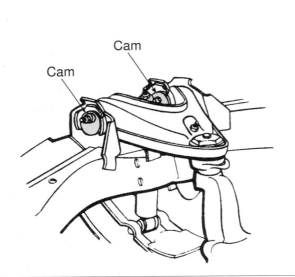

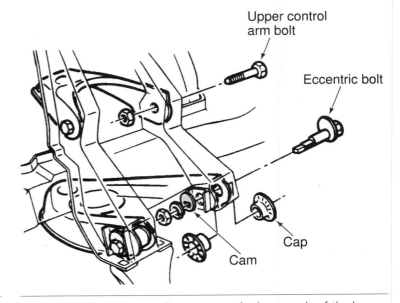

Figure 16-5 Cam adjustment on the inner ends of the upper control arms. (Courtesy of Hunter Engineering Company.)

Figure 16-6 Cam adjustment on the inner ends of the lower control arms. (Courtesy of Hunter Engineering Company.)

Figure 16-7 Upper control arm shaft kit provides an extra 1 1/2° camber adjustment. (Courtesy of Cooper Industries.)

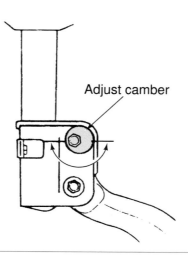

Figure 16-8 Eccentric steering-knuckle-to-strut camber adjusting bolt, MacPherson strut front suspension. (Courtesy of Hunter Engineering Company.)

kits that provide 2 1/2° of camber adjustment are available for many MacPherson strut suspension systems (Figure 16-9 and Figure 16-10). Similar camber adjustment kits are available for many cars with MacPherson strut front suspension systems. On some double wishbone front suspension systems a graduated cam on the inner end of the lower control arm provides a camber adjustment (Figure 16-11).

Slotted Strut Mounts and Frames

On some MacPherson strut front suspension systems the upper strut mounts may be loosened and moved inward or outward to adjust camber (Figure 16-12). Other MacPherson strut front suspen-

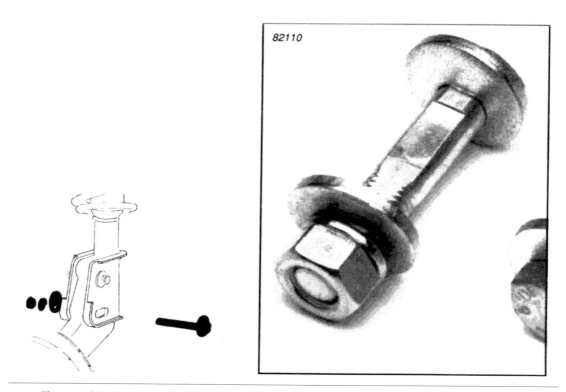

Figure 16-9 Camber adjustment kits provide 2 1/2° camber adjustment on some General Motors cars. (Courtesy of Specialty Products Company.)

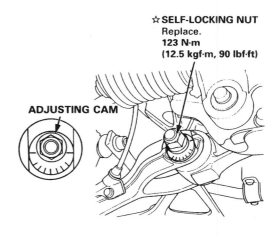

☆ **SELF-LOCKING NUT**
Replace.
123 N·m
(12.5 kgf·m, 90 lbf·ft)

ADJUSTING CAM

☆ : **Corrosion resistant bolt/nut**

Figure 16-10 Camber adjustment kits provide 2 1/2° camber adjustment on some General Motors and American Motors Corporation Products. (Courtesy of Specialty Products Company)

Figure 16-11 Camber adjustment double wishbone front suspension. (Courtesy of American Honda Motor Company, Inc.)

sion systems do not provide a camber adjustment. For these cars, aftermarket manufacturers sell slotted upper strut bearing plates that provide camber adjustment capabilities (Figure 16-13). If the front wheel camber adjustment is not correct, check for worn components such as ball joints or upper strut mounts. Also check the engine cradle mounts. Since the inner ends of the lower control

Figure 16-12 The upper strut mounting bolts may be loosened and the mount moved inward or outward to adjust camber on some MacPherson strut front suspension systems. (Courtesy of Hunter Engineering Company)

Figure 16-13 Slotted upper strut bearing plates provide camber adjustment on some MacPherson strut front suspension systems. (Courtesy of Specialty Products Company)

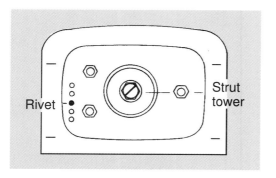

Figure 16-14 Upper strut mount with camber plate and rivet, modified MacPherson strut front suspension. (Courtesy of Hunter Engineering Company)

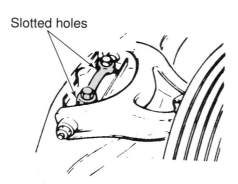

Figure 16-15 Elongated bolt holes for upper control arm shaft bolts provide camber adjustment on some short-and-long arm front suspension systems. (Courtesy of Hunter Engineering Company)

arms are attached to the engine cradle on many front-wheel-drive cars, a cradle that is out of position may cause improper camber readings.

A camber plate is locked in place with a pop rivet on some upper strut mounts. The pop rivet may be removed to release the camber plate (Figure 16-14). After the upper strut mount bolts are loosened, the mount may be moved inward or outward to adjust the camber. When the camber is adjusted to the vehicle manufacturer's specification, the strut mount bolts should be tightened to the correct torque and a new pop rivet installed.

On some short-and-long arm front suspension systems, the bolt holes in the frame are elongated for the inner shaft on the upper control arm. When these bolts are loosened, the upper control arm may be moved inward or outward to adjust camber (Figure 16-15).

Eccentric Bushing or Ball Joint Eccentric

In some short-and-long arm front suspension systems an eccentric bushing in the upper end of the steering knuckle may be used to adjust camber (Figure 16-16). On other short-and-long arm front suspension systems a ball joint eccentric may be rotated to adjust camber (Figure 16-17).

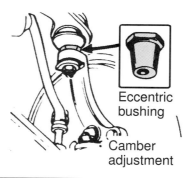

Figure 16-16 Eccentric bushing in upper end of the steering knuckle may be used to adjust camber on some short-and-long arm front suspension systems. (Courtesy of Hunter Engineering Company)

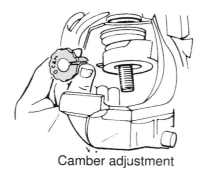

Figure 16-17 Upper ball joint eccentric may be used to adjust camber. (Courtesy of Hunter Engineering Company)

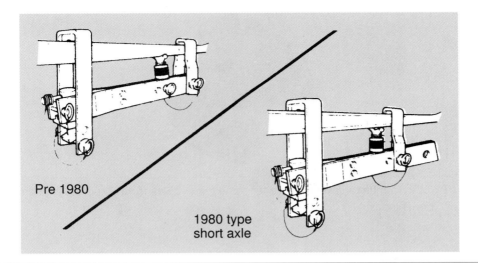

Figure 16-18 Adjusting camber to a more positive setting, twin I-beam front suspension. (Courtesy of Hunter Engineering Company)

Camber Adjustment on Twin I-Beam Front Syspension System

On twin I-beam front suspension systems a special clamp-type bending tool with a hydraulic jack is used to bend the I-beams and change the camber setting. On a twin I-beam front suspension the clamp and hydraulic jack must be positioned properly to adjust the camber to a more positive setting (Figure 16-18) or a more negative setting (Figure 16-19). On some twin I-beam front suspension systems camber sleeves are available to adjust camber. These sleeves are located in the top hole in the outer end of the I-beams, and the upper ball joint studs extend through the openings in the camber sleeves (Figure 16-20).

Special Tools

Twin I-beam bending tool

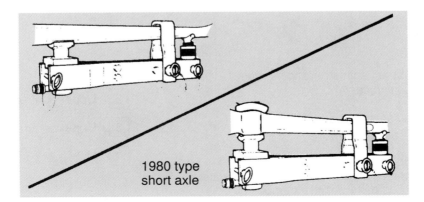

Figure 16-19 Adjusting camber to a more negative setting, twin I-beam front suspension. (Courtesy of Hunter Engineering Company)

Figure 16-20 Camber adjusting sleeve for some twin I-beam front suspension systems. (Courtesy of Cooper Industries)

Caster Adjustment Procedure

Shims

On a short-and-long arm front suspension system the same shims are used for camber and **caster adjustment.** If the shims are located between the upper control arm shaft and the inside of the frame, subtract shims from the front mounting bolt and add shims to the rear mounting bolt to increase positive caster (Figure 16-21). After the caster is adjusted, the camber should be rechecked.

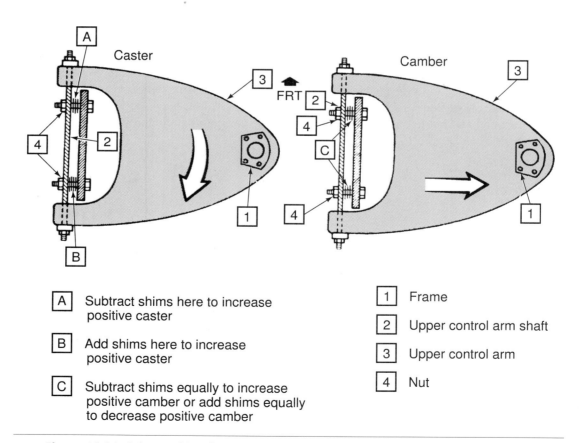

A	Subtract shims here to increase positive caster
B	Add shims here to increase positive caster
C	Subtract shims equally to increase positive camber or add shims equally to decrease positive camber

1	Frame
2	Upper control arm shaft
3	Upper control arm
4	Nut

Figure 16-21 Subtract shims from the front upper control arm mounting bolt and add shims to the rear mounting bolt to increase positive caster. (Courtesy of Chevrolet Motor Division, General Motors Corp.)

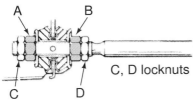

Unscrew B then tighten
A to reduce caster

A   B

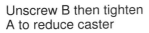

C, D locknuts

C   D

Unscrew A then tighten B
to increase caster

Figure 16-22 Strut rod nuts may be adjusted to lengthen or shorten the strut rod and adjust caster. (Courtesy of Hunter Engineering Company)

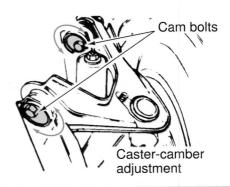

Cam bolts

Caster-camber adjustment

Figure 16-23 Eccentric cams on the inner ends of the upper or lower control arms may be used to adjust camber and caster. (Courtesy of Hunter Engineering Company)

Strut Rod Length Adjustment

On some suspension systems the nuts on the forward end of the strut rod may be adjusted to lengthen or shorten the strut rod, which changes the caster setting (Figure 16-22). Shorten the strut rod to increase the positive caster.

Eccentric Cams

The same **eccentric cams** on the inner ends of the upper or lower control arms may be used to adjust camber or caster (Figure 16-23). If an eccentric bushing in the outer end of the upper control arm is rotated to adjust camber, this same eccentric also adjusts caster. On some double wishbone front suspension systems, the pivot adjuster mounting bolt nuts must be loosened under the compliance pivot and the graduated cam may then be rotated to adjust caster (Figure 16-24).

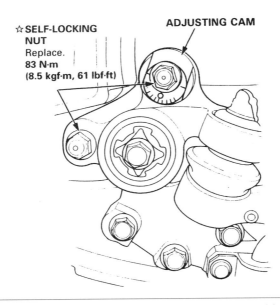

☆ SELF-LOCKING
NUT
Replace.
83 N·m
(8.5 kgf·m, 61 lbf·ft)

ADJUSTING CAM

Figure 16-24 Graduated cam for caster adjustment, double wishbone front suspension. (Courtesy of American Honda Motor Company, Inc.)

Slotted Strut Mounts and Frames

If a caster adjustment is required and the suspension does not have an adjustment for this purpose, check for worn components such as upper strut mounts and ball joints. Also check the engine cradle mounts. Since the inner ends of the lower control arms are attached to the engine cradle on many front-wheel-drive cars, a cradle that is out-of-position may cause improper caster readings.

The slots in the frame at the upper control arm shaft mounting bolts provide camber and caster adjustment. If the upper strut mount is adjustable on a MacPherson strut suspension, the mount retaining bolts may be loosened and the mount moved forward or rearward to adjust caster. In some suspension systems, one of the upper strut mount bolt holes has to be elongated with the proper size drill and a round file prior to caster adjustment (Figure 16-25).

> **CAUTION:** After any alignment angle adjustment, always be sure the adjustment bolts are tightened to the vehicle manufacturer's specified torque. Loose suspension adjustment bolts will result in improper alignment angles, steering pull, and tire wear. Loose suspension adjustment bolts may also cause reduced directional stability, which could result in a vehicle accident and possible personal injury.

Caster Adjustment on Twin I-Beam Front Suspension Systems

> **CAUTION:** Improper use of bending equipment may result in personal injury. Always follow the equipment and vehicle manufacturer's recommended procedures.

Special Tools

Twin I-beam radius rod bending tool

The same bending tool and hydraulic jack may be used to adjust camber and caster on twin I-beam front suspension systems. This equipment is installed on the radius rods connected from the I-beams to the chassis to adjust caster. If the hydraulic jack is placed near the rear end of the radius rod and the clamp arm is positioned over the center area of this rod, the rod is bent downward in the center when the jack is operated (Figure 16-26). This bending action tilts the I-beam and caster line rearward to increase positive caster.

If the hydraulic jack is placed in the center area of the radius arm and the clamps are positioned at the ends of the arm, the arm is bent upward when the jack is operated (Figure 16-27). When this bending action occurs, the I-beam and caster line are tilted forward to decrease positive caster.

On some twin I-beam front suspension systems combination camber and caster sleeves are available to provide a camber and caster adjustment (Figure 16-28). These sleeves are located in the top hole in the outer end of the I-beams. The upper ball joint studs extend through the openings in the camber sleeves. Aftermarket parts manufacturers supply a special radius arm bushing to provide a caster adjustment on some twin I-beam front suspension systems (Figure 16-29).

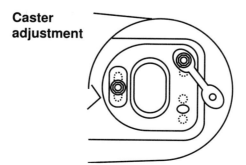

Caster adjustment

Figure 16-25 Some adjustable upper strut mounts may be moved forward or rearward to adjust caster. (Courtesy of Hunter Engineering Company)

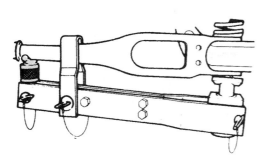

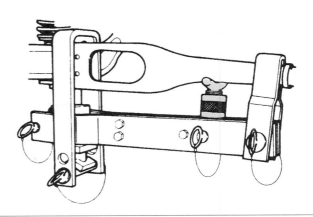

Figure 16-26 Hydraulic jack and bending tool positioned on the radius rod to increase positive caster on a twin I-beam front suspension. (Courtesy of Hunter Engineering Company)

Figure 16-27 Hydraulic jack and clamp positioned on radius rod to decrease positive caster on a twin I-beam front suspension system. (Courtesy of Hunter Engineering Company)

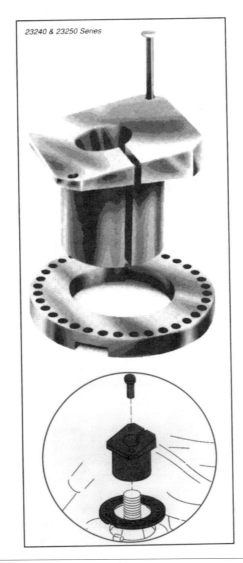

Figure 16-28 Combination camber/caster sleeves for some twin I-beam front suspension systems. (Courtesy of Specialty Products Company)

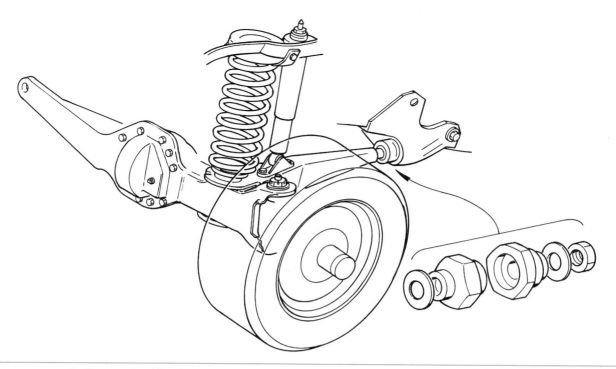

Figure 16-29 Radius arm bushing for caster adjustment on some twin I-beam suspension systems. (Courtesy of Cooper Industries)

Setback Measurement and Correction Procedure

Computer wheel aligners have **setback** measuring capabilities on the front wheels. These aligners automatically display setback angles with the other front suspension alignment angles. If a front-wheel-drive car has experienced collision damage on the left front, the left side of the cradle or sub-frame may be driven rearward (Figure 16-30). This type of collision damage also moves the bottom of the left front strut rearward to a setback condition and shifts the caster to a negative position.

⚠️ **WARNING:** If the engine cradle has to be removed or loosened, the engine support fixture must hold the engine in the normal position to prevent damage to the fan shroud, power steering hoses, and other components.

🔲 **CAUTION:** Prior to any service operation where the body is going to be moved in relation to the cradle or subframe, the intermediate shaft on the rack and pinion gear must be disconnected. If this shaft is not disconnected, it may be damaged, resulting in loss of steering control, serious vehicle damage, and personal injury.

With this type of collision damage, it may be possible to adjust the caster back to the specified setting, but that does not correct the setback problem. If the caster is adjusted and the setback problem is ignored, the steering may pull to one side while driving straight ahead, especially if the setback condition is severe. The cradle must be straightened and the unitized body must be pulled back to the original measurements to correct this problem. A severely bent, distorted cradle should be replaced.

Collision damage may push a front cradle sideways. This condition may increase the positive camber on one side of the car and move the camber toward a negative position on the opposite side. When this condition is encountered, always inspect and correct the cradle position. If the camber is adjusted in this situation, the steering will not provide proper directional control.

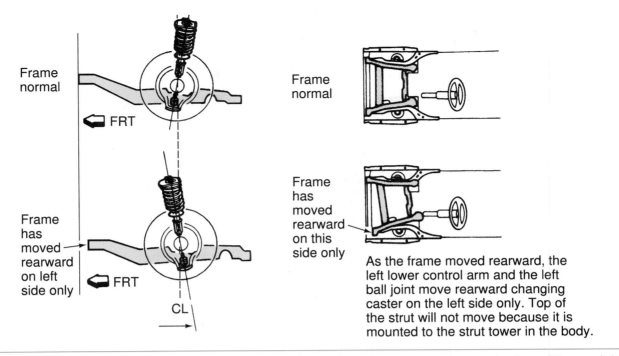

As the frame moved rearward, the left lower control arm and the left ball joint move rearward changing caster on the left side only. Top of the strut will not move because it is mounted to the strut tower in the body.

Figure 16-30 Collision damage on the left front moves the cradle and strut rearward to a setback condition, and shifts the caster to a negative position. (Courtesy of Oldsmobile Motor Division, General Motors Corp.)

Prior to cradle removal, an engine support fixture must be installed to support the engine and transaxle (Figure 16-31). The engine support fixture must hold the engine in the original position to prevent damage to the power steering hoses or other components.

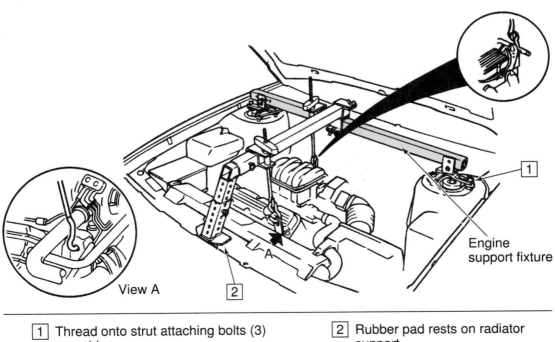

| 1 | Thread onto strut attaching bolts (3) per side | 2 | Rubber pad rests on radiator support |

Figure 16-31 Engine support fixture installed to support engine and transaxle prior to cradle removal. (Courtesy of Oldsmobile Motor Division, General Motors Corp.)

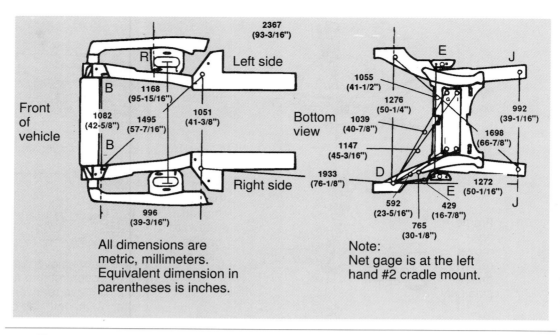

Figure 16-32 Front and rear underbody measurements. (Courtesy of Oldsmobile Motor Division, General Motors Corp.)

The steering axis inclination (SAI) line is an imaginary line through the centers of the upper and lower ball joints, or through the center of the lower ball joint and the upper strut mount.

The SAI angle is the angle between the SAI line and the true vertical centerline of the tire and wheel.

The included angle is the sum of the SAI angle and the camber angle if the camber is positive. If the camber is negative, the camber setting must be subtracted from the SAI angle to obtain the included angle.

Before the body is moved in relation to the cradle or subframe, the intermediate shaft on the rack and pinion steering gear must be disconnected to prevent damage to this shaft. Remove all the engine and transaxle mount bolts in the cradle, plus the cradle retaining bolts and insulators to remove the cradle. Perform the front underbody measurements with a tram gauge (Figure 16-32), and pull the unitized body back to these specifications. This pulling operation is usually done in an autobody shop rather than an automotive repair shop.

The cradle must be restored to the specified measurements (Figure 16-33). Check all cradle insulators and replace any that are damaged, oil soaked, or worn. When the cradle is replaced, all the insulators must be installed properly, and all the cradle bolts should be aligned with the bolt holes in the body. Alignment hole A in the cradle must be perfectly aligned with the matching alignment hole in the body (Figure 16-34). Observing this alignment hole is a quick way to check cradle position.

Steering Axis Inclination (SAI) Correction Procedure

Steering axis inclination (SAI) is not considered adjustable. If the SAI is not within specifications, the upper strut tower may be out of position, the lower control arm may be bent or the center crossmember, or cradle, could be shifted. In most cases, these defects are caused by collision damage. When the SAI angle is correct but the camber and included angle are less than specified, the strut or spindle may be bent. The difference in the included angles on the front wheels must not exceed 1 1/2°.

When the SAI is not within specifications, steering pull may occur while braking or accelerating and steering may be erratic on road irregularities.

Toe Adjustment

⚠️ **WARNING:** Do not heat tie-rod sleeves to loosen them. This action may weaken the sleeves, resulting in sudden sleeve failure, loss of steering control, personal injury, and vehicle damage.

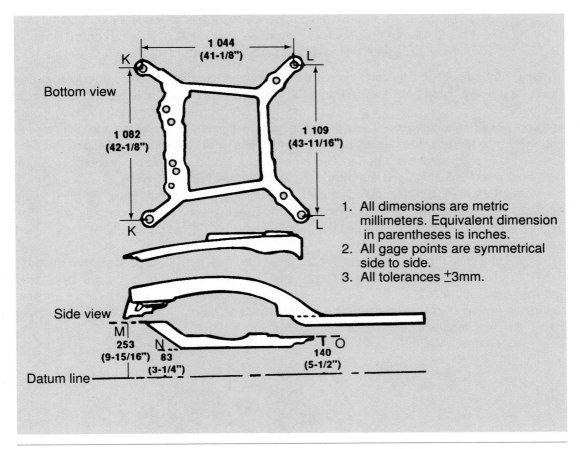

Figure 16-33 Front cradle measurements. (Courtesy of Oldsmobile Motor Division, General Motors Corp.)

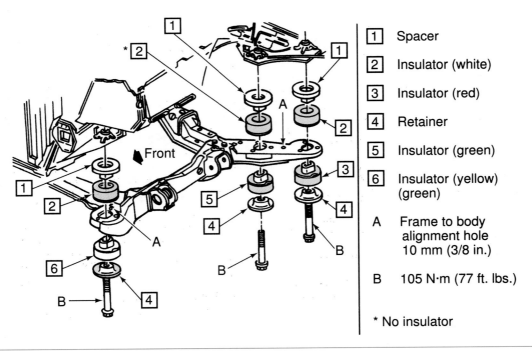

1	Spacer
2	Insulator (white)
3	Insulator (red)
4	Retainer
5	Insulator (green)
6	Insulator (yellow) (green)
A	Frame to body alignment hole 10 mm (3/8 in.)
B	105 N·m (77 ft. lbs.)
*	No insulator

Figure 16-34 Cradle insulators and alignment holes in cradle and body. (Courtesy of Oldsmobile Motor Division, General Motors Corp.)

Tie-rod sleeve rotating
tool

Toe-in occurs when the
distance between the
front edges of the front
wheels is less than the
distance between the
rear edges of the front
wheels.

Toe-out is present
when the distance
between the front
edges of the front
wheels is more than
the distance between
the rear edges of the
front wheels.

Total toe is the sum of
the toe settings on the
front wheels.

⚠ **WARNING:** Do not use a pipe wrench to loosen tie rod sleeves. This tool may crush and weaken the sleeve, causing sleeve failure, loss of steering control, personal injury, and vehicle damage. A tie rod rotating tool must be used for this job.

When a front wheel **toe** adjustment is required, apply penetrating oil to the tie-rod adjusting sleeves and sleeve clamp bolts. Loosen the tie-rod adjusting sleeve clamp bolts enough to allow the clamps to partially rotate. One end of each tie-rod sleeve contains a right-hand thread; the opposite end contains a left-hand thread. These threads match the threads on the tie rod and outer tie-rod end. When the tie-rod sleeve is rotated the complete tie rod, sleeve, and tie-rod end assembly is lengthened or shortened. Use a tie-rod sleeve rotating tool to turn the sleeves until the toe-in on each front wheel is equal to one-half the total toe specification (Figure 16-35).

After the **toe-in** adjustment is completed, the adjusting sleeve clamps must be turned so the openings in the clamps are positioned away from the slots in the adjusting sleeves (Figure 16-36). Replace clamp bolts that are rusted, corroded, or damaged and tighten these bolts to the specified torque.

On some rack and pinion steering gears, the tie rod lock nut is loosened and the tie rod is rotated to adjust toe on each front wheel (Figure 16-37). Prior to rotating the tie rod, the small outer bellows boot clamp should be removed to prevent it from twisting during tie rod rotation. After the toe-in adjustment, the tie rod nut must be torqued to specifications, and the small bellows boot clamp must be reinstalled.

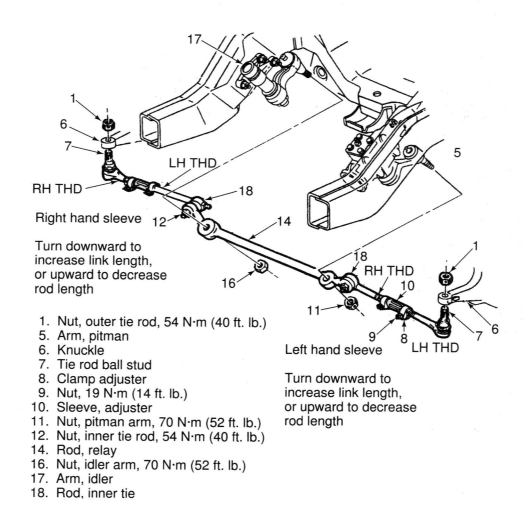

RH THD

Right hand sleeve

Turn downward to
increase link length,
or upward to decrease
rod length

LH THD

Left hand sleeve

Turn downward to
increase link length,
or upward to decrease
rod length

1. Nut, outer tie rod, 54 N·m (40 ft. lb.)
5. Arm, pitman
6. Knuckle
7. Tie rod ball stud
8. Clamp adjuster
9. Nut, 19 N·m (14 ft. lb.)
10. Sleeve, adjuster
11. Nut, pitman arm, 70 N·m (52 ft. lb.)
12. Nut, inner tie rod, 54 N·m (40 ft. lb.)
14. Rod, relay
16. Nut, idler arm, 70 N·m (52 ft. lb.)
17. Arm, idler
18. Rod, inner tie

Figure 16-35 Rotating tie-rod sleeves to adjust front wheel toe. (Courtesy of Chevrolet Motor Division, General Motors Corp.)

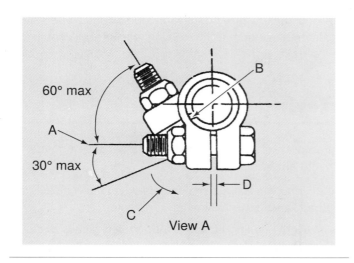

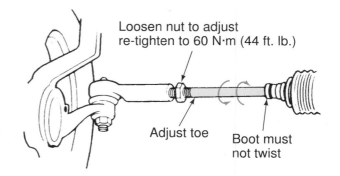

Figure 16-36 Proper tie rod adjusting sleeve clamp position. (Courtesy of Chevrolet Motor Division, General Motors Corp.)

Figure 16-37 Rotating tie rod to adjust the toe on a rack and pinion steering gear. (Courtesy of Chevrolet Motor Division, General Motors Corp.)

On other rack and pinion steering gears, the tie-rod and outer tie-rod end have internal threads and a threaded adjuster is installed in these threads. One end of the adjuster has a right-hand thread and the opposite end has a left-hand thread. Matching threads are located in the outer tie-rod end and tie rod. Clamps on the outer tie-rod end and tie rod secure these components to the adjuster. A hex-shaped nut is designed into the center of the adjuster (Figure 16-38). After the clamps are loosened, an open end wrench is placed on this hex nut to rotate the adjuster and change the toe setting.

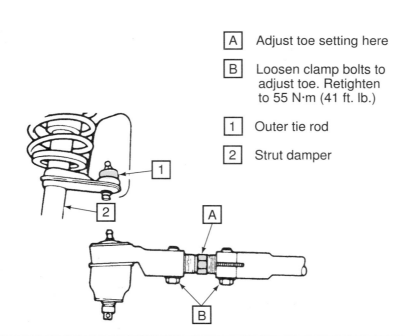

Figure 16-38 Rack and pinion steering gear with externally threaded toe adjuster and internal threads on the tie rod and outer tie-rod end. (Courtesy of Oldsmobile Motor Division, General Motors Corp.)

When the toe-in adjustment is completed, the toe must be set within the manufacturer's specifications with the steering wheel in the centered position. If the steering wheel is not centered, a centering adjustment is required.

Steering Wheel Centering Procedure

 SERVICE TIP: The most accurate check of a properly centered steering wheel is while driving the vehicle straight ahead during a road test.

Road test the vehicle and determine if the steering wheel spoke is centered when the vehicle is driven straight ahead. If steering wheel centering is necessary, follow this procedure:

1. Lift the front end of the vehicle with a hydraulic jack and position safety stands under the lower control arms. Lower the vehicle onto the safety stands and place the front wheels in the straight-ahead position.

2. Use a piece of chalk to mark each tie-rod sleeve in relation to the tie rod, and loosen the sleeve clamps (Figure 16-39).

3. Position the steering wheel spoke in the position it was in while driving straight ahead during the road test. Turn the steering wheel to the centered position and note the direction of the front wheels.

4. If the steering wheel spoke is low on the left side while driving the vehicle straight ahead, use a tie-rod sleeve rotating tool to shorten the left tie rod and lengthen the right tie rod (Figure 16-40). One-quarter turn on a tie-rod sleeve moves the steering wheel position approximately one inch. Turn the tie-rod sleeves the proper amount to bring the steering wheel to the centered position. For example, if the steering wheel spoke is two inches off-center, turn each tie-rod sleeve one-half turn.

5. If the steering wheel spoke is low on the right side while driving the vehicle straight ahead, lengthen the left tie rod and shorten the right tie rod.

6. Mark each tie-rod sleeve in its new position in relation to the tie rod. Be sure the sleeve clamp openings are positioned properly as indicated earlier in this chapter. Tighten the clamp bolts to the specified torque.

7. Lift the front chassis with a floor jack and remove the safety stands. Lower the vehicle onto the shop floor and check the steering wheel position during a road test.

Classroom Manual
Chapter 16, page 354

When all the front suspension alignment angles are adjusted within the manufacturer's specifications, road test the vehicle. When the vehicle is driven on a relatively smooth, straight road, there should be directional stability with no drift to the right or left, and the steering wheel must be centered.

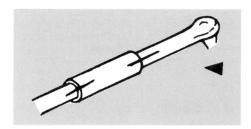

Figure 16-39 Marking tie-rod sleeves in relation to the tie rods and outer tie-rod ends. (Courtesy of Ammco Tools, Inc.)

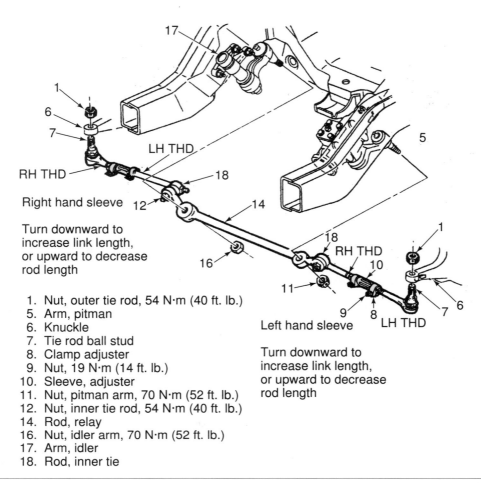

1. Nut, outer tie rod, 54 N·m (40 ft. lb.)
5. Arm, pitman
6. Knuckle
7. Tie rod ball stud
8. Clamp adjuster
9. Nut, 19 N·m (14 ft. lb.)
10. Sleeve, adjuster
11. Nut, pitman arm, 70 N·m (52 ft. lb.)
12. Nut, inner tie rod, 54 N·m (40 ft. lb.)
14. Rod, relay
16. Nut, idler arm, 70 N·m (52 ft. lb.)
17. Arm, idler
18. Rod, inner tie

Figure 16-40 Rotating tie-rod sleeves to center steering wheel. (Courtesy of Chevrolet Motor Division, General Motors Corp.)

Causes of Improper Rear Wheel Alignment

The following suspension or chassis defects will cause incorrect rear wheel alignment.

1. Collision damage that results in a bent frame or distorted unitized body.

2. A leaf-spring eye that is unwrapped or spread open.

3. Leaf-spring shackles that are broken, bent, or worn.

4. Broken leaf springs or leaf-spring center bolts.

5. Worn rear upper or lower control arm bushings.

6. Worn trailing arm bushings or dislocated trailing arm brackets.

7. Bent components such as radius rods, control arms, struts, and rear axles.

Rear Suspension Adjustments

Rear Wheel Camber Adjustment

⚠ **WARNING:** Always follow the vehicle manufacturer's recommended rear suspension adjustment procedure in the service manual to avoid improper alignment angles and reduced steering control.

Improper **rear wheel camber** causes excessive wear on the edges of the rear tire treads. Cornering stability may be affected by improper rear wheel camber, especially while cornering at high speeds. Improper rear wheel camber on a front-rear-drive car may change the understeer or oversteer characteristics of the vehicle. Since a tilted wheel rolls in the direction of the tilt, rear wheel camber may cause steering pull. For example, excessive positive camber on the right rear wheel causes the rear suspension to drift to the right and steering to pull to the left.

Rear suspension adjustments vary depending on the type of suspension system. Some manufacturers of wheel alignment equipment and tools provide detailed diagrams for front and rear wheel alignment adjustments on all makes of domestic and imported vehicles. Our objective is to show a few of the common methods of adjusting rear wheel alignment angles on various types of rear suspension systems.

On some semi-independent rear suspension systems, camber and toe are adjusted by inserting different sizes of shims between the rear spindle and the spindle mounting surface. These shims are retained by the spindle mounting bolts. The shim thickness is changed between the top or bottom of the spindle to adjust camber (Figure 16-41).

Many rear camber shims are now circular and are available in a wide variety of configurations to fit various rear wheels. Some computer wheel aligners indicate the thickness of shim required and the proper shim position. These same shims also adjust rear wheel toe (Figure 16-42).

On some transverse leaf rear suspension systems such as those on a later model Oldsmobile Cutlass Supreme, the spindle must be removed from the strut, and the lower strut bolt hole elongated to allow a camber adjustment (Figure 16-43). Once this bolt hole is elongated, assemble the knuckle and strut and leave the retaining bolts loose. The top of the tire and the spindle may be moved outward or inward to adjust camber. After the camber is adjusted to specifications, the strut-to-knuckle nuts must be tightened to the specified torque.

On some independent rear suspension systems with a lower control arm and ball joint, the strut-to-knuckle bolts are loosened and a special camber tool must be installed on the lower strut-to-knuckle bolt and the rear side of the strut (Figure 16-44). Rotate the adjusting bolt in the back of the tool to adjust the camber, then tighten the strut-to-knuckle nuts to the specified torque.

On other rear suspension systems, a camber adjustment wedge may be installed between the top of the knuckle and the strut to adjust camber (Figure 16-45). The nuts on the strut-to-knuckle

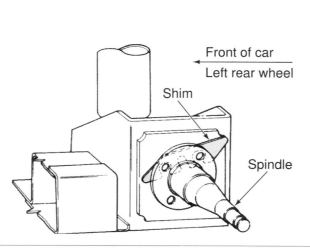

Figure 16-41 Shim adjustment for rear wheel camber, semi-independent rear suspension system. (Courtesy of Chrysler Corp.)

Figure 16-42 Circular shim for rear wheel camber and toe adjustment. (Courtesy of Cooper Industries.)

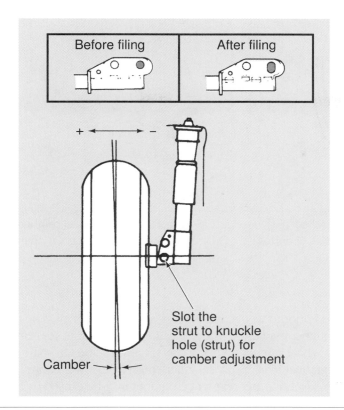

Figure 16-43 Elongating the lower strut bolt hole provides camber adjustment. (Courtesy of Oldsmobile Motor Division, General Motors Corp.)

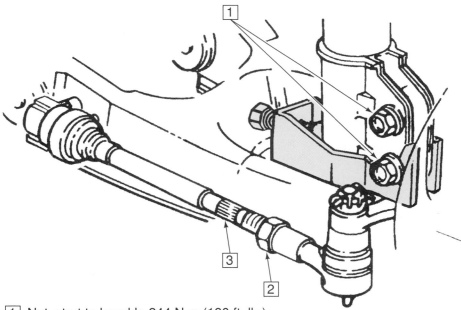

1 Nut, strut to knuckle 244 N·m (180 ft. lb.)

2 Lock nut

3 Inner tie rod

Figure 16-44 Camber adjustment with special tool inserted on rear lower strut-to-knucle bolt. (Courtesy of Oldsmobile Motor Division, General Motors Corp.)

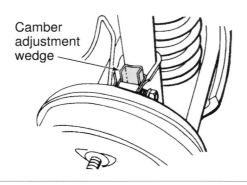

Figure 16-45 Camber adjusting wedge installed between the top of the rear knuckle and strut. (Courtesy of Hunter Engineering Company)

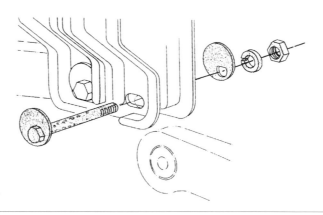

Figure 16-46 Rear wheel camber adjusting kit. (Courtesy of Specialty Products Company)

A camber wedge may be installed between the strut and knuckle to adjust camber on some rear suspension systems.

bolts must be loosened prior to wedge installation, and these nuts must be tightened to the specified torque after the camber is adjusted to specifications. Aftermarket adjustment cam kits are available to adjust the rear wheel camber on many cars (Figure 16-46). On some double wishbone rear suspension systems, graduated cams on the inner ends of the lower control arms provide camber adjustment (Figure 16-47). Improper rear wheel camber may be caused by sagged rear springs on some cars or station wagons (Figure 16-48). If the improper rear wheel camber is caused by sagged rear springs, the springs should be replaced and the camber adjusted to specifications.

Rear Wheel Toe Adjustment

Improper **rear wheel toe** moves the thrust line away from the geometric centerline, which causes steering pull and feathered tire tread wear. If the left rear wheel has excessive toe-out, the **thrust line** is moved to the left of the geometric centerline. Excessive toe-in on the left rear wheel moves the thrust line to the right of the geometric centerline. When toe-in or toe-out is excessive on the

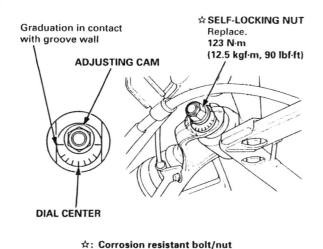

Figure 16-47 Graduated cam for rear camber adjustment, double wishbone suspension. (Courtesy of American Honda Motor Company, Inc.)

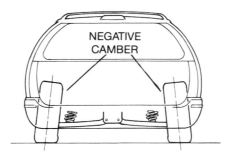

Figure 16-48 Improper rear wheel camber may be caused by sagged rear springs on independent rear suspension systems. (Courtesy of Moog Automotive, Inc.)

right rear wheel it has the opposite effect on the thrust line compared to the left rear wheel. When the thrust line is moved to the left of the geometric centerline, the steering pulls to the right, whereas a thrust line positioned to the right of the **geometric centerline** results in steering pull to the left.

Improper rear wheel toe adjustment on a non-driving rear axle causes feathered tire tread wear (Figure 16-49). The toe may be adjusted on many rear suspension systems used on front-wheel-drive cars by changing the tapered shim thickness between the front and rear edges of the rear spindle mounting flange (Figure 16-50). As mentioned previously, these shims are also used to adjust rear wheel camber.

On some transverse leaf rear suspension systems, a special toe adjusting tool is inserted in openings in the jack pad and the rear lower control rod (Figure 16-51). This tool has a turnbuckle in the center of the tool. When the nut on the inner end of rear lower control rod is loosened, the turnbuckle is rotated to lengthen or shorten the tool and move the rear lower control rod to adjust the rear wheel toe (Figure 16-52). After the toe is adjusted to specification, the nut on the rear lower control rod bolt must be tightened to the specified torque.

On some independent rear suspension systems an eccentric star wheel on the rear lower control rod may be rotated to adjust the rear wheel toe (Fiture 16-53).

On some rear suspension systems with a lower control arm and ball joint, the nut on the tie rod is loosened and the tie rod is rotated to adjust the rear wheel toe (Figure 16-54). After the toe is adjusted to specifications, the tie-rod nut must be tightened to the specified torque.

On some front-wheel-drive cars, the rear wheel toe is adjusted by loosening the toe link bolt and moving the wheel in or out to adjust the toe (Figure 16-55). After the rear wheel toe is properly adjusted, the toe link bolt must be tightened to the specified torque. On other front-wheel-drive cars the rear wheel toe is adjusted by loosening the nut on the inner adjustment link cam nut (Figure 16-56). The cam bolt is rotated to adjust the rear wheel toe. After the rear wheel toe is properly adjusted, the adjustment link cam nut must be tightened to the specified torque.

A variety of aftermarket bushing kits are available to adjust rear wheel toe. Some of these kits also adjust rear wheel toe (Figure 16-57).

On some transverse leaf rear suspension systems the toe is adjusted with a special turnbuckle tool installed in openings in the jack pad and the rear lower control rod. When the inner nut on the lower control rod is loosened, the tool is lenghthened or shortened to adjust the rear wheel toe.

A star wheel is located on the inner end of the lower control rod to adjust the toe on some rear suspension systems.

On other rear suspension systems, the tie-rod nut may be loosened and the tie rod rotated to adjust rear wheel toe.

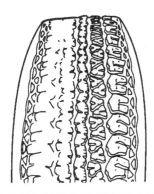

Figure 16-49 Tire tread wear caused by improper rear wheel toe adjustment (Courtesy of General Motors Corp.)

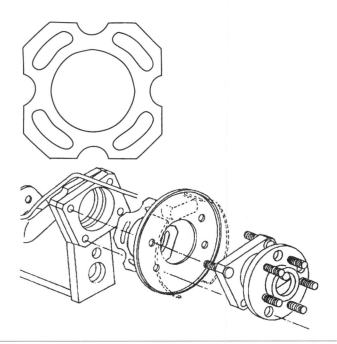

Figure 16-50 Adjusting rear wheel toe by changing shim thickness between the rear spindle and mounting flange. (Courtesy of Chrysler Corp.)

Check and set alignment with a full fuel tank.

Vehicle must be jounced 3 times before checking alignment to eliminate false readings.

Toe left and right side to be set separately per wheel to achieve specified total toe and thrust angle.

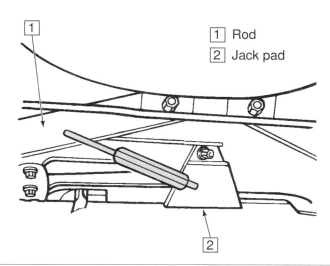

1 Rod
2 Jack pad

Figure 16-51 Special toe adjusting tool for transverse leaf-spring rear suspension. (Courtesy of Oldsmobile Motor Division, General Motors Corp.)

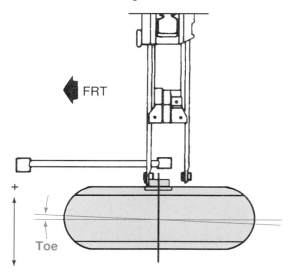

Figure 16-52 Moving lower rear control rod position to adjust rear wheel toe. (Courtesy of Oldsmobile Motor Division, General Motors Corp.)

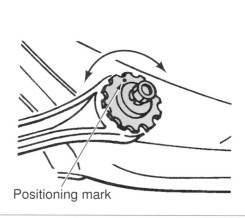

Positioning mark

Figure 16-53 Eccentric star wheel on inner end of rear lower control rod for toe adjustment. (Courtesy of Hunter Engineering Company)

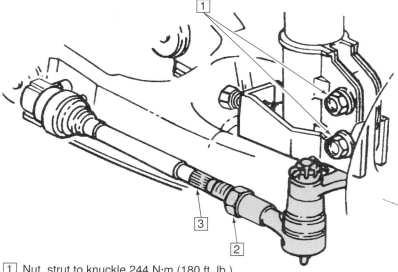

1 Nut, strut to knuckle 244 N·m (180 ft. lb.)
2 Lock nut
3 Inner tie rod

Figure 16-54 Rotating rear tie rod to change tie rod length and adjust rear wheel toe. (Courtesy of Oldsmobile Motor Division, General Motors Corp.)

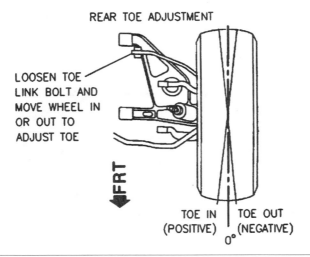

Figure 16-55 Loosening toe link bolt to adjust rear wheel toe. (Courtesy of General Motors Corp.)

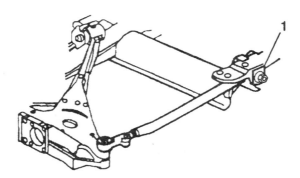

Figure 16-56 Cam bolt for rear wheel toe adjustment. (Courtesy of General Motors Corp.)

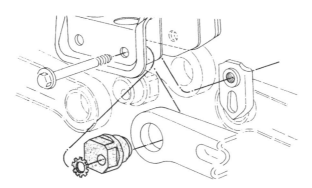

Figure 16-57 Aftermarket bushing kit to adjust rear wheel toe and camber. (Courtesy of Specialty Products Company)

Photo Sequence 14   Perform Front and Rear Suspension Alignment Adjustments

A 1993-1995 Ford Taurus or Mercury Sable and a computer wheel aligner are required for this sequence.

P16-1 Position vehicle properly on alignment rack with front wheels centered on turn tables and rear wheels on slip plates.

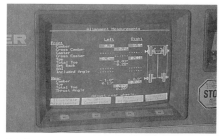

P16-2 After prealignment inspection and front and rear ride height measurement, measure front and rear suspension angles with computer wheel aligner.

P16-3 Drill spot welds to remove alignment plate on top of the strut tower. After strut upper mount retaining nuts and alignment plate are removed, shift the strut inward or outward to adjust front wheel camber. Shift the upper strut mount forward or rearward to adjust front wheel caster.

P16-4 Loosen the lock nuts and adjust the tie rod length to adjust front wheel toe.

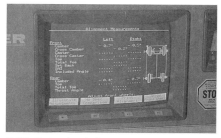

P16-5 Recheck front suspension alignment readings on the monitor screen.

P16-6 Adjust left rear wheel camber and toe by rotating the cam bolt on the inner end of the lower control arm. If the rear wheel camber and toe have not been adjusted previously, a rear camber adjusting kit may have to be installed in the inner end of the lower control arm.

P16-7 Adjust the left rear wheel camber and toe by rotating the cam bolt on the inner end of the lower control arm. If the rear wheel camber and toe have not been adjusted previously, a rear camber adjusting kit may have to be installed in the inner end of the lower control arm.

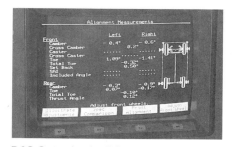

P16-8 Recheck all front and rear alignment readings on the monitor screen. Be sure all readings are within specifications including the thrust angle.

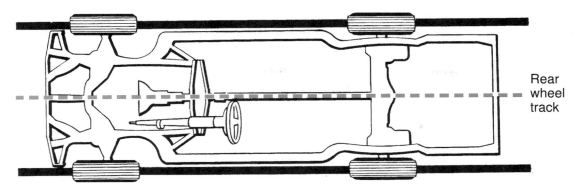

Figure 16-58 When all four wheels are parallel to the geometric centerline, the thrust line is positioned at the geometric centerline and the rear wheels track directly behind the front wheels. (Courtesy of Hunter Engineering Company)

Rear Wheel Tracking Measurement with a Track Guage

Prior to the introduction of computer alignment systems, wheel alignment equipment basically performed a two wheel alignment on the front wheels and a track gauge was used to measure **rear wheel tracking**. A **track gauge** is connected between the front and rear wheels to determine the rear wheel tracking or thrust line. When all four wheels are parallel to the geometric centerline of the vehicle, the thrust line is positioned at the geometric centerline. Under this condition, the rear wheels track directly behind the front wheels (Figure 16-58).

If the left front of the vehicle is involved in a severe collision, the left front wheel may be driven rearward to a setback condition, and the rear wheels may be moved straight sideways. Under this condition, the rear wheels no longer track directly behind the front wheels (Figure 16-59). The term sideset is applied to the condition where the rear axle is moved straight sideways.

Special Tools

Track gauge

When all four wheels are parallel to the geometric centerline and the thrust line is positioned on the geometric centerline, the rear wheels track directly behind the front wheels.

If the track gauge measurements are the same on both sides of the vehicle, the rear wheel tracking is satisfactory.

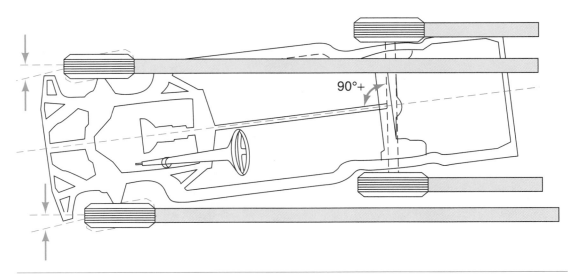

90°+

Figure 16-59 Collision damage causes front wheel setback and improper rear wheel tracking with rear axle sideset. (Courtesy of Hunter Engineering Company)

A track gauge is a straight metal bar graduated in inches or millimeters. The scale on the outside edge of the track bar is similar to a tape measure. Three straight metal pointers are mounted in brackets on the track gauge. These pointers are also graduated much like a tape measure. When the wing nuts in the brackets are loosened, the brackets may be moved lengthwise on the track bar. The metal pointers may be adjusted inward or outward in the brackets.

To check rear wheel tracking, one of the track gauge pointers is positioned against the rear outside edge of the front wheel rim at spindle height. The other two pointers are positioned on the outside edges of the rear wheel rim at spindle height. The track gauge must be kept parallel to the side of the vehicle (Figure 16-60).

Rear Wheel Track Measurement Procedure

Follow these steps to measure rear wheel tracking:

1. Lift one side of the front suspension with a floor jack and lower the suspension onto a safety stand. Be sure the tire is lifted off the shop floor.

2. Place a lateral runout gauge against the outside edge of the rim lip and rotate the wheel to check wheel runout. (Refer to Chapter 4 for wheel runout measurement.) Place a chalk mark at the location of the maximum wheel runout. If the wheel runout is excessive, replace the wheel.

3. Rotate the chalk mark to the exact top or bottom of the wheel, and remove the safety stand and floor jack to lower the wheel onto the shop floor.

4. Repeat steps 1, 2, and 3 at the other three wheels.

5. Hold the single pointer on the track gauge so it contacts the rear edge of the front rim.

6. Adjust the other two pointers so they contact the outer edges of the rear wheel rim.

7. Adjust the pointers on the rear wheel so the track gauge is parallel to the side of the vehicle. Be sure all three pointers are clamped securely.

8. Move the track gauge to the opposite side of the vehicle and position the pointers in the same location on these wheels. If the vehicle has proper tracking, all three pointers on the track gauge will contact the front and rear rims at exactly the same location on both sides of the vehicle, and the track gauge will be parallel to the sides of the vehicle.

Figure 16-60 Track bar positioned to measure rear wheel tracking.

Examples of Improper Rear Wheel Tracking Measured with a Track Gauge

Rear Axle Offset. When the track gauge is positioned on the left side of the vehicle, the rear pointer on the rear wheel must be moved further inward compared to the pointer at the front edge of the rear rim. On the right side of the vehicle the front pointer is positioned forward, ahead of the rim edge, when the two pointers are positioned on the edges of the rear rim. This indicates the wheelbase is less between the wheels on the right side of the vehicle compared to the left side. The front pointer on the rear rim is some distance from the rim face (Figure 16-61). These pointer positions indicate the right rear wheel has moved forward and the left rear wheel has moved rearward. This rear axle offset condition moves the thrust line to the left of the geometric centerline.

Setback. If a vehicle has been subjected to a severe right front sideways impact, the left front wheel may be driven outward and rearward. The track gauge and pointers are positioned on the right side of the vehicle and then moved to the left side. The front pointer is positioned ahead of the rim edge on the left side, and the track gauge is not parallel to the side of the vehicle (Figure 16-62).

The track gauge must be moved inward on the front pointer to position the bar parallel to the side of the vehicle. When this adjustment is made the distance on the front pointer is less on the left side of the vehicle compared to this distance on the right side of the vehicle. These track gauge measurements indicate the left front wheel is moved rearward to a setback position, as well as being outward.

Diamond-Shaped Frame Condition. When a vehicle experiences severe collision damage on the left front, the entire left side of the vehicle may be driven rearward in relation to the right side of the chassis. Under this condition the left front wheel is moved to a setback condition, and the rear axle is offset, which moves the thrust line to the left of the geometric centerline. This particular type of damage is most likely to occur on a rear-wheel-drive vehicle with a frame and a one-piece rear axle.

Since the entire left side of the vehicle is driven rearward, the wheel base is the same on both sides of the vehicle. Therefore the track gauge pointers indicate the wheelbase is the same between the front and rear wheels on both sides of the vehicle. However, the two rear track gauge pointers must be adjusted to different lengths to contact the rear rim edges. On the left side of the vehicle the

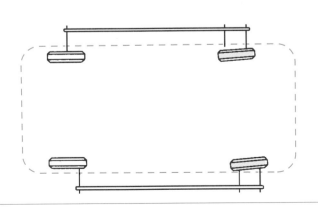

Figure 16-61 Track bar measurement of an offset rear axle.

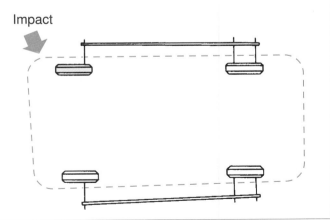

Figure 16-62 Track bar measurement of a left front wheel that is moved outward and rearward to a setback position.

567

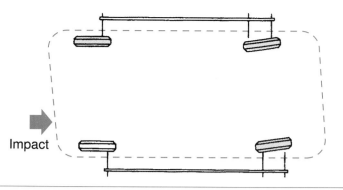

Figure 16-63 A severe left front collision impact moves the entire left side of the vehicle rearward resulting in front wheel setback and rear axle offset, which provides a diamond shaped chassis.

rear track gauge pointer is some distance from the rear rim edge, if these two pointers are set at the same measurement (Figure 16-63).

When the track gauge is positioned on the right side of the vehicle, there is some distance between the front pointer and the edge of the rear rim if both rear pointers are set at the same distance.

Measuring vehicle tracking with a track gauge is a two-person operation and requires a considerable amount of time to perform the wheel runout and track bar measurements. The track gauge also lacks precision accuracy. Computer alignment systems perform the same checks by measuring the thrust line angle, setback, and all the other front and rear suspension alignment angles with speed and accuracy.

CUSTOMER CARE: Those individuals involved in the automotive service industry are like everyone else: we do make mistakes. If you make a mistake that results in a customer complaint, always be willing to admit your mistake and correct it. Do not try to cover up the mistake or blame someone else. Customers are usually willing to live with an occasional mistake that is corrected quickly and efficiently.

Guidelines for Four Wheel Alignment Adjustments

1. Different methods for adjusting camber include shims, eccentric cams, slotted strut mounts and frames, eccentric bushings or ball joints, and upper strut mount camber plates.

2. A special tool is available to bend the I-beams and adjust camber on some twin I-beam front suspensions.

3. Caster adjustments on various suspension systems include shims, strut rod lenth, slotted upper strut mounts and frames, and eccentric cams.

4. On some twin I-beam front suspension systems the radius rods may be bent with a special tool to adjust caster.

5. SAI is not considered adjustable.

6. If the SAI is not within specifications the upper strut tower may be out of position, the lower control arm may be bent, or the front cradle may be out of position.

7. The difference between the included angles on the two front wheels must not exceed 1 1/2°.

8. If the SAI angle is not correct on one front wheel, steering pull may occur while braking or during hard acceleration, and steering may be erratic.

9. The camber and caster must be within specifications before checking the front wheel toe.

10. While checking the front wheel toe, the wheels must be straight ahead with the steering wheel locked and the brakes applied.

11. The tie-rod sleeves must be rotated with a sleeve rotating tool to adjust front wheel toe.

12. If heat is used to loosen tie-rod sleeves, the sleeves may break causing loss of steering control.

13. When a pipe wrench is used to rotate tie-rod sleeves, the sleeves may be crushed and weakened resulting in sleeve failure and loss of steering control.

14. Before tightening the tie-rod sleeve clamps, the opening in the clamps must be positioned away from the slots in the sleeves.

15. The small outer bellows boot clamp must be removed to prevent the boot from twisting before the tie rod is rotated to adjust toe on a rack and pinion steering system.

16. Steering wheel centering may be necessary after a toe adjustment.

17. On many rear suspension systems on front-wheel-drive cars, shims are positioned between the spindle and the spindle mounting surface to adjust rear wheel camber and toe.

18. On some transverse leaf rear suspension systems the lower bolt hole in the rear strut must be elongated to provide a camber adjustment.

19. On some rear suspension systems with a lower control arm and ball joint, a special tool is installed on the strut and lower strut bolt to adjust camber.

20. A camber wedge may be installed between the strut and knuckle to adjust camber on some rear suspension systems.

21. On some transverse leaf rear suspension systems the toe is adjusted with a special turnbuckle tool installed in openings in the jack pad and the rear lower control rod. When the inner nut on the lower control rod is loosened, the tool is lengthened or shortened to adjust the rear wheel toe.

22. A star wheel is located on the inner end of the lower control rod to adjust the toe on some rear suspension systems.

23. On other rear suspension systems the tie rod nut may be loosened and the tie rod rotated to adjust rear wheel toe.

24. When all four wheels are parallel to the geometric centerline and the thrust line is positioned on the geometric centerline, the rear wheels track directly behind the front wheels.

CASE STUDY

A customer complained that the left rear tire on his Cutlass Calais had suffered severe tread wear in the last 2,000 miles. The technician examined this tire and found the customer was not exaggerating. All the tread wear indicators were showing and the other tires showed very little wear. The customer indicated that the left rear tire tread looked like the tread on the other tires 2,000 miles previously. While checking the tread wear on the left rear tire, the technician noticed the car had been painted recently. Further discussion with the customer revealed the car had been purchased recently, and there was no warranty on the vehicle. During a road test the technician found that the steering pulled to the left while driving

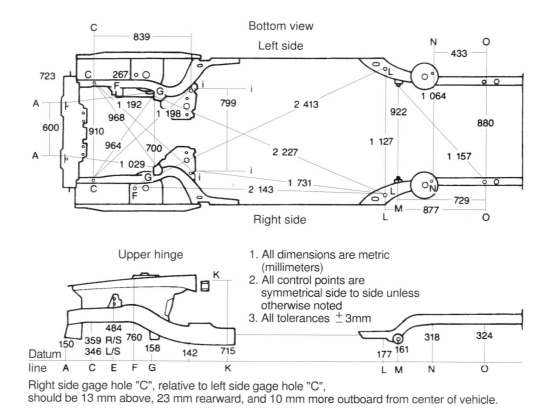

Figure 16-64 Underbody measurement specifications and locations. (Courtesy of Oldsmobile Motor Division, General Motors Corp.)

straight ahead. The technician suggested a thorough prealignment inspection and complete four wheel alignment with a computer alignment system.

During the prealignment inspection all the steering and suspension components except the left rear tire appeared to be in normal condition. However, the technician noticed the left rear quarter panel and other body components had been replaced. Apparently the car had suffered severe collision damage in this area.

The four wheel alignment results were satisfactory except the left rear wheel, which had a 5° toe-in and a thrust angle of 8°. This severe rear toe-in condition had moved the thrust line to the right of the geometric centerline. When the technician referred to the manufacturer's service manual, this information stated the rear wheel toe is not adjustable, and if the rear wheel camber and toe are not within specifications the rear underbody and suspension should be inspected for damage. The technician located the underbody dimension specifications in the service manual (Figure 16-64 and Figure 16-65).

The technician began checking the rear suspension underbody measurements with a tram gauge. When dimension M was measured between the center of the bolt heads in the trailing arm brackets, this measurement was 1 in (25.4 mm) less than specified. This measurement indicated the left rear subframe had been forced inward toward the center of the vehicle, which resulted in the improper toe and thrust angle. Further inspection of the rear axle channel with a long straightedge indicated this channel was bent rearward in the center. The car was sent to the body shop to pull the rear subframe and unitized body back to the original position and provide the specified distance between the center of bolt heads on the trailing arms. Since the rear axle channel is a critical component in providing rear

REF.	HORIZONTAL	VERTICAL	LOCATION
A	Leading edge of 24 mm gage hole*	Lower surface at gage hole to datum line	Front end lower tie bar
B	Center of 16 mm gage hole	None	Front end upper tie bar
C	Leading edge of 20 mm gage hole	Lower surface at gage hole to datum line	Motor compartment side rail, forward of transaxle anchor support plate on left side rail, and at rear of engine mounting support on right side rail. For access on right side with air conditioning, remove air compressor
D	Center of 16 mm gage hole	None	Front upper surface of motor compartment side rail
E	Center of lower attaching hole in transaxle support anchor plate	Datum line to horizontal centerline of transaxle support anchor plate lower attaching hole	Transaxle support anchor plate on left side rail
F	Center of shock tower strut front attaching hole	Upper surface at shock tower strut front attaching hole	Shock tower
G	Leading edge of oblong master gage hole	Datum line to horizontal leading edge of oblong master gage hole	Front suspension lower control arm mount support
H	Center of 16 mm gage hole	None	Gage hole in upper surface of motor compartment side rail
I	Center of 19 mm gage hole	Datum line to lower surface at gage hole	Outboard 19 mm gage hole in front suspension control arm mounting reinforcement support
J	Center of hood hinge pivot pin head	None	Hood hinge
K	Center of front upper hinge pin hole	Upper surface at hinge pin hole	Front upper door hinge, body side
L	Leading edge of 22 mm gage hole	Lower surface at leading edge of gage hole to datum line	Compartment pan longitudinal rail, forward of rear suspension spring seat support
M	Center of attaching bolt head	Datum line to horizontal centerline of attaching bolt head of rear suspension control arm support. Dimension is calculated by placing side surface of pointer at center of bolt head	Rear suspension control arm support forward of rear suspension spring seat support
N	Center of 16 mm gage hole	Datum line to lower surface at gage hole	Outboard gage hole of rear suspension spring seat support. (Not accessible with rear suspension assembly in place.)
O	Leading edge of 21 mm gage hole	Lower surface at gage hole to datum line	Compartment pan longitudinal rail. For access on right side, remove muffler
P (Right)	Center of lower stud attaching hole of right side steering unit rack and pinion support mount plate	Centerline of lower stud attaching hole of right side steering unit rack and pinion support mount plate to datum line	Steering unit rack and pinion support mount plate attached to lower right side of dash panel
P (Left)	Center of lower stud attaching hole of left side steering unit rack and pinion support mount plate	Centerline of lower stud attaching hole of left side steering unit rack and pinion support mount plate to datum line	Steering unit rack and pinion support mount plate attached to lower left side of dash panel
Q (Right)	Center of upper stud attaching hole to right side steering unit rack and pinion support mount plate	Centerline of upper stud attaching hole of right side steering unit rack and pinion support mount plate to datum line	Steering unit rack and pinion support mount plate attached to lower right side of dash panel
Q (Left)	Center of upper stud attaching hole to left side steering unit rack and pinion support mount plate	Centerline of upper stud attaching hole of left side steering unit rack and pinion support mount plate to datum line	Steering unit rack and pinion support mount plate attached to lower left side of dash panel
R	Center of outboard attaching hole for front suspension — lower control arm support	Lower surface at gage hole	Front end lower tie bar assembly

*Leading edge refers to the most forward edge of the hole

Figure 16-65 Explanation of underbody horizontal and vertical measurement locations. (Courtesy of Oldsmobile Motor Division, General Motors Corp.)

suspension ride quality and alignment, the customer was advised this component should be replaced. A new rear tire was also installed.

When the rear suspension was reassembled, all the alignment angles were rechecked and the rear wheel toe and thrust line were within specifications. During a road test there was no evidence of steering pull or any other steering problems.

From this experience the technician learned the importance of four wheel alignment. The technician also discovered that accurate underbody measurement specifications are absolutely essential.

Terms to Know

Camber adjustment	Rear wheel camber	Thrust line
Caster adjustment	Rear wheel toe	Toe
Eccentric cams	Rear wheel tracking	Toe-in
Geometric centerline	Setback	Track gauge
Rear axle offset	Steering axis inclination (SAI)	

ASE Style Review Questions

1. While discussing camber adjustments on a MacPherson strut front suspension system:
   Technician A says an eccentric cam on the strut-to-steering-knuckle bolt may be used to adjust camber.
   Technician B says an eccentric on the lower ball joint stud may be used to adjust camber.
   Who is correct?
   A. A only
   B. B only
   C. Both A and B
   D. Neither A nor B

2. While discussing camber adjustment on a short-and-long arm front suspension system with adjusting shims positioned between the upper control arm mounting shaft and the outside of the frame:
   Technician A says adding equal shim thickness on both mounting bolts increases negative camber.
   Technician B says adding more shim thickness on the front bolt decreases positive camber.
   Who is correct?
   A. A only
   B. B only
   C. Both A and B
   D. Neither A nor B

3. While diagnosing and adjusting front wheel camber:
   A. front wheel camber is measured with the front wheels turned 20°.
   B. the steering pulls to the side with the least amount of positive camber.
   C. the right front wheel may have more positive camber to compensate for road crown.
   D. the upper strut mount may be moved inward or outward to adjust camber.

4. When diagnosing and adjusting front wheel caster:
   A. on some suspensions the radius rod should be shortened to decrease positive caster.
   B. on some twin I-beam suspensions the I-beams may be bent to adjust caster.
   C. on some double wishbone suspensions a graduated cam on the pivot adjuster is rotated to adjust caster.
   D. the steering pulls to the side of the suspension that has the most positive caster.

5. The camber adjustment is within specifications, but the SAI angle is more than specified on the left side of a MacPherson strut front suspension.
   Technician A says the camber adjustment bolt may be adjusted to correct the problem.
   Technician B says the left front wheel may have a setback condition.
   Who is correct?
   A. A only
   B. B only
   C. Both A and B
   D. Neither A nor B

6. Unequal SAI angles on the left and right sides of the front suspension may cause:
   A. tread wear on the front tires.
   B. brake pull during sudden stops.
   C. bump steer on a rough road.
   D. steering wander while driving straight ahead.

7. On many front suspension systems the maximum variation between the left and right SAI is:
   A. 0.75°
   B. 1°
   C. 1.5°
   D. 2.5°

8. All these statements about adjusting front wheel toe are true EXCEPT:
   A. If the positive caster is increased on the right front wheel, this wheel moves toward a toe-out position.
   B. The tie-rod sleeves may be heated with an acetylene torch to loosen them.
   C. The opening in the sleeve clamp must be positioned away from the slots in the tie-rod sleeve.
   D. The front wheel toe should be measured with the front wheels straight ahead.

9. A rear independent suspension system has a tie rod connected from the steering knuckle to the lower control arm.
   Technician A says if the tie rod is lengthened, the rear wheel is moved toward a toe-out position.
   Technician B says if the tie rod is shortened the positive camber is increased on the rear wheel.
   Who is correct?
   A. A only
   B. B only
   C. Both A and B
   D. Neither A nor B

10. While discussing a vehicle on which the front suspension alignment angles are within manufacturer's specifications, but the thrust line is 7° to the right of the vehicle centerline:
    Technician A says this problem may be caused by excessive toe-out on the right rear wheel.
    Technician B says this vehicle will tend to steer to the left when driven straight ahead.
    Who is correct?
    A. A only
    B. B only
    C. Both A and B
    D. Neither A nor B

ASE Challenge Questions

1. A car with a MacPherson strut front suspension has 1.5° positive camber on the left front wheel and 1.5° negative camber on the right front wheel. The most likely cause of this problem is:
   A. worn upper strut mounts.
   B. improperly positioned steering gear.
   C. engine cradle shifted to the right.
   D. bent strut rods.

2. A pickup truck with a MacPherson strut front suspension pulls to the left. Preliminary inspection shows that the SAI is correct, but the camber of the right front wheel is −1 1/4°.
   Technician A says a bent spindle could be the cause.
   Technician B says the problem may be a worn out strut.
   Who is correct?
   A. A only
   B. B only
   C. Both A and B
   D. Neither A nor B

3. A car seems to "push" the nose of the car into the turn when making a right turn.
   Technician A says the probable cause is there is not enough left front wheel toe-out on turns.
   Technician B says the probable cause is too much positive camber on the left rear wheel.
   Who is correct?
   A. A only
   B. B only
   C. Both A and B
   D. Neither A nor B

4. "Feathering" type wear of a rear tire is likely caused by:
   A. improper rear wheel camber alignment.
   B. improper rear tire inflation.
   C. improper rear wheel balance.
   D. improper rear wheel toe alignment.

5. The owner of an older pickup says the truck has begun to "wander and shimmy" in the past few months, especially when the bed of the truck is empty.
   Technician A says the twin I-beam front suspension caster is probably out of spec.
   Technician B says the eyes of the rear leaf springs are probably worn.
   Who is correct?
   A. A only
   B. B only
   C. Both A and B
   D. Neither A nor B

Table 16-1 NATEF and ASE TASK

Check for front wheel setback; determine needed repairs or adjustments.

Problem Area	Symptoms	Possible Causes	Classroom Manual	Shop Manual
STEERING CONTROL	Steering pull while driving straight ahead	Severe front wheel setback condition	358	550

Table 16-2 NATEF and ASE TASK

Check front cradle (subframe) alignment; determine needed repairs or adjustments.

Problem Area	Symptoms	Possible Causes	Classroom Manual	Shop Manual
STEERING CONTROL	Steering pull while driving straight ahead	1. Improper camber caused by improperly positioned cradle	353	552
		2. Improper setback caused by improperly positioned cradle	358	552

Table 16-3 NATEF and ASE TASK

Check SAI/KPI (steering axis inclination/king pin inclination); determine needed repairs.

Problem Area	Symptoms	Possible Causes	Classroom Manual	Shop Manual
STEERING CONTROL	Steering pull during hard acceleration	Improper SAI on one side of front suspension	354	552
	Steering pull while braking	Improper SAI on one side of front suspension	356	552
	Erratic steering on road irregularities	Improper SAI on one side of front suspension	356	552

Table 16-4 NATEF and ASE TASK

Check included angle; determine needed repairs.

Problem Area	Symptoms	Possible Causes	Classroom Manual	Shop Manual
STEERING CONTROL	Steering pull while braking or accelerating	Improper SAI on one side of front suspension	355	552
	Erratic steering on road irregularities	Improper SAI on one side of front suspension	355	552

Table 16-5 NATEF and ASE TASK

Check and adjust front wheel toe.

Problem Area	Symptoms	Possible Causes	Classroom Manual	Shop Manual
TIRE LIFE	Feathered wear on front tire treads	Improper front wheel toe	359	554

Table 16-6 NATEF and ASE TASK

Check and adjust front and rear wheel camber on suspension systems with a camber adjustment.

Problem Area	Symptoms	Possible Causes	Classroom Manual	Shop Manual
TIRE LIFE	Wear on the outside edges of rear tire treads	Excessive negative or positive camber	363	558
STEERING CONTROL	Steering pull	Improper rear wheel camber	365	558
	Oversteer or understeer while cornering at high speed	Improper rear wheel camber	365	558

Table 16-7 NATEF and ASE TASK

Check front and rear wheel camber on non-adjustable suspension systems; determine needed repairs.

Problem Area	Symptoms	Possible Causes	Classroom Manual	Shop Manual
TIRE LIFE	Wear on outside edges of rear tire treads	Excessive negative or positive camber	355	543
STEERING CONTROL	Steering pull	Improper rear wheel camber	355	543

Table 16-8 NATEF and ASE TASK

Check rear wheel toe; determine needed repairs or adjustments.

Problem Area	Symptoms	Possible Causes	Classroom Manual	Shop Manual
TIRE LIFE	Feathered wear on rear tires	Improper rear wheel toe setting	365	559
STEERING CONTROL	Steering pull	Improper toe setting on one rear wheel	365	559

Table 16-9 NATEF and ASE TASK

Check rear wheel thrust angle; determine needed repairs or adjustments.

Problem Area	Symptoms	Possible Causes	Classroom Manual	Shop Manual
STEERING CONTROL	Steering pull	Thrust line positioned to the left or right of the geometric centerline	365	567

Job Sheet 47

47

Name _____ Date _____

Center Steering Wheel

NATEF and ASE Correlation

This job sheet is related to NATEF and ASE Automotive Suspension and Steering Task List content area: C. Wheel Alignment Diagnosis, Adjustment, and Repair. Task 8: Center steering wheel.

Tools and Materials

Tie-rod sleeve rotating tool

Vehicle make and model year _____

Vehicle VIN _____

Procedure

1. List the suspension and steering service procedures that may require steering wheel centering after these procedures are completed.

   _____
   _____
   _____
   _____

2. The steering wheel is centered properly with the front wheels straight ahead in the shop, but the steering wheel is not centered when driving the vehicle straight ahead. List the causes of this problem.

   _____
   _____
   _____
   _____

3. Lift the front end of the vehicle with a hydraulic jack and position safety stands under the lower control arms. Lower the vehicle onto the safety stands and place the front wheels in the straight-ahead position.

   Are the lower control arms securely supported on safety stands? yes _____ no _____

   Are the front wheels straight ahead?

   Instructor check _____

4. Use a piece of chalk to mark each tie-rod sleeve in relation to the tie rod, and loosen the sleeve clamps.

   Are the tie-rod sleeves marked in relation to the tie rods? yes _____ no _____

   Instructor check _____

5. Position the steering wheel spoke in the position it was in while driving straight ahead during the road test. Turn the steering wheel to the centered position and note the direction of the front wheels.

Direction the front wheels are turned with the steering wheel centered:
right _____ left _____

6. If the steering wheel spoke is low on the left side while driving the vehicle straight ahead, use a tie-rod sleeve rotating tool to shorten the left tie rod and lengthen the right tie rod. A one-quarter turn on a tie-rod sleeve moves the steering wheel position approximately one inch. Turn the tie-rod sleeves the proper amount to bring the steering wheel to the centered position. For example, if the steering wheel spoke is two inches off-center, turn each toe-rod sleeve one-half turn.

Left tie-rod sleeve: lengthened _____ shortened _____

Right tie-rod sleeve: lengthened _____ shortened _____

Amount of left tie-rod sleeve rotation _____

Amount of right tie-rod sleeve rotation _____

Is the steering wheel centered with front wheel straight ahead? yes _____ no _____

Instructor check _____

7. If the steering wheel spoke is low on the right side while driving the vehicle straight ahead, lengthen the left tie rod and shorten the right tie rod.

8. Mark each tie rod sleeve in its new position in relation to the tie rod. Be sure the sleeve clamp openings are positioned properly as indicated previously in this chapter. Tighten the clamp bolts to the specified torque.

Are the tie-rod sleeves marked in relation to tie rods? yes _____ no _____

Are the tie-rod sleeve clamps properly installed? yes _____ no _____

Tie-rod sleeve bolt specified torque _____

Tie-rod sleeve bolt actual torque _____

Instructor check _____

9. Lift the front chassis with a floor jack and remove the safety stands. Lower the vehicle onto the shop floor and check the steering wheel position during a road test.

Steering wheel position while driving straight ahead during a road test:
satisfactory _____ unsatisfactory _____

Instructor Check _____

Job Sheet 48

48

Name _____ Date _____

Adjust Front Wheel Alignment Angles

NATEF and ASE Correlation

This job sheet is related to NATEF and ASE Automotive Suspension and Steering Task List content area: C. Wheel Alignment Diagnosis, Adjustment, and Repair.

Task 3: Check and adjust front and rear wheel camber on suspension systems with a camber adjustment.

Task 5: Check and adjust caster on suspension systems with a caster adjustment.

Task 7: Check and adjust front wheel toe.

Tools and Materials

A vehicle with improperly adjusted front wheel camber, caster, and toe.

Vehicle make and model year _____

Vehicle VIN _____

Procedure

1. Position the vehicle properly on the alignment ramp with the front wheels properly positioned on the turn tables and the rear wheels on slip plates.

   Are the front wheel properly positioned on turn tables? yes _____ no _____

   Are the rear wheels properly positioned on slip plates? yes _____ no _____

   Is the parking brake applied? yes _____ no _____

   Instructor check _____

2. Perform a prealignment inspection and correct any defective conditions or components found during the inspection. List the defective conditions or components and explain the reasons for your diagnosis.

   _____

   _____

   _____

3. Measure front and rear ride height and correct it if necessary.

   Specified front suspension ride height _____

   Actual front suspension ride height _____

   Specified rear suspension ride height _____

   Actual rear suspension ride height _____

List the necessary repairs to correct the front and rear suspension ride height and explain the reasons for your diagnosis.

_____

_____

_____

_____

4. Measure and adjust front wheel camber.

Specified front wheel camber _____

Actual front wheel camber: left _____ right _____

Front wheel camber after adjustment: left _____ right _____

Camber adjustment method _____

5. Measure SAI and included angle.

Specified SAI _____

Actual SAI: left _____ right _____

Specified included angle _____

Actual included angle: left _____ right _____

If the SAI and/or included angle are not within specifications, state the necessary repairs and explain the reasons for your diagnosis.

_____

_____

_____

_____

6. Measure and adjust front wheel caster.

Specified front wheel caster _____

Actual front wheel caster: left _____ right _____

Front wheel caster after adjustment: left _____ right _____

Caster adjustment method _____

7. Recheck camber measurements and adjust if necessary.

Camber adjustment: satisfactory _____ unsatisfactory _____

Left front camber readjusted to _____

Right front camber readjusted to _____

8. Measure and adjust front wheel toe.

Specified front wheel toe _____

Actual front wheel toe: left _____ right _____

Total toe _____

Front wheel toe after adjustment: left _____ right _____

Total front wheel toe _____

9. Check steering wheel centering with front wheels straight ahead and center the wheel if necessary.

Is steering wheel centered? yes _____ no _____

Instructor check _____

10. Measure turning angle.

Specified turning angle _____

Actual turning angle left front wheel _____

Actual turning angle right front wheel _____

Turning angle correction required: yes _____ no _____

Left front wheel turned inward 20°, turning angle on right front wheel _____

Right front wheel turned inward 20°, turning angle on left front wheel _____

If the turning angle is not within specifications, state the necessary repairs and explain the reasons for your diagnosis.

_____

_____

_____

_____

✓ **Instructor Check** _____

Job Sheet 49

49

Name _____ Date _____

Adjust Rear Wheel Alignment Angles

NATEF and ASE Correlation

This job sheet is related to NATEF and ASE Automotive Suspension and Steering Task List content area: C. Wheel Alignment Diagnosis, Adjustment, and Repair.

Task 3:   Check and adjust front and rear wheel camber on suspension systems with a camber adjustment.

Task 12: Check rear wheel toe; determine needed repairs or adjustments.

Tools and Materials

A vehicle with improperly adjusted rear wheel camber and toe.

Vehicle make and model year _____

Vehicle VIN _____

Procedure

1. Position the vehicle properly on the alignment ramp with the front wheels properly positioned on the turn tables and the rear wheels on lsip plates.

   Are the front wheels properly positioned on turn tables? yes _____ no _____

   Are the rear wheels properly positioned on slip plates? yes _____ no _____

   Is the parking brake applied?

   Instructor check _____

2. Perform a prealignment inspection and correct any defective conditions or components found during the inspection. List the defective conditions or components and explain the reasons for your diagnosis.

   _____

   _____

   _____

3. Measure front rear ride height and correct if necessary.

   Specified front suspension ride height _____

   Actual front suspension ride height _____

   Specified rear suspension ride height _____

   Actual rear suspension ride height _____

   If the rear ride height is not within specifications, state the necessary repairs and explain the reasons for your diagnosis.

   _____

   _____

   _____

   _____

4. Measure and adjust camber on left rear wheel.

Specified left rear wheel camber _____

Actual left rear wheel camber _____

Left rear wheel camber after adjustment camber _____

Camber adjustment method _____

5. Measure left rear wheel toe.

Specified left rear wheel toe _____

Actual left rear wheel toe _____

Left rear wheel toe after adjustment _____

Toe adjustment method _____

6. Measure and adjust camber on right rear wheel.

Specified right rear wheel camber _____

Actual right rear wheel camber _____

Right rear wheel camber after adjustment _____

Camber adjustment method _____

7. Measure right rear wheel toe.

Specified right rear wheel toe _____

Actual right rear wheel toe _____

Right rear wheel toe after adjustment _____

Toe adjustment method _____

8. Check thrust angle.

Specified thrust angle _____

Actual thrust angle _____

If the thrust angle is not within specifications, state the necessary repairs and explain the reasons for your diagnosis.

_____
_____
_____
_____

9. Road test vehicle for satisfactory steering and suspension operation.

Steering directional control: satisfactory _____ unsatisfactory _____

Complete steering operation: satisfactory _____ unsatisfactory _____

Suspension operation: satisfactory _____ unsatisfactory _____

If the steering or suspension operation is unsatisfactory, state the necessary repairs and explain the reasons for your diagnosis.

_____
_____
_____
_____

✔ **Instructor Check** _____

APPENDIX A

ASE PRACTICE EXAMINATION

1. After new tires and new alloy rims are installed on a sports car, the owner complains about steering wander and steering pull in either direction while braking.
   Technician A says there may be brake fluid on the front brake linings.
   Technician B says the replacement rims may have a different offset than the original rims.
   Who is correct?
   A. Technician A
   B. Technician B
   C. Both A and B
   D. Neither A nor B

2. *Technician A* says when a vehicle pulls to one side, the problem will not be caused by the manual steering gear.
   Technician B says when an unbalanced power steering gear valve causes a vehicle to pull to one side the steering effort will be very light in the direction of the pull and normal or heavier in the opposite direction.
   Who is correct?
   A. Technician A
   B. Technician B
   C. Both A and B
   D. Neither A nor B

3. The outside edge of the left front tire on a rear wheel drive car is badly scalloped.
   Technician A says the cause could be worn ball joints.
   Technician B says the cause could be incorrect tire pressure.
   Who is correct?
   A. Technician A
   B. Technician B
   C. Both A and B
   D. Neither A nor B

4. The owner of a large rear wheel drive sedan says the front tires squeal loudly during low-speed turns. The most probable cause of this condition is:
   A. excessive positive camber.
   B. negative caster adjustment.
   C. improper steering axis inclination (S.A.I.).
   D. improper turning angle.

5. A mini pickup has a severe shudder when the vehicle is started from a stop with a load in the bed.
   Technician A says the problem may be worn spring eyes.
   Technician B says the problem may be axle torque wrap-up.
   Who is correct?
   A. Technician A
   B. Technician B
   C. Both A and B
   D. Neither A nor B

6. A cyclic noise ("moaning," "whining," or "howling") that changes pitch with road speed and is present whenever the vehicle is in motion may be caused by any of the following EXCEPT:
   A. worn differential gears.
   B. rear axle bearings.
   C. incorrect driveshaft runout.
   D. off-road tire tread pattern.

7. *Technician A* says hard steering may be caused by low hydraulic pressure due to a stuck flow control valve in the pump.
   Technician B says hard steering may be caused by low hydraulic pressure due to a worn steering gear piston ring or housing bore.
   Who is correct?
   A. Technician A
   B. Technician B
   C. Both A and B
   D. Neither A nor B

8. Wheels and tires on a pickup truck were changed from a standard 14-inch to standard 15-inch light truck rims. The first time the brakes were applied, the truck shook and shuddered. When the 15-inch wheels were replaced by the 14-inch wheels, braking was uneventful.
   Technician A says the 15-inch rim is one inch wider which causes the brakes to grab.
   Technician B says the additional inch diameter increases braking leverage overloading worn suspension bushings.
   Who is correct?
   A. Technician A
   B. Technician B
   C. Both A and B
   D. Neither A nor B

9. While discussing tire tread wear, *Technician A* says a scalloped pattern of tire wear indicates an out-of-round wheel or tire.

Technician B says uneven wear on one side of a tire may indicate radial force variation.

Who is correct?

A. Technician A
B. Technician B
C. Both A and B
D. Neither A nor B

10. A front wheel drive car makes chattering noises only during a hard right turn.

Technician A says the problem is probably caused by an upper spring seat binding.

Technician B says the problem is probably caused by a defective strut bearing.

Who is correct?

A. Technician A
B. Technician B
C. Both A and B
D. Neither A nor B

11. A customer says he equipped the rear of his 1988 minivan with inflatable air shocks to handle heavier trailer hitch tongue weights, but the van's rear end still drags when he hooks up his ski boat trailer.

Technician A says the problem is the air shocks are not heavy duty.

Technician B says the problem is heavy duty springs are also needed with the air shocks.

Who is correct?

A. Technician A
B. Technician B
C. Both A and B
D. Neither A nor B

12. All of the following could cause shock absorber noise EXCEPT:

A. a bent piston rod.
B. worn shock absorbers.
C. shock fluid leaks.
D. extreme temperatures.

13. When a vehicle pulls to one side, any of the following problems may be the cause EXCEPT:

A. worn ball joints.
B. reduced curb height.
C. bent strut rod.
D. improper turning angle.

14. A customer says when he applied the brakes hard on his front wheel drive car, "the whole car shook."

Technician A says the problem could be worn ball joints.

Technician B says the problem could be worn or loose strut rod bushings.

Who is correct?

A. Technician A
B. Technician B
C. Both A and B
D. Neither A nor B

15. A customer with a sport utility vehicle says after an off-road outing over the weekend, his vehicle pulls to the left on acceleration. Which of the following could cause this problem?

A. Broken leaf spring center bolt
B. Loose steering unit
C. Stuck brake pad
D. Stuck in 4WD

16. While discussing electronic air suspension.

Technician A says the compressor must be running when a corner of the vehicle is lifted from the ground.

Technician B says on a car with electronic air suspension the switch should be in the "on" position.

Who is correct?

A. Technician A
B. Technician B
C. Both A and B
D. Neither A nor B

17. Front wheel "shimmy"—a side-to-side movement of the front wheels that is felt in the steering wheel— may be caused by any of the following EXCEPT:

A. worn tie rod ends.
B. wheel/tire imbalance.
C. rack bushings and rack alignment.
D. tight sector shaft adjustment.

18. A customer says her front wheel drive car is hard to steer because on the steering wheel no longer returns to center. After the turn, she has to bring it back to center.

Technician A says a corroded or stuck strut bearing plate could be the cause of the problem.

Technician B says a bent tension rod could be the cause of the problem.

Who is correct?

A. Technician A
B. Technician B
C. Both A and B
D. Neither A nor B

19. While discussing 4-wheel steering (4WS),
Technician A says only a fault in the main steering angle sensor will cause a DTC in the Honda Prelude 4WS main processing unit.
Technician B says temporary codes in the Honda Prelude 4WS system are stored in memory and recalled each time the ignition key is turned on.
Who is correct?
   A. Technician A   C. Both A and B
   B. Technician B   D. Neither A nor B

20. A customer says her rear wheel drive car "nearly shakes the wheel out of my hands" on rough roads. All of the following could cause or contribute to this problem EXCEPT:
   A. worn ball joints.
   B. damaged axle housing.
   C. worn steering damper.
   D. worn or loose gear box mounts.

21. While discussing suspension height,
Technician A says raising the suspension height in the rear of a vehicle will affect the front suspension geometry.
Technician B says raising the suspension height in the rear of a vehicle could lead to premature rear spring failure.
Who is correct?
   A. Technician A   C. Both A and B
   B. Technician B   D. Neither A nor B

22. A vehicle with recirculating ball steering has excessive steering wheel freeplay.
Technician A says a loose worm bearing preload adjustment may cause the problem.
Technician B says loose column U-joints may cause the problem.
Who is correct?
   A. Technician A   C. Both A and B
   B. Technician B   D. Neither A nor B

23. The engine has recently been replaced in a front wheel drive car with power rack and pinion steering. The customer now complains about excessive steering effort. A preliminary check of the steering revealed the fluid level was OK and the belts were not slipping.
Technician A says perhaps the rack was knocked out of alignment when the engine was installed.
Technician B says perhaps the pressure line was bent or pinched when the engine was installed.
Who is correct?
   A. Technician A   C. Both A and B
   B. Technician B   D. Neither A nor B

24. While discussing rear suspension systems,
Technician A says semi-independent rear suspension systems momentarily have a slight negative camber and toe-in when the wheel goes over a bump.
Technician B says constantly carrying a lot of weight in the rear of a car with semi-independent rear suspension may cause the rear tires to wear on the inside edge.
Who is correct?
   A. Technician A   C. Both A and B
   B. Technician B   D. Neither A nor B

25. The owner of a 4WD sport utility vehicle complains about wheel thumping and vibration on tight turns at low speed. The most probable cause of this condition is:
   A. excessive positive scrub radius.
   B. improper steering axis inclination (SAI).
   C. transfer case is in 4WD.
   D. bent steering knuckle.

26. If a power steering system has proper pump pressure, but steering effort is more than specs, all of the following could cause or contribute to the problem EXCEPT:
   A. worn front tires.
   B. cold fluid.
   C. hose restrictions.
   D. steering gear malfunction.

27. The customer says his vehicle has a rapid "thumping" noise and a vibration in the steering wheel that is most noticeable when the vehicle is at a steady speed in a long curve.
Technician A says the problem could be caused by a tire defect.
Technician B says the problem could be improper dynamic wheel balance.
Who is correct?
   A. Technician A   C. Both A and B
   B. Technician B   D. either A nor B

28. While discussing power steering fluids,
Technician A says a foamy, milky power steering fluid is caused by mixing automatic transmission fluid with hydraulic fluid intended for power steering use.
Technician B says using automatic transmission fluid instead of power steering hydraulic fluids will lower pump pressure and increase steering effort.
Who is correct?
   A. Technician A   C. Both A and B
   B. Technician B   D. Neither A nor B

29. On a car with power rack and pinion steering, the steering suddenly swerves to the right or left when a front wheel strikes a road irregularity. The most likely cause of the problem is:
A. a loose rack bearing adjustment.
B. worn front struts.
C. loose steering gear mounting bushings.
D. bent steering arm.

30. A customer returns to your shop and says she just had steering work done on the car the day before and it has been pulling to the left since then. A check of the records shows that lower control arm ball joints and both rubber encapsulated tie rod ends were replaced.
Technician A says the tie rod end may have been installed with the wheels turned slightly away from center.
Technician B says the alignment should have been checked after installing the ball joints.
Who is correct?
A. Technician A
B. Technician B
C. Both A and B
D. Neither A nor B

31. A front wheel drive car tends to drift to the right except when brakes are applied. All of the following could contribute to this problem EXCEPT:
A. uneven positive front wheel camber.
B. dragging left rear wheel brake.
C. misadjusted right front wheel bearing.
D. worn right front tie rod end.

32. While discussing excessive vehicle noise and vibration, *Technician A* says suspension and driveline vibrations may be amplified by the resonance of the body or steering wheel.
Technician B says noises that phase in and out of hearing range may be two separate noises that alternately amplify then cancel each other.
Who is correct?
A. Technician A
B. Technician B
C. Both A and B
D. Neither A nor B

33. If the Honda Prelude 4WS system detects a problem in the system while the vehicle is being driven, all of the following events will occur EXCEPT:
A. the 4WS control unit energizes the damper relay.
B. the control unit stores the DTC.
C. the actuator motor quickly centers the rear wheels.
D. the 4WS indicator light illuminates.

34. While discussing unibody and frame problems, *Technician A* says an indication of possible frame damage is excessive tire wear when the alignment angles are correct.
Technician B says a worn strut lower ball joint is often an indicator of unibody torsional damage.
Who is correct?
A. Technician A
B. Technician B
C. Both A and B
D. Neither A nor B

35. While discussing frame damage, *Technician A* says wrinkles in the upper flange of a truck frame indicate a sag.
Technician B says diamond shape distortion of a 4WD sport utility vehicle frame was possibly caused by towing or being towed from a frame corner.
Who is correct?
A. Technician A
B. Technician B
C. Both A and B
D. Neither A nor B

36. Power steering fluid is leaking from the high pressure outlet fitting on a nonsubmerged power steering pump. To correct this problem, you should:
A. replace the cover seal.
B. replace the O ring.
C. replace the fitting.
D. replace the check valve.

37. The lock cylinder on a light duty GM truck with an automatic transmission is very hard to turn between "Off" and "Off-Lock." All of the following could cause this problem EXCEPT:
A. broken lock-bolt spring.
B. distorted lock rack.
C. burr on tang of shift gate.
D. shift linkage not adjusted.

38. The steering wheel of a GM car is loose in every other tilt position. Of the following possible causes, which is the most likely to cause this problem?
A. Housing support screws loose
B. Column misaligned
C. Lock shoe springs broken
D. Loose fit of lock shoe to lock pivot pin

39. While discussing unibody and frame damage, *Technician A* says a vehicle that has been rear ended could have later problems with steering and tracking. *Technician B* says towing a 2,000 pound trailer on a class II hitch bolted to the frame of a minivan will cause diamond distortion of the frame.
Who is correct?
- **A.** Technician A
- **B.** Technician B
- **C.** Both A and B
- **D.** Neither A nor B

40. While discussing a Honda Prelude 4WS system, *Technician A* says that, before diagnosing the system, the idle speed has to be steady at the specified rpm.
Technician B says before diagnosing the 4WS system the rear steering center lock pin must be installed.
Who is correct?
- **A.** Technician A
- **B.** Technician B
- **C.** Both A and B
- **D.** Neither A nor B

41. A customer says the steering wheel of her front wheel drive car does not return to the center position after a turn. A test drive reveals that the steering wheel is stiff and only returns to within approximately 90° of center after a 180° turn.
Technician A says the problem is memory steer and could be caused by binding in the steering shaft universal joints.
Technician B says memory steer in a front wheel drive car may be also caused by binding upper strut mounts.
Who is correct?
- **A.** Technician A
- **B.** Technician B
- **C.** Both A and B
- **D.** Neither A nor B

42. A customer says his front wheel drive car shakes and moans when decelerating at low speeds. The most likely cause of the problem is:
- **A.** worn upper strut bearings.
- **B.** worn inner drive axle joints.
- **C.** loose wheel bearings.
- **D.** excessive tire inflation pressure.

43. A sport utility vehicle with a power recirculating ball steering gear makes a rattling noise when the steering wheel is turned. This noise is noticeable whether the car is in motion or standing still.
Technician A says the cause of the problem could be a loose pitman arm or shaft.
Technician B says the cause of the problem could be the high pressure power steering hose touching some part of the vehicle.
Who is correct?
- **A.** Technician A
- **B.** Technician B
- **C.** Both A and B
- **D.** Neither A nor B

44. A rear wheel drive car has a steering wander problem. The cause of this problem could be:
- **A.** tire tread wear.
- **B.** loose outer tie-rod ends.
- **C.** worn idler arm.
- **D.** all of the above.

45. A customer purchased P-metric tires to replace the 6.00–15 tires on his pickup truck.
Technician A says P-metric tires should not be installed on a vehicle designed to ride on bias ply tires.
Technician B says radial tires and belted bias tires should never be mixed on the same axle.
Who is correct?
- **A.** Technician A
- **B.** Technician B
- **C.** Both A and B
- **D.** Neither A nor B

46. Many vehicles have too much play in the steering wheel, usually caused by loose steering components. A check of these components should be:
- **A.** done with vehicle weight on the tires.
- **B.** done with a computerized alignment system.
- **C.** completed after a front wheel alignment.
- **D.** performed with the vehicle raised and the wheels unsupported.

47. While discussing shock absorber diagnosis, *Technician A* says the "bounce test" is the sure way to quickly pinpoint bad shocks and struts.
Technician B says oil film on the lower chamber of a shock or strut indicates leakage and requires replacement.
Who is correct?
- **A.** Technician A
- **B.** Technician B
- **C.** Both A and B
- **D.** Neither A nor B

48. While discussing front suspension angles,
Technician A says turning angle is fixed by the
steering and suspension system design.
Technician B says toe-out on turns may be adjusted
with the steering stopper bolts.
Who is correct?
A. Technician A   **C.** Both A and B
B. Technician B   **D.** Neither A nor B

49. While discussing rear suspension alignment,
Technician A says in straight ahead driving the rear
wheels must track exactly behind the front wheels or
the vehicle will not handle correctly.
Technician B says a dynamic tracking rear
suspension that lets the rear wheels turn in the same
direction as the fronts makes curve negotiation and
lane changes much quicker and safer.
Who is correct?
A. Technician A   **C.** Both A and B
B. Technician B   **D.** Neither A nor B

50. While discussing rear wheel tracking,
Technician A says a simple tracking check is to
compare the distance between the centers of front
and rear wheels on both sides of the vehicle with the
distance diagonally between front and rear.
Technician B says if diagonal measurements are
different, but front to rear measurements are the
same, the frame is diamond shaped.
Who is correct?
A. Technician A   **C.** Both A and B
B. Technician B   **D.** Neither A nor B

APPENDIX B

Metric Conversions

	to convert these	to these,	multiply by:
TEMPERATURE	Centigrade Degrees	Fahrenheit Degrees	1.8 then + 32
	Fahrenheit Degrees	Centigrade Degrees	0.556 after − 32
LENGTH	Millimeters	Inches	0.03937
	Inches	Millimeters	25.4
	Meters	Feet	3.28084
	Feet	Meters	0.3048
	Kilometers	Miles	0.62137
	Miles	Kilometers	1.60935
AREA	Square Centimeters	Square Inches	0.155
	Square Inches	Square Centimeters	6.45159
VOLUME	Cubic Centimeters	Cubic Inches	0.06103
	Cubic Inches	Cubic Centimeters	16.38703
	Cubic Centimeters	Liters	0.001
	Liters	Cubic Centimeters	1000
	Liters	Cubic Inches	61.025
	Cubic Inches	Liters	0.01639
	Liters	Quarts	1.05672
	Quarts	Liters	0.94633
	Liters	Pints	2.11344
	Pints	Liters	0.47317
	Liters	Ounces	33.81497
	Ounces	Liters	0.02957
WEIGHT	Grams	Ounces	0.03527
	Ounces	Grams	28.34953
	Kilograms	Pounds	2.20462
	Pounds	Kilograms	0.45359
WORK	Centimeter-Kilograms	Inch-Pounds	0.8676
	Inch-Pounds	Centimeter-Kilograms	1.15262
	Meter-Kilograms	Foot-Pounds	7.23301
	Foot-Pounds	Newton-Meters	1.3558
PRESSURE	Kilograms/Square Centimeter	Pounds/Square Inch	14.22334
	Pounds/Square Inch	Kilograms/Square Centimeter	0.07031
	Bar	Pounds/Square Inch	14.504
	Pounds/Square Inch	Bar	0.06895

APPENDIX C

Automotive Suspension and Steering Systems Special Tool Suppliers

Automotive Group
Kent-Moore Division, SPX Corporation
Roseville, MI

Bear Automotive Service Equipment Company
Milwaukee, WI

Easco/KD Tools
Lancaster, PA

Hennessy Industries, Inc.
LaVerge, TN

Hunter Engineering Company
Bridgeton, MO

Mac Tools, Inc.
Washington Courthouse, OH

OTC Division, SPX Corporation
Owatonna, MN

Sears Industrial Sales
Cincinnati, OH

Snap-On Tools Corporation
Kenosha, WI

Specialty Products Company
Longmont, CO

Sun Electric Corporation
Crystal Lake, IL

GLOSSARY

Note Terms are highlighted in color, followed by Spanish translation in bold.

Accidental air bag deployment An unintended air bag deployment caused by improper service procedures.
Despliegue accidental del Airbag Despliegue imprevisto del Airbag ocasionado por procedimientos de reparación inadecuados.

Adjustment screen A display on a computer wheel aligner that allows the technician to see the results of certain suspension adjustments.
Pantalla para la visualización de ajustes Representación visual en una computadora para alineación de ruedas que le permite al técnico ver los resultados de ciertos ajustes en la suspensión.

Air bag deployment module The air bag and deployment canister assembly that is mounted in the steering wheel for the driver's side air bag, or in the dash panel for the passenger's side air bag.
Unidad de despliegue del Airbag El conjunto del Airbag y elemento de despliegue montado en el volante de dirección para proteger al conductor, o en el tablero de instrumentos para proteger al pasajero.

Air bleeding The process of bleeding air from a hydraulic system such as the power steering system.
Muestra de aire Proceso a través del cual se extrae el aire de un sistema hidráulico, como por ejemplo un sistema de dirección hidráulica.

Air conditioning programmer (ACP) A control unit that may contain the computer, solenoids, motors, and vacuum diaphragms for air conditioning control.
Programador del acondicionamiento de aire Unidad de control que puede incluir la computadora, los solenoides, los motores y los diafragmas de vacío para el control del acondicionamiento de aire.

Alignment ramp A metal ramp positioned on the shop floor on which vehicles are placed during wheel alignment procedures.
Rampa de alineación Rampa de metal ubicada en el suelo del taller de reparación de automóviles sobre la que se colocan los vehículos durante los procedimientos de alineación de ruedas.

Anhydrous calcium grease A special lubricant used in manual rack and pinion steering gears.
Grasa de calcio anhidro Lubricante especial utilizado en mecanismos de dirección de cremallera y piñón manuales.

Anti-theft lug nuts Locking lug nuts that help to prevent wheel theft.
Tuercas de orejetas anti-robo Tuercas de orejetas autobloqueantes que ayudan a evitar el robo de las ruedas.

Anti-theft wheel covers Locking wheel covers that help to prevent wheel theft.
Cubrerruedas anti-robo Cubrerruedas autobloqueantes que ayudan a evitar el robo de las ruedas.

Asbestos parts washer A parts washer designed to clean brake and clutch components without releasing asbestos into the shop air.
Lavador de partes de amianto Un lavador de partes diseñado para lavar los componentes de freno e embrague sin soltar el amianto al aire del taller.

ASE blue seal of excellence An ASE logo displayed by automotive service shops that employ ASE certified technicians.

Sello azul de excelencia de la ASE Logotipo exhibido en talleres de reparación de automóviles donde se emplean mecánicos certificados por la ASE.

ASE technician certification ASE provides certification of automotive technicians in eight different areas of expertise.
Certificación de mecánico de la ASE Certificación de mecánico de automóviles otorgada por la ASE en ocho áreas diferentes de especialización.

Auto/manual diagnostic check A test procedure in diagnosing a Ford computer-controlled suspension system.
Revisión diagnóstica automática/manual Procedimiento de prueba llevado a cabo para diagnosticar un sistema de suspensión controlado por computadora de la Ford.

Auto/manual test A test procedure in diagnosing a Ford computer-controlled suspension system.
Prueba auto-manual Un procedimiento de pruebas en diagnosticar un sistema de suspensión Ford controlado por computadora.

Auto test A test procedure in diagnosing a Ford computer-controlled suspension system.
Prueba automática Procedimiento de prueba llevado a cabo para diagnosticar un sistema de suspensión controlado por computadora de la Ford.

Auto test diagnostic mode A test procedure in diagnosing a Ford computer-controlled suspension system.
Modo diagnóstico de autoprueba Un procedimiento de prueba en diagnosticar un sistema de suspensión Ford controlado por computadora.

Automatic ride control (ARC) system The ARC system on some four-wheel drive vehicles has a front torsion bar suspension and a rear leaf spring suspension and four air springs. The air suspension system operates only when additional weight is added to the vehicle, or if the driver selects four-wheel-drive high or four-wheel-drive low.
Sistema de control de marcha automática (ARC) El sistema ARC en algunos vehículos de tracción en cuatro ruedas tiene una suspensión delantera de barra de torsión y una suspensión trasera de muelle de hojas con cuatro resortes neumáticos. El sistema de suspensión neumático opera solamente cuando se añade peso adicional al vehículo, o si el conductor selecciona marcha alta con tracción en cuatro ruedas o marcha baja con tracción en cuatro ruedas.

Axle offset A condition where the complete rear axle assembly has turned so one rear wheel has moved forward, and the opposite rear wheel has moved rearward.
Eje descentrado Una condición en la cual la asamblea total del eje trasero se ha girado para que una rueda se ha movido hacia afrente, y la rueda opuesta trasera se ha movido hacia atrás.

Axle pullers A special puller required for axle shaft removal.
Extractores del eje Extractor especial requerido para la remoción del árbol motor.

Backup power supply A voltage source, usually located in the air bag computer, that is used to deploy an air bag if the battery cables are disconnected in a collision.

Alimentación de reserva Fuente de tensión, por lo general localizada en la computadora del Airbag, que se utiliza para desplegar el Airbag si se desconectan los cables de la batería a consecuencia de una colisión.

Ball joint axial movement Vertical movement in a ball joint because of internal joint wear.

Movimiento axial de la junta esférica Movimiento vertical en una junta esférica ocasionado por el desgaste interno de la misma.

Ball joint radial movement Horizontal movement in a ball joint because of internal joint wear.

Movimiento radial de la junta esférica Movimiento horizontal en una junta esférica ocasionado por el desgaste interno de la misma.

Ball joint removal and pressing tools Special tools required for ball joint removal and replacement.

Herramientas para la remoción y el ajuste de la junta esférica Herramientas especiales requeridas para la remoción y el reemplazo de la junta esférica.

Ball joint unloading Removing the ball joint tension supplied by the vehicle weight prior to ball joint diagnosis.

Descarga de la junta esférica Remoción de la tensión de la junta esférica provista por la carga del vehículo antes de llevarse a cabo la diagnosis de la junta esférica.

Ball joint vertical movement Vertical movement in a ball joint because of internal joint wear.

Movimiento vertical de la junta esférica Un movimiento vertical en la junta esférica debido al desgaste interior de la junta.

Ball joint wear indicator A visual method of checking ball joint wear.

Indicador de desgaste de la junta esférica Método visual de revisar el desgaste interior de una junta esférica.

Bearing abrasive roller wear Fine scratches on the bearing surface.

Desgaste abrasivo del rodillo del cojinete Rayados finos en la superficie del cojinete.

Bearing abrasive step wear A fine circular wear pattern on the ends of the rollers.

Desgaste abrasivo escalonado del cojinete Desgaste circular fino en los extremos de los rodillos.

Bearing brinelling Straight line indentations on the races and rollers.

Acción de Brinnell en un cojinete Hendiduras en línea recta en los anillos y los rodillos.

Bearing etching A loss of material on the bearing rollers and races.

Corrosión del cojinete Pérdida de material en los rodillos y los anillos del cojinete.

Bearing fatigue spalling Flaking of the surface metal on the rollers and races.

Escamación y fatiga del cojinete Condición que ocurre cuando el metal de la superficie de los rodillos y los anillos comienza a escamarse.

Bearing frettage A fine, corrosive wear pattern around the races and rollers, with a circular pattern on the races.

Cinceladura del cojinete Desgaste corrosivo y fino alrededor de los anillos y los rodillos que se hace evidente a través de figuras circulares en los anillos.

Bearing galling Metal smears on the ends of the rollers.

Desgaste por rozamiento en un cojinete Ralladuras metálicas en los extremos de los rodillos.

Bearing heat discoloration A dark brown or bluish discoloration of the rollers and races caused by excessive heat.

Descoloramiento del cojinete ocasionado por el calor Descoloramiento marrón oscuro o azulado de los rodillos y los anillos ocasionado por el calor excesivo.

Bearing preload A tension placed on the bearing rollers and races by an adjustment or assembly procedure.

Carga previa del cojinete Tensión aplicada a los rodillos y los anillos del cojinete a través de un procedimiento de ajuste o de montaje.

Bearing pullers Special tools designed for bearing removal.

Extractores de cojinetes Herramientas especiales diseñadas para la remoción del cojinete.

Bearing smears Metal loss from the races and rollers in a circular, blotched pattern.

Ralladuras en los cojinetes Pérdida del metal de los anillos y los rodillos que se hace evidente a través de una figura circular oxidada.

Bearing stain discoloration A light brown or black discoloration of the rollers and races caused by incorrect lubricant or moisture.

Descoloramiento del cojinete Descoloramiento marrón claro o negro de los rodillos y los anillos ocasionado por la humedad o la utilización de un lubricante incorrecto.

Bellows boots Accordion-style boots that provide a seal between the tie rods and the housing on a rack and pinion steering gear.

Botas de fuelles Botas en forma de acordeón que proveen una junta de estanqueidad entre las barras de acoplamiento y el alojamiento en un mecanismo de dirección de cremallera y piñón.

Belt tension gauge A special gauge used to measure drive belt tension.

Calibrador de tensión de la correa de transmisión Calibrador especial que se utiliza para medir la tensión de una correa de transmisión.

Body sway Excessive body movement from side to side.

Oscilación de la carrocería Movimiento lateral excesivo de la carrocería.

Brake pedal depressor A special tool installed between the front seat and the brake pedal to apply the brakes during certain wheel alignment measurements.

Depresor del pedal de frenos Una herramienta especial instalada entre el asiento de afrente y el pedal de frenos para aplicar los frenos durante ciertas medidas de alineación de las ruedas.

Brake pedal jack A special tool installed between the front seat and the brake pedal to apply the brakes during certain wheel alignment measurements.

Gato del pedal de freno Herramienta especial instalada entre el asiento delantero y el pedal de freno que se utiliza para aplicar los frenos durante ciertas medidas de alineación de ruedas.

Bump steer The tendency of the steering to veer suddenly in one direction when one or both front wheels strike a bump.

Cambio de dirección ocacionado por promontorios en el terreno Tendencia de la dirección a cambiar repentinamente de sentido cuando una o ambas ruedas delanteras golpea un promontorio.

Cable reel A conductive ribbon mounted on top of the steering column to maintain electrical contact between the air bag inflator module and the air bag electrical system. This component may be called a clock spring electrical connector.

Bobina de cable Cinta conductiva montada sobre la columna de dirección que se utiliza para mantener el contacto eléctrico entre la unidad infladora y el sistema eléctrico del Airbag. Dicho componente se conoce también como conector eléctrico de cuerda de reloj.

Camber The inward or outward tilt of a line through the center of a front or rear tire in relation to the true vertical centerline of the tire and wheel.

Comba La inclinación, hacia adentro o hacia afuera, de una línea que atraviesa el centro de un neumático delantero, o trasero, en relación al eje mediano verdadero del neumático y la rueda.

Camber adjustment A method of adjusting the inward or outward tilt of a front or rear wheel in relation to the true vertical centerline of the tire.

Ajuste de la combadura Método de ajustar la inclinación hacia adentro o hacia afuera de una rueda delantera o trasera con relación a la línea central vertical real del neumático.

Camber angle The inward or outward tilt of a line through the center of a front or rear tire in relation to the true vertical centerline of the tire and wheel.

Ángulo de combadura Inclinación hacia adentro o hacia afuera de una línea a través del centro de una rueda delantera o trasera con relación a la línea central vertical real del neumático y la rueda.

Carbon monoxide A poisonous gas present in vehicle exhaust in small quantities.

Monóxido de carbono Gas mortífero presente en pequeñas cantidades en el escape de los motores.

Caster A line through the center of the upper and lower ball joints, or lower ball joint and upper strut mount, in relation to the true vertical centerline of the tire and wheel viewed from the side.

Ángulo de inclinación Una línea que atraviesa el centro de las juntas esféricas superiores e inferiores, o por la montadura del tirante, en relación al eje mediano vertical verdadero del neumático y la rueda vistos de un lado.

Caster adjustment A method of adjusting the forward or rearward tilt of a line through the center of the upper and lower ball joints, or lower ball joint and upper strut mount, in relation to the true vertical centerline of the tire and wheel viewed from the side.

Ajuste de comba de eje Método de ajustar la inclinación hacia adelante o hacia atrás de una línea a través del centro de las juntas esféricas superior e inferior, o la junta esférica inferior y el montaje del montante superior, con relación a la línea central vertical real del neumático y la rueda vista desde la parte lateral.

Caster angle A line through the center of the upper and lower ball joints, or lower ball joint and upper strut mount, in relation to the true vertical centerline of the tire and wheel viewed from the side.

Ángulo de comba de eje Línea a través del centro de las juntas esféricas superior e inferior, o la junta esférica inferior y el montaje del montante superior, con relación a la línea central vertical real del neumático y la rueda vista desde la parte lateral.

Circlip A round circular clip used as a locking device on components such as front drive axles.

Grapa circular Grapa circular esférica que se utiliza como dispositivo de bloqueo en algunos componentes, como por ejemplo ejes de mando de tracción delantera.

Claw washer A special locking washer used to retain the tie rods to the rack in some rack and pinion steering gears.

Arandela de garra Arandela de bloqueo especial que se utiliza para sujetar las barras de acoplamiento a la cremallera en algunos mecanismos de dirección de cremallera y piñón.

Climate control center (CCC) The air conditioning controls in the instrument panel.

Centro de control climático (CCC) Los controles del acondicionador del aire en el tablero de instrumentos.

"C" locks A thick metal locking device used to lock components in place, such as the rear drive axles.

Retenedores en C Un dispositivo de cerrojo de metal grueso que sirve para enclavar los componentes en su lugar, tal como los ejes propulsores traseros.

Clock spring electrical connector A conductive ribbon in a plastic container mounted on top of the steering column that maintains electrical contact between the air bag inflator module and the air bag electrical system.

Conector eléctrico de cuerda de reloj Cinta conductiva envuelta en una cubierta plástica montada sobre la parte superior de la columna de dirección, que mantiene el contacto eléctrico entre la unidad infladora y el sistema eléctrico del Airbag.

Coil spring compressing tool A special tool required to compress a coil spring prior to removal of the spring from the strut.

Herramienta para la compresión del muelle helicoidal Herramienta especial requerida para comprimir un muelle helicoidal antes de remover el muelle del montante.

Collapsible steering column A steering column that is designed to collapse when impacted by the driver in a collision to help reduce driver injury.

Columna de dirección plegable Columna de dirección diseñada para plegarse al ser impactada por el conductor durante una colisión con el propósito de reducir las lesiones que el conductor pueda recibir.

Compressing tool A tool used to compress components during disassembly and assembly procedures.

Herramienta compresor Una herramienta que sirve para comprimir los componentes durante los procedimientos de desmontaje y montaje.

Computer command ride (CCR) system A computer-controlled suspension system that controls strut or shock absorber firmness in relation to driving and road conditions.

Sistema de viaje ordenado por computadora Sistema de suspensión controlado por computadora que controla la firmeza de los montantes o de los amortiguadores de acuerdo a las condiciones de viaje o del camino.

Computer wheel aligner A type of wheel aligner that uses a computer to measure wheel alignment angles at the front and rear wheels.

Computadora para alineación de ruedas Tipo de alineador de ruedas que utiliza una computadora para medir los ángulos de alineación de las ruedas delanteras y traseras.

Continual strut cycling A test procedure for a computer command ride suspension system that allows continual cycling of the strut armatures.

Funcionamiento cíclico continuo del montante Procedimiento de prueba para un sistema de suspensión de viaje ordenado por computadora que permite el funcionamiento cíclico continuo de las armaduras de los montantes.

Continuous DTCs A diagnostic trouble code that represents a defect that is always present in a computer-controlled system.

DTCs contínuos Un código diagnóstico que representa un defecto que siempre está presente en un sistema controlado por computadora.

Control arm bushing tools Special tools required for control arm bushing removal and replacement.

Herramientas para el buje del brazo de mando Herramientas especiales requeridas para la remoción y el reemplazo del buje del brazo de mando.

Control arm movement monitor A monitor available in some computer wheel aligners. On short-and-long arm front suspensions this feature indicates the required shim thickness to provide the specified camber and caster .

Monitor del movimiento del brazo de mando Un monitor disponible en algunos alineadores de ruedas computerizadas. En las suspensiones delanteras tipo brazo largo-y-corto este elemento indica el espesor de la cuña requerido para proveer el ángulo de la inclinación especificado.

Cornering angle The turning angle of one front wheel in relation to the opposite front wheel during a turn. This angle may be called turning radius.

Ángulo de viraje Ángulo de giro de una rueda delantera con relación a la rueda delantera opuesta durante un viraje. Dicho ángulo se conoce también como radio de giro.

Crocus cloth A very fine paper for polishing or removing small abrasions from metal.

Tela fina de esmeril Papel sumamente fino que se utiliza para pulir o remover pequeñas abrasiones del metal.

Curb riding height The distance between the vehicle chassis and the road surface measured at specific locations.

Altura del cotén del viaje Distancia entre el chasis del vehículo y la superficie del camino medida en puntos específicos.

Current code A fault code in a computer that is present at all times.

Código actual Código de fallo siempre presente en una computadora.

Cylinder end stopper A circular bushing in the end of a rack and pinion steering gear housing.

Tapador en el extremo del cilindro Buje circular en el extremo del alojamiento de un mecanismo de dirección de cremallera y piñón.

Data link connector (DLC) An electrical connector for computer system diagnosis mounted under the instrument panel or in the engine compartment.

Conector de enlace de datos Conector eléctrico montado debajo del tablero de instrumentos o en el compartimiento del motor, que se utiliza para la diagnosis del sistema informático.

Datum line A straight reference line such as the top of a dedicated bench system.

Línea de datos Línea recta de referencia, como por ejemplo la parte superior de un sistema de banco dedicado.

Dedicated bench system A heavy steel bed with special fixtures for aligning unitized bodies.

Sistema de banco dedicado Asiento pesado de acero con aparatos especiales que se utiliza para la alineación de carrocerías unitarias.

Diagnostic drawing and text screen A display on a computer wheel aligner that provides illustrations and written instructions for the technicians.

Pantalla para la visualización de textos y diagramas diagnósticos Representación visual en una computadora para alineación de ruedas que le provee a los mecánicos ilustraciones e instrucciones escritas.

Diagnostic trouble codes (DTCs) Codes displayed in digital form that represent faults in a computer-controlled system.

Códigos diagnósticos de averías (DTCs) Los códigos manifestados en forma digital que representan las averías en un sistema controlado por computadora.

Dial indicator A precision measuring device with a stem and a rotary pointer.

Indicador de cuadrante Dispositivo para medidas precisas con vástago y aguja giratoria.

Diamond-shaped chassis A vehicle chassis that is shaped like a diamond from collision damage.

Chasis en forma de diamante Chasis que ha adquirido la forma de un diamante a causa del impacto recibido durante una colisión.

Diesel particulates Carbon particles emitted in diesel engine exhaust.

Partículas de diesel Partículas de carbón presentes en el escape de un motor diesel.

Digital adjustment photos Photos that are available in some computer-controlled wheel aligners to inform the technician regarding suspension adjustment procedures.

Fotos digitales de ajuste Las fotos que son disponibles en algunos alineadores de ruedas controlados por computadoras para informar a los técnicos en cuanto a los procedimientos de ajustes de la suspensión.

Digital signal processor An electronic device in some wheel sensors on computer-controlled wheel aligners.

Procesor de señales digitales Un dipositivo electrónico en algunos sensores de ruedas en los alineadores de ruedas controlado por computadora.

Directional stability The tendency of a vehicle steering to remain in the straight-ahead position when driven straight ahead on a smooth, level road surface.

Estabilidad direccional Tendencia de la dirección del vehículo a permanecer en línea recta al ser así conducido en un camino cuya superficie es lisa y nivelada.

Drive cycle diagnosis A test procedure when diagnosing computer-controlled suspension.

Diagnosis del ciclo propulsor Procedimiento de prueba llevado a cabo durante la diagnosis de una suspensión controlada por computadora.

Dynamic imbalance The imbalance of a wheel in motion.

Desequilibrio dinámico Desequilibrio de una rueda en movimiento.

Eccentric camber bolt A bolt with an out-of-round metal cam on the bolt head that may be used to adjust camber.

Perno de combadura excéntrica Perno con una leva metálica con defecto de circularidad en su cabeza, que puede utilizarse para ajustar la combadura.

Eccentric cams Out-of-round metal cams mounted on a retaining bolt with the shoulder of the cam positioned against a component. When the cam is rotated the component position is changed.

Levas eccéntricas Las levas de metal ovaladas montadas en un perno retenedor con lo saliente de la leva posicionado contra un componente. Cuando se gira la leva cambia la posición del componente.

Eccentric cams or bushings Out-of-round metal cams mounted on a retaining bolt with the shoulder of the cam positioned against a component. When the cam is rotated, the component position is changed.

Bujes o levas excéntricas Levas metálicas con defecto de circularidad montadas sobre un perno de retenida; el apoyo de la leva está colocado contra un componente. Al girar la leva, cambia la posición del componente.

Electronically erasable programmable read only memory (EEPROM) A chip in a computer that may be erased easily with special equipment.

Memoria de solo lectura borrable y programable electrónicamente (EEPROM) Pastilla en una computadora que puede borrarse fácilmente con equipo especial.

Electronic neutral check A check while servicing an electronically controlled four wheel steering system.

Revisión electrónica neutra Revisión llevada a cabo durante la reparación de un sistema de dirección en las cuatro ruedas controlado electrónicamente.

Electronic wheel balancers A computer-controlled balancer that provides static and dynamic wheel balance.

Equilibrador de ruedas electrónico Un equilibrador controlado por computadora que provee la equilibración estática y dinámico.

Expanding tool A tool used to expand certain components, such as Teflon rings, during the assembly procedure.

Expandidor Una herramienta que sirve para ensanchar algunos componentes, tal como los anillos de teflón, durante el proceso de montaje.

Fail-safe function A mode entered by a computer if the computer detects a fault in the system.

Función de autoprotección Modo adaptado por una computadora si la misma detecta un fallo en el sistema.

Fail safe mode A mode entered by a computer if the computer detects a fault in the system.

Modo de seguridad Un modo iniciado por la computadora si ésta descubre un fallo en el sistema.

Fault codes Numeric codes stored in a computer memory representing specific computer system faults.

Códigos de fallo Códigos numéricos almancenados en la memoria de una computadora que representan fallos específicos en el sistema informático.

Floor jack A hydraulically operated lifting device for vehicle lifting.

Gato de pie Dispositivo activado hidráulicamente que se utiliza para levantar un vehículo.

Flow control valve A special valve that controls fluid movement in relation to system demands.

Válvula de control de flujo Válvula especial que controla el movimiento del fluido de acuerdo a las exigencias del sistema.

Frame flange The upper or lower horizontal edge on a vehicle frame.

Brida del armazón Borde horizontal superior o inferior en el armazón del vehículo.

Frame web The vertical side of a vehicle frame.

Malla del armazón Lado vertical del armazón del vehículo.

Front and rear wheel alignment angle screen A display on a computer wheel aligner that provides readings of the front and rear wheel alignment angles.

Pantalla para la visualización del ángulo de alineación de las ruedas delanteras y traseras Representación visual en una computadora para alineación de ruedas que provee lecturas de los ángulos de alineación de las ruedas delanteras y traseras.

Front main steering angle sensor An input sensor mounted in the front steering gear in an electronically controlled four wheel steering system.

Sensor principal del ángulo de la dirección delantera Sensor de entrada montado en el mecanismo de dirección delantera en un sistema de dirección en las cuatro ruedas controlado electrónicamente.

Front sub steering angle sensor An input sensor mounted in the front steering gear in an electronically controlled four wheel steering system.

Sensor auxiliar del ángulo de la dirección delantera Sensor de entrada montado en el mecanismo de dirección delantera en un sistema de dirección en las cuatro ruedas controlado electrónicamente.

Functional test 211 A diagnostic procedure in a computer-controlled suspension system.

Prueba funcional 211 Procedimiento diagnóstico en un sistema de suspensión controlado por computadora.

Functional tests A diagnostic procedure in a computer-controlled suspension system.

Pruebas de función Un procedimiento diagnóstico en un sistema de suspensión controlado por computadora.

Gear backlash, or lash Movement between gear teeth that are meshed with each other.

Contragolpe o juego en los engranajes Movimiento entre los dientes del engranaje que están endentados entre sí.

Geometric centerline An imaginary line through the exact center of the front and rear wheels.

Línea central geométrica Línea imaginaria a través del centro exacto de las ruedas delanteras y traseras.

Hard fault A fault code in a computer that represents a fault that is present at all times.

Fallo consistente Un código de averías en la computadora que representa una avería que siempre está presente.

Hard fault code A fault code in a computer that represents a fault that is present at all times.

Código de fallo duro Código de fallo en una computadora que representa un fallo que está siempre presente.

Heavy spots A spot in a tire casing that is heavier than the rest of the tire.

Puntos pesados Punto en la cubierta del neumático más pesado que el resto del neumático.

High frequency transmitter An electronic device that sends high frequency voltage signals to a receiver. Some wheel sensors on computer-controlled wheel aligners send high frequency signals to a receiver in the wheel aligner.

Transmisor de alta frecuencia Un dispositivo electrónico que manda las señales de voltaje de alta frecuencia a un receptor. Algunos sensores de ruedas en los alineadores controlados por computadora mandan señales de alta frecuencia a un receptor en el alineador de ruedas.

History code A fault code in a computer that represents an intermittent defect.

Código histórico Código de fallo en una computadora que representa un defecto intermitente.

Hydraulic press A hydraulically operated device for disassembling and assembling components that have a tight press-fit.

Prensa hidráulica Dispositivo activado hidráulicamente que se utiliza para desmontar y montar componentes con un fuerte ajuste en prensa.

Included angle The sum of the camber and steering axis inclination (SAI) angles.

Ángulo incluido La suma del ángulo de inclinación de la comba y de ángulo de inclinación del eje de dirección (SAI).

Instrument panel cluster (IPC) A display in the instrument panel that may include gauges, indicator lights, or digital displays to indicate system functions.

Instrumentos agrupados en el tablero de instrumentos Representación visual en el tablero de instrumentos que puede incluir calibradores, luces indicadoras, o visualizaciones digitales para indicar las funciones del sistema.

Integral reservoir A reservoir that is joined with another component such as a power steering pump.

Tanque integral Tanque unido a otro componente, como por ejemplo una bomba de la dirección hidráulica.

Intermittent fault A fault code in a computer memory that is caused by a defect that is not always present.

Fallo intermitente Un código de fallos en la memoria de la computadora causado por un defecto que no siempre está presente.

Intermittent fault code A fault code in a computer memory that is caused by a defect that is not always present.

Código de fallo intermitente Código de fallo en la memoria de una computadora ocasionado por un defecto que no está siempre presente.

International system (SI) A system of weights and measures.

Sistema internacional Sistema de pesos y medidas.

Jack stand A metal stand that may be used to support a vehicle.

Soporte de gato Soporte de metal que puede utilizarse para apoyar un vehículo.

Lateral axle sideset The amount that the rear axle is moved straight sideways in relation to the front axle.

Desplazamiento lateral del eje La cantidad que el eje trasero está desplazado directamente hacia un lado en relación al eje delantero.

Lateral movement Movement from side to side.

Movimiento lateral Movimiento de un lado a otro.

Lateral runout The variation in side-to-side movement.

Desviación lateral Variación del movimiento de un lado a otro.

Light emitting diode (LED) A special diode that emits light when current flows through the diode.

Diodo emisor de luz (LED) Diodo especial que emite luz cuando una corriente fluye a través del mismo.

Lithium-based grease A special lubricant that is used on the rack bearing in manual rack and pinion steering gears.

Grasa con base de litio Lubricante especial utilizado en el cojinete de la cremallera en mecanismos de dirección de cremallera y piñón manuales.

Lower insulator An insulator positioned between the lower end of the coil spring and the spring seat to reduce noise.

Aislador inferior Un aislador posicionado entre la extremidad inferior de un resorte espiral y el asiento del resorte para disminuir el ruido.

Machinist's rule A steel ruler used for measuring short distances. These rulers are available in USC or metric measurements.

Regla para mecánicos Regla de acero que se utiliza para medir distancias cortas. Dichas reglas están disponibles en medidas del USC o en medidas métricas.

Magnetic gauges Wheel alignment gauges that are held on each front hub with a magnet.

Calibradores magnéticos Los calibradores de alineación de las ruedas que se sostienen en cada cubo delantero por medio de un imán.

Magnetic wheel alignment gauge A wheel alignment gauge that is held on each front hub with a magnet.

Calibrador magnético para la alineación de ruedas Calibrador para la alineación de ruedas que se fija a cada uno de los cubos delanteros con un imán.

Main menu A display on a computer wheel aligner from which the technician selects various test procedures.

Menú principal Representación visual en una computadora para alineación de ruedas de la que el mecánico puede eligir varios procedimientos de prueba.

Memory steer The tendency of the vehicle steering not to return to the straight-ahead position after a turn, but to keep steering in the direction of the turn.

Dirección de memoria Tendencia de la dirección del vehículo a no regresar a la posición de línea recta después de un viraje, sino a continuar girando en el sentido del viraje.

Molybdenum disulphide lithium-based grease A special lubricant containing molybdenum disulphide and lithium that may be used on some steering components.

Grasa de bisulfuro de molibdeno con base de litio Lubricante especial compuesto de bisulfuro de molibdeno y litio que puede utilizarse en algunos componentes de la dirección.

Multi-pull A hydraulically operated system that pulls in more than one location when straightening unitized bodies.

Enderezador de puntos múltiples Un sistema operado hidráulicamente que jala en más de un punto al enderezar las carrocerías monocasco.

Multipull unitized body straightening system A hydraulically operated system that pulls in more than one location when straightening unitized bodies.

Sistema de tiro múltiple para enderezar la carrocería unitaria Sistema activado hidráulicamente que tira hacia más de una dirección al enderezar carrocerías unitarias.

National Institute for Automotive Service Excellence (ASE) An organization that provides voluntary automotive technician certification in eight areas of expertise.

Instituto Nacional para la excelencia en la reparación de automóviles Organización que provee una certificación voluntaria de mecánico de automóviles en ocho áreas diferentes de especialización.

On demand DTCs Diagnostic trouble codes that represent intermittent faults in some computer-controlled systems.

DTCs a solicitud Los códigos de fallos diagnósticos que representan los fallos intermitentes en algunos sistemas controlados por computadora.

Optical toe gauge A piece of equipment that uses a light beam to measure front wheel toe.

Calibrador óptico del tope Equipo que utiliza un rayo de luz para medir el tope de la rueda delantera.

Oversteer The tendency of a vehicle to turn sharper than the turn selected by the driver.

Sobreviraje Tendencia de un vehículo a girar más de lo que el conductor desea.

Parameter identification screen (PID) A scan tool display that may be used to determine if automatic ride control (ARC) system components are operating electrically.

Pantalla de identificación de parámetros (PID) Una presentación de un explorador que se puede usar para determinar si los componentes del sistema automático de control de viaje están operando electrónicamente.

Part finder database Information available in some computer wheel aligners that allows the technician to access part numbers and prices from many undercar parts manufacturers.

Datos para localización de partes La información disponible en algunos alineadores de ruedas electrónicos que permite que el técnico tenga acceso a los números y precios de partes de muchos de los fabricantes de partes del carro inferior.

Pilot bearing A bearing that supports the end of a shaft, such as a pinion shaft in a rack and pinion steering gear.

Cojinete piloto Cojinete que apoya el extremo de un árbol, como por ejemplo el árbol de piñón en un mecanismo de dirección de cremallera y piñón.

Pinion and bearing assembly The pinion gear and supporting bearings in a rack and pinion steering gear.

Conjunto del piñón y cojinete Engranaje del piñón y cojinetes de soporte en un mecanismo de dirección de cremallera y piñón.

Pin point test A special test procedure to locate the specific cause of a diagnostic trouble code (DTC) in a computer-controlled system.

Prueba específica Un procedimiento especial de prueba para localizar la causa específica de un código diagnóstico de fallos (DTC) en un sistema controlado por computadora.

Pilot bearing A bearing that supports the end of a shaft, such as a pinion shaft in a rack and pinion steering gear.

Cojinete piloto Un cojinete que sostiene la extremidad de una flecha, tal como el eje piñón en la dirección de piñón y cremallera.

Pinpoint tests Specific test procedures to locate the exact cause of a fault code in a computer system.

Pruebas llevadas a cabo con precisión Procedimientos de prueba específicos que se llevan a cabo para localizar la causa exacta de un código de fallo en un sistema informático.

Pitman arm puller A special puller required to remove a pitman arm.

Extractor del brazo pitman Extractor especial requerido para la remoción de un brazo pitman.

Plumb bob A weight with a sharp, tapered point that is suspended and centered on a string.

Plomada Balanza con un extremo cónico puntiagudo, suspendida y centrada en una hilera.

Ply separation A parting of the plies in a tire casing.

Separación de estrías División de las estrías en la cubierta de un neumático.

Pneumatic tools Tools operated by air pressure.

Herramientas neumáticas Herramientas activadas por aire a presión.

Power steering pressure gauge A gauge and valve with connecting hoses for checking power steering pump pressure.

Calibrador de dirección hidráulica Un calibrador y una válvula con mangueras conectas para revisar la presión de la bomba de la dirección hidráulica.

Power steering pump pressure gauge A gauge and valve with connecting hoses for checking power steering pump pressure.

Calibrador de presión de la bomba de la dirección hidráulica Calibrador y válvula con mangueras de conexión utilizadas para verificar la presión de la bomba de la dirección hidráulica.

Powertrain control module (PCM) A computer that controls engine and possibly transmission functions.

Unidad de control del tren transmisor de potencia Computadora que controla las funciones del motor y posiblemente las de la transmisión.

Prealignment inspection A check of steering and suspension components prior to a wheel alignment.

Inspección antes de una alineación Verificación de los componentes de la dirección y de la suspensión antes de llevarse a cabo una alineación de ruedas.

Preliminary inspection screen A display on a computer wheel aligner that allows the technician to check and record the condition of many suspension and steering components.

Pantalla para la visualización durante una inspección preliminar Representación visual en una computadora para alineación de ruedas que le permite al mecánico revisar y anotar la condición de un gran número de componentes de la suspensión y de la dirección.

Pressure gauge A gauge that may be connected to a pressure source, such as a power steering pump, to determine pump condition.

Indicador de presión Un manómetro que se puede conectar a una fuente de presión, tal como la bomba de la dirección hidráulica, para determinar la condición de la bomba.

Pressure relief valve A valve designed to limit pump pressure, such as a power steering pump.

Válvula de alivio de presión Válvula diseñada para limitar la presión de una bomba, como por ejemplo una bomba de la dirección hidráulica.

Programmed ride control (PRC) system A computer-controlled suspension system in which the computer operates an actuator in each strut to control strut firmness.

Sistema de control programado del viaje Sistema de suspensión controlado por computadora en el que la computadora opera un accionador en cada uno de los montantes para controlar la firmeza de los mismos.

Rack bushing A bushing that supports the rack in the rack and pinion steering gear housing.

Buje de la cremallera Buje que apoya la cremallera en el alojamiento del mecanismo de dirección de cremallera y piñón.

Radial runout The variations in diameter of a round object such as a tire.

Desviación radial Variaciones en el diámetro de un objeto circular, como por ejemplo un neumático.

Real time damping (RTD) The real time damping module controls the road-sensing suspension system.

Amortiguamiento en tiempo real La unidad del amortiguamiento en tiempo real controla el sistema de suspensión con equipo sensor.

Rear axle offset A condition in which the complete rear axle assembly has turned so one rear wheel has moved forward and the opposite rear wheel has moved rearward.

Desviación del eje trasero Condición que ocurre cuando todo el conjunto del eje trasero ha girado de manera que una de las ruedas traseras se ha movido hacia adelante y la opuesta se ha movido hacia atrás.

Rear axle sideset A condition in which the rear axle assembly has moved sideways from its original position.

Resbalamiento lateral del eje trasero Condición que ocurre cuando el conjunto del eje trasero se ha movido lateralmente desde su posición original.

Rear main steering angle sensor An input sensor in the rear steering actuator of an electronically controlled four wheel steering system.

Sensor principal del ángulo de la dirección trasera Sensor de entrada en el accionador de la dirección trasera de un sistema de dirección en las cuatro ruedas controlado electrónicamente.

Rear steering center lock pin A special pin used to lock the rear steering during specific service procedures in an electronically controlled four wheel steering system.

Pasador de cierre central de la dirección trasera Pasador especial que se utiliza para bloquear la dirección trasera durante procedimientos de reparación específicos en un sistema de dirección en las cuatro ruedas controlado electrónicamente.

Rear sub steering angle sensor An input sensor in the rear steering actuator of a four wheel steering system.

Sensor auxiliar del ángulo de la dirección trasera Sensor de entrada en el accionador de la dirección trasera de un sistema de dirección en las cuatro ruedas.

Rear wheel camber The tilt of a line through the center of a rear tire and wheel in relation to the true vertical centerline of the tire and wheel.

Comba de la rueda trasera La inclinación de una línea que atraviesa el centro del neumático y la rueda en relación al eje mediano verdadero del neumático y la rueda.

Rear wheel toe The toe setting on the rear wheels.

Ángulo de inclinación de la rueda trasera El ajuste del ángulo de inclinación en las ruedas traseras.

Rear wheel tracking Refers to the position of the rear wheels in relation to the front wheels.

Encarrilamiento de las ruedas traseras Se refiere a la posición de las ruedas traseras con relación a las ruedas delanteras.

Rebound bumper A rubber stop that prevents metal-to-metal contact between the suspension and the chassis when a wheel strikes a road irregularity and the suspension moves fully upward.

Paragolpes de rebotes Un tope de caucho que previene el contacto de metal a metal entre la suspensión y el chasis cuando una rueda golpea una irregularidad en el camino y la suspensión se mueva totalmente hacia arriba.

Remote reservoir A container containing fluid that is mounted separately from the fluid pump.

Tanque remoto Recipiente lleno de líquido que se monta separado de la bomba del fluido.

Ribbed V-belt A belt containing a series of small "V" grooves on the underside of the belt.

Correa nervada en V Correa con una serie de ranuras pequeñas en forma de "V" en la superficie inferior de la misma.

Ride height screen A display on a computer wheel aligner that illustrates the locations for ride height measurement.

Pantalla para la visualización de la altura del viaje Representación visual en una computadora para alineación de ruedas que muestra los puntos donde se mide la altura del viaje.

Rim clamps Special clamps designed to clamp on the wheel rims and allow the attachment of alignment equipment such as the wheel units on a computer wheel aligner.

Abrazaderas de la llanta Abrazaderas especiales diseñadas para sujetar las llantas de las ruedas y así permitir la fijación del equipo de alineación, como por ejemplo las unidades de la rueda en una computadora para alineación de ruedas.

Road feel A feeling experienced by a driver during a turn when the driver has a positive feeling that the front wheels are turning in the intended direction.

Sensación del camino Sensación experimentada por un conductor durante un viraje cuando está completamente seguro de que las ruedas delanteras están girando en la dirección correcta.

Road sensing suspension (RSS) A computer-controlled suspension system that senses road conditions and adjusts suspension firmness to match these conditions in a few milliseconds.

Suspensión con equipo sensor Sistema de suspensión controlado por computadora que advierte las condiciones actuales del camino y ajusta la firmeza de la suspensión para equiparar dichas condiciones en un período de milisegundos.

Scan tester A digital computer system tester used to read trouble codes and perform other diagnostic functions.

Verificador de exploración Verificador digital del sistema informático que se utiliza para leer códigos indicativos de problemas y llevar a cabo otras funciones diagnósticas.

Scan tools A digital computer system tester used to read trouble codes and perform other diagnostic functions.

Exploradores Un probador del sistema de computadora digital que sirve para leer los códigos de errores y para executar otras funciones diagnósticas.

Seal drivers Designed to maintain even contact with a seal case and prevent seal damage during installation.

Empujadores para sellar Diseñados para mantener un contacto uniforme con el revestimiento de la junta de estanqueidad y evitar el daño de la misma durante el montaje.

Section modulus The measurement of a frame's strength based on height, width, thickness, and the shape of the side rails.

Coeficiente de sección Medida de la resistencia de un armazón, basada en la altura, el ancho, el espesor y la forma de las vigas laterales.

Selective shim Shims of different thicknesses that are used to provide an adjustment such as bearing preload.

Laminillas selectivas Laminillas de diferentes espesores que se utilizan para proveer un ajuste, como por ejemplo la carga previa del cojinete.

Self-locking nut A special nut designed so it will not loosen because of vibration.

Tuerca de cierre automático Tuerca con un diseño especial que evita su aflojamiento a causa de la vibración.

Serpentine belt A ribbed V-belt drive system in which all the belt-driven components are on the same vertical plane.

Correa serpentina Sistema de transmisión con correa nervada en V en el que todos los componentes accionados por una correa se encuentran sobre el mismo plano vertical.

Service bay diagnostics A diagnostic system supplied by Ford Motor Company to their dealers that diagnoses vehicles and provides communication between the dealer and manufacturer.

Diagnóstico para el puesto de reparación Sistema diagnóstico que les suministra la Ford a sus distribuidores de automóviles; dicho sistema diagnostica vehículos y permite una buena comunicación entre el distribuidor y el fabricante.

Service check connector A diagnostic connector used to diagnose computer systems. A scan tester may be connected to this connector.

Conector para la revisión de reparaciones Conector diagnóstico que se utiliza para diagnosticar sistemas informáticos. A dicho conector se le puede conectar un verificador de exploración.

Setback Occurs when one front wheel is driven rearward in relation to the opposite front wheel.

Retroceso Condición que ocurre cuando una de las ruedas delanteras se mueve hacia atrás con relación a la rueda delantera opuesta.

Sheared injected plastic Is inserted into steering column shafts and gear shift tubes to allow these components to collapse if the driver is thrown against the steering column in a collision.

Plástico cortado inyectado Insertado en los árboles de la columna de dirección y en los tubos del cambio de engranajes de velocidad para permitir que estos componentes se pleguen si la columna de dirección es impactada por el conductor durante una colisión.

Shim display screen A display on a computer wheel aligner that shows the technician the thickness and location of the proper shims for rear wheel alignment.

Pantalla de laminillas de ajuste Una presentación en un alineador de ruedas que demuestra al técnico el espesor y localización de las laminillas apropiadas para alineación de las ruedas traseras.

Shim select screen A display on a computer wheel aligner that shows the technician the thickness and location of the proper shims for rear wheel alignment.

Pantalla para la visualización durante la selección de laminillas Representación visual en una computadora para alineación de ruedas que le muestra al mecánico el espesor y la ubicación de las laminillas correctas para la alineación de las ruedas traseras.

Shock absorber manual test A test in which the lower end of the shock absorber is disconnected and the shock absorber is operated manually to determine its condition.

Prueba manual del amortiguador Prueba en la que se desconecta el extremo inferior del amortiguador con el fin de poder operarlo manualmente y lograr determinar su condición.

Shock absorber or strut bounce test A shock absorber test performed by manually pushing downward and releasing the vehicle bumper.

Prueba de rebote de los amortiguadores o los tirantes Una prueba de los amortiguadores que se efectúa empujando el parachoques del vehículo hacia abajo con las manos.

Shop layout The location of all shop facilities including service bays, equipment, safety equipment, and offices.

Arreglo del taller de reparación Ubicación de todas las instalaciones del taller, incluyendo los puestos de reparación, el equipo, el equipo de seguridad, y las oficinas.

Single-pull Hydraulically operated equipment that pulls in one location while straightening unitized bodies.

Jalado simple El equipo operado hidráulicamente que jala en un sólo lugar al enderezar los bastidores monocasco.

Single-pull unitized body straightening system Hydraulically operated equipment that pulls in one location while straightening unitized bodies.

Sistema enderezador de tiro único para la carrocería unitaria Sistema activado hidráulicamente que tira hacia una dirección al enderezar carrocerías unitarias.

Slip plates Plates that are placed under the rear wheels during a wheel alignment to allow rear wheel movement during rear suspension adjustments.

Placas de patinaje Las placas que se colocan debajo de las ruedas traseras durante un alineación de ruedas traseras para permitir el movimiento de las ruedas traseras durante el ajuste de la suspensión trasera.

Soft-jaw vise A vise equipped with soft metal, such as copper, in the jaws.

Tornillo con tenacilla maleable Tornillo equipado con un metal blando, como por ejemplo cobre, en las tenacillas.

Specifications menu A display on a computer wheel aligner that allows the technician to select, enter, alter, and display vehicle specifications.

Menú de especificaciones Representación visual en una computadora para alineación de ruedas que le permite al mecánico seleccionar, introducir datos, cambiar e indicar especificaciones referentes al vehículo.

Spiral cable A conductive ribbon mounted in a plastic container on top of the steering column that maintains electrical contact between the air bag inflator module and the air bag electrical system. A spiral cable may be called a clock spring electrical connector.

Cable espiral Cinta conductiva montada en un recipiente plástico sobre la columna de dirección que mantiene el contacto eléctrico entre la unidad infladora y el sistema eléctrico del Airbag. El cable espiral se conoce también como conector eléctrico de cuerda de reloj.

Spring compressing tools Tools that are used to compress a coil spring during the removal and replacement procedure.

Herramientas de compresión de resortes Las herramientas que se usan para comprimir un resorte helicoidal durante el procedimiento de remoción y refacción.

Spring fill diagnostics A diagnostic procedure in a computer-controlled air suspension system.

Diagnóstico con relleno para muelles Procedimiento diagnóstico en un sistema de suspensión de aire controlado por computadora.

Spring insulators, or silencers Rings usually made from plastic and positioned on each end of a coil spring to reduce noise and vibration transfer to the chassis.

Aisladores de muelles, o silenciadores Anillos, por lo general fabricados de plástico y colocados a ambos extremos de un muelle helicoidal, que se utilizan para disminuir la transferencia de ruido y de vibración al chasis.

Spring sag Occurs when a spring becomes weak, and the curb riding height is reduced compared to the original height.

Agotamiento de los muelles Condición que ocurre cuando el muelle se debilita; la altura del cotén del viaje disminuye si se la compara con la altura orginal.

Spring silencers Plastic spacers placed between spring leaves to reduce noise.

Amortiguadores de los resortes Las arandelas de plástico puestas entre los muelles de hojas para disminuir el ruido.

Stabilizer bar, or sway bar A round steel bar connected between the front or rear lower control arms that reduces body sway.

Barra estabilizadora o barra de oscilación lateral Barra circular de acero contectada entre los brazos de mando delantero o trasero que disminuye la oscilación lateral de la carrocería.

Static imbalance Refers to the imbalance of a wheel and tire at rest.

Desequilibrio estático Se refiere al desequilibrio de una rueda y de un neumático cuando se ha detenido la marcha del vehículo.

Steering axis inclination (SAI) The tilt of a line through the center of the upper and lower ball joints or through the center of the lower ball joint and the upper strut mount in relation to the true vertical centerline of the tire and wheel.

Inclinación del eje de dirección (SAI) La inclinación de una línea que atraviesa el centro de las juntas esféricas superiores e inferiores o por el centro de la junta esférica inferior y el montaje del tirante superior en relación al eje mediano verdadero del neumático y de la rueda.

Steering effort The amount of effort required by the driver to turn the steering wheel.

Esfuerzo de dirección Amplitud de esfuerzo requerido por parte del conductor para girar el volante de dirección.

Steering effort imbalance Occurs when more effort is required to turn the steering wheel in one direction than in the opposite direction.

Desequilibrio del esfuerzo en dirección Ocurre cuando se requiere más esfuerzo para girar la rueda en una dirección que en la dirección opuesta.

Steering pull The tendency of the steering to pull to the right or left when the vehicle is driven straight ahead on a smooth, straight road surface.

Tiro de la dirección Tendencia de la dirección a desviarse hacia la derecha o hacia la izquierda mientras se conduce el vehículo en línea recta en un camino cuya superficie es lisa y nivelada.

Steering wheel freeplay The amount of steering wheel movement before the front wheels begin to turn.

Juego libre del volante de dirección Amplitud de movimiento del volante de dirección antes de que las ruedas delanteras comiencen a girar.

Steering wheel locking tool A special tool used to lock the steering wheel during certain wheel alignment procedures.

Herramienta para el cierre del volante de dirección Herramienta especial que se utiliza para bloquear el volante de dirección durante ciertos procedimientos de alineación de ruedas.

Stethoscope A special tool that amplifies sound to help diagnose noise location.

Estetoscopio Herramienta especial que amplifica el sonido para ayudar a diagnosticar la procedencia de los ruidos.

Strikeout, or rebound, bumper A rubber block that prevents the control arm from striking the chassis when the wheel hits a large road irregularity.

Parachoques de rebote Bloque de caucho que evita que el brazo de mando choque contra el chasis cuando la rueda golpea una irregularidad en el camino.

Strut cartridge The inner components in a strut that may be replaced rather than replacing the complete strut.

Cartucho del montante Componentes internos de un montante que pueden ser reemplazados en vez de tener que reemplazarse todo el montante.

Strut chatter A chattering noise as the steering wheel is turned often caused by a binding upper strut mount.

Vibración del montante Rechinamiento producido mientras se gira el volante de dirección. A menudo ocasionado por el trabamiento del montaje del montante superior.

Strut rod, or radius rod A rod connected from the lower control arm to the chassis to prevent forward and rearward control arm movement.

Varilla del montante o varilla radial Varilla conectada del brazo de mando inferior al chasis que se utiliza para evitar el movimiento hacia adelante o hacia atrás del brazo de mando.

Strut tower A circular, raised, reinforced area inboard of the front fenders that supports the upper strut mount and strut assembly.

Torre del montante Área circular, elevada y reforzada en la parte interior de los guardafangos delanteros que apoya el conjunto del montante y montaje del montante superior.

Sulfuric acid A very corrosive acid used in automotive batteries.

Ácido sulfúrico Ácido sumamente corrosivo utilizado en las baterías de automóviles.

Sunburst cracks Cracks that radiate outward from an opening in the vehicle frame.

Grietas radiales Las grietas que radian hacia afuera desde una apertura en el armazón del vehículo.

Supplemental inflatable restraint (SIR) An air bag system.

Sistema de seguridad inflable suplementario Sistema del Airbag.

Suspension adjustment link An adjustable link connected horizontally from the suspension to the chassis.

Biela de ajuste de la suspensión Una biela ajustable conectada horizontalmente de la suspensión al chasis.

Symmetry angle measurements On some computer-controlled wheel aligners help the technician determine if out-of-specifications readings may have been caused by collision or frame damage.

Medidas de simetría de los ángulos En algunos alineadores de ruedas controlados por computadora ayuda al técnico en determinar si las medidas fuera de especificación puedan haber sido causadas por una colisión o un daño al bastidor.

Tapered-head bolts A bolt with a taper on the underside of the head.

Pernos de cabeza cónica Perno con una cabeza cuya superficie inferior es cónica.

Teflon ring compressing tool A special tool used to compress the Teflon rings on the pinion prior to installation in a rack and pinion steering gear.

Herramienta para la compresión de anillos de teflón Herramienta especial que se utiliza para comprimir los anillos de teflón en el piñón antes de ser instalados en un mecanismo de dirección de cremallera y piñón.

Teflon ring expander A special tool required to expand the Teflon rings prior to installation on the pinion in a rack and pinion steering gear.

Expansor para anillos de teflón Herramienta especial requerida para expandir los anillos de teflón antes de ser instalados en un mecanismo de dirección de cremallera y piñón.

Thrust line A line positioned at a 90° angle to the rear axle and projected toward the front of the vehicle.

Línea de empuje Línea colocada a un ángulo de 90° con relación al eje trasero y proyectada hacia la parte frontal del vehículo.

Tie rod end and ball joint puller A special puller designed for removing tie rod ends and ball joints.

Extractor para la junta esférica y el extremo de la barra de acoplamiento Extractor especial diseñado para la remoción de las juntas esféricas y los extremos de barras de acoplamiento.

Tie rod sleeve adjusting tool A special tool required to rotate tie rod sleeves without damaging the sleeve.

Herramienta para el ajuste del manguito de la barra de acoplamiento Herramienta especial requerida para girar los manguitos de las barras de acoplamiento sin estropearlos.

Tire changer Equipment used to demount and mount tires on wheel rims.

Cambiador de neumáticos Equipo que se utiliza para desmontar y montar los neumáticos en las llantas de las ruedas.

Tire condition screen A display on a computer wheel aligner that allows the technician to check and enter tire condition.

Pantalla para la visualización de la condición del neumático Representación visual en una computadora para alineación de ruedas que le permite al mecánico revisar e introducir datos referentes a la condición del neumático.

Tire conicity Occurs when the tire belt is wound off-center in the manufacturing process creating a cone-shaped belt, which results in steering pull.

Conicidad del neumático Condición que ocurre cuando la correa del neumático se devana fuera del centro durante el proceso de fabricación creando así una correa en forma cónica y trayendo como resultado el tiro de la dirección.

Tire inspection screen A screen on a computer-controlled wheel aligner that allows the technician to enter the condition of each vehicle tire.

Pantalla de inspección de los neumáticos Una pantalla en un alineador de ruedas controlado por computadora que permite que el técnico anote la condición de cada neumático del vehículo.

Tire rotation Involves moving each wheel and tire to a different location on the vehicle to increase tire life.

Rotación de los neumáticos Involucra moviendo cada rueda y neumático a una posición diferente en el vehículo para prolongar la vida del neumático.

Tire thump A pounding noise as the tire and wheel rotate usually caused by improper wheel balance.

Ruido sordo del neumático Ruido similar al de un golpe pesado que se produce mientras el neumático y la rueda están girando. Por lo general este ruido lo ocasiona el desequilibrio de la rueda.

Tire tread depth gauge A special tool required to measure tire tread depth.

Calibrador de la profundidad de la huella del neumático Herramienta especial requerida para medir la profundidad de la huella del neumático.

Tire vibration Vertical or sideways tire oscillations.

Vibración del neumático Oscilaciones verticales o laterales del neumático.

Toe Refers to the distance between the front edges on a pair of opposite wheels in relation to the distance between the rear edges of these wheels.

Ángulo de inclinación Refiere a la distancia entre las orillas delanteras en un par de ruedas opuestas en relación a la distancia entre las orillas traseras de esas ruedas.

Toe gauge A special tool used to measure front or rear wheel toe.

Calibrador del tope Herramienta especial que se utiliza para medir el tope de las ruedas delanteras o traseras.

Toe-in A condition in which the distance between the front edges of the tires is less than the distance between the rear edges of the tires.

Convergencia Condición que ocurre cuando la distancia entre los bordes frontales de los neumáticos es menor que la distancia entre los bordes traseros de los mismos.

Toe-out A condition in which the distance between the front edges of the tires is more than the distance between the rear edges of the tires.

Divergencia Condición que ocurre cuando la distancia entre los bordes delanteros de los neumáticos es mayor que la distancia entre los bordes traseros de los mismos.

Toe-out on turns The turning angle of the wheel on the inside of a turn compared to the turning angle of the wheel on the outside of the turn.

Divergencia durante un viraje El ángulo de giro de la rueda en el interior de un viraje comparado con el ángulo de giro de la rueda en el exterior del viraje.

Torque steer The tendency of the steering to pull to one side during hard acceleration on front wheel drive vehicles with unequal length front drive axles.

Dirección de torsión Tendencia de la dirección a desviarse hacia un lado durante una aceleración rápida en un vehículo de tracción delantera con ejes de mando desiguales.

Total toe The sum of the toe angles on both wheels.

Tope total Suma de los ángulos del tope en ambas ruedas.

Track gauge A long straight bar with adjustable pointers used to measure rear wheel tracking in relation to the front wheels.

Calibrador del encarrilamiento Barra larga y recta con agujas ajustables que se utiliza para medir el encarrilamiento de las ruedas traseras con relación a las ruedas delanteras.

Traction control system (TCS) A computer-controlled system that prevents one drive wheel from spinning more than the opposite drive wheel on acceleration.

Sistema para el control de la tracción Sistema de suspensión controlado por computadora que evita que una rueda motriz gire más que la opuesta durante la aceleración del vehículo.

Tram gauge A long, straight bar with adjustable pointers used to measure unitized bodies.

Calibrador del tram Barra larga y recta con agujas ajustables que se utiliza para medir carrocerías unitarias.

Tread wear indicators Raised portions near the bottom of the tire tread that are exposed at a specific tread wear.

Indicadores del desgaste del neumático Secciones elevadas cerca de la parte inferior de la huella del neumático que quedan expuestas cuando la huella alcanza una cantidad de desgaste específica.

Trim height The normal chassis riding height on a computer-controlled air suspension system.

Altura equilibrada Altura normal de viaje del chasis en un sistema de suspensión controlado por computadora.

Turning angle screen A display on a computer wheel aligner that displays the turning angle.

Pantalla para la visualización del ángulo de giro Representación visual en una computadora para alineación de ruedas que muestra el ángulo de giro.

Turning imbalance A condition in which more effort is required to turn the steering wheel in one direction compared to the opposite direction.

Desequilibrio de giro Condición que ocurre cuando se requiere más esfuerzo para girar el volante en una dirección que en la dirección opuesta.

Turning radius The turning angle on the front wheel on the inside of a turn compared to the front wheel turning angle on the outside of the turn.

Radio de giro, o círculo de giro Ángulo de giro de la rueda delantera en el interior de un viraje con relación al ángulo de giro de la rueda delantera en el exterior del viraje.

Turning radius gauge A gauge with a degree scale mounted on turn tables under the front wheels.

Calibrador del radio de giro Calibrador con una escala de grados montado sobre plataformas giratorias debajo de las ruedas delanteras.

Turning radius gauge turn tables A gauge with a degree scale mounted on turn tables under the front wheels.

Plataforma giratoria calibradora del radio de viraje Un calibrador que tiene una escala de grados montado en las plataformas giratorias debajo de las ruedas delanteras.

Turn tables Are placed under the front wheels during a wheel alignment to allow the front wheels to be turned in each direction to measure various front suspension angles.

Plataforma giratoria Se coloca debajo de las ruedas delanteras durante la alineación de ruedas para permitir girar las ruedas delanteras en cada dirección para medir varios ángulos de la suspensión delantera.

Understeer The tendency of a vehicle not to turn as much as desired by the driver.

Dirección pobre Tendencia de un vehículo a no girar tanto como lo desea el conductor.

United States Customary (USC) A system of weights and measures patterned after the British system.

Sistema usual estadounidense (USC) Sistema de pesos y medidas desarrollado según el modelo del Sistema Imperial Británico.

Upper insulator An insulator positioned between the top of the coil spring and the upper strut mount to reduce noise.

Aislante superior Un aislante posicionado entre la parte superior de un resorte helicoidal y el montaje del tirante superior para disminuir el ruido.

Vacuum hand pump A mechanical pump with a vacuum gauge and hose used for testing vacuum-operated components.

Bomba de vacío manual Bomba mecánica con un calibrador de vacío y una manguera que se utiliza para probar componentes a depresión.

V-belt A drive belt with a V-shape.

Correa en V Correa de transmisión en forma de V.

Vehicle lift A hydraulically or air-operated mechanism for lifting vehicles.

Levantamiento del vehículo Mecanismo activado hidráulicamente o por aire que se utiliza para levantar vehículos.

Vehicle wander The tendency of the steering to pull to the right or left when the vehicle is driven straight ahead on a straight road.

Desviación de la marcha del vehículo Tendencia de la dirección a desviarse hacia la derecha o hacia la izquierda cuando se conduce el vehículo en línea recta en un camino cuya superficie es lisa.

Wear indicator A device that visually indicates ball joint wear.

Indicador de desgaste Dispositivo que indica de manera visual el desgaste de una junta esférica.

Wheel balancer Equipment for wheel balancing that is usually computer controlled.

Equilibrador de ruedas Equipo para la equilibración de ruedas que por lo general es controlado por computadora.

Wheel runout compensation screen A display on a computer wheel aligner that is shown during the wheel runout compensation procedure.

Pantalla para la visualización de la compensación de desviación de la rueda Representación visual en una computadora para alineación de ruedas que aparece durante el procedimiento de compensación de desviación de la rueda.

Woodruff key A half-moon shaped metal key used to retain a component, such as a pulley, on a shaft.

Chaveta woodruff Chaveta de metal en forma de media luna que se utiliza para sujetar un componente, como por ejemplo una roldana, en un árbol.

Yield strength A measurement of the material strength from which a frame is manufactured.

Límite para fluencia Medida de la resistencia del material del que está fabricado un armazón.

INDEX